Benchmark Papers in Geology

Editor: Rhodes W. Fairbridge, Columbia University

A selection from the published volumes in this series

STATISTICAL ANALYSIS IN GEOLOGY / *John M. Cubitt and Stephen Henley*

SEAFLOOR SPREADING CENTERS: Hydrothermal Systems / *Peter A. Rona and Robert P. Lowell*

MEGACYCLES: Long-Term Episodicity in Earth and Planetary History / *G. E. Williams*

RIFT VALLEYS: Afro-Arabian / *A. M. Quennell*

OROGENY / *John G. Dennis*

OPHIOLITIC AND RELATED MELANGES / *G. J. H. McCall*

MANTLE FLOW AND PLATE THEORY / *Zvi Garfunkel*

Related titles of interest

THE ENCYCLOPEDIA OF GEOCHEMISTRY AND ENVIRONMENTAL SCIENCES, IVA / *Rhodes W. Fairbridge*

THE ENCYCLOPEDIA OF MINERALOGY / *Keith Frye*

VOLCANOES OF THE WORLD / *T. Simkin, L. Siebert, L. McClelland, D. Bridge, C. Newhall, and J. H. Latter*

BASALTS

Edited by

PAUL C. RAGLAND
The Florida State University

and

JOHN J. W. ROGERS
The University of North Carolina at Chapel Hill

A Hutchinson Ross Benchmark® Book

VNR VAN NOSTRAND REINHOLD COMPANY

Copyright © 1984 by **Van Nostrand Reinhold Company Inc.**
Benchmark Papers in Geology, Volume 71
Library of Congress Catalog Card Number: 84-7445
ISBN: 0-442-27769-5

Manufactured in the United States of America.

Published by Van Nostrand Reinhold Company Inc.
135 West 50th Street
New York, New York 10020

Van Nostrand Reinhold Company Limited
Molly Millars Lane
Wokingham, Berkshire RG11 2PY, England

Van Nostrand Reinhold
480 Latrobe Street
Melbourne, Victoria 3000, Australia

Macmillan of Canada
Division of Gage Publishing Limited
164 Commander Boulevard
Agincourt, Ontario MIS 3C7, Canada

15 14 13 12 11 10 9 8 7 6 5 4 3 2 1

Library of Congress Cataloging in Publication Data
Main entry under title:
Basalts.

  (Benchmark papers in geology ; 71)
  "A Hutchinson Ross Benchmark book."
  Bibliography; p.
  Includes index.

  1. Basalt—Addresses, essays, lectures. I. Ragland,
Paul C., 1936–   . II. Rogers, John J. W. (John James
William), 1930–   . III. Series.
QE462.B3B29 1984   552'.26   84-7445
ISBN 0-442-27769-5

CONTENTS

Series Editor's Foreword   xi
Preface   xiii
Contents by Author   xvii

Introduction   1

PART I: CLASSIFICATION OF BASALTS

Editors' Comments on Papers 1 Through 18   6

1   **KENNEDY, W. Q.:**   Trends of Differentiation in Basaltic Magmas   13
Am. Jour. Sci. **25:**239–242 (1933)

2   **TILLEY, C. E.:**   Some Aspects of Magmatic Evolution (Abstract)   17
Geol. Soc. London Quart Jour. **106:**37 (1950)

3   **MACDONALD, G. A., and T. KATSURA:**   Chemical Composition of Hawaiian Lavas   18
Jour. Petrology **5:**84–89 (1964)

4   **SCHWARZER, R. R., and J. J. W. ROGERS:**   Erratum to "A Worldwide Comparison of Alkali Olivine Basalts and Their Differentiation Trends"   25
Earth and Planetary Sci. Letters **25:**91 (1975)

5   **KUNO, H.:**   Differentiation of Basalt Magmas   26
Basalts: The Poldervaart Treatise on Rocks of Basaltic Composition, vol. 2,
H. H. Hess and A. Poldervaart, eds., Wiley-Interscience, New York, 1968,
pp. 625–628

6   **JAKEŠ, P., and J. GILL:**   Rare Earth Elements and the Island Arc Tholeiitic Series   31
Earth and Planetary Sci. Letters **9:**18 (1970)

7   **MIYASHIRO, A., F. SHIDO, and M. EWING:**   Diversity and Origin of Abyssal Tholeiite from the Mid-Atlantic Ridge near 24° and 30° North Latitude (Abstract)   32
Contr. Mineralogy and Petrology **23:**38 (1969)

8   **PEACOCK, M. A.:**   Classification of Igneous Rock Series   33
Jour. Geology **39:**57 (1931)

9   **DONNELLY, T. W., and J. J. W. ROGERS:**   Igneous Series in Island Arcs: The Northeastern Caribbean Compared With Worldwide Island-Arc Assemblages (Abstract)   34
Bull. Volcanol. **43-2:**347 (1980)

Contents

10 GILL, J. B.: Geochemistry of Viti Levu, Fiji, and Its Evolution as an
Island Arc   35
Contr. Mineralogy and Petrology **27**:195-196 (1970)

11 MIYASHIRO, A.: Volcanic Rock Series in Island Arcs and Active
Continental Margins   38
Am. Jour. Sci. **274**:325 (1974)

12 GUNN, B. M., M. J. ROOBOL, and A. L. SMITH: Petrochemistry of the
Peléan-Type Volcanoes of Martinique   39
Geol. Soc. America Bull. **85**:1023-1024 (1974)

13 IRVINE, T. N., and W. R. A. BARAGAR: A Guide to the Chemical
Classification of the Common Volcanic Rocks   41
Canadian Jour. Earth Sci. **8**:524 (1971)

14 NICHOLLS, J., and I. S. E. CARMICHAEL: A Commentary on the
Absarokite-Shoshonite-Banakite Series of Wyoming, U.S.A.   42
Schweizer. Mineralog. u. Petrog. Mitt. **49(1)**:61-63 (1969)

15 JOPLIN, G. A.: The Shoshonite Association: A Review   46
Geol. Soc. Australia Jour. **15(2)**:284-288 (1968)

16 VILJOEN, M. J., and R. P. VILJOEN: The Geology and Geochemistry of
the Lower Ultramafic Unit of the Onverwacht Group and a
Proposed New Class of Igneous Rocks   51
Geol. Soc. South Africa Spec. Pub. No. 2: "Upper Mantle Project," 1969,
pp. 80-84

17 BROOKS, C., and S. R. HART: On the Significance of Komatiite   58
Geology **2**:107-108 (1974)

18 ARNDT, N. T., A. J. NALDRETT, and D. R. PYKE: Komatiitic and Iron-Rich
Tholeiitic Lavas of Munro Township, Northeast Ontario   62
Jour. Petrology **18(2)**:320 (1977)

PART II: ORIGIN OF BASALTIC MAGMAS

Editors' Comments on Papers 19 Through 33   64

19 BOWEN, N. L.: The Crystallization of Haplobasaltic, Haplodioritic,
and Related Magmas   73
Am. Jour. Sci., ser. 4, **40(256)**:174-185 (1915)

20 FENNER, C. N.: The Crystallization of Basalts   85
Am. Jour. Sci., ser. 5, **18(105)**:238-253 (1929)

21 OSBORN, E. F.: Role of Oxygen Pressure in the Crystallization and
Differentiation of Basaltic Magma   101
Am. Jour. Sci. **257**:609-610, 632-643, 647 (1959)

22 YODER, H. S., Jr., and C. E. TILLEY: Origin of Basalt Magmas: An
Experimental Study of Natural and Synthetic Rock Systems   114
Jour. Petrology **3**:342, 504-509 (1962)

23 GREEN, D. H., and A. E. RINGWOOD: The Genesis of Basalt Magmas   121
Contr. Mineralogy and Petrology **15**:104-105, 159-167 (1967)

24 O'HARA, M. J., and H. S. YODER, JR.: Formation and Fractionation of
Basic Magmas at High Pressures   133
Scottish Jour. Geology **3**:67, 88, 99-115, 116, 117 (1967)

25 O'HARA, M. J.:   The Bearing of Phase Equilibria Studies in Synthetic and Natural Systems on the Origin and Evolution of Basic and Ultrabasic Rocks 153
Earth-Sci. Rev. **4:**126–128 (1968)

26 WYLLIE, P. J.: 156
The Dynamic Earth
Wiley, New York, 1971, pp. 198, 199

27 KUSHIRO, I.:   The System Forsterite-Diopside-Silica with and without Water at High Pressures 159
Am. Jour. Sci. **267A:**280–294 (1969)

28 KUSHIRO, I.:   Effect of Water on the Composition of Magmas Formed at High Pressures (Abstract) 173
Jour. Petrology **13:**311 (1972)

29 EGGLER, D. H.:   Effect of CO_2 on the Melting of Peridotite 174
Carnegie Inst. Washington Year Book 73, pp. 215–224 (1974)

30 MYSEN, B. O., and A. L. BOETTCHER: Melting of a Hydrous Mantle: II. Geochemistry of Crystals and Liquids Formed by Anatexis of Mantle Peridotite at High Pressures and High Temperatures as a Function of Controlled Activities of Water, Hydrogen, and Carbon Dioxide 185
Jour. Petrology **16:**549–550, 586–590 (1975)

31 PRESNALL, D. C., J. R. DIXON, T. H. O'DONNELL, and S. A. DIXON: Generation of Mid-ocean Ridge Tholeiites 193
Jour. Petrology **20:**3, 20–30, 32, 33, 34, 35 (1979)

32 WALKER, D., T. SHIBATA, and S. E. DELONG: Abyssal Tholeiites from the Oceanographer Fracture Zone. II. Phase Equilibria and Mixing 208
Contr. Mineralogy and Petrology **70:**111–125 (1979)

33 STOLPER, E.:   A Phase Diagram for Mid-Ocean Ridge Basalts: Preliminary Results and Implications for Petrogenesis 223
Contr. Mineralogy and Petrology **74:**13–27 (1980)

PART III: BASALTS AND MANTLE EVOLUTION

Editors' Comments on Papers 34 Through 37 240

34 JAMIESON, B. G., and D. B. CLARKE: Potassium and Associated Elements in Tholeiitic Basalts 242
Jour. Petrology **11:**193–200, 203, 204 (1970)

35 HART, S. R., C. BROOKS, T. E. KROGH, G. L. DAVIS, and D. NAVA: Ancient and Modern Volcanic Rocks: A Trace Element Model (Abstract) 251
Earth and Planetary Sci. Letters **10:**17 (1970)

36 JAHN, B.-M.:   Trace Element Geochemistry of Archean Volcanic Rocks and Its Implication for the Chemical Evolution of the Upper Mantle 252
Soc. Géol. France Bull. **19:**1259–1269 (1977)

37 SUN, S. S., and G. N. HANSON: Evolution of the Mantle: Geochemical Evidence from Alkali Basalt 263
Geology **3:**297–302 (1975)

PART IV: OCEANIC BASALTS

Editors' Comments on Papers 38 Through 45   **270**

38   **KAY, R., N. J. HUBBARD, and P. W. GAST:**   Chemical Characteristics and Origin of Oceanic Ridge Volcanic Rocks   **274**
Jour. Geophys. Research **75:**1585, 1606–1611 (1970)

39   **WHITE, W. M., and W. B. BRYAN:**   Sr-isotope, K, Rb, Cs, Sr, Ba, and Rare-Earth Geochemistry of Basalts from the FAMOUS Area   **281**
Geol. Soc. America Bull. **88:**571–576 (1977)

40   **LANGMUIR, C. H., J. F. BENDER, A. E. BENCE, G. N. HANSON, and S. R. TAYLOR:**   Petrogenesis of Basalts from the FAMOUS Area: Mid-Atlantic Ridge (Abstract)   **287**
Earth and Planetary Sci. Letters **36:**133 (1977)

41   **SHIBATA, T., G. THOMPSON, and F. A. FREY:**   Tholeiitic and Alkali Basalts from the Mid-Atlantic Ridge at 43°N   **288**
Contr. Mineralogy and Petrology **70:**127, 136–141 (1979)

42   **SUN, S.-S., R. W. NESBITT, and A. Ya. SHARASKIN:**   Geochemical Characteristics of Mid-Ocean Ridge Basalts (Abstract)   **293**
Earth and Planetary Sci. Letters **44:**119 (1979)

43   **THOMPSON, G., W. B. BRYAN, F. A. FREY, and C. M. SUNG:**   Petrology and Geochemistry of Basalts and Related Rocks from Sites 214, 215, 216, DSDP Leg 22, Indian Ocean   **294**
Initial Reports of the Deep Sea Drilling Project **22:**459–468 (1974)

44   **SCHILLING, J.-G.:**   Azores Mantle Blob: Rare-Earth Evidence (Abstract)   **304**
Earth and Planetary Sci. Letters **25:**103 (1975)

45   **CLAGUE, D. A., and M. H. BEESON:**   Trace Element Geochemistry of the East Molokai Volcanic Series, Hawaii   **305**
Am. Jour. Sci. **280-A:**820, 838–843 (1980)

PART V: BASALTS OF SUBDUCTION ZONES

Editors' Comments on Papers 46 Through 49   **314**

46   **KUNO, H.:**   Lateral Variation of Basalt Magma Type Across Continental Margins and Island Arcs   **318**
Bull. Volcanol. **29:**196, 217 (1966)

47   **DICKINSON, W. R.:**   Potash-Depth (K-H) Relations in Continental Margin and Intra-Oceanic Magmatic Arcs   **320**
Geology **3:**53–56 (1975)

48   **GREEN, T. H.:**   Island Arc and Continent-Building Magmatism—A Review of Petrogenic Models Based on Experimental Petrology and Geochemistry (Abstract)   **324**
Tectonophysics **63:**367–368 (1980)

49   **CAWTHORN, R. G.:**   Petrological Aspects of the Correlation between Potash Content of Orogenic Magmas and Earthquake Depth   **325**
Mineralog. Mag. **41:**173–182 (1977)

PART VI: CONTINENTAL BASALTS

Editors' Comments on Papers 50 Through 53   336

50   WATERS, A. C.:   Basalt Magma Types and Their Tectonic Associations:
                 Pacific Northwest of the United States   340
         Crust of the Pacific Basin, Geophysical Monograph No. 6, 1962, pp. 158-170

51   HOOPER, P. R.:   The Columbia River Basalts   353
                 Science **215:**1463-1468 (1982)

52   LIPMAN, P. W., and H. H. MEHNERT: Late Cenozoic Basaltic Volcanism
                 and Development of the Rio Grande Depression in the
                 Southern Rocky Mountains   359
         Geol. Soc. America Mem. **144:**119-120, 137-147, 148, 149, 150, 151, 152,
                 153, 154 (1975)

53   IRVING, A. J., and D. H. GREEN: Geo-chemistry and Petrogenesis of
                 the Newer Basalts of Victoria and South Australia   374
         Geol. Soc. Australia Jour. **23**(Pt. 1):45-47, 57-66 (1976)

PART VII: CORRELATION OF BASALT TYPE
AND TECTONIC ENVIRONMENT

Editors' Comments on Papers 54 Through 58   388

54   SCHMIDT, R. G.:   Petrology of the Volcanic Rocks   393
                 U.S. Geol. Survey Prof. Paper 280-B, 1957, p. 128

55   JAKEŠ, P., and J. R. WHITE: Major and Trace Element Abundances in
                 Volcanic Rocks of Orogenic Areas   394
                 Geol. Soc. America Bull. **83:**29-39 (1972)

56   ROGERS, J. J. W.:   Criteria for Recognizing Environments of Formation
                 of Volcanic Suites; Application of These Criteria to
                 Volcanic Suites in the Carolina Slate Belt   405
         Geol. Soc. America Spec. Paper 191, 1982, pp. 99, 100, 101, 102

57   PEARCE, J. A., and J. R. CANN: Tectonic Setting of Basic Volcanic
                 Rocks Determined Using Trace Element Analyses   408
                 Earth and Planetary Sci. Letters **19:**295 (1973)

58   PEARCE, J. A., and G. H. GALE: Identification of Ore-Deposition
                 Environment from Trace-Element Geochemistry of Associated
                 Igneous Host Rocks   410
         Volcanic Processes in Ore Genesis, Inst. Mining and Metallurgy; Geol. Soc.
                 London, 1977, pp. 14-24

Author Citation Index   421
Subject Index   426
About the Editors   430

SERIES EDITOR'S FOREWORD

The philosophy behind the Benchmark Papers in Geology is one of collection, sifting, and rediffusion. Scientific literature today is so vast, so dispersed, and, in the case of old papers, so inaccessible for readers not in the immediate neighborhood of major libraries that much valuable information has been ignored by default. It has become just so difficult, or so time consuming, to search out the key papers in any basic area of research that one can hardly blame a busy person for skimping on some of his or her "homework."

This series of volumes has been devised, therefore, as a practical solution to this critical problem. The geologist, perhaps even more than any other scientist, often suffers from twin difficulties—isolation from central library resources and immensely diffused sources of material. New colleges and industrial libraries simply cannot afford to purchase complete runs of all the world's earth science literature. Specialists simply cannot locate reprints or copies of all their principal reference materials. So it is that we are now making a concerted effort to gather into single volumes the critical materials needed to reconstruct the background of any and every major topic of our discipline.

We are interpreting "geology" in its broadest sense: the fundamental science of the planet Earth, its materials, its history, and its dynamics. Because of training in "earthy" materials, we also take in astrogeology, the corresponding aspect of the planetary sciences. Besides the classical core disciplines such as mineralogy, petrology, structure, geomorphology, paleontology, and stratigraphy, we embrace the newer fields of geophysics and geochemistry, applied also to oceanography, geochronology, and paleoecology. We recognize the work of the mining geologists, the petroleum geologists, the hydrologists, and the engineering and environmental geologists. Each specialist needs a working library. We are endeavoring to make the task of compiling such a library a little easier.

Each volume in the series contains an introduction prepared by a specialist (the volume editor)—a "state of the art" opening or a summary of the object and content of the volume. The articles, usually some twenty to fifty reproduced either in their entirety or in significant extracts, are selected in an attempt to cover the field, from the key papers of the last century to fairly recent work. Where the original works are in foreign languages, we have endeavored to locate or commission translations. Geologists, because of their global subject, are often acutely aware of the oneness of our world. The selections cannot therefore be restricted to any one country, and whenever possible an attempt is made to scan the world literature.

To each article, or group of kindred articles, some sort of "highlight commentary" is usually supplied by the volume editor. This commentary should serve to bring that article into historical perspective and to emphasize its particular role in the growth of the field. References, or citations, wherever possible, will be reproduced in their entirety—for by this means the observant reader can assess the background material available to that particular author, or, if desired, he or she too can double check the earlier sources.

A "benchmark," in surveyor's terminology, is an established point on the ground that is recorded on our maps. It is usually anything that is a vantage point, from a modest hill to a mountain peak. From the historical viewpoint, these benchmarks are the bricks of our scientific edifice.

RHODES W. FAIRBRIDGE

PREFACE

Basalts are probably the most widely studied of all igneous rocks. As an example, at least 3,500 papers were published on basalts from 1970 to 1979. Aside from the obvious fact that basalts comprise a major portion of the earth's crust, they provide us with a window through which we can examine, albeit with difficulty, a myriad of earth processes. Modern marine geology, volcanology, and plate tectonics depend in part on a knowledge of basalt genesis.

Basalt petrologists have played an important role in developing our current understanding about the nature of the earth's lithosphere and relationships between the upper mantle and crust. Moreover, recent space expeditions and probes have revealed that basalts roughly similar to those on earth are apparently abundant crustal components of our moon and the other inner planets; hence, basalts are not only significant with regard to the internal evolution of the earth but also to the origin of our solar system. We realize increasingly that basalts are genetically related to certain sulfide ore deposits and are being considered actively as repositories for nuclear wastes. Thus, on purely academic, as well as more practical, grounds, the study of basalts probably touches on as many, or more, aspects of geology than that of any other igneous rock. It would seem that the potential rewards amply justify the time spent on exploring these fascinating but enigmatic rocks.

In this book, we depart slightly from the format followed in the majority of other Benchmark series books. Our aim is twofold: (1) to recall the most significant excerpts from some classic papers that are widely recognized to have had a profound effect on our thinking about basalts and (2) to present the critical parts of some papers, most of which are quite recent, that are probably not yet classics but that we believe to be the wave of the future. We have decided to integrate excerpts from a large number of papers rather than to reproduce a few papers in toto, which leaves us open to criticisms of taking material out of context. We are particularly vulnerable in Part I and with relatively short segments from some exceptionally long articles in other parts. We strongly believe, however, that this overall format is absolutely essential to provide a sufficiently broad coverage but to keep the book to a reasonable length. Moreover, because of this style we hope this book can be used as a textbook for an advanced course on basalt petrology. At the very least, it should be a useful supplement for a standard text or a valuable reference book.

Because of the broad nature of this topic and the proliferation of published information, we have found it necessary to exclude certain more specialized subjects. We will not consider basaltic rocks such as gabbros or diabases (dolerites) except in the context of their derivation from basaltic magmas. The so-called andesite and spilite problems must await other Benchmark or similar efforts. Secondary alteration of basalt, whether metamorphic, hydrothermal, or by weathering, is only peripherally considered, as are economic deposits associated with basaltic rocks. Only basalts on earth are examined in any detail. Physical parameters that characterize basalt magmas and emplacement mechanisms are not amply treated. Finally, petrographic descriptive material is minimized.

If these topics are either not treated or inadequately covered, a fair question to ask is what subjects are included. Part I discusses the classification of basalts, with emphasis on tectonic and/or quasi-chemical classifications. Part II deals with the genesis of basaltic magmas, organized around a more-or-less chronological treatment of those experimental studies that we believe have had the most impact. In Part III we attempt to examine the role of basalts in mantle evolution. Parts IV through VI deal with basalts in three major tectonic environments: ocean basins, subduction zones, and continents. Finally, Part VII is an effort to draw much of this information together to explore ways in which basalt petrology affects interpretation of tectonic environments, particularly paleoenvironments.

Finally, perhaps parenthetically, we should comment briefly on our being chosen to edit this book. The senior author has co-authored several papers on basaltic rocks (principally diabases; e.g., Ragland, Rogers, and Justus, 1968; Weigand and Ragland, 1970; Ragland, Brunfelt, and Weigand, 1971; Steele and Ragland, 1976; Ragland, Hatcher, and Whittington, 1983) and is continuing to work on Mesozoic diabases from eastern North America. The junior author has frequently published on basalts and andesites (Leeman and Rogers, 1970; Schwarzer and Rogers, 1974; Rogers and Novitsky-Evans, 1977; Paper 9; Rogers and Ragland, 1980). Neither of us, however, would characterize himself as a basalt petrologist. We have primarily used the geochemistry of basaltic rocks as a tool to aid in our regional geologic interpretations. An argument could be put forth readily that a person with, for example, an extensive background in the experimental petrology of basalts could better cull the Benchmark-quality papers from the chaff. It is certainly true that such a person would be far more knowledgeable about the intricacies of basalt petrogenesis than we are. Our only defense is that occasionally a generalist can provide a more even-handed outlook than a specialist, particularly with regard to the impact of a particular paper upon the profession as a whole. Some worthy papers and authors have undoubtedly been omitted; to those individuals we apologize. We have attempted, however, to be even-handed in our choice of papers; the reader must determine to what degree we have been successful.

ACKNOWLEDGMENTS

First and foremost, I (P.C.R.) wish to thank Joan Ragland for her tireless efforts in doing numerous boring jobs such as helping compile the index and editing the text. Sharon Reeves helped with the manuscript, and Robert Loucks provided advice and useful discussion. I am also indebted to David Green, who many years ago introduced me to the complexities of basalt petrogenesis. Norman Tilford furnished the opportunity for me to continue working in this area, and Jelle de Boer gave encouragement and stimulation. Finally, I am grateful to a number of graduate students down through the years who have persistently forced me to learn at least something about basaltic rocks; in particular, thanks are due Don Hermes, Phil Justus, Peter Weigand, Ken Steele, Joe Wooden, Greg McHone, Paul Drez, Marc Defant, David Whittington, and Laura Cummins.

I (J.J.W.R.) would specifically like to thank Barbara Rogers for her continued assistance in assembly of materials and other logistical efforts connected with publication of this work. Harry Posey and Cathy Cash also provided great assistance in this work. Throughout the years I have greatly benefitted from discussions with numerous colleagues, including Nick Donnelly, Rudy Schwarzer, Joyce Novitsky-Evans, Paul Mueller, and Bill Leeman.

We both would like to specify that, despite all of this assistance, the mistakes in this book are solely the result of our own efforts.

PAUL C. RAGLAND
JOHN J. W. ROGERS

REFERENCES

Leeman, W. P., and J. J. W. Rogers, 1970, Late Cenozoic Alkali-Olivine Basalts of the Basin-Range Province, USA, *Contr. Mineralogy and Petrology* **25**:1–24.

Ragland, P. C., A. O. Brunfelt, and P. W. Weigand, 1971, Rare Earth Abundances in Mesozoic Dolerite Dikes from Eastern United States, in *Activation Analysis in Geochemistry and Cosmochemistry*, A. O. Brunfelt and E. Steinnes, eds., Oslo Universitets for laget, Oslo, Norway, pp. 272–284.

Ragland, P. C., R. D. Hatcher, Jr., and D. Whittington, 1983, Juxtaposed Mesozoic Diabase Dike Sets from the Carolinas, *Geology* **11**:394–399.

Ragland, P. C., J. J. W. Rogers, and P. S. Justus, 1968, Origin and Differentiation of Triassic Dolerite Magmas, North Carolina, USA, *Contr. Mineralogy and Petrology* **20**:57–80.

Rogers, J. J. W., and J. E. Novitsky-Evans, 1977, The Clarno Formation of Central Oregon, U.S.A.—Volcanism on a Thin Continental Margin, *Earth and Planetary Sci. Letters* **34**:56–66.

Rogers, J. J. W., and P. C. Ragland, 1980, Trace Elements in Continental-Margin Magmatism; Part I. Trace Elements in the Clarno Formation of Central Oregon and the Nature of the Continental Margin on which Eruption Occurred, *Geo. Soc. America Bull.* **91:**196–198, Part I; p. 1217–1292, Part II, Card 3.

Schwarzer, R. R., and J. J. W. Rogers, 1974, A Worldwide Comparison of Alkali Olivine Basalts and Their Differentiation Trends, *Earth and Planetary Sci. Letters* **23:**286–296.

Steele, K. F., and P. C. Ragland, 1976, Model for the Closed-System Fractionation of a Dike Formed by Two Pulses of Dolerite Magma, *Contr. Mineralogy and Petrology* **57:**307–316.

Weigand, P. W. and P. C. Ragland, 1970, Geochemistry of Mesozoic Dolerite Dikes from Eastern North America, *Contr. Mineralogy and Petrology* **29:**195–214.

CONTENTS BY AUTHOR

Arndt, N. T., 62
Baragar, W. R. A., 41
Beeson, M. H., 305
Bence, A. E., 287
Bender, J. F., 287
Boettcher, A. L., 185
Bowen, N. L., 73
Brooks, C., 58, 251
Bryan, W. B., 281, 294
Cann, J. R., 408
Carmichael, I. S. E., 42
Cawthorn, R. G., 325
Clarke, D. B., 242
Clague, D. A., 305
Davis, G. L., 251
DeLong, S. E., 208
Dickinson, W. R., 320
Dixon, J. R., 193
Dixon, S. A., 193
Donnelly, T. W., 34
Eggler, D. H., 174
Ewing, M., 32
Fenner, C. N., 85
Frey, F. A., 288, 294
Gale, G. H., 410
Gast, P. W., 274
Gill, J., 31, 35
Green, D. H., 121, 374
Green, T. H., 324
Gunn, B. M., 39
Hanson, G. N., 263, 287
Hart, S. R., 58, 251
Hooper, P. R., 353
Hubbard, N. J., 274
Irvine, T. N., 41

Irving, A. J., 374
Jahn, B.-M., 252
Jakeš, P., 31, 394
Jamieson, B. G., 242
Joplin, G. A., 46
Katsura, T., 18
Kay, R., 274
Kennedy, W. Q., 13
Krogh, T. E., 251
Kuno, H., 26, 318
Kushiro, I., 159, 173
Langmuir, C. H., 287
Lipman, P. W., 359
MacDonald, G. A., 18
Mehnert, H. H., 359
Miyashiro, A., 32, 38
Mysen, B. O., 185
Naldrett, A. J., 62
Nava, D., 251
Nesbitt, R. W., 293
Nicholls, J., 42
O'Donnell, T. H., 193
O'Hara, M. J., 133, 153
Osborn, E. F., 101
Peacock, M. A., 33
Pearce, J. A., 408, 410
Presnall, D. C., 193
Pyke, D. R., 62
Ringwood, A. E., 121
Rogers, J. J. W., 25, 34, 405
Roobol, M. J., 39
Schilling, J.-G., 304
Schmidt, R. G., 393
Schwarzer, R. R., 25
Sharaskin, A. Ya., 293

Contents by Author

Shibata, T., 208, 288
Shido, F., 32
Smith, A. L., 39
Stolper, E., 223
Sun, S. S., 263, 293
Sung, C. M., 294
Taylor, S. R., 287
Thompson, G., 288, 294
Tilley, C. E., 17, 114

Viljoen, M. J., 51
Viljoen, R. P., 51
Walker, D., 208
Waters, A. C., 340
White, A. J. R., 394
White, W. M., 281
Wyllie, P. J., 156
Yoder, H. S., Jr., 114, 133

INTRODUCTION

Speculation about basalts began early. Pliny's *Natural History* was published about A.D. 77 (see Adams, 1938) and classified basalt as black marble, but he referred to most dense building stones as marble, so probably was implying no genetic link. Although some ancient Greeks and Romans correctly guessed that basalts were related to volcanism, at the end of the Dark Ages basalt was thought to be of sedimentary origin, deposited in the Great Deluge described in the Book of Genesis (Adams, 1938). This concept persisted into the early 1700s, when one popular idea held that heat necessary for volcanic eruptions was caused by the combustion of oils and fats of land animals and fishes, presumably at least in part trapped in sediments during the Great Flood. In Gesner's *De Rerum Fossilium*, published in 1565, a fascinating sketch of hexagonal basalt columns exhibits pyramidal terminations, implying that the columns were crystals (see Adams, 1938, Fig. 25, p. 128). That basalt columns were, in fact, crystals precipitated from sea water was a common idea in those times. Spallanzani (1798) and Raspe (1776) were among the first to identify correctly the origins, if not the specifics, of columnar jointing in basalt.

In the latter 30 years of the 1700s, geologic thought in Europe was dominated by the charismatic Abraham Werner and his Neptunist school, which also held that basalt was of sedimentary origin (Werner, 1788). Ironically, most of Werner's Neptunist theories came to grief over the origin of basalt, with the Plutonists eventually ruling the day.

Perhaps the first person to undertake a systematic field study of basalts and to theorize correctly concerning their origins was Gutteard, who in 1751 recognized extinct basaltic volcanoes in the Auvergne Mountains of southern France. It is interesting that 1,700 years earlier the Romans settled a village in this area and named it Volvic (a contraction from Latin for "volcanic village"), so they clearly recognized that ancient basalts were of volcanic origin. Desmarest (1774) extended the volcanic orgin for the Auvergne basalts to those in many other European areas, although we should emphasize that Gutteard's and Desmarest's views were not widely held because of the popularity of

Werner's Neptunist theories. Werner's conclusions on the origin of basalt were eventually doomed, and the Plutonists prevailed when, first, Playfair published his rewrite of Hutton's classic work, *Illustrations of the Huttonian Theory of the Earth,* in 1802 and, second, when three of Werner's best students (D'Aubisson, von Buch, and von Humboldt) turned against their old mentor on this issue during the first 10 years of the 1800s. Thereafter, basalt has been considered by most geologists to be of volcanic origin.

Other early studies of note include those by Hall (1805) and Cordier (1816), who attempted some of the first experimental work by melting natural basalts and cooling them to reproduce basaltic textures. Although Darwin obviously spent more of his efforts in other fields, in 1844 he recognized crystal settling of plagioclase in basalt and suggested that it could cause the process we now call *crystal fractionation.* Von Bunsen (1853), more famed for his early exploits in chemistry, chemically analyzed basalts from Iceland for some major oxides and concluded that chemical trends could be explained by crystal fractionation or magma mixing, a forerunner to modern petrochemistry. Finally, Fouque and Michel-Levy (1878) artificially produced basalts by melting appropriate mixtures of natural minerals and then cooling them.

Other early workers, such as Scrope, Dana, Sorby, von Richthofen, Iddings, and Daly should be mentioned as well. N. L. Bowen's initial work in the early 1900s, however, clearly marked the beginning of the modern era for all types of igneous petrology, especially basalt petrology. For this reason, Part II begins with part of one of Bowen's best papers. This early work culminated in the publication of Bowen's classic book, *The Evolution of the Igneous Rocks,* in 1928.

Few materials from any of the several excellent books on basalts or other igneous rocks are reproduced in this book. Thus, we must credit some of the books, or chapters in books, of special significance. They range in sophistication from undergraduate texts to professional treatises. As in the brief survey given here on early theories about basalts, it is most straightforward to cite these books in chronological order. Note the rapid increase in the number of titles since the 1970s.

Daly, 1914, *Igneous Rocks and the Depths of the Earth* (Chapter 16);
Bailey et al., 1924, *Tertiary and Post-Tertiary Geology of Mull, Loch Aline, and Oban;*
Wager and Deer, 1939, *The Petrology of the Skaergaard Intrusion, Kangerdlugssuaq, East Greenland;*
Wahlstrom, 1950, *Theoretical Igneous Petrology;*
Turner and Verhoogen, 1960, *Igneous and Metamorphic Petrology,* 2nd ed. (Chapters 6–8);

Barth, 1962, *Theoretical Petrology,* 2d. (Part III);
Hess and Poldervaart, 1967, 1968, *Basalts: The Poldervaart Treatise on Rocks of Basaltic Composition,* 2 vols.;
Wyllie, 1971, *The Dynamic Earth* (Chapters 7 and 8);
Hyndman, 1972, *Petrology of Igneous and Metamorphic Rocks* (Chapters 3 and 4);
Carmichael, Turner, and Verhoogen, 1974, *Igneous Petrology* (Chapters 8-10);
Ringwood, 1975, *Composition and Petrology of the Earth's Mantle;*
Yoder, 1976, *Generation of Basaltic Magma;*
Cox, Bell, and Pankhurst, 1979, *The Interpretation of Igneous Rocks* (Chapters 9 and 10);
Morse, 1980, *Basalts and Phase Diagrams;*
Sood, 1981, *Modern Igneous Petrology* (Chapters 2, 6, 7);
Best, 1982, *Igneous and Metamorphic Petrology* (Chapters 5 and 6);
Barker, 1983, *Igneous Rocks* (Chapter 11).

We also must mention a recent tribute to N. L. Bowen, edited by H. S. Yoder, Jr. (1979) and entitled *The Evolution of the Igneous Rocks,* in honor of Bowen's original book by the same name. The book deals directly or indirectly with basalts. Chapter 6, by I. Kushiro, provides a particularly useful state of the art summary up to 1979.

REFERENCES

Adams, F. D., 1938, *The Birth and Development of the Geological Sciences,* Dover Publications, New York, 506p.

Bailey, E. B., C. T. Clough, W. B. Wright, J. E. Richey, and G. V. Wilson, 1924, *Tertiary and Post-Tertiary Geology of Mull, Loch Aline, and Oban,* Geological Survey of Scotland Memoir, Edinburgh, 445p.

Barker, D. S., 1983, *Igneous Rocks,* Prentice-Hall, Englewood Cliffs, N.J., 417p.

Barth, T. F. W., 1962, *Theoretical Petrology, 2nd ed.,* John Wiley and Sons, New York, 416p.

Best, M. G., 1982, *Igneous and Metamorphic Petrology,* W. H. Freeman and Co., San Francisco, 630p.

Bowen, N. L., 1928, *The Evolution of the Igneous Rocks,* Princeton University Press, Princeton, N.J., 334p.

von Bunsen, R. W. E., 1853, Über die Processe der Bulkanische Gesteinsbildungen Islands, *Annales de Chimie et de Physique* **38:**215-300.

Carmichael, I. S. E., F. J. Turner, and J. Verhoogen, 1974, *Igneous Petrology,* McGraw-Hill, New York, 739p.

Cordier, P. L. A., 1816, Sur les Substances Minérales, Dites en Masse, qui Servent Debase aux Roches Volcaniques, *Jour de Physique* **84:**135-161.

Cox, K. G., J. D. Bell, and R. J. Pankhurst, 1979, *The Interpretation of Igneous Rocks,* George Allen and Unwin, London, 450p.

Daly, R. A., 1914, *Igneous Rocks and Their Origin*, McGraw-Hill, New York, 563p.

Darwin, C. R., 1844, *Geological Investigations on the Volcanic Islands during the Voyage of the H.M.S. Beagle*, Smith, Elder and Co., London, 175p.

Desmarest, N., 1774, Memoires sur l'Origine et la Nature du Basalte à Grandes Colones Polygones, Determines par l'Histoire Naturelle de cette Pierre Observée en Auvergne, *Memoires de l'Academie Royale des Sciences (An. 1771)*, Paris, 705–775.

Fouque, F. A., and A. Michel-Levy, 1878, Reproduction des Feldspathes par Fusion et par Maintien Prolongé à Une Temperature Voisine de Celle de Fusion, *Acad. Sci. Comptes Rendus* **87:**700–702.

Hall, J., 1805, Results of the Slow Cooling of Melted Rock, *Royal Soc. Edinburgh Trans.* **5:**43–48.

Hess, H. H., and A. Poldervaart, 1967, *Basalts: The Poldervaart Treatise on Rocks of Basaltic Composition*, vol. 1, Wiley-Interscience Publishers, New York, 482p.

Hess, H. H., and A. Poldervaart, 1968, *Basalts: The Poldervaart Treatise on Rocks of Basaltic Composition*, vol. 2, Wiley-Interscience Publishers, New York, 380p.

Hyndman, D. W., 1972, *Petrology of Igneous and Metamorphic Rocks*, McGraw-Hill, New York, 533p.

Morse, S. A., 1980, *Basalts and Phase Diagrams*, Springer-Verlag, New York, 493p.

Playfair, J., 1802, *Illustrations of the Huttonian Theory of the Earth*, 1956 reproduction by Dover Publications, New York, 528p.

Raspe, R. E., 1776, Columnar Jointing in Basaltic Lava: An Account of Some German Volcanoes, in K. F. Mather and S. L. Mason, eds., 1939, *Source Book in Geology*, McGraw-Hill, New York, pp. 11–13.

Ringwood, A. E., 1975, *Composition and Petrology of the Earth's Mantle*, McGraw-Hill, New York, 618p.

Sood, M. K., 1981, *Modern Igneous Petrology*, Wiley-Interscience, New York, 244p.

Spallanzani, L., 1798, Observations which have an Immediate Relation with Volcanization of the Eolian Isles—Inquiries Relative to the Origin of Basalts, *in* L. Spallanzani, *Travels in the Two Sicilies and Some Parts of the Appenines* (Anonymous English Translation of Unpublished Manuscript in Italian), G. G. and J. Robinson Publishers, London, **3:**152–215.

Turner, F. J., and J. Verhoogen, 1960, *Igneous and Metamorphic Petrology*, 2nd ed., McGraw-Hill, New York, 694p.

Wager, L. R., and W. A. Deer, 1939, Geological Investigations in East Greenland: Part III. The Petrology of the Skaergaard Intrusion, Kangerdlugssuaq, East Greenland, *Medd. Grönland 105*, 353p.

Wahlstrom, E. E., 1950, *Theoretical Igneous Petrology*, John Wiley and Sons, New York, 365p.

Werner, A. G., 1788, Bekanntmachung einer am Scheibenberger Hügel über die Entstehung des Basaltes Gemachten Entdeckung, *Allgemeiner Literaturzeitung, Intelligenzblatt* **57:**484–485.

Wyllie, P. J., 1971, *The Dynamic Earth*, John Wiley and Sons, New York, 416p.

Yoder, H. S., Jr., 1976, *Generation of Basaltic Magma*, National Academy of Sciences, Washington, D.C., 265p.

Yoder, H. S., Jr., ed., 1979, *The Evolution of the Igneous Rocks: Fiftieth Anniversary Perspectives*, Princeton University Press, Princeton, N.J., 588p.

Part I

CLASSIFICATION OF BASALTS

Editors' Comments
on Papers 1 Through 18

1 KENNEDY
Excerpt from *Trends of Differentiation in Basaltic Magmas*

2 TILLEY
Excerpt from *Some Aspects of Magmatic Evolution*

3 MACDONALD and KATSURA
Excerpt from *Chemical Composition of Hawaiian Lavas*

4 SCHWARZER and ROGERS
Erratum to *A Worldwide Comparison of Alkali Olivine Basalts and Their Differentiation Trends*

5 KUNO
Excerpt from *Differentiation of Basalt Magmas*

6 JAKEŠ and GILL
Excerpt from *Rare Earth Elements and the Island Arc Tholeiitic Series*

7 MIYASHIRO, SHIDO, and EWING
Abstract from *Diversity and Origin of Abyssal Tholeiite from the Mid-Atlantic Ridge near 24° and 30° North Latitude*

8 PEACOCK
Excerpt from *Classification of Igneous Rock Series*

9 DONNELLY and ROGERS
Abstract from *Igneous Series in Island Arcs: The Northeastern Caribbean Compared with Worldwide Island-Arc Assemblages*

10 GILL
Excerpt from *Geochemistry of Viti Levu, Fiji, and Its Evolution as an Island Arc*

11 MIYASHIRO
Excerpt from *Volcanic Rock Series in Island Arcs and Active Continental Margins*

12   GUNN, ROOBOL, and SMITH
Excerpt from *Petrochemistry of the Peléan-Type Volcanoes of Martinique*

13   IRVINE and BARAGAR
Excerpt from *A Guide to the Chemical Classification of the Common Volcanic Rocks*

14   NICHOLLS and CARMICHAEL
Excerpt from *A Commentary on the Absarokite-Shoshonite-Banakite Series of Wyoming, USA*

15   JOPLIN
Excerpt from *The Shoshonite Association: A Review*

16   VILJOEN and VILJOEN
Excerpt from *The Geology and Geochemistry of the Lower Ultramafic Unit of the Onverwacht Group and a Proposed New Class of Igneous Rocks*

17   BROOKS and HART
Excerpt from *On the Significance of Komatiite*

18   ARNDT, NALDRETT, and PYKE
Excerpt from *Komatiitic and Iron-rich Tholeiitic Lavas of Munro Township, Northeast Ontario*

Geologists have applied different names to different types of basalts for many years. Much of current usage began to be formalized in the 1920s and 1930s. One of the first major efforts was the distinction between *olivine* (or *alkali olivine*) *basalts,* which yield an alkaline differentiation sequence, and *tholeiites,* which yield a subalkaline trend. The paper by W. Q. Kennedy (1933), an excerpt from which is reprinted as Paper 1, is a classic in this effort. Tilley (1950, the abstract from which appears as Paper 2) provided a post-World War II summary of thought along these lines.

The concept of separate alkali and subalkali *primary* basalts was explored in depth by Yoder and Tilley (1962). Figure 1 is their classification, the well-known *basalt tetrahedron.* Another representation of the basalt tetrahedron (Fig. 2) was given by Green and Ringwood (1967). Bass (1972) further clarified the concept of basalts transitional between alkali and tholeiitic. A chemical rather than normative distinction between alkali basalts and tholeiites was proposed by MacDonald and Katsura (1964) for Hawaiian lavas (Paper 3). However, MacDonald and Katsura's classification system was challenged later by McBirney and Williams (1969, pp. 128–129). A summary of various chemical characteristics of alkali olivine basalts was given by Schwarzer and Rogers (1974, Paper 4).

In addition to alkali and tholeiitic basalts as potential primary magmas, Kuno (1968) introduced the term *high-alumina basalt.* The diagrams that distinguish high-alumina basalt, plus other relationships of basalt sequences, are shown in Paper 5. This concept was also discussed by Kushiro and Kuno (1963).

The term *tholeiite* has been modified in various ways. For example, Jakes and Gill (1970) distinguished tholeiites of mid-ocean ridges (MORB) from the early island-arc tholeiites (Paper 6). Miyashiro, Shido, and Ewing (1969) discussed compositional variability of MORB

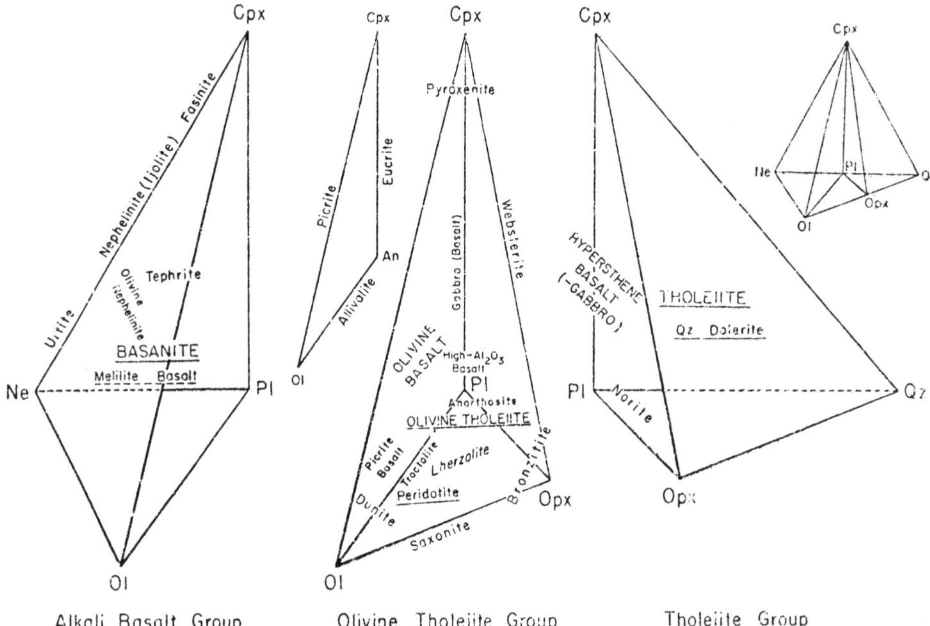

Figure 1. Exploded view of generalized simple basalt system illustrated in upper right inset. Entered are names of rocks whose major normative phases are contained in the tetrahedra. Names underlined are within tetrahedra. Names in faces are written parallel to base. Lherzolite is in Ol-Cpx-Opx face. An additional inset, Ol-Cpx-An, gives alternative nomenclature when plagioclase is rich in An. *(From H. S. Yoder, Jr., and C. E. Tilley, 1962, Origin of Basalt Magmas: An Experimental Study of Natural and Synthetic Rock Systems, Jour. Petrology 3:352; copyright © 1962.)*

and described high-alumina and low-alumina varieties (Paper 7). Further detailed work on MORB has been reported by a number of workers; we present the abstract (Paper 42) of an important paper by Sun, Nesbitt, and Sharaskin (1979) in Part IV. We also call attention to extensive investigations by J. G. Schilling and co-workers of normal oceanic ridge segments and areas affected by mantle plumes (e.g., Schilling et al., 1983).

In addition to discrimination of various types of basalts, a great deal of effort has been expended on recognition of different igneous rock series. At one time, it was assumed that each series evolved from a different type of primary magma. Hence, there has been an unfortunate tendency to refer, for example, not only to tholeiitic basalts but also to tholeiitic differentiation sequences. This tendency to use the same term both for rock types and rock series complicates the already noticed tendency for different writers to define the same term in

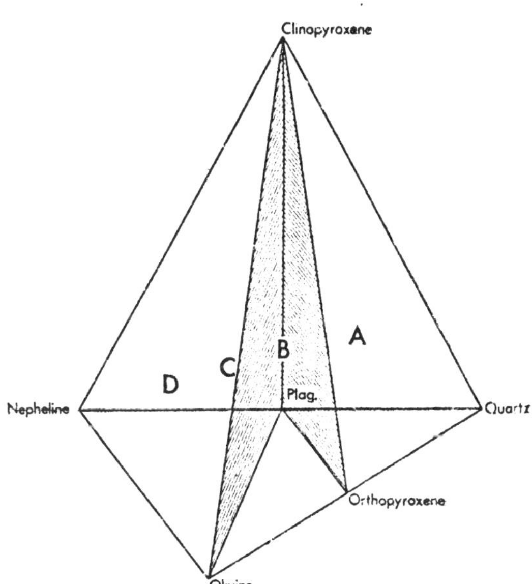

Figure 2. Diagrammatic representation of the major mineralogy of basalts using the "basalt tetrahedron" (Yoder and Tilley, 1962). The plane olivine—clinopyroxene—plagioclase is referred to as the "critical plane of undersaturation". *A* field of quartz tholeiites; *B* field of olivine tholeiites; *C* field of alkali olivine basalts; *D* field of olivine basanites. *(From D. H. Green and A. E. Ringwood, 1967, The Genesis of Basaltic Magmas, Contr. Mineralogy and Petrology 15:106; copyright © 1967.)*

different ways. Geologists who seek a tidy consistency in terminology will not find it in the petrologic literature.

One of the earliest, still used, classifications of igneous sequences (Paper 8) was given by Peacock (1931). Peacock defined various terms, including *calc-alkalic,* which has also been written as *calcalkaline* and *calc-alkali.* The term has been a magnet for new definitions. Kuno, for example, originally identified *pigeonitic* and *hypersthenic* series in Japan, but he later reclassified these terms as *tholeiitic* and *calcalkaline* respectively (Paper 5). Kuno's *critical iron enrichment line,* separating tholeiitic (iron-enriched) from calcalkaline (not iron-enriched) suites on an AFM (alkali-iron oxide-magnesia) diagram, is the middle line on the AFM diagram (Paper 4) reproduced from Schwarzer and Rogers (1975). Donnelly and Rogers (1980, abstract given as Paper 9) recognized two presumably subduction-related sequences in the Caribbean: (1) a primitive island-arc sequence roughly equivalent to Jakeš and Gill's (1970) island-arc tholeiites and (2) a calcalkaline sequence having many of the characteristics of Kuno's hypersthenic series. Many writers have assumed that calcalkaline and high-alumina suites are more or less equivalent, although Kuno (1968) specifically indicated that high-alumina basalts could yield either pigeonitic or hypersthenic suites by fractionation.

Other classifications of igneous suites, re-using the terms *tholeiitic* and *calcalkaline,* include those of Gill (Paper 10) and Miyashiro (Paper 11). Gunn, Roobol, and Smith (Paper 12) used a different set of terms for these rock suites. A suite that is, say, calcalkaline by one definition is commonly also calcalkaline by other definitions, although that consistency is not always found. It is noteworthy that most of these classifications are based, at least partly, on variations in abundances of alkali elements (e.g., Gill; Gunn, Roobol, and Smith), Fe/Mg ratios (e.g., Miyashiro), or all of these elements (e.g., Kuno, 1968). A thorough effort to describe types of basalts and their differentiation sequences was given by Irvine and Baragar (Paper 13).

We must comment on two terms that have been used extensively in recent years—*shoshonite* and *komatiite.* Shoshonites were originally described by Iddings (1895) as porphyritic rocks with a set of characteristics more recently summarized by Nicholls and Carmichael (Paper 14). Joplin (1968) adopted the term for certain high-K rocks (Paper 15); her paper also distinguished among high-K shoshonites, high-K members of alkali basalt suites, and high-Na members of alkali basalts. Gill (Paper 10) identified shoshonites as a subduction-related suite separate from calcalkaline and tholeiitic suites. Rogers et al. (1974) hesitantly used the term *shoshonite* for certain high-K rocks in the western United States. On a K_2O-SiO_2 diagram, Joplin's shoshonites either show little change in K_2O with SiO_2 or a decrease in K_2O as SiO_2 increases; Gill's shoshonites show a rapid increase in K_2O as SiO_2 increases. Rogers et al. found no relationship between K_2O and SiO_2.

Having clarified that problem, we now can turn to komatiites. These ultramafic flow rocks were originally identified by Viljoen and Viljoen (1969, Paper 16) in the Barberton Mountain belt of southern Africa. Characteristics of different types of komatiites have been described in more detail by Brooks and Hart (1974, Paper 17) and Arndt, Naldrett, and Pyke (1977, Paper 18).

REFERENCES

Bass, M. N., 1972, Ocurrence of Transitional Basalt, *Lithos* **5:**57–67.

Donnelly, T. W., and J. J. W. Rogers, 1980, Igneous Series in Island Arcs: The Northeastern Caribbean Compared with Worldwide Island-arc Assemblages, *Bull. Volcanol.* **43-2:**347–382.

Green, D. H., and A. E. Ringwood, 1967, The Genesis of Basaltic Magmas, *Contr. Mineralogy and Petrology* **15:**103–190.

Jakeš, P., and J. Gill, 1970, Rare Earth Elements and the Island Arc Tholeiitic Series, *Earth and Planetary Sci. Letters* **9:**17–28.

Joplin, G. A., 1968, The Shoshonite Association: A Review, *Geol. Soc. Australia Jour.* **15**(2):275–294.

Iddings, J. P., 1895, Absarokite-Shoshonite-Banakite Series, *Jour. Geology* **3:**935–959.

Kennedy, W. Q., 1933, Trends of Differentiation in Basaltic Magmas, *Am. Jour. Sci.* **25:**239–256.

Kuno, H., 1968, Differentiation of Basalt Magmas, in *The Poldervaart Treatise on Rocks of Basaltic Composition,* vol. 2, *Basalts,* H. H. Hess and A. Poldervaart, eds., Wiley–Interscience, New York, pp. 623–688.

Kushiro, I., and H. Kuno, 1963, Origin of Primary Basalt Magmas and Classification of Basaltic Rocks, *Jour. Petrology* **4:**75–89.

McBirney, A. R., and H. Williams, 1969, Geology and Petrology of the Galapagos Islands, *Geol. Soc. America Mem.* **118,** 197p.

MacDonald, G. A., and T. Katsura, 1964, Chemical Composition of Hawaiian Lavas, *Jour. Petrology* **5:**82–133.

Miyashiro, A., F. Shido, and M. Ewing, 1969, Diversity and Origin of Abyssal Tholeiite from the Mid-Atlantic Ridge Near 24° and 30° North Latitude, *Contr. Mineralogy and Petrology* **23:**38–52.

Peacock, M. A., 1931, Classification of Igneous Rock Series, *Jour. Geology* **39:**54–67.

Rogers, J. J. W., B. C. Burchfield, E. W. Abbott, J. K. Anepohl, A. E. Ewing, P. J. Koehnken, J. M. Novitsky-Evans, and S. C. Talukdar, 1974, Paleozoic and Lower Mesozoic Volcanism and Continental Growth in the Western United States, *Geol. Soc. America Bull.* **85:**1913–1924.

Schilling, J. G., M. Zajac, R. Evans, T. Johnston, W. White, J. D. Devine, and R. Kingsley, 1983, Petrologic and Geochemical Variations along the Mid-Atlantic Ridge from 29 N° to 73 N°, *Am. Jour. Sci.* **283:**510–586.

Schwarzer, R. R., and J. J. W. Rogers, 1974, A Worldwide Comparison of Alkali

Olivine Basalts and Their Differentiation Trends, *Earth and Planetary Sci. Letters* **23:**286–296.

Sun, S. -S., R. W. Nesbitt, and A. Ya. Sharaskin, 1979, Geochemical Characteristics of Mid-Ocean Ridge Basalts, *Earth and Planetary Sci. Letters* **44:**119–138.

Tilley, C. E., 1950, Some Aspects of Magmatic Evolution, *Quart. Jour. Geol. Sci.* **106:**37–61.

Yoder, H. S., Jr., and C. E. Tilley, 1962, Origin of Basalt Magmas: An Experimental Study of Natural and Synthetic Rock Systems, *Jour. Petrology* **3:**342–532.

1

TRENDS OF DIFFERENTIATION IN BASALTIC MAGMAS.*

W. Q. KENNEDY.

INTRODUCTION.

The origin of the two main lines of magmatic descent, the alkaline series on the one hand and the calc-alkaline on the other, constitutes a problem which is of fundamental importance in modern petrogenesis and has long occupied a conspicuous position in petrological thought.

To-day there are two main theories concerning the evolution of alkaline magmas both of which have certain facts in common. Daly and his followers regard "primary basalt magma" as the universal parent from which all the different rock associations have been derived but believe that alkaline types arise by the desilication of normal calc-alkaline magmas when the latter assimilate limestone or other calcic sediments.[1] Bowen, on the other hand, derives both alkaline and calc-alkaline types from primary basalt magma by processes of fractional crystallization but considers that *normally* basalt magma gives only quartzose late differentiates.[2]

In both theories there is a tendency to regard the alkaline magmas as abnormal and due to the operation of exceptional factors. The geographical and geotectonic relationships of the two magma series, however, seem to the author to be at variance with the conception of abnormality in either case, a conception which arises from the attempt to explain all igneous rocks as products of a single parent magma subjected to varying conditions of cooling. It seems much more probable that some factor other than the physical conditions of crystallization is involved and, in the present paper, an attempt is made to show that the alkaline and calc-alkaline rock associations belong to two distinct lines of descent, each of which has its origin in a separate parent basaltic magma.

THE PRIMARY BASALT MAGMAS.

The body of evidence, both experimental and geological, is undoubtedly such as to justify the conclusion that primary

---

* Published by permission of the Director, H. M. Geological Survey.

[1] Daly, R. A., Igneous Rocks and their Origin, New York, 1913, pp. 410-445. Shand, S. J., Geol. Mag., **67**, 415-427, 1930. This paper gives many references to other literature on the subject.

[2] Bowen, N. L., The Evolution of the Igneous Rocks, Princeton, 1928, p. 235.

13

basaltic magma is to be regarded as the parent of the majority
of igneous rock series.   At the same time some confusion
appears to exist regarding the nature of the primary basalt,
and recent investigations have demonstrated the existence of
two distinct types of basaltic magma, both of which possess
characters which indicate that they cannot be derivatives of
some pre-existing basic magma.[3]

Primary basalt magma must possess certain characters.   It
must represent a true liquid and, in addition, show world-wide
distribution, uniformity of composition and great aggregate
bulk.   If we exclude porphyritic rocks on the ground that they
do not represent original liquids,[4] there remain two basaltic
types which fulfil the conditions required in a *primary* magma;
these are (a) the olivine-basalts and dolerites and (b) the
olivine-free basalts, tholeiites and quartz-dolerites.

Both types were recognized among the plateau basalts of
the Hebrides, and Bailey and Thomas, who made the first
attempt to give a close definition of their characters,[5] showed
that in petrological discussion it is necessary to distinguish
between them.   In the Mull memoir they were designated as
the *Plateau Magma-Type* and *Non-Porphyritic Central
Magma-Type* respectively, but as these terms possess only a
local significance the more widely applicable terms, *Olivine-
basalt Magma-Type* and *Tholeiitic Magma-Type,* will be used
in the present paper.

In a recent paper the author showed that the Tholeiitic
Magma occurred in enormous bulk and was represented by the
great majority of world plateau basalts.[6]   This fact at once
elevated it to the status of a primary magma-type, a conclusion which, unknown to the writer, had previously been recognized by Wahl.[7]   We are led, therefore, to the conclusion that,
instead of the single primary basalt magma postulated by

[3] This statement is relative and does not exclude the possibility that the
two magmas may be complementary differentiates of some deep-seated original melt.   It seems evident, however, that so far as petrogenesis is concerned they must be regarded as primary, for their origin is probably to be
referred to cosmical rather than petrological processes.

[4] It has been demonstrated by Tsuboi and Bowen that the bulk composition of a porphyritic rock does not possess any genetical significance.   It
may or may not represent some original liquid and as doubt is constantly
present on this account it is better to include only aphanitic rocks in the
present discussion.

[5] Bailey, E. B. and Thomas, H. H., The Tertiary and Post-Tertiary
Geology of Mull, etc., Mem. Geol. Sur., 13-28, 1924.

[6] Kennedy, W. Q., The Parent Magma of the British Tertiary Province,
Summary of Progress for 1930, Pt. 2, Mem. Geol. Sur., 62-67, 1931.

[7] Wahl, W., Fenna, **24**, No. 3, 69, 1908.

Daly,[8] it is necessary to recognize two primary basalt types and to consider the part that each may play in the evolution of igneous rocks.

The mineralogical and chemical characters of the olivine-basalt and tholeiitic magma-types have been described in detail by Bailey and Thomas[9] and by the writer[10] so that only a short statement is necessary here.

Olivine-Basalt Magma-Type. The essential minerals are olivine, augite, basic plagioclase and iron ore. The pyroxene is a diopsidic or basaltic augite, often a titaniferous variety. A little residual, interstitial material may be present and this is of alkaline nature, without free quartz. The composition assigned to the magma-type by Bailey and Thomas is shown in Table I.

Rocks belonging to this magma-type are very widespread. They constitute the common rock-type of the oceanic islands and in alkaline provinces generally. As plateau lavas they show their greatest development in the Thulean province and form, in addition, many of the great shield volcanoes. Many of the Patagonian plateau basalts appear to be of this type[11] and some, at least, of the Siberian Traps.[12]

Tholeiitic Magma-Type. The essential minerals are pyroxene, basic plagioclase and iron ore. Olivine is either completely absent or present in very subordinate amount. Characteristically an interstitial, acid residuum is developed which may be glassy but is dominantly quartzo-feldspathic. The pyroxene belongs typically to the enstatite-augite (pigeonite) series of lime-poor pyroxenes. The composition is given in Table I.

TABLE I.

	Olivine-basalt Magma-Type	Tholeiitic Magma-Type
SiO_2	45	50
Al_2O_3	15	13
Fe_2O_3 $+FeO$	13	13
MgO	8	5
CaO	9	10
Na_2O	2.5	2.8
K_2O	0.5	1.2

[8] Daly, R. A., Igneous Rocks and their Origin, New York, 1913.
[9] Bailey, E. B. and Thomas, H. H., Mull Memoir. pp. 13-18.
[10] Kennedy, W. Q., op. cit., pp. 62-67.
[11] Baeckström, O., Bull. Geol. Inst. Upsala, **13**, 115-182, 1915. Backlund, H., Geol. Mag., **63**, 418, 1926.
[12] Backlund, H., Acta Acad. Abo., Math. et Phys., **1**, 1-29, 1922.

Basalts and dolerites belonging to the tholeiitic magma-type show a great regional distribution. The Deccan Traps and the majority of the plateau basalts which have been studied are of tholeiitic composition, and we must include also the quartz-dolerites of the great sill swarms such as the Karroo dolerites of South Africa and the Parana sills of South America. To this group belong the massive sills of Sudbury type in North America. It is probable, also, that many of the so-called andesites of the Cordilleras are actually olivine-free basalts, in which case they would also be reckoned among the tholeiitic types. To sum up, the tholeiitic magma occurs in volume which is at least of the same order of magnitude as that of the olivine-basalt magma.

The enormous volume, wide distribution in space and time and relative uniformity of composition which mark the two magma-types seem to preclude the possibility that either is a derivative of some other magma. It is necessary, therefore, to consider the origin and differentiation of igneous rocks in the light of the existence of two possible basaltic parents, and to determine in what degree the original differences of composition shown by the two magmas are maintained throughout their subsequent cooling history.

[*Editors' Note:* Material has been omitted at this point.]

2

SOME ASPECTS OF MAGMATIC EVOLUTION

C. E. Tilley

THE ANNIVERSARY ADDRESS OF THE PRESIDENT, PROFESSOR CECIL EDGAR TILLEY, PH.D. F.R.S., AT THE ANNUAL GENERAL MEETING OF THE SOCIETY, 26 APRIL, 1950.

SUMMARY

The address is concerned with some aspects of magmatic evolution in volcanic successions, and includes particularly a discussion of the nature and genetic relations of primary and derivative basic and intermediate magmas.

Among the volcanic associations of the oceanic areas, those of the Hawaiian archipelago provide data of critical importance in the study of basaltic evolution. The established chronological succession of magmas in these islands makes it probable that the alkali-poor tholeiites to be recognized in the lavas of the primitive shield volcanoes form the closest approach to the primary magma of the ocean basins, and from them derive by fractional crystallization processes the Hawaiian alkali magma series of alkali olivine-basalt, mugearite (so-called andesite) and trachyte.

The course of volcanic variation there exhibited provides a new orientation to the study of the genetic evolution of the Tertiary igneous Brito-Icelandic province, and leads to the conception that the Hebridean alkali magma series derives through tholeiites as an alternative line of descent.

The course of fractional crystallization of tholeiite magma in the great layered intrusion with its final product of granophyre is illustrative of the great flexibility in line of descent which fractionation processes permit. Derivation by crystallization differentiation in a convecting system comparable to that demonstrated in layered intrusions seems, indeed, the most likely mechanism for the genesis of the primitive sialic layer of the crust itself.

The problem of intermediate magma in the volcanic successions of the orogenic belts is critical in petrogenic theory. Though fractionation processes appear most competent to give a liquid line of descent through andesite to rhyolite, normal andesitic magma does not reflect in its chemistry the course of basaltic (tholeiitic) differentiation.

Consideration is devoted to the conception that andesites, though ultimately of basaltic parentage, have derived in part by sialic contamination.

The testing of the genetic relations of assemblages within the basalt-andesite range of the volcanic suites of the orogens remains one of the most important problems of volcanic petrogenesis.

Systematic attack through intimate phase analysis and geochemical study may provide the data which will go far to solve, not only the status of the voluminous andesite, but of intermediate magma in general in igneous rock evolution.

17

3

CHEMICAL COMPOSITION OF HAWAIIAN LAVAS

G. A. Macdonald and T. Katsura

[*Editors' Note:* In the original, material precedes and follows this excerpt.]

NOMENCLATURE OF BASALTIC ROCKS

During the past decade the terms tholeiite and alkali olivine basalt have been
in common usage as defined by Kennedy (1933), with the substitution of alkali
olivine basalt for Kennedy's term 'plateau basalt' (Tilley, 1950). As thus defined,
tholeiite is a rock essentially saturated in silica, in which magnesian olivine bears
a reaction relationship with Ca-poor pyroxene, whereas alkali olivine basalt is
an undersaturated rock, in which magnesian olivine and calcic clinopyroxene
crystallize simultaneously and in equilibrium with each other. Various other
characteristics of the two rock types have been described (Kuno *et al.*, 1957), but
the essential difference has been considered to be the presence or absence of the
reaction relationship between olivine and Ca-poor pyroxene. Thus, it has been
recognized that some rocks that are closely related to the tholeiites are considerably undersaturated, but that nevertheless the reaction relationship holds, and
the pyroxene is predominantly Ca-poor. A fact that appears to have received
too little emphasis is that the alkali olivine basalts, as defined by the mineralogical relationships, actually are alkalic, in the sense of containing a larger
proportion of alkalis than tholeiitic rocks with the same silica content—a fact
that was presumably implicit in the original choice of the name.

Recently, Yoder and Tilley (1962, p. 352) have proposed the division of
basaltic rocks into the following five groups, on the basis of their normative
composition:

1. Tholeiite (oversaturated): normative hypersthene and quartz.
2. Tholeiite (saturated; hypersthene basalt): normative hypersthene.
3. Olivine tholeiite (undersaturated): normative hypersthene and olivine.
4. Olivine basalt: normative olivine.
5. Alkali basalt: normative olivine and nepheline.

The division between tholeiite and olivine tholeiite is taken as the plane of
silica saturation (En–Di–Ab), and that between olivine tholeiite and alkali

18

basalt as the critical plane of silica undersaturation (Fo–Di–Ab), in their Di–Fo–Ne–Qz tetrahedron. Olivine basalt lies directly on the Fo–Di–Ab plane, and contains neither hypersthene nor nepheline in the norm. They emphasize, however, that modal composition should take precedence over normative composition in classifying rocks, and they emphasize, also, the uninterrupted gradation, chemically, from one type of basalt to another.

The above classification limits the term *alkali basalt* to only a portion of the rocks formerly included under that name, and the term *olivine basalt* to only a small portion of the basaltic rocks containing olivine. Indeed, olivine basalt is reduced almost to the vanishing-point! There are very, very few basalts that contain neither normative hypersthene nor nepheline! Both restrictions seem to us undesirable. Basalt and olivine basalt are very useful general terms in field work and general petrology. An attempt to restrict the term olivine basalt to only a very small portion of the olivine-bearing basaltic rocks will inevitably lead only to partial acceptance, and consequently to confusion. We shall have 'olivine basalt' used in two different senses, one specific and one general, just as we now have two meanings for the term 'granite'; and certainly most petrologists will agree that this duality of usage is to be avoided if possible! If any name is needed for this minute group of basalts containing olivine but no nepheline or hypersthene in the norm, which seems doubtful, it should be a new one, not 'olivine basalt', which has long been established in a much broader usage.

We propose, therefore, to retain *olivine basalt* as a general term for all basaltic rocks containing a significant proportion of modal olivine. The amount of olivine to be regarded as significant is arbitrary, but there is no apparent reason for abandoning the figure of 5 percent used previously (Macdonald, 1949, p. 1544). Where chemical or microscopic analysis indicates the rock to be tholeiitic, the terms tholeiite or tholeiitic basalt, and olivine tholeiite or tholeiitic olivine basalt, should be used where the greater precision is desirable.

However, if olivine basalt is retained as a general term, some other name is needed for the rocks called 'olivine basalt' by Yoder and Tilley. The fundamental characteristics of these rocks include not only undersaturation in silica, evidenced by the presence of normative olivine, but also a greater abundance of alkalis, at any given silica percentage, than in the corresponding rocks of the tholeiitic suite. Indeed, their undersaturated character is more the result of more abundant alkalis than of less abundant silica. The essential factor, i.e. richness in alkalis, is called to immediate attention by the designation *alkalic basalt*, formerly used by Tilley (1950), and we propose to retain that designation.

For the more strongly undersaturated rocks that contain a significant amount of nepheline in the norm, well-established names exist. The amount of nepheline regarded as significant is again arbitrary. A figure of 5 percent has previously been suggested (Macdonald, 1949, p. 1544), and there appears to be no adequate reason to change it. We propose, therefore, to call rocks of this group containing less than 5 percent normative nepheline, *alkalic basalt*; and those that contain 5 percent or more normative nepheline *basanite* if the nepheline is also modal, and *basanitoid* if it is occult.

The presence or absence of hypersthene in the norm depends on factors other than merely the degree of alkali richness and silica poorness. One of the principal of these is the state of oxidation of the iron—a factor not necessarily closely related to the degree of oxidation of the magma as it rises from depth. This is clearly shown by the fact that rocks having the mineral composition of alkalic basalts or the closely related hawaiites, and previously classified in the alkalic suite, may contain up to several percent of normative hypersthene (Yoder & Tilley, 1957, p. 158). The presence or absence of normative hypersthene does not, therefore, seem to be an appropriate basis for separating the tholeiitic from the alkalic basalts.

It is pointed out on a later page (p. 91) that classification on the basis of modal composition is sometimes difficult even in rocks that are very largely or wholly crystalline, and in glassy rocks it is impossible. Some chemical scheme for separating the rocks of the tholeiitic and alkalic suites is, therefore, very desirable even though the modal composition continues to be the main basis for the recognition of the two suites. The presence or absence of normative hypersthene, as used by Yoder and Tilley, constituted such a scheme. If the hypersthene criterion is removed, some other criterion is necessary for separating the tholeiitic and alkalic basalts on the basis of chemical composition. The best seems to be the proportion of alkalis to silica. Yoder and Tilley (1962, p. 355) comment that in alkalic basalts the total alkalis usually exceed 3 percent. However, the amount of alkalis must be considered in relation to the amount of silica, as appears to have been recognized, though not explicitly stated, by Yoder and Tilley (1962, Table 2, p. 361). In some siliceous tholeiites the alkalis exceed 3 percent, whereas some silica-poor alkalic basalts contain less than 3 percent total alkalis.

Fig. 1 is an alkali:silica diagram in which are plotted all the analysed Hawaiian rocks that fall within the compositional limits of the diagram. Alkalic rocks of the nephelinic suite (Macdonald & Katsura, 1961, p. 360) have been included with those of the major alkalic suite, but ankaramites have been indicated by a different symbol. As pointed out by Kuno and his associates (1957, p. 203), there is a remarkably clean separation of rocks that have the respective mineral compositions of the two suites. The field of the tholeiitic rocks can be separated from that of the alkalic rocks by a diagonal line drawn through the region of low population that lies between the respective population concentrations of the two suites. H. S. Yoder, Jr. has suggested (personal communication, July 1963) that the line approximates the trace of the critical plane of silica undersaturation—a suggestion that appears very plausible, since it is the abundance of alkalis that largely determines whether or not the available silica is adequate to produce saturated minerals in the norm calculation. It is proposed herein to use this empirical line, whatever its cause, as the boundary between the two suites.

All the clearly alkalic rocks fall above the line, except for three ankaramites.

The latter rocks belong mineralogically with the alkalic suite, but because of their great enrichment in augite and olivine (presumably by accumulation of phenocrysts) they have a very small salic fraction, and consequently low alkali content. No rocks with definite tholeiitic mineral composition have been found to lie above the line. A plot of a tholeiitic basalt in the region above the line in a previously-published diagram (Macdonald & Katsura, 1962, p. 189) has been found, on re-study of the rock, to have been in error. The rocks close below the line appear to contain typical tholeiitic minerals, and are transitional toward the alkalic basalts only in the sense of containing more alkalis than other tholeiitic rocks of the same silica content.

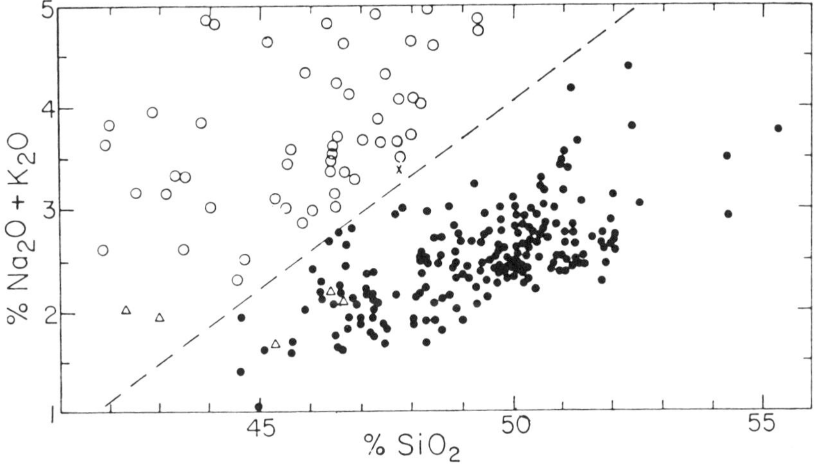

FIG. 1. Alkali:silica diagram of Hawaiian basaltic rocks, showing the boundary (diagonal dashed line) between the tholeiitic and alkalic fields. Rocks with tholeiitic mineral composition are shown by solid dots, rocks with alkalic mineral composition by open circles, and ankaramites by triangles. The liquid that rose into the drill hole through the crust on the Kilauea Iki lake is shown by an *X*.

Thus, the rocks of the two suites as separated by the line in Fig. 1 correspond modally with Yoder and Tilley's criterion of presence or absence of hypersthene or its near equivalent, pigeonite, even although normative hypersthene is present in some of the alkalic rocks.

A rock from Niihau described and analysed by Washington and Keyes (1926, p. 347, analysis 4), which would be expected from its associations in the post-erosional lavas of that island to be alkalic olivine basalt, falls just below the line in the field of the tholeiitic suite. However, the rock is not available for re-examination and its mineralogical affinities are unknown. It has been omitted from Fig. 1.

The liquid that rose into the drill hole through the crust on the Kilauea Iki lava lake is unquestionably derived from tholeiitic magma (Macdonald & Katsura, 1961), but it falls just above the line, in the alkalic field. The rock contains unresorbed olivine phenocrysts, and microphenocrysts of augite with large 2V, but the groundmass is largely glassy. Its position is shown on Fig. 1, in the alkalic field just above the boundary line. Its alkalic affinity is suggested also by its high titania content (see p. 105). If this rock is truly alkalic, its theoretical importance is obvious.

Thus, the diagonal line in Fig. 1 gives a very satisfactory division between the tholeiitic and alkalic suites in the Hawaiian region. To what extent it can be applied more widely is not yet certain, but an examination of the literature indicates that the same boundary line probably is generally applicable in other regions. Alkalic basalt of lower Eocene age in western Washington and Oregon lies very close to the line (P. D. Snavely, Jr., personal communication, May 1963), as also does one tholeiitic basalt from the Deccan region, and one from the Okonjeje complex in south-west Africa (Simpson, 1954, p. 169). One specimen of the late Yakima–Ellensburg type of basalt (Waters, 1961, p. 594) lies just above the line, and another lies well within the alkalic field, whereas two others lie in the tholeiitic field. This suggests that the late Yakima basalts may represent a gradation into the alkalic suite—a suggestion supported to some degree by their high content of titania.

The pegmatite of the tholeiitic Whin sill (Tomkeieff, 1929; cited by Walker & Poldervaart, 1941, p. 444) lies just on the alkalic side of the line, but those of the Palisade (Walker, 1940) and Dillsburg (Hotz, 1953) sills lie on the tholeiitic side. Some tholeiitic rocks of Mull (Bailey *et al.*, 1924, p. 17) lie on or barely above the line, and one alkalic basalt (op. cit., p. 15) falls below the line in the tholeiitic field, but these rocks are somewhat weathered (Tilley & Muir, 1962) and perhaps affected by regional alteration, and probably should be disregarded.

Only a few rocks with the mineralogic characteristics of the alkalic suite (see next section) have been found to contain less than 5 percent modal olivine. These we refer to as alkalic basalt, whereas those containing 5 percent or more modal olivine we term alkalic olivine basalt. However, alkalic basalt is used also as a non-specific term for all basaltic rocks with alkalic mineral composition or for those lying in the alkalic part of the field in Fig. 1.

In summary, the classification of basaltic rocks used in the following pages is as follows:

(1) Tholeiitic suite: rocks with tholeiitic mineral composition, falling below the boundary line in Fig. 1.

(*a*) Tholeiitic basalt—containing less than 5 percent modal olivine.
(*b*) Tholeiitic olivine basalt—containing 5 percent or more modal olivine.
(*c*) Oceanite (picrite–basalt of oceanite type)—containing very abundant phenocrysts of olivine, and less than 30 percent feldspar.

(2) Alkalic suite: rocks with alkalic mineral composition, falling above the boundary line in Fig. 1.

(*a*) Alkalic basalt—containing less than 5 percent modal olivine.

(*b*) Alkalic olivine basalt—containing 5 percent or more modal olivine, and less than 5 percent normative nepheline.

(*c*) Basanite—containing more than 5 percent normative nepheline, and with both modal nepheline and feldspar.

(*d*) Basanitoid—containing more than 5 percent normative, but no modal nepheline.

(*e*) Ankaramite (picrite–basalt of ankaramite type)—containing very abundant phenocrysts of olivine and augite, and less than 30 percent total feldspar.

(*f*) Hawaiite—a rock with moderate to high color index and frequently basaltic habit, in which the normative and modal feldspar is andesine, and with soda:potash ratio greater than 2:1 (Macdonald, 1960, p. 175).

(*g*) Mugearite—a rock similar to hawaiite but in which the feldspar is oligoclase.

The terms nepheline basalt, trachyte, and rhyodacite have their usual significance.

REFERENCES

[*Editors' Note:* Only the references cited in the preceding excerpt are reproduced here.]

BAILEY, E. B., & others, 1924. Tertiary and post-Tertiary geology of Mull, Loch Aline, and Oban. *Mem. Geol. Surv. Scotland,* 445 pp.

HOTZ, P. E., 1953. Petrology of granophyre in diabase near Dillsburg, Pennsylvania. *Bull. Geol. Soc. Amer.* **64,** 675-704.

KENNEDY, W. Q., 1933. Trends of differentiation in basaltic magmas. *Amer. J. Sci.,* ser. 5, **25,** 239-56.

KUNO, H., YAMASAKI, K., IIDA, C., & NAGASHIMA, K., 1957. Differentiation of Hawaiian magmas. *Jap. J. Geol. Geograph.* **28,** 179-218.

MACDONALD, G. A., 1949. Hawaiian petrographic province. *Bull. geol. Soc. Amer.* **60,** 1541-96.

———— DAVIS, D. A., & Cox, D. C., 1960. Geology and ground-water resources for the island of Kauai, Hawaii. *Hawaii Div. Hydrog. Bull.* **13,** 212 pp.

———— 1960. Dissimilarity of continental and oceanic rock types. *J. Petrol.* **1,** 172-7.

———— & KATSURA, T., 1961. Variations in the lava of the 1959 eruption in Kilauea Iki. *Pacific Sci.* **15,** 358-69.

———— ———— 1962. Relationship of petrographic suites in Hawaii. *Amer. Geoph. Un. Monog.* **6,** 187-95.

SIMPSON, E. S. W., 1954. The Okonjeje igneous complex, south-west Africa. *Trans. geol. Soc. S. Afr.,* **57,** 125-72.

TILLEY, C. E., 1950. Some aspects of magmatic evolution. *Jour. geol. Soc. London,* **106,** 37-61.

TILLEY, C. E., & MUIR, I. D., 1962. The Hebridean plateau magma type. *Trans. geol. Soc. Edinb.* **19,** pt. 2, 208-15.

WALKER, F., 1940. Differentiation of the Palisade diabase, New Jersey. *Bull. geol. Soc. Amer.* **51,** 1059–106.

———— & POLDERVAART, A., 1941. The Hangnest dolerite sill, S.A. *Geol. Mag.* **78,** 429–50.

WASHINGTON, H. S. & KEYES, M. G., 1926. Petrology of the Hawaiian Islands: V. The Leeward Islands. *Amer. J. Sci.* **12,** 336–52.

WATERS, A. C., 1961. Stratigraphic and lithologic variations in the Columbia River basalt. *Ibid.* **259,** 583–611.

YODER, H. S., Jr., & TILLEY, C. E., 1957. Basalt magmas. *Ann. Rept. Geophys. Lab., Carnegie Instn. Wash. Yearb.* **56,** 156–61.

———— ———— 1962. Origin of basalt magmas: An experimental study of natural and synthetic rock systems. *J. Petrol.* **3,** 342–532.

Reprinted from *Earth and Planetary Sci. Letters* **25**:91 (1975)

ERRATUM TO "A WORLDWIDE COMPARISON OF ALKALI OLIVINE BASALTS AND THEIR DIFFERENTIATION TRENDS"

R. R. Schwarzer and J. J. W. Rogers

R.R. Schwarzer and J.J.W. Rogers, A worldwide comparison of alkali olivine basalts and their differentiation trends, Earth Planet. Sci. Lett. 23 (1974) 286–296.

Page 292. Figs. 3 and 4 did not reproduce properly. As these figures are important to the proper understanding of the data presented in the article, the correct figures along with their captions are shown below.

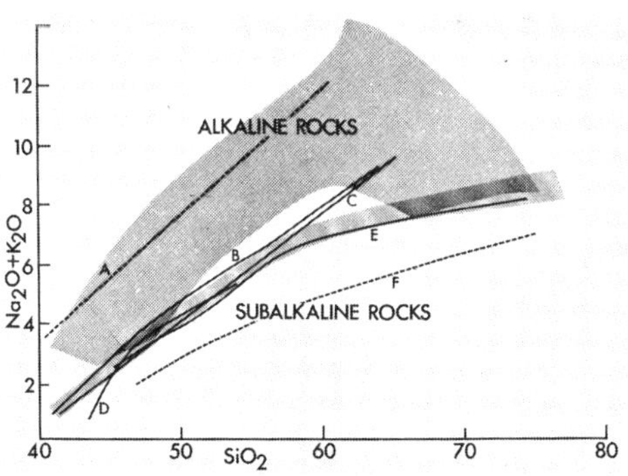

Fig. 3. ($Na_2O + K_2O$) vs. SiO_2 diagram showing the area (stippled) occupied by differentiation trends of all suites listed in Table 1. No discrimination is possible between continental, island-arc, and oceanic suites on this diagram (see text). The proposed boundary between the Alkaline Rock Series and the Subalkaline Rock Series is shown by a diagonal pattern. Curves A, B, C, D, E, and F are from Fig. 1.

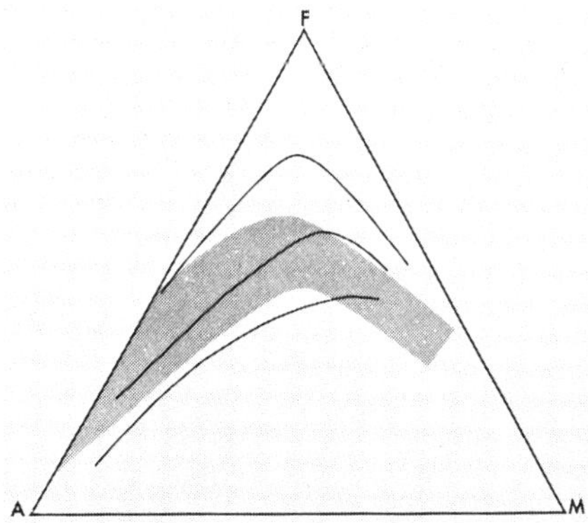

Fig. 4. AMF plot showing the differentiation trends (stippled area) of the alkali olivine basalt series listed in Table 1. No discrimination is possible between continental, island-arc, and oceanic suites on this diagram (see text). Kuno's [2] tholeiitic and calcalkaline differentiation trends of Japan are also shown.

5

DIFFERENTIATION OF BASALT MAGMAS

H. Kuno

[*Editors' Note:* In the original, material precedes and follows this excerpt.]

PARENTAL BASALT MAGMAS

Three types of basalt magmas—tholeiite, high-alumina basalt, and alkali olivine basalt—are distinguished in this paper. The distinction can be made more clearly in volcanic rocks than in plutonic rocks because the latter are greatly affected by selective crystal accumulation.

Tholeiite is characterized by the existence of a reaction relation between Mg olivine and Ca-poor pyroxenes, namely orthopyroxene and pigeonite (Tilley, 1950; Kuno and others, 1957). Therefore, Mg olivine is usually absent from the groundmass of volcanic rocks; where present, it is surrounded by reaction rims of pigeonite provided the groundmass is reasonably crystalline. In some tholeiite representing a late stage of fractionatoin, Fe olivine may appear in the groundmass. Such olivine has no reaction relation to pyroxene. Most tholeiite contains either silica minerals in the groundmass or siliceous glass.

Alkali olivine basalt is characterized by the absence of a reaction relation between olivine and pyroxene (Tilley, 1950; Kuno and others, 1957); olivine is invariably present in the groundmass. The pyroxene is usually Ca-rich clinopyroxene, although Ca-poor pyroxenes are sometimes also found. Alkali feldspar and/or zeolite, sometimes accompanied by feldspathoid, take the place of the silica minerals of the tholeiite groundmass.

The distinction of high-alumina basalt is based on two characteristics: Al_2O_3 is higher than 16.5% in aphyric rocks, and the $Na_2O + K_2O$ contents lie between those of the other two basalt types for a given SiO_2 content. Mineralogically, high-alumina basalt is transitional between the other two basalt types (Kuno, 1960). It may or may not show a reaction relation between olivine and Ca-poor pyroxene. Silica minerals may or may not be present, but alkali feldspar is usually found in the groundmass. These characteristics depend on whether the composition of rock is close to that of tholeiite or of alkali olivine basalt.

The mineralogical criteria alone are not sufficient to distinguish the three basalt types. We must use some chemical criteria, too. Because the three types are completely gradational in chemical composition, their classification is naturally arbitrary. Figure 1 is a diagram for such classification (Kuno, 1960) constructed in the following way:

Tholeiite and alkali olivine basalt whose analyses were available were identified by their mineralogy. High-alumina basalt was identified partly by its Al_2O_3 content being higher than 16.5%, using only aphyric rocks, and partly by mineralogy. Aphyric tholeiite, aphyric high-alumina basalt, and aphyric

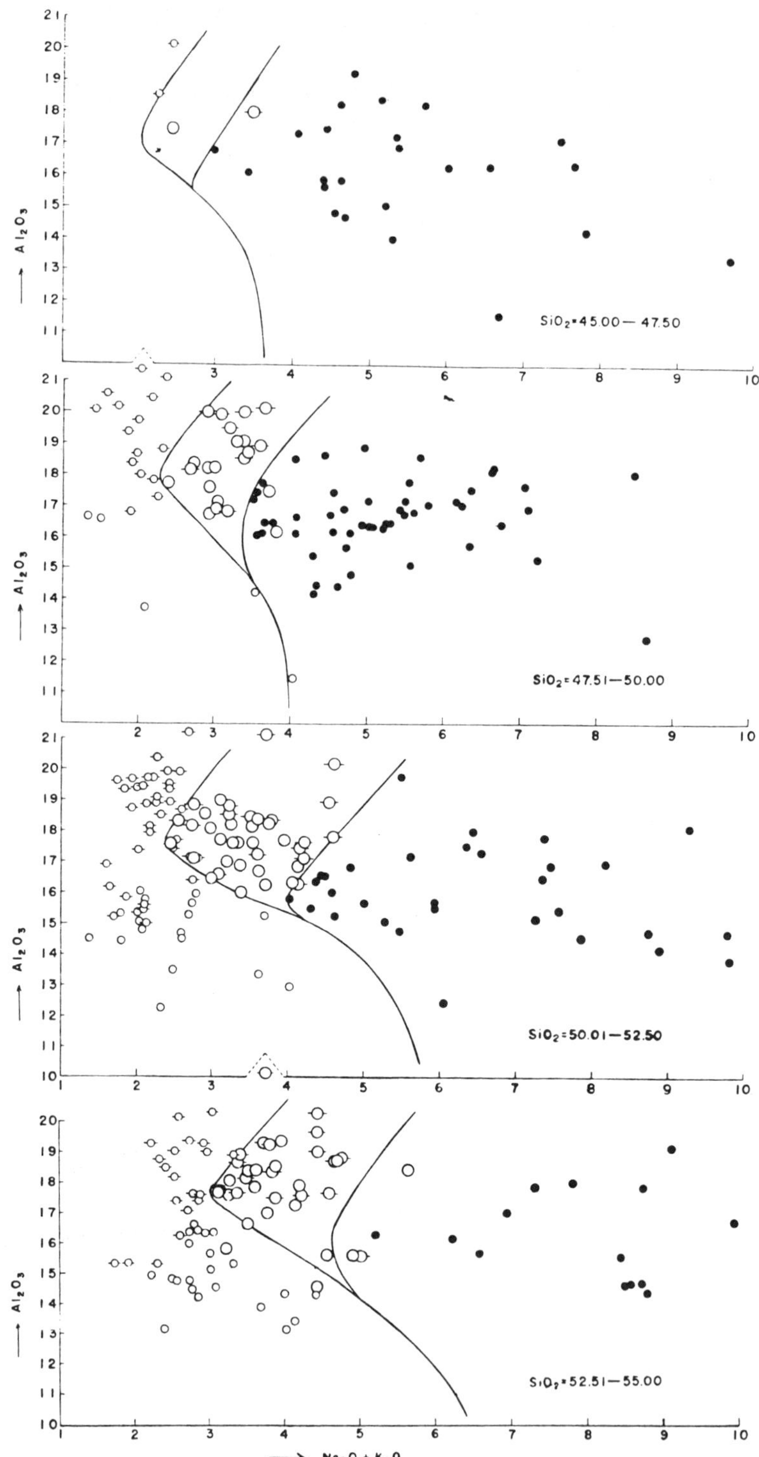

Figure 1. Al$_2$O$_3$–Na$_2$O + K$_2$O–SiO$_2$ relation in tholeiite, high-alumina basalt, and alkali olivine basalt from Japan, Korea, Manchuria, and Sakhalin. Small open circles without stripes: aphyric tholeiite; small open circles with stripes: porphyritic tholeiite; large open circles without stripes: aphyric high-alumina basalt; large open circles with stripes: porphyritic high-alumina basalt; solid circles: alkali olivine basalt, trachybasalt, and mugearite. Arbitrary boundary lines are drawn to separate the fields of the three basalt types (Kuno, 1960).

28

and porphyritic alkali olivine-basalt from central Japan, Korea, and Manchuria were classified into four groups according to their SiO_2 content, and analyses of the rocks of each group were plotted in an Al_2O_3 versus $Na_2O + K_2O$ diagram. Boundary lines between the fields of the three basalt types were drawn.

In Figure 1, in which these boundary lines are reproduced, analyses of aphyric and porphyritic basalts from the above-mentioned regions are plotted. Even the porphyritic tholeiites and high-alumina basalts generally fit within the boundary lines.

It has also been shown (Kuno, 1960) that basaltic rocks from other regions of the world classified by previous authors as tholeiite, high-alumina basalt, and alkali olivine basalt plot within their respective fields of Figure 1, except for some iron-rich basaltic rocks representing advanced stages of fractionation.

A simpler way to distinguish the three types of basalt is to plot their $Na_2O + K_2O$ contents against SiO_2. In Figure 2, analyses of aphyric and porphyritic tholeiite and high-alumina basalt from the Cenozoic volcanic province of central Japan, and alkali olivine basalt of the same age from central and south-

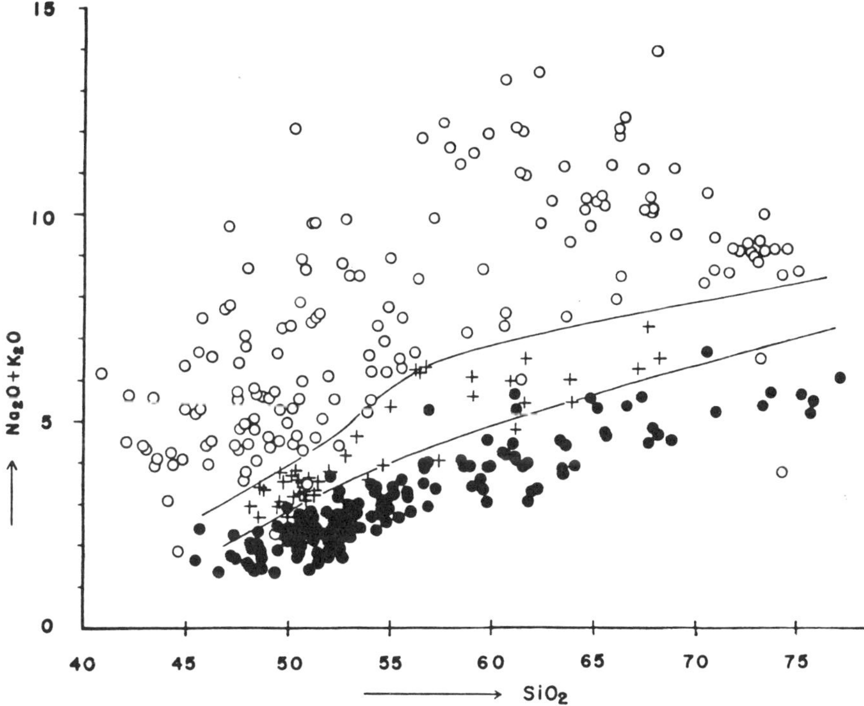

Figure 2. Alkali–silica diagram for tholeiite (solid circles) and high-alumina basalt (crosses) of central Japan and alkali olivine basalt (open circles) of central to southwestern Japan, Korea, and Manchuria, all of Cenozoic age. Their fractionation products in the same provinces—namely andesite, dacite, and rhyolite derived from the tholeiite (the pigeonitic rock series) and those derived from the high-alumina basalt (the high-alumina basalt series) and mugearite, trachyandesite, trachyte, and alkali rhyolite derived from the alkali olivine basalt—are also plotted by the respective marks. Both aphyric and porphyritic rocks are included. General boundaries between the fields of the three types of basalt and their fractionation products are shown by curves.

western Japan, Korea, and Manchuria, are plotted together with their fractionation products. So far as the alkali–silica relation is concerned, the high-alumina basalt is transitional between the tholeiite and alkali olivine basalt. It should be noted that, if all of the points for the high-alumina basalt were removed from the diagram, a distinct gap should appear between the fields of the other two basalt types. The fractionation products of high-alumina basalt (andesite and dacite) lie midway between those of the tholeiite and alkali olivine basalt. General boundary lines between the fields of these three basalt types and their fractionation products are drawn in the figure. It is not implied that these boundary lines separate compositional fields of fractionation products of the three basalt types in all other provinces; they serve merely as references whenever the alkali–silica relations of rocks of other provinces are compared with those of Japan, Korea, and Manchuria.

Figures 1 and 2 illustrate that there is a fairly wide variation in composition even among the basalts of each type. It may be supposed therefore that basalts of each type have been derived from their own parental magma or from magmas of more limited compositions. These parental magmas themselves may grade into one another.

REFERENCES

[*Editors' Note:* Only the references cited in the preceding excerpt are reproduced here.]

Kuno, H., 1960, High-alumina basalt: Jour. Petrology, **1,** 121–145

Kuno, H., 1962a, Frequency distribution of rock types in oceanic, orogenic, and kratogenic volcanic associations: The Crust of the Pacific Basin: Am. Geophys. Union Geophys. Mon. **6,** 133–139

Kuno, H., Yamasaki, K., Iida, C., and Nagashima, K., 1957, Differentiation of Hawaiian magmas: Japan Jour. Geology and Geography, **28,** 179–218

Tilley, C. E., 1950, Some aspects of magmatic evolution: Quart. Jour. Geol. Soc. London, **106,** 37–61

6

RARE EARTH ELEMENTS AND THE ISLAND ARC THOLEIITIC SERIES

P. Jakeš and J. Gill

[*Editors' Note:* Only Figure 2 is reproduced here.]

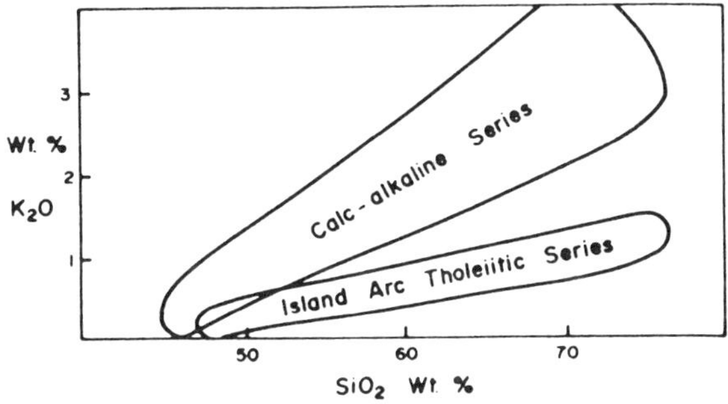

Fig. 2. Generalized K_2O-SiO_2 relationships in selected island arcs. Data from Gill [8], (fig. 8). Analyses from Tonga, Fiji, S. Kurile and S. Sandwich islands, Saipan-Guam, and the pigeonitic series of Japan and the Izu-N. Mariana islands all fall within the island arc tholeiitic series field. Analyses from the Aleutian and N. Kurile islands, central New Guinea, Bougainville, the New Hebrides, and central Viti Levu, Fiji fall within the calc-alkaline series field. Note that analyses from St. Kitts, Lesser Antilles also plot within the island arc tholeiitic field but have no iron-enrichment. Similarly, those from northern New Guinea and New Britain have characteristic iron enrichment but nevertheless plot within the calc-alkaline field.

[8] J.B. Gill, Geochemistry of Viti Levu, Fiji and its evolution as an island arc, Contr. Mineral. Petrol. 27 (1970) 179.

DIVERSITY AND ORIGIN OF ABYSSAL THOLEIITE FROM THE MID-ATLANTIC RIDGE NEAR 24° and 30° NORTH LATITUDE

Akiho Miyashiro, Fumiko Shido, and Maurice Ewing

Abstract. On cursory examination of hand specimens and thin sections, the abyssal tholeiite in a dredge haul may appear to be uniform in composition. Chemical analyses of a considerable number of fragments, however, have always revealed the existence of regular compositional variation in them. The MgO content decreases with increasing SiO_2. In abyssal tholeiites with relatively low Al_2O_3 contents, the SiO_2, total iron, Na_2O and P_2O_5 contents tend to increase and the MgO content tends to decrease with increasing iron/magnesia ratio, probably owing to crystallization differentiation.

In a certain dredge haul, high-alumina abyssal tholeiites (with Al_2O_3 contents near or over 17%) occur in association with low-alumina abyssal tholeiites. The magma of high-alumina abyssal tholeiites would be generated from that of low-alumina abyssal tholeiites by differentiation at a depth around 30 km.

In pillow lavas of abyssal tholeiite free from weathering and metamorphism, the chilled rim of the pillow usually has virtually the same chemical composition as the more crystalline core except for a decrease of K_2O content toward the rim. On the other hand, the weathered rim of pillow lavas shows marked compositional change. The Fe_2O_3/FeO ratio of unweathered abyssal tholeiite is in the range of 0.1 to 0.3. This ratio and the H_2O^- and H_2O^+ contents increase with advancing weathering.

8

CLASSIFICATION OF IGNEOUS ROCK SERIES

M. A. Peacock

[*Editors' Note:* Only Figure 1 is reproduced here.]

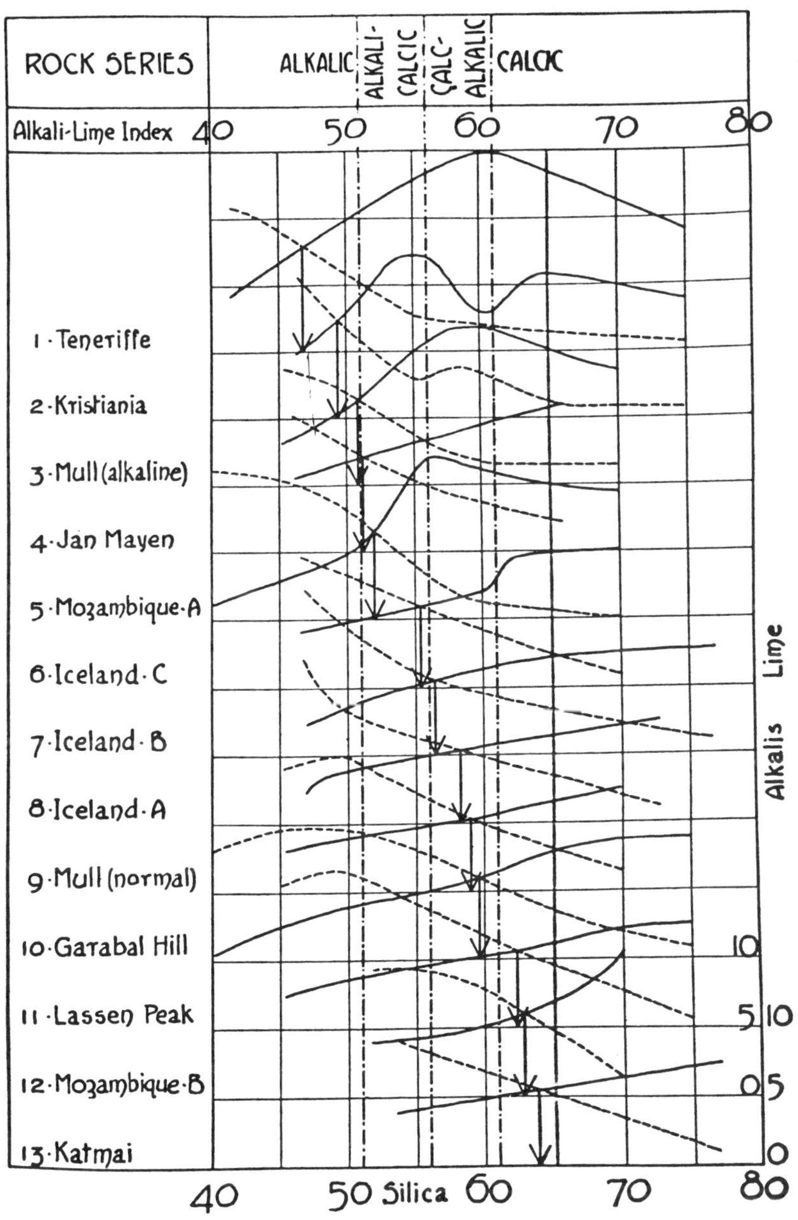

Fig. 1.—Curves for $Na_2O + K_2O$ (full lines) and for CaO (broken lines) for thirteen rock series arranged in four groups according to the "alkali-lime index" (SiO_2 value at which $Na_2O + K_2O = CaO$).

Igneous Series in Island Arcs: The Northeastern Caribbean Compared With Worldwide Island-Arc Assemblages

T.W. DONNELLY   *Department of Geological Sciences, State University of New York at Binghamton, Binghamton, N.Y. 13901, U.S.A.*

J.J.W. ROGERS   *Department of Geology, University of North Carolina at Chapel Hill, Mitchell Hall 029A, Chapel Hill, N.C. 27514, U.S.A.*

ABSTRACT

Two major magma series are recognised in the northeastern Carribean island arcs and correlated with numerous examples elsewhere, mainly in the western Pacific. The Primitive Island Arc (PIA) series is apparently the oldest in each of its island-arc occurrences and is followed by the calcalkaline (CA) series, which locally co-occurs with a high-potassium (HK) variety, especially in topographically larger island areas. Also occurring in the northeastern Caribbean are tectonically emplaced examples of metamorphosed mid-ocean ridge basalt (MORB) and a diabase dike swarm possibly related to a rifting event.

PIA series is distinguished from the CA by its lower content of LIL elements, including REE, Th, U, Zr, K, and Ba. The difference is most striking in the abundant siliceous differentiates of this series, in which these elements are scarcely higher in abundance than in mafic examples. The HK series is considered to be a subseries of the CA which is higher in K, Ba, Sr, Mg, and Ni, but has REE, Th, and other element levels very similar to the CA. The PIA series has been confused with MORB but has lower REE (with flatter patterns generally), much lower Ni, lower Ti, Zr, and Mg, and appears to be intrinsically hydrated. The abundance of siliceous differentiates also separates this series from MORB.

PIA magma is believed to be generated by massive melting of hydrated mantle which rises in front of the descending snout of a newly subducting slab at the initial stage of island-arc genesis; i.e., before the island-arc platform has formed. CA magma has a different source and is probably the partially fused material of the slab itself, more or less modified by volcanic-sedimentary materials dragged to depth during subduction.

10

GEOCHEMISTRY OF VITI LEVU, FIJI, AND ITS EVOLUTION AS AN ISLAND ARC

J. B. Gill

[*Editors' Note:* In the original, material precedes and follows this excerpt.]

For comparison, the "ferro-femic index" of Nockolds' (1954) average calcalkaline series is 42; of the Cascades, 54—70; of the Thingmuli trend, 80; and of the Skaergaard, 95 (see Coats, 1968). Fig. 8, adapted from Dickinson (1968b) gives estimated K_2O—SiO_2 curves for the areas summarized in Table 4.

"Island arc tholeiites" are characteristic of Tonga, the Mariana islands, the Izu islands and peninsula, the southern Kuriles, and the Scotia arc. They have in common average ferro-femic indices between 74—78, low K_2O and K_2O—SiO_2 slopes, and high Na_2O/K_2O ratios. They are similar to Viti Levu tholeiites in some aspects of trace element chemistry. Three Saipan samples reported by Taylor *et al.* (1969) and several Izu Peninsula samples (Mazuda, 1966, 1968) all have rare earth abundance patterns subparallel to those of chondrites and La/Yb ratios of 1—2.5. (Samples of the hypersthene rock series (Taylor and White, 1966) have typical calcalkaline rare earth contents and abundance patterns.) Th and U concentrations are variable but <0.1 ppm in one Saipan andesite and 0.1—0.3 ppm in most Izu analyses (Tatsumoto, 1966). Cs, Ba, and Sr are all low in the Saipan samples. K/Rb ratios of 500—870 in Saipan and 450 in the Scotia arc (Baker, 1968b) are lower than those of Viti Levu.

In contrast, calcalkaline suites are characterized by varying iron enrichment (ferro-femic indices between 50—70), moderate K_2O—SiO_2 slopes, Na_2O/K_2O ratios of 2—3, and trace element abundances as summarized by Taylor and White (1966) and Taylor (1968). These are found in the New Hebrides, the Solomon islands, eastern Papua in New Guinea, the northern Kuriles, and the Aleutians.

Thus all island arcs are not necessarily or even principally calcalkaline provinces. Instead, the calcalkaline suite may represent but one stage in arc evolution (Jakes and White, 1970).

Although the rocks of the Lesser Antilles are generally considered calcalkaline, they are characterized by "tholeiitic" K_2O—SiO_2 variations and more iron enrichment than other calcalkaline provinces (Nockolds and Allen, 1953; Baker, 1968a; Hatherton and Dickinson, 1969). Relative to the "average andesite" of Taylor (1968), St. Kitts samples of comparable SiO_2 have less Ni and Cr as do Viti Levu tholeiites.

Rocks of the coastal regions of the New Guinea and New Britain are considered tholeiitic by Morgan (1966) and Jakes and White (1970) due to their relative iron

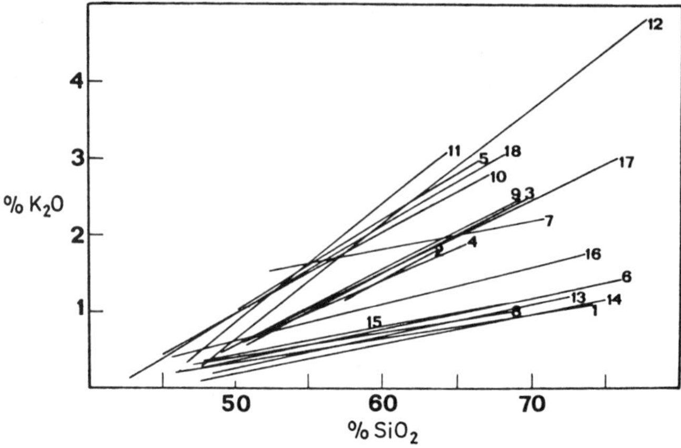

Fig. 8. Estimated K$_2$O—SiO$_2$ variation diagram for regions summarized in Table 4. Based on Dickinson (1968). Note both the continuum of slopes but also the three rather distinct populations of approximately 0.2%, 0.5%, and 0.7% K$_2$O per 5% SiO$_2$ respectively. Data as follows: *1* First period, Fiji; *2* Second period, Fiji; *3* New Britain (Lowder and Carmichael, 1970); *4* North coast New Guinea (Morgan, 1966); *5* East Papua, New Guinea; *6* Mariana Islands; *7* Bougainville, Solomon Islands; *8* St. Kitts, Lesser Antilles; *9* Malekula, New Hebrides; *10* Central Islands, New Hebrides; *11* Semispochnoi, Aleutian Islands (U.S.G.S. Bull. 1028-0); *12* Umnak, Aleutian Islands (Byers, 1961); *13* Izu Islands and Peninsula; *14* Tonga; *15* S. Kuriles; *16* C. Kuriles; *17* Aleutian Islands; *18* N. Kuriles. References not listed are found in Table 4

enrichment. They also have several trace element characteristics of "island arc tholeiites": K/Rb ratios around 1000 (Jakes and White, in prep.), U and Th generally less than 0.5 ppm with Th/U ratios around 1, and low Ni and Cr contents (Ewart and Gill, unpublished data). In contrast, however, they have more K$_2$O and associated trace elements and a steeper K$_2$O—SiO$_2$ slope. Rocks of the Central Kuriles are intermediate between tholeiitic and calcalkaline trends in both iron enrichment and K$_2$O—SiO$_2$ variations.

In all island arc regions where shoshonites are known (Indonesia, New Guinea, Fiji, Kamchatka), they represent the last volcanic event and are furthest from the trench. In this respect they are similar to the alkali basalts of Japan and the Aleutians, but differ from them principally in Na/K relationships and corresponding mineralogy.

Of all the areas examined, however, only in Viti Levu is there a clear secular evolution from tholeiitic to calcalkaline to shoshonitic volcanism. Donnelly *et al.* (in prep.) describe a transition from submarine spilites and keratophyres to subaerial calcalkaline rocks in the Lesser Antilles. The earliest rocks are characterized by low Th and U contents (< 0.5 ppm) and Th/U ratios around 1, whereas all are higher in samples of the later period. A similar change from submarine to subaerial assemblages accompanied by a decrease in iron enrichment and increase in potassium is reported from the western Aleutians (Wilcox, 1959). Baker (1968b) and Jakes and White (1970) also suggest that early arc development is tholeiitic in character, giving way to subsequent calcalkaline activity.

REFERENCES

Baker, P. E.: Petrology of Mt. Misery volcano, St. Kitts, West Indies. Lithos **1,** 124–150 (1968a)
——— Comparative volcanology and petrology of the Atlantic island-arcs. Bull. Volcanol. **32,** 189–206 (1968b)
Byers, F. M. Jr.: Petrology of three volcanic suites, Umnak and Bogoslof Islands, Aleutian Islands, Alaska. Bull. Geol. Soc. Am. **72,** 93–128 (1961).
Coats, R. R.: Magmatic differentiation in Tertiary and Quaternary volcanic rocks from Adak and Kanaga Islands, Aleutian Islands, Alaska. Bull. Geol. Soc. Am. **63,** 485–514 (1952).
——— Geologic reconnaissance of Semisopochnoi Island, western Aleutian Islands, Alaska. U.S. Geol. Surv. Bull. **1028-0,** 477–519 (1959)
Dickinson, W. R.: Circum-Pacific andesite types. J. Geophys. Res. **73,** 2261–2269 (1968b)
Donnelly, T. W., Rogers, J. J. W., Pushkar, P., Armstrong, R. L.: Chemical evolution of the igneous rocls of the eastern West Indies—An investigation of thorium, uranium, and potassium distributions, and lead and strontium isotopic ratios. Geol. Soc. Amer. Memoir **130,** 181–224 (1971).
Hatherton, T., Dickinson, W. R.: The relationship between andesitic volcanism and seismicity in Indonesia, the Lesser Antilles, and other island arcs. J. Geophys. Res. **74,** 5301–5310 (1969).
Jakes, P., White, A. J. R.: Structure of the Melanesian Arcs and correlation with distribution of magma types. Tectonophysics **8,** 223–236 (1970).
Jakes, P., White, A. J. R.: K/Rb ratios of rocks from island arcs (in prep.).
Lowder, G. G., Carmichael, I. S. E.: The volcanoes and caldera of Talasea, New Britain. Bull. Geol. Soc. Am. **81,** 17–38 (1970)
Masuda, A.: Lanthanides in basalts of Japan with three distinct types. Geochem. J. **1,** 11–26 (1966).
——— Geochemistry of lanthanides in basalts of Central Japan. Earth Planet. Sci. Letters **4,** 284–292 (1968).
Morgan, W. R.: A note on the petrology of some lava types from East New Guinea. J. Geol. Soc. Australia **13,** 583–591 (1966).
Nockolds, S. R., Allen, R.: The geochemistry of some igneous rock series. Geochim. Cosmochim. Acta **4,** 105–142 (1953).
Tatsumoto, M.: Isotopic composition of lead in volcanic rocks from Hawaii, Iwo Jima, and Japan. J. Geophys. Res. **71,** 1721–1733 (1966).
Taylor, S. R.: Geochemistry of andesites. In: Origin and distribution of the elements (ed. L. H. Ahrens), p. 559–583. Oxford: Pergamon Press 1968.
——— Capp, A. C., Graham, A. L., Blake, D. H.: Trace element abundances in andesites. II Saipan, Bougainville and Fiji. Contr. Mineral. and Petrol. **23,** 1–26 (1969).
——— White, A. J. R.: Trace element abundances in andesites. Bull. Volcanol. **29,** 177–194 (1966).
Wilcox, R. E.: Igneous rocks of the Near Islands, Aleutian Islands, Alaska. Int. Geol. Congr. XX Sess., Mexico **11-A,** *365–378 (1959).*

11

VOLCANIC ROCK SERIES IN ISLAND ARCS AND ACTIVE CONTINENTAL MARGINS

Akiho Miyashiro

[*Editors' Note:* Only Figure 1 is reproduced here.]

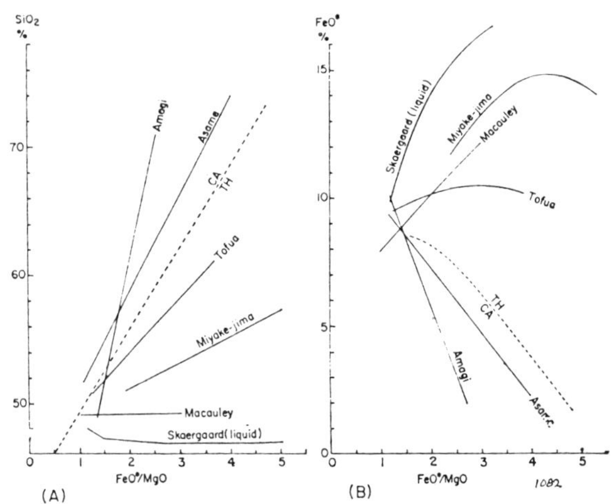

Fig. 1. Comparison of non-alkalic volcanic rock series of some provinces. The location of the provinces and the sources of data are given in table 1.

12

Petrochemistry of the Peléan-Type Volcanoes of Martinique

BERNARD M. GUNN *Department of Geology, Université de Montréal, Montréal, Québec, Canada*
MICHAEL JOHN ROOBOL *Department of Geology, University of the West Indies, Mona, Jamaica*
ALAN LEWIS SMITH *Department of Geology, University of Puerto Rico, Mayaguez, Puerto Rico*

ABSTRACT

The northern part of the island of Martinique in the Lesser Antilles consists of three overlapping volcanic cones: Mont Pelée, Pitons du Carbet, and Morne Jacob. The southern half of the island consists of a complex of older, smaller, more deeply dissected and weathered centers. The deposits of the island are largely volcaniclastic, originating as nuée ardente pyroclastic flows, air-fall material, explosion breccia, mudflow, and lahars. Many of the deposits cannot be grouped as above because of extensive reworking shortly after their eruption and deposition in an extremely friable and uncompacted state. Lava flows, domes, spines, and pitons compose only a small part of the exposed volume of the island. Chemically, the majority of rocks are close to 60 percent ±3 percent SiO_2, but distinctly different abundances of alkali elements are present in the different centers. The older Morne Jacob lavas have more than double the K, Rb, Ba, and Th of the Mont Pelée rocks, whereas the latter have higher Na and Sr. Other element abundances are similar. Mont Pelée is thus one of the low-K andesite series typical of island arcs, whereas Morne Jacob tends toward the Andean type in alkali content, although still composed of calcic andesite. *Key words: geochemistry, volcanology, orogeny.*

INTRODUCTION

Mont Pelée, at the northern end of the island of Martinique in the Lesser Antilles (Fig. 1), achieved a certain notoriety with the violent eruption of 1902 that annihilated the city of St. Pierre. The detailed description of Lacroix (1904) of this eruptive event has lead to the adoption into the study of volcanology of the term "Peléan type," signifying the multiple eruption of small-volume pyroclastic flows termed "nuées ardentes" ("glowing clouds"). These eruptions consist of white-hot gas, ash, scoria, and lava blocks, up to several meters in diameter, which cascade down the flanks of the volcano. Perret (1937) recorded 185 nuées ardentes in a 29-day period during January and February 1930. During the 1902 and 1929 eruptions of Mont Pelée, the activity was accompanied by and closed with tholoidlike dome and spine building, but usage of the term "Peléan type" includes eruptions which lack this character.

Peléan-type eruptions appear to be characteristic not only of Mont Pelée but also of other centers on the island of Martinique and of many of the other islands of the Lesser Antilles. The volcano of Soufrière, on St. Vincent, which erupted simultaneously with Mont Pelée in 1902 and with greater violence, nevertheless produces a much more basic lava, averaging about 54 percent SiO_2, compared with an average of almost 60 percent for Mont Pelée. The explosive nature of the eruptions, while undoubtedly related to the water content of the magma, does not appear to be restricted to any narrow range of chemical composition. In Martinique and adjacent islands, recent and ancient exposures contain many sheets of unsorted nuée ardente material, usually intercalated with air-fall pyroclastic deposits. The deposits of different eruptions are separated by paleosoils and horizons of redistributed volcaniclastic material. Andesite lava flows are not usually the dominant component; rather, the lava tends to occur as domes, spines, or pitons at or near the site of the vent. The ready erosion of the enclosing uncompacted pyroclastic sheets, which dip away from the centers at angles of rest close to 10° or 20°, leaves the conical "pitons" as characteristic features of the Antillean landscape.

The major-element chemistry of Mont Pelée was well documented by Lacroix (1904) with 64 analyses, 21 from dated nuées ardentes and the rest of older lavas. Grunevald (1965) has presented 21 more analyses of Mont Pelée and of the older volcanic centers of Martinique. There are some discrepancies between the two sets of data: those of Lacroix are systematically higher in K_2O and lower in CaO. As described below, the ratio between the alkali elements is of some importance in determining the petrogenesis of a volcanic series, and increasing stress is placed upon the abundance of certain trace elements, especially Rb, Sr, Ba, Th, Cr, Ni, and Co, which have not been previously determined for this island. Irvine and Baragar (1971) have suggested that andesitic volcanic lavas may be differentiated into high-K and low-K series. Based on a study of the 8,000 analyses of volcanic rocks in the GEOKEM Data File at the Université de Montréal (of which more than a third have been carried out in the same laboratory), we propose that this classification should be extended into:

1. An "Andean" trend, which is followed by apparently all continental orogenic basalt-andesite-dacite-rhyolite suites (Chile, Peru, Guatemala, Costa Rica, Mexico, Cascade province, Japan, New Zealand), as well as some near-continental island-arc series (Aleutians, Kuriles). Most alkali basalt–trachybasalt–trachyte series follow a similar trend (for example, the Terlingua-Solitario province of west Texas, the McMurdo Volcanics of Antarctica and the Canary, Cape Verde, and Azores Islands). Baragar (1972) has referred to this as the "circumpacific trend," but island arcs of the western Pacific are all more sodic.

2. A "low-K" trend followed by the island-arc associations of the western Pacific margin, including Tonga, the New Hebrides, the pigeonite series of Japan, and the Deception Island and Scotia Arc lavas. Most of the islands of the Antilles follow this trend, including the Mount Misery series of St. Kitts (Baker, 1968) and the St. Vincent lavas. Some hawaiite-mugearite-benmoreiite series follow the same trend.

3. The "ophiolite" trend forms an extremely K-depleted series, as in the Whalesback ophiolites of Newfoundland, the Troodos ophiolites, the Archaean metabasalts of Canada, Australia, and South Africa, and, though higher differentiates are rare, apparently by the ocean-floor and oceanic-ridge tholeiites. Spilites and keratophyres also belong here, but albitization of feldspar, ion exchange with sea water, and metasomatic effects are common in this group.

4. The "high-K" group is a much less consistent series of potassic, often leucite-bearing, basalts, potassic trachybasalts and leucite trachytes, jumillites, and shoshonites. Possibly, the shoshonites in

39

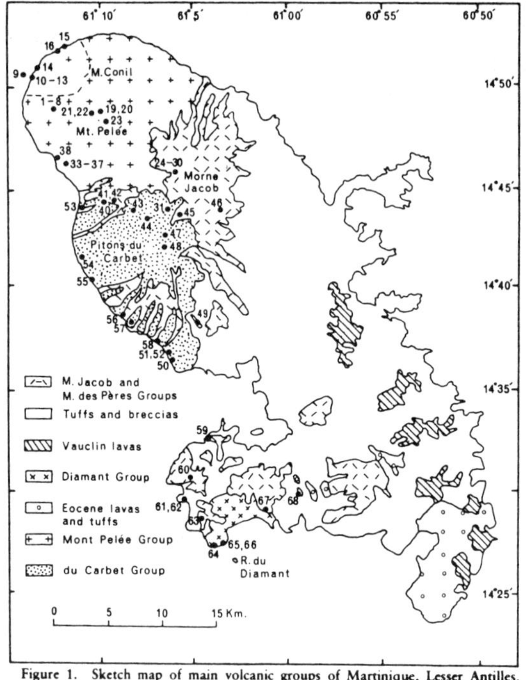

Figure 1.   Sketch map of main volcanic groups of Martinique, Lesser Antilles, showing sample locations.

which K slightly exceeds Na should be separated from the jumillites with K/Na ratios >2, but high-potassium lavas are not the subject of this paper and will not be discussed further. Full references, means, and computer listings of the numerous data cited above are available from the Geology Department, Université de Montréal.

The low-K andesites, restricted as they are to the younger island arcs, have other characteristics, notably a low Sr abundance (<300 ppm compared to 400 to 600 ppm for continental andesites) and low Rb, Ba, and Th. These and other characteristics have led Jakeš and Gill (1970) to term similar rocks in Fiji "island-arc tholeiite." This term is perhaps unfortunate, as these rocks, or at least those of the Antilles, have typical orogenic andesite petrography, contain orthopyroxene, follow distinctive andesitic fractionation trends, and form distinctive andesite composite cones, in strong contrast to the pillowed tholeiitic flood basalts of the ocean floor or to Hawaiian tholeiites. Similar trends are followed by trondhjemite plutonic suites. Whether all the centers of the Lesser Antilles are uniformly of the low-K andesite type or whether greater compositional differences exist is not yet known. This paper is concerned with the variations within the island of Martinique.

[*Editors' Note:* Material has been omitted at this point.]

REFERENCES

[*Editors' Note:* Only the references cited in the preceding excerpt are reproduced here.]

Baker, P. E., 1968, Petrology of Mt. Misery volcano, St. Kitts, West Indies: Lithos, v. 1, p. 124-150.

Baragar, W. R. A., 1972, Some physical and chemical aspects of Precambrian volcanic belts of the Canadian Shield: Ottawa, Canada Earth. Phys. Br. Pub., v. 42, no. 3, p. 129-140.

Irvine, T. N., and Baragar, W. R. A., 1971, A guide to the chemical classification of the common volcanic rocks: Canadian Jour. Earth Sci., v. 8, no. 5, p. 523-548.

Jakeš, P., and Gill, J. B., Rare earth elements and island arc tholeiitic series: Earth and Planetary Sci. Letters, v. 9, p. 17-28.

Lacroix, A., 1904, La Montagne Pelée et ses éruptions: Paris, Masson, 662 p.

Perret, F. A., 1937, The eruption of Mt. Pelee, 1929-32: Carnegie Inst. Washington Pub. 458, 126 p.

A GUIDE TO THE CHEMICAL CLASSIFICATION
OF THE COMMON VOLCANIC ROCKS

T. N. Irvine and W. R. A. Baragar

[*Editors' Note:* Only Figure 1 is reproduced here.]

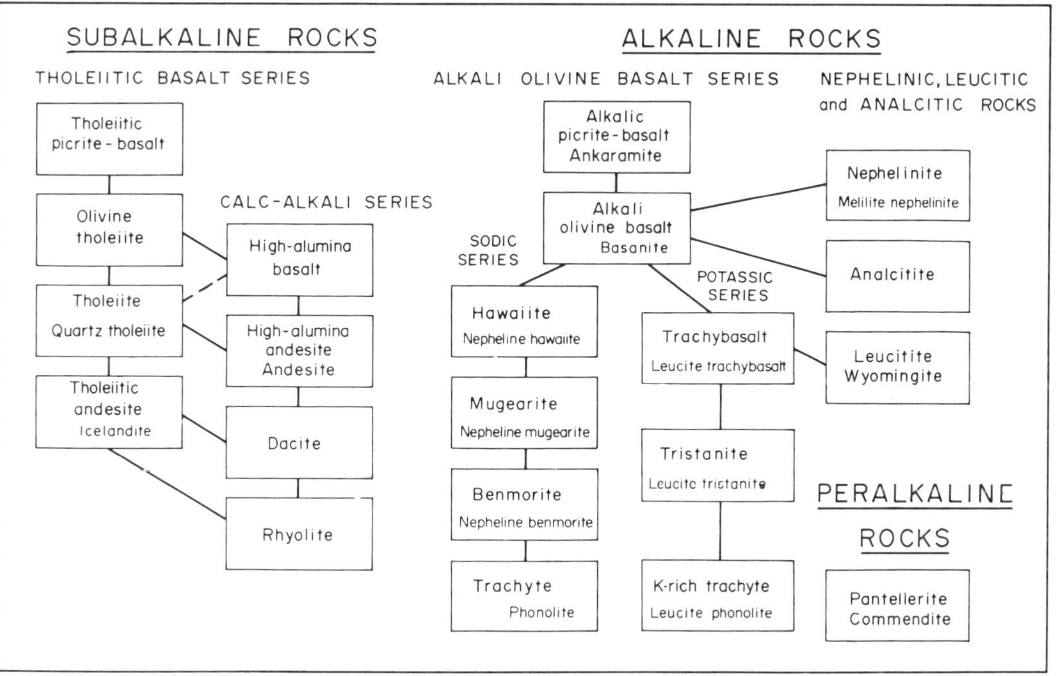

Figure 1. General classification scheme for the common volcanic rocks. The lines joining boxes serve to outline common associations. The rocks indicated by small print within the boxes are variants of the main rock.

14

Reprinted from pages 61–63 of *Schweizer. Mineralog. u. Petrog. Mitt.*
49(1):47–64 (1969)

A COMMENTARY ON THE ABSAROKITE-SHOSHONITE-BANAKITE SERIES OF WYOMING, U.S.A.

J. Nicholls and I. S. E. Carmichael

[*Editors' Note:* In the original, material precedes and follows this excerpt.]

The average latite (Table 6) is less rich in calcium (lower ratio of $CaO/Na_2O + K_2O$) than the shoshonites, and accordingly has a normative feldspar richer in alkali components.

The results of X-ray fluorescence analysis of the minor and trace elements in the five rocks of Table 5 are given in Table 7. In common with the generality of potassic lavas (TURNER and VERHOOGEN, 1960; CARMICHAEL, 1967c), the representatives of the absarokite-shoshonite-banakite series (Table 7) contain notable amounts of P, Sr, Ba, La and the light rare earths, lesser amounts of Cr and Ni, and are rather poor in Zr especially in comparison to the average latite of the Sierra Nevada (NOCKOLDS and ALLEN, 1954), the only comparable rock-type with trace-element data.

Table. 6. *Chemical analyses (recalculated H_2O and CO_2 free)*

| | Absarokite | Shoshonites | | | | | | | | | | ? Banakite | |
	225	A.	M.	114	118 A	109 A	S.	T.A.	Doreite	Latite	119	B.
SiO_2	53.53	51.29	53.90	53.08	54.16	55.11	54.68	55.0	56.52	54.45	61.04	55.37
TiO_2	0.81	0.98	1.43	0.92	0.97	0.98	0.84	0.95	1.30	1.19	0.64	0.73
Al_2O_3	13.02	13.18	14.61	17.06	16.32	16.64	18.25	17.5	16.97	17.36	15.26	18.87
Fe_2O_3	3.52	4.54	3.44	4.00	3.40	4.22	4.62	4.8	3.77	3.86	1.38	4.44
FeO	4.78	4.38	4.38	3.59	5.06	3.98	3.12	3.0	4.40	4.01	4.01	2.49
MnO	0.17	0.11	—	0.13	0.17	0.11	0.07	0.14	0.13	0.12	0.11	0.11
MgO	7.97	10.23	7.28	5.66	4.93	4.60	3.70	4.5	3.42	3.90	5.29	2.82
CaO	7.64	8.14	6.93	8.26	8.60	7.24	6.58	6.8	6.93	6.81	5.81	4.77
BaO	0.25	0.39	—	0.24	0.17	0.20	0.25	—	—	—	0.16	0.26
Na_2O	2.34	2.13	2.23	3.33	3.56	3.69	3.49	3.5	3.59	3.35	3.73	4.36
K_2O	5.29	3.94	5.78	3.04	2.15	2.59	3.84	3.2	2.62	4.46	2.30	5.16
P_2O_5	0.62	0.69	—	0.53	0.37	0.50	0.56	0.60	0.33	0.49	0.18	0.62

A:	average of Wyoming absarokites (IDDINGS, 1899, p. 329).
M:	average (recalculated) of 64 minettes (MÉTAIS and CHAYES, 1963, p. 157).
S:	average of shoshonites (IDDINGS, 1899, p. 340).
TA:	average of 10 trachyandesite lavas, Absaroka Mountains (NELSON and PIERCE, 1968).
Doreite and latite:	average analyses (recalculated) from NOCKOLDS (1954).
B:	average Wyoming banakite and quartz-banakite (IDDINGS, 1899, p. 347).

Table 7. *X-ray fluorescence analyses of trace and minor elements of the rocks in Table 5*

	225	114	118 A	109 A	119
Ga	15	20	15	20	20
Cr	415	90	145	100	370
V	160	170	185	175	100
Ni	70	60	75	45	190
Cu	70	85	75	60	40
Zn	95	90	115	110	95
Zr	110	135	120	145	125
Y	20	15	20	20	10
La	60	80	70	80	40
Ce	50	110	80	100	50
Pr	<10	10	<10	10	<10
Nd	30	40	30	40	20
Sm	<10	<10	<10	<10	<10
Sr	485	1245	1075	1115	725
Ba	2130	2140	1460	1670	1360
Rb	180	80	55	50	35

Analyses by R. N. JACK.

THE ABSAROKITE-SHOSHONITE-BANAKITE SERIES

This series was considered by IDDINGS (1899) to be genetically related, but the nature of the relationship was not treated, except that banakites were suggested to be the highly feldspathic modification of shoshonite magma, and are complementary to absarokite, which represents the least feldspathic modification of the same magma. Shoshonite was also considered to be related to basalt, and petrographic transitions between the two rock-types were described by IDDINGS. HOLMES (HOLMES and HARWOOD, 1937) re-interpreted Iddings' data, and suggested that each of the three rock-types was itself a series, and some factor, other than differentiation (crystal fractionation), controlled the series. HOLMES later (op. cit., p. 147) suggested that the "olivine-rich absarokites of Montana may be syntectic products due to the action of alkali ultrabasic magma on sediments which were themselves transfused by alakli emanations from magma".

The new analytical and mineralogical data presented in this paper do not provide the solution to the genetic link between each of these rock-types. There must be considerable doubt over the original composition of most of these volcanics, some of which contain over three percent of water and carbon dioxide. However, if it is assumed that the absarokite (225) (Tables 5 and 6) has not been greatly changed in composition subsequently to its eruption, then no combination of phases present as phenocrysts in this volcanic series can be found, which by addition and/or subtraction to the absarokite (or the average absarokite (Table 6)) will give a close approximation to the composition of the shoshonites. Similarly, there seems to be no combination of phenocrysts either added to or subtracted from the shoshonites which will give either the absarokite (225) or the average absarokite (Table 6).

In porphyritic volcanics such as the absarokite-shoshonite-banakite series, where the liquid line of descent (if it exists) is unknown, the accumulative nature, or otherwise, of one rock-type cannot be established. Therefore there cannot be an unequivocal conclusion about the dominant control in the genesis of this series, especially as it would depend on accepting the analyses of Table 5 to be representative of the lavas in a pristine condition.

THE SHOSHONITE SERIES

This magma series, recently defined and proposed by JOPLIN (1965, 1966) is analogous to, and parallel with, the alkaline olivine-basalt magma series. The shoshonite magma series as conceived by JOPLIN (1965, 1966) has high potash as the dominant characteristic, and has two branches, one rich in feldspathoids, and highly undersaturated, and the other less undersaturated and which includes the absarokites, shoshonites and banakites, and the potassic trachytes and all their plutonic equivalents. The potassic lamprophyres are also considered representatives of the shoshonite magma series.

It should be clear that it is not our intention to question the existence of the postulated shoshonite magma series, or its implied genetic thread, although the series would seem to include a wide variety of rock-types, but rather to offer data on the rock-type (possibly accumulative) which gives its name to this alkaline magma series. Whatever the shoshonite series is, or is conceived to be, it has little or no affiliation with, or similarity to, the tholeiitic or the (calc-alkaline) orogenic series. Only in exceptional geological circumstances is JOPLIN (1966, p. 42) prepared to concede that orthopyroxene can crystallize in a representative of the shoshonitic magma series. It is difficult to encourage the name shoshonite, which, in the type locality of Wyoming, is a lava of tholeiitic aspect, containing calcium-rich and calcium-poor pyroxenes, and a siliceous glassy residuum, for the title of an alkaline magma series characterized largely of what it is not. We therefore advocate abandoning the title of shoshonite for this magma series. CHAYES (1966) made an identical proposal concerning things tholeiitic, but the consequent challenge (TILLEY and MUIR, 1967) depended largely on preservation of the status quo, and the present widespread usage of the term. We hope therefore that this commentary will forestall shoshonite coming into common petrological usage.

REFERENCES

CARMICHAEL, I. S. E. (1967c): The mineralogy and petrology of the volcanic rocks from the Leucite Hills, Wyoming. Contr. Mineral. Petrol. *15*, 24–66.

CHAYES, F. (1966): Alkaline and subalkaline basalts. Amer. J. Sci. *264*, 128–145

HOLMES, A. and HARWOOD, H. F. (1937): The petrology of the volcanic area of Bufumbira. Mem. Geol. Survey Uganda *3*.

IDDINGS, J. P. (1899): HAGUE, et al., Geology of the Yellowstone National Park. Part II. Monograph of the U.S. Geol. Survey *32,* chapter 9.

JOPLIN, G. A.(1965): The problem of the potash-rich basaltic rocks. Mineral. Mag. *34,* 266–275.

——— (1966): On lamprophyres. J. Roy. Soc. New South Wales (Australia) *99,* 37–44.

METAIS, D., and CHAYES, F. (1963): Varieties of lamprophyre. Ann. Rept. Geophys. Lab. Yearbook *62,* 156–157.

NELSON, W. H. and PIERCE, W. G. (1968): Wapiti formation and Trout Peak trachyandesite, northwestern Wyoming. U.S. Geol. Survey Bull. 1254-H.

NOCKOLDS, S. R. and ALLEN, R. (1954): The geochemistry of some igneous rock series. Part II. Geochim. Cosmochim. Acta *5,* 245–285.

TILLEY, C. E. and MUIR, I. D. (1967): Tholeiite and tholeiitic series. Geol. Mag. *104,* 337–343.

TURNER, F. J. and VERHOOGEN, J. (1960): Igneous and metamorphic petrology. 2nd ed., McGraw-Hill Book Co., New York, 694 p.

15

Reprinted from pages 284–288 of *Geol. Soc. Australia Jour.* **15**(2):275–294 (1968)

THE SHOSHONITE ASSOCIATION: A REVIEW

Germaine A. Joplin

[*Editors' Note:* In the original, material precedes and follows this excerpt.]

RELATION OF THE SHOSHONITE ASSOCIATION TO OTHER IGNEOUS ASSOCIATIONS

(i) *Relation to the Tholeiite Association*

The shoshonitic rocks differ in containing much higher aluminium and showing relatively little iron enrichment (see Fig. 1). The alkalis (Fig. 2) are much higher in rocks of the shoshonite association.

(ii) *Relation to the Alkali Basalt Association*

Like the alkali basalt association, members of the shoshonite association contain high alkalis, but sodium predominates over potassium in the alkali basalts and their differentiates (Fig. 2). Also like the alkali basalts there appears to be two separate trends one towards silicon enrichment with the devlopment of comendites in the alkali-basalt association

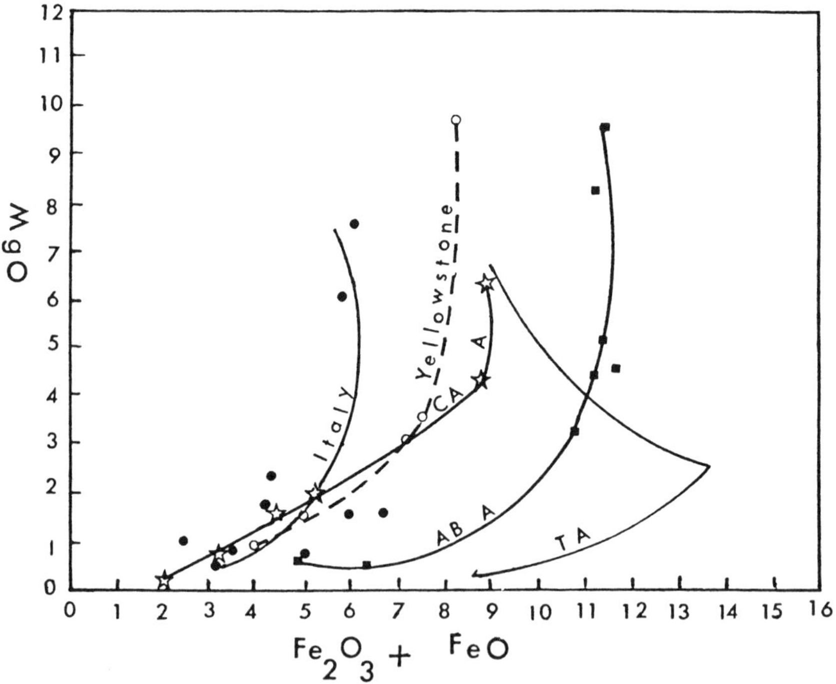

Fig. 1. Iron enrichment of shoshonite association at Yellowstone Park and Italy compared with calcalkaline association, alkali basalt association and tholeiite association.
CA A = calcalkaline association; AB A = alkali basalt association illustrated by rocks from Hawaii and from Central Victoria; TA = tholeiite association illustrated by Tasmanian tholeiites.

and liparites in the shoshonite association, and the other towards alkali enrichment and the development of trachytes. Trachytes occurring with the alkali basalts commonly contain phenocrysts of anorthoclase and sodium predominates over potassium, whereas trachytes in the shoshonite association contain phenocrysts of sanidine and potassium is predominant over sodium.

In both associations clinopyroxene is the dominant mafic mineral, but orthopyroxene occurs in some members of the shoshonite

potassium enrichment, the former being the more common. I have shown this diagrammatically in Figure 2.

Alkali Basalt
|
Hawaiite

Na   K
Mugearite   Trachybasalt
Benmoreite   Tristanite
Trachyte   K-trachyte

The potassium-rich suite derived from the alkali basalt is quite different from the shosho-

TABLE IX

Suggested relation between tectonic environment and rock association

TECTONIC ENVIRONMENT			ROCK ASSOCIATIONS			
Mobile Belt	Basin within the Hedreocraton	Rift Valley	Calcalkaline	Shoshonite	Tholeiite	Alkali Basalt
Eugeosynclinal Stage			Andesites, etc.			
Early stage of consolidation and welding			Adamellite and granodiorite intrusions	Lamprophyre dykes Small bosses of monzonite		
Later stage of consolidation and welding	Geosyncline or basin forming		K-rich calc-alkaline lavas and granite ring-dykes	Lavas and high level intrusions		
Faulting and tilting	Faulting and tilting	Faulting and tilting		Shoshonitic lavas and leucite rocks		
Relative stability	Relative stability	Relative stability			Lavas and high-level intrusions	Lavas and high-level intrusions

association. Furthermore, biotite is typical in shoshonitic rocks and very rare in alkali-basalt types. Normative hypersthene occurs in most members of the shoshonite association with the exception of the trachytes. Again iron-enrichment is marked in the alkali-basalt association but shoshonite association shows little enrichment (Fig. 1).

Shoshonitic rocks occur in all of the well known places where leucite-bearing rocks occur and leucite may be present in the more mafic members of the shoshonite association, but undersaturated members of the alkali-basalt association commonly contain the sodium-rich feldspathoids, nepheline or analcime. Recently, however, it has been shown that leucite may also occur with the alkali basalts (Le Maitre, 1962; Le Maitre & Gass, 1963; Baker, Gass, Harris & Le Maitre, 1964) and Tilley & Muir (1964) believe the alkali basalt may trend towards either sodium or

nite association, though some trachybasalts and some shoshonites may be chemically similar. Some tristanites may contain leucite. A comparison of the curves of sodium and for potassium in Figure 2 shows that though potassium is high in the Tristan da Cunha-Nightingale suite, it is lower than in the shoshonite association and that sodium is far higher in the Tristan da Cunha-Nightingale rocks.

(iii) *Relation to the Calcalkaline Volcanic Association*

A close field association, high aluminium contents, the common occurrence of biotite and orthopyroxene, a low iron-enrichment and an association with ignimbrites links the shoshonite association with the calcalkaline volcanic association, but they differ markedly in other respects. Trachytes and leucite-bearing rocks are not commonly associated with the andesitic suite except in those areas where saturated

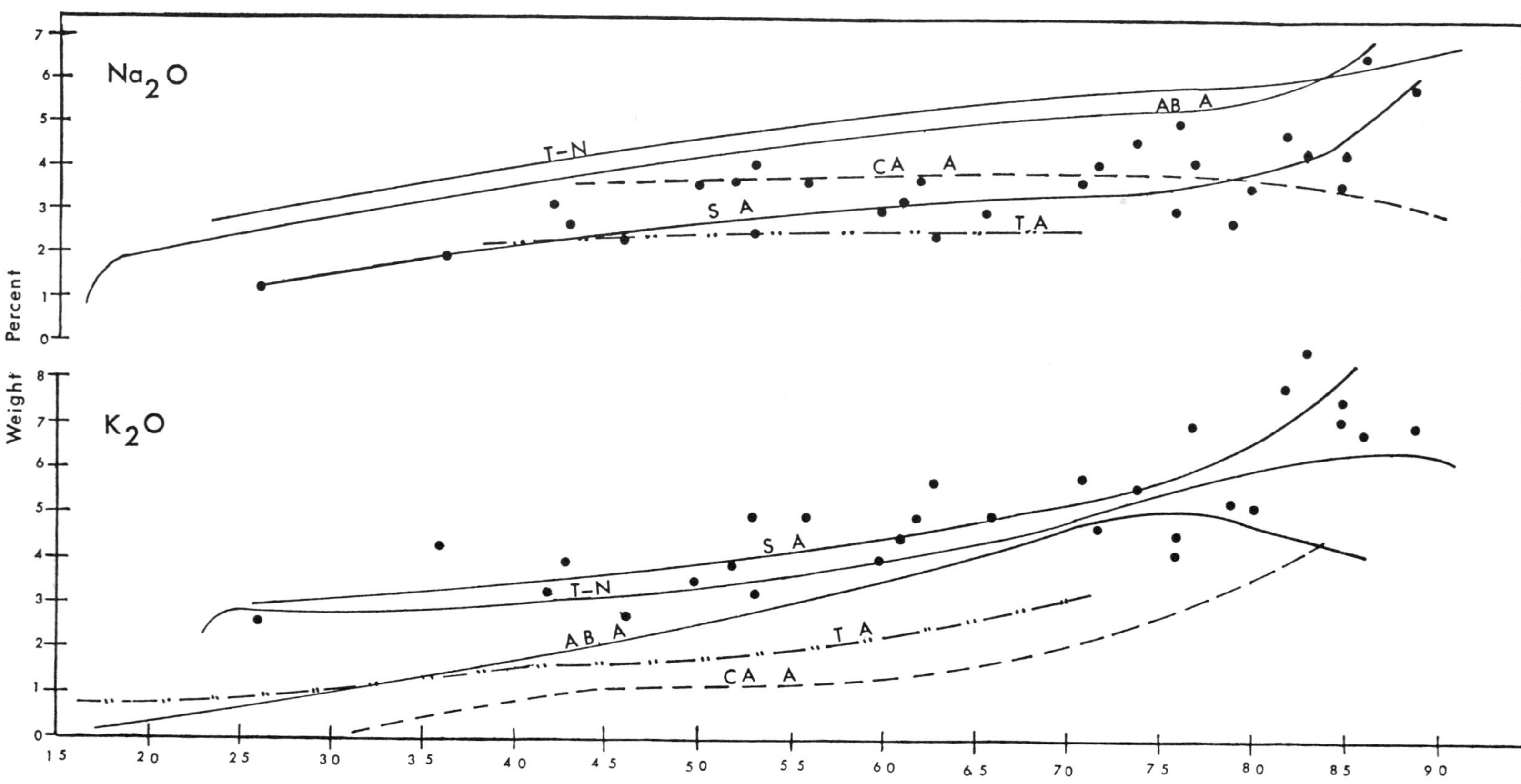

Fig. 2. Alkali trends in the shoshonite association (S A), in the Tristan da Cunha-Nightingale lavas (T-N); in the alkali basalts as shown by those from Hawaii and Central Victoria (AB A); in the tholeiite association as shown in Tasmania (TA); and in the calcalkaline lavas using Nockolds's averages (CA A).

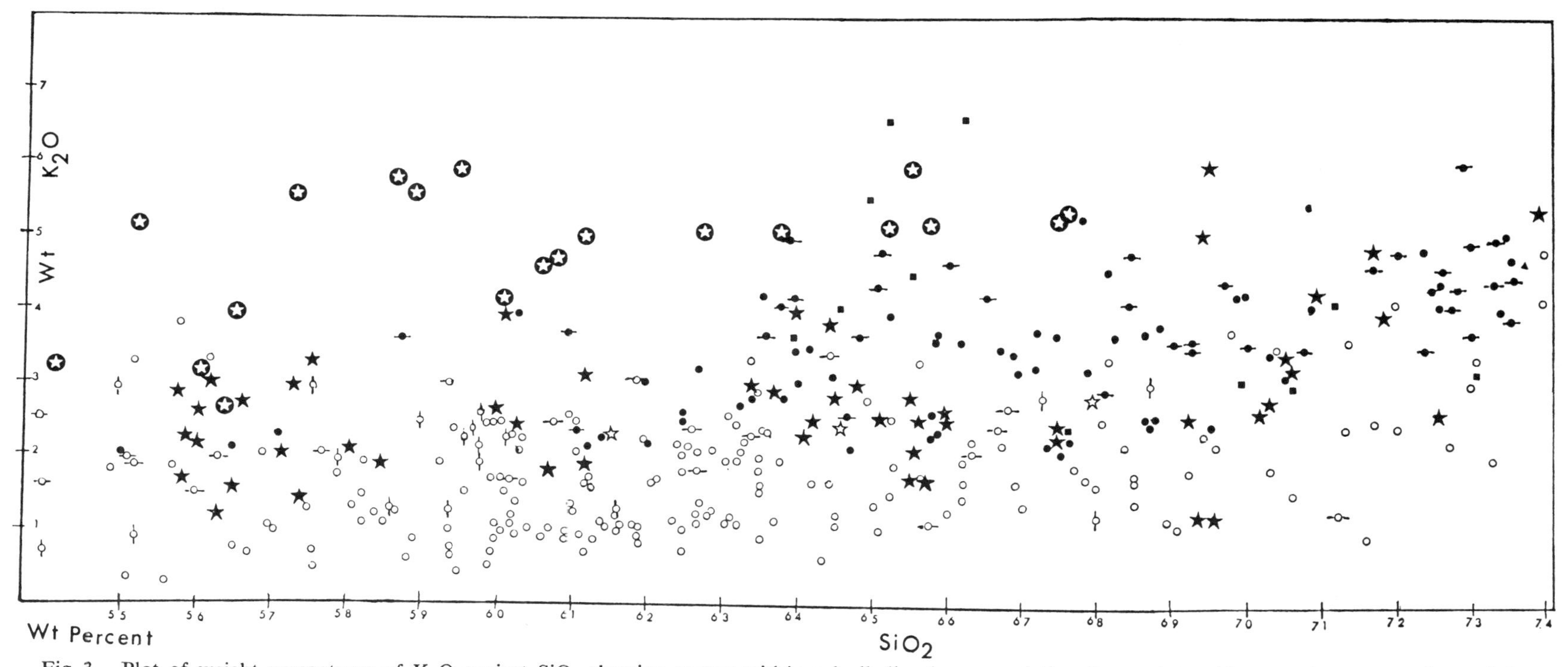

Fig. 3. Plot of weight percentages of K_2O against SiO_2 showing scatter within calcalkaline lava association. Large dots with open star are members of the shoshonite association with K_2O higher than most other rocks shown. Solid dots without bar are 'toscanites', dacites, rhyodacites from Central Victoria. Solid dots with bar are the same types from Burrinjuck, N.S.W., and solid squares are from the Georgetown area in Queensland. Solid stars are andesites, etc., from Yellowstone Park, and three open stars are breccias from the same locality.
Open circles are calcalkaline lavas from eugeosynclinal areas. Open circles with vertical bar from New Guinea; with horizontal bar from Indonesia; and without bar from West Indies and Andes.

49

members of the shoshonite association also occur. Potassium is high only in the more felsic members of the calcalkaline suite whereas it persists throughout the shoshonite suite. Reference to Figure 3 shows that potassium is considerably higher and fairly constant in the shoshonite association whereas it is lower and very variable among the calcalkaline volcanic rocks.

(iv) *Relation to the Carbonatites*

The relation of the shoshonite association to the carbonatites is puzzling. Both are found in close proximity in the Rift Valley of Africa and possibly in the Rhine Valley of West Germany and certain rock-types commonly found in carbonatite complexes occur with monzonites at Bufumbira, at Mt Dromedary in New South Wales, in the Tyrol, and in Montana.

Though no leucite rocks occur in the carbonatite complexes of the African Rift, they appear to be associated at the Kaiserstuhl (Wimmenauer, 1964), and at both localities highly potassic trachytes and sanidinites occur with carbonatites (Sutherland, 1965, 1967), but these also occur with leucite rocks in the shoshonite association.

Pyroxenites and biotite pyroxenites were described by Holmes from Bufumbira and they occur with monzonites at Mt Dromedary, in the Tyrol and in Montana, yet these rocks are typical of some carbonatite complexes.

Some rocks at Mt Dromedary (Brown, 1930) were likened to ijolites and melteigites from Fen, to covites from Magnet Cove and to jacupirangites from Jacupiranga and all of these three areas have since proved to be carbonatite complexes. Shonkinites appear to be associated with monzonites at Mt Dromedary and also in Montana, but they also occur with many carbonatites.

Thus, the question of the relation between the carbonatites and the shoshonite association must remain an open one for the present.

[*Editors' Note:* Only the references cited in the preceding excerpt are reproduced here.]

REFERENCES

BAKER, P. E., GASS, I. G., HARRIS, P. G., & LE MAITRE, R. W., 1964: The volcanological report of the Royal Society expedition to Tristan da Cunha, 1962. *Phil. Trans. R. Soc.*, 256A, pp. 439–575.

BROWN, IDA A., 1930: The Geology of the South Coast of New South Wales. Part iii. The Monzonitic Complex of the Mount Dromedary District. *Proc. Linn. Soc. N.S.W.*, 55, pp. 637–698.

LE MAITRE, R. W., 1962: Petrology of Volcanic Rocks, Gough Island, South Atlantic. *Bull. geol. Soc. Am.*, 73, pp. 1309–1340.

———, & GASS, I. G., 1963: Occurrence of leucite in volcanic rocks from Tristan da Cunha. *Nature, Lond.*, 198, pp. 779–780.

SUTHERLAND, D. S., 1965: Potash-trachytes and ultra-potassic rocks associated with the carbonatite complex of the Toror Hills, Uganda. *Mineralog. Mag.*, 35, pp. 363–378.

———, 1967: A note on the occurrence of potassium-rich trachytes in the Kaiserstuhl carbonatite complex, West Germany. *Mineralog. Mag.*, 36, pp. 334–341.

TILLEY, C. E., & MUIR, I. D., 1964: Intermediate members of the oceanic basalt-trachyte association. *Geol. For. Stockh. Forh.*, 85, pp. 436–444.

WIMMENAUER, W., 1966: Carbonatites of the Kaiserstuhl (West Germany) and their magmatic environment. *Pap. Proc. int. mineralog. Assoc., 4th Gen. Meeting, New Delhi* (1964), pp. 54–57. Mineralog. Soc. India.

16

THE GEOLOGY AND GEOCHEMISTRY OF THE LOWER ULTRAMAFIC UNIT OF THE ONVERWACHT GROUP AND A PROPOSED NEW CLASS OF IGNEOUS ROCKS

M. J. Viljoen and R. P. Viljoen

[*Editors' Note:* In the original, material precedes and follows this excerpt.]

C. Chemical Comparison of the Mafic and Ultramafic Rocks of the Lower Ultramafic Unit with other Well-established Classes of Mafic and Ultramafic Rock

1. Introduction

In an accompanying paper the unique chemical characteristics of the extrusive peridotitic magma of the Komati formation (peridotitic komatiite) are illustrated by a comparison with well-established classes of peridotitic and picritic rocks. In the present section the chemistry of all the ultramafic and mafic rocks of the lower formations of the Onverwacht Group are also compared with the chemistry of well-established classes of peridotitic and picritic rocks in addition to basaltic and extra-terrestrial rocks. The classes of peridotitic and picritic rocks used are the same as those employed in the comparison of the peridotitic komatiites of the Komati formation (Paper No. 5). They include the class of so-called high temperature peridotite intrusions (Green, 1967) and the average of the peridotite inclusions in kimberlite pipes and in alkali basalts. The classes of picrite used can be divided into two groups, viz. basaltic picrites and picrites of the picritic minor intrusions (see Paper No. 5). Other classes of igneous rock employed for comparative purposes are limburgites. In addition to the above classes which were all used in the accompanying paper (No. 5) several other additional rock classes have been employed for comparative purposes in the present paper. These include the well-known basalt classes, viz. oceanic tholeiites, continental tholeiites, alkali basalts and olivine basalts (Table V). In addition, the average chemistry of the meta-tholeiites from the upper formations of the Onverwacht Group as well as of Archean metatholeiites from the Superior Province of the Canadian Shield have also been included for comparison. Of the extra-terrestrial material the average chondrite and achondrite as well as the average lunar sample from the Apollo 11 landing site on the Sea of Tranquillity (recalculated to 100% after subtracting 8.3% TiO_2 and 8.3% FeO to eliminate excessive ilmenite) have been included (see Table V).

2. Presentation of Results by Means of Variation Diagrams

(1) CaO vs Al_2O_3

Two of the most distinctive components of both the peridotitic and basaltic komatiites, viz. calcium and aluminium, have been plotted against each other in Fig. 4a. This plot serves to distinguish komatiites of both the peridotitic and basaltic varieties from most well-established classes of basaltic, picritic and peridotitic rocks. The ankaramite picrites of Madagascar and Hawaii have Ca/Al ratios approaching those of the mafic rocks under consideration, with those from Madagascar plotting close to the field of the Badplaas type and those from Hawaii close to the field of the Barberton type (Fig. 4a). The higher alkali and lower silica content of the ankaramites, however, readily distinguishes them from the basaltic komatiites. The calcium-rich (omphacite) eclogites from kimberlite pipes also have a high Ca/Al ratio but are distinguished from basaltic komatiites by their higher alkali (particularly Na_2O) content which is ascribed to the jadeite molecule in the pyroxene. Recent chemical data from the Apollo 11 lunar sample (L.S.P.E.T., 1969) indicates that this material has a Ca/Al ratio approaching that of the basaltic komatiites of the Barberton type. The high iron and titanium content of the lunar material, however, renders it unlike the Barberton or any other earth rocks.

TABLE V

	Komatiites						Ave. perid- otite	Picrites						Lim- burg- ites	Basalts								Extraterrestrial Material		
	Peridotitic		Basaltic																						
	1	2	3	4	5	6	7	8	9	10	11	12	13	14	15	16	17	18	19	20	21	22	23	24	25
SiO_2	47.49	45.94	53.75	53.29	50.02	51.78	44.52	46.00	42.60	46.41	41.26	44.10	45.07	50.23	46.25	45.36	50.24	48.35	49.94	51.50	48.16	49.83	46.86	48.51	49.96
TiO_2	0.55	0.34	0.87	0.57	0.49	0.45	0.15	0.61	0.29	1.98	2.66	2.74	1.30	3.25	0.47	0.75	0.16	2.77	1.51	1.20	2.91	0.94	0.13	0.48	2.46
Al_2O_3	3.45	2.98	10.02	5.50	7.17	3.95	2.75	10.86	8.98	8.53	10.31	11.60	7.85	9.60	15.79	7.49	7.71	13.18	16.69	16.30	18.31	14.64	3.08	13.04	13.17
Fe_2O_3	6.39	6.23	1.25	1.00	1.25	11.58	1.74	3.98	1.50	2.47	5.58	2.84	2.42	2.54	1.06	5.20	1.02	2.35	2.01	2.80	4.24	3.03		1.11	
FeO	5.86	4.80	9.90	9.06	8.53	n.d.	6.74	6.86	8.25	9.82	8.62	9.91	6.44	8.24	9.60	9.72	4.64	9.08	6.90	7.90	5.89	8.77	15.34	15.90	12.79
MnO	0.20	0.18	0.22	0.72	0.23	0.18	0.04	0.14	0.14	0.15	0.14	0.16	–	0.15	–	0.31	0.11	0.14	0.17	0.17	0.16	0.21	0.30		
MgO	26.92	33.79	10.30	15.56	21.53	21.08	40.98	22.15	32.12	20.81	14.00	15.13	18.39	12.97	14.36	15.55	17.79	9.72	7.28	5.90	4.87	7.36	29.37	7.87	9.84
CaO	7.40	4.73	10.18	13.09	8.78	10.01	2.26	7.88	5.48	7.38	11.65	10.69	14.29	7.79	9.12	10.92	16.66	10.34	11.86	9.80	8.79	10.46	2.40	11.00	12.15
Na_2O	0.52	0.15	2.70	1.23	0.41	0.10	0.24	1.07	0.56	1.58	3.19	1.66	1.31	2.40	2.10	2.66	0.97	2.42	2.76	2.50	4.05	2.02	1.21	0.68	0.66
K_2O	0.05	0.03	0.47	0.09	0.06	0.02	0.70	0.04	0.04	0.32	0.88	0.54	1.19	1.92	0.30	0.00	0.11	0.58	0.16	0.86	1.69	0.23	0.21	0.25	0.17

N.B. All analyses re-calculated anhydrous

1 *Peridotitic komatiite* (average of Sandspruit formation).
2 *Peroditttic komatiite* (average of Komati formation).
3 *Basaltic komatiite* (average of Barberton type).
4 *Basaltic komatiite* (average of Badplaas type).
5 *Basaltic komatiite* (average of Geluk type).
6 *Basaltic komatiite* (average of Geluk type–low alumina variety).
7 *Peridotite* (average of high temperature intrusions and nodules in basalt and kimberlite, see accompanying Paper No. 5).
8 *Picritic minor intrusion*-chilled margin of dyke – Skye (Drever and Johnson, 1961. p. 81).
9 *Picritic minor intrusion* – centre of sill – Skye (Drever and Johnson, 1967, p. 81).
10 *Oceanite* type of picrite basalt – Hawaii (Mac Donald and Katsura, 1964, p. 19).
11 *Mimosite* type of picrite basalt – Hawaii (Mac Donald, 1949, p. 1571).
12 *Ankaramite* type of picrite basalt – Hawaii (Mac Donald and Katsura, 1964).
13 *Ankaramite* type of picrite basalt – Ankaramy, Madagascar (Johannsen, 1937).
14 *Limburgite* – average from Naunetsi Igneous Province, Rhodesia (Cox et al., 1965, p. 145).
15 *Eclogite* nodule – typical kimberlitic ecolgite from Roberts Victor Mine (new analysis, analysts – N.I.M. = Johannesburg).
16 *Eclogite* nodule with abundant omphacite relative to garnet – Roberts Victor Mine (Williams, 1932).
17 *Eclogite*, Roberts Victor Mine (Wagner, 1928).
18 *Olivine basalt*-average Hawaiian (Mac Donald, 1949).
19 *Oceanic tholeiite* – average given by Engel et al. (1965).
20 *Continental tholeiite*-average given by Mansen (1967).
21 *Alkali Basalt* – average from East Pacific rise (Engel et al., 1965).
22 *Archaean Metabasalt* of Superior Province of Canadian Shield – average given by Wilson et al. (1965).
23 *Chondrite* – average of superior analyses given by Urey and Craig, 1953; recalculated without Fe and FeS.
24 *Basaltic Achondrite* – average given by Urey and Craig, (1953), quoted in Engel et al. (1965).
25 *Lunar Sample* minus excessive ilmenite (8.31% TiO_2 and 8.31% FeO) from Apollo 11 landing site on the Sea of Tranquillity (L.S.P.E.T., 1969).

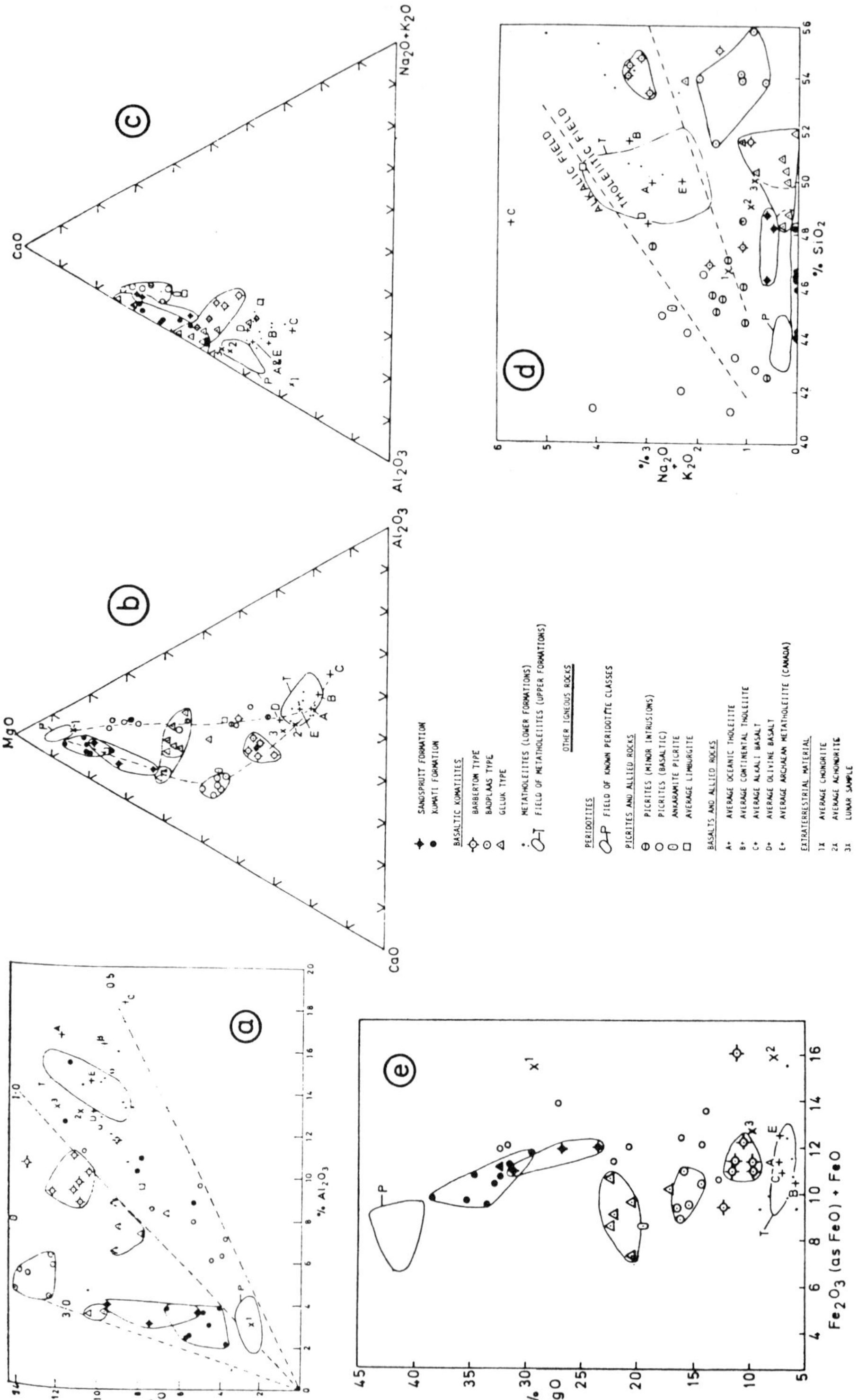

Fig. 4. Chemical comparison of the mafic and ultramafic rocks of the lower formations of the Onverwacht Group with established classes of peridotites, picrites, basalts and allied rocks, as well as extra-terrestrial materials.

The meta-tholeiites of the lower formations of the Onverwacht Group have normal Ca/Al ratios (± 0.70) and are comparable to most classes of tholeiite, including those from the upper formations of the Onverwacht Group, as well as the Archaean meta-tholeiites from the Canadian Shield. Two additional variation diagrams, making use of the other diagnostic variables of the distinctive mafic and ultramafic rocks of the lower formations, viz. magnesium and the alkalies, are depicted in Figs. 4b and 4c.

(ii) MgO—CaO—Al$_2$O$_3$

In this plot the peridotitic rocks of the Onverwacht Group fall in a field towards the magnesium corner of the diagram close to the fields of known peridotite classes and olivine-rich picrites (Fig. 4b). They are removed from these fields, however, as a result of their higher Ca/Al ratio, and form an elongate field starting with the peridotites of the Komati formation and extending through the field of peridotites from the Sandspruit formation towards the calcium corner of the diagram (Fig. 4b). Beyond this cluster a trend through the fields of the basaltic komatiites of the Geluk (low-alumina variety) and Badplaas types (at which point it swings towards the aluminium corner of the diagram) towards the field of the basaltic komatiites of the Barberton type and eventually into the field of tholeiitic basalts can be followed (Fig. 4b). A complementary trend joining most of the well-known rock classes employed in this comparison starts at the field of known peridotite classes and the average chondrite at the magnesium corner of the diagram and passes through various types of picrites to the field of basalts. The separation of this and the above trend is mainly a function of calcium and aluminium abundances. When the tholeiite field is reached both trends converge and all the basalt classes mentioned above (except for alkali basalts) cluster in a small, fairly well-defined area (Fig. 4b). The field of one class of rock, viz. basaltic komatiites of the Geluk type, is spread between and partly overlaps the two trends mentioned above, occurring towards the magnesium corner of the diagram (Fig. 4b).

The only rock types falling on or close to the komatiite trend are basaltic kimberlites (discussed in Paper No. 5), ankaramites from Madagascar, certain high calcium (omphacite-rich) eclogites (not plotted) and the average lunar sample. Some genetic connection could exist between the ankaramites and omphacite-rich eclogites and the basaltic komatiites of the Barberton and Badplaas types but, as pointed out previously, other differences clearly separate them. The lunar sample plots on the komatiite trend between the field of the Barberton type and the tholeiite field.

(iii) K$_2$O + Na$_2$O—CaO—Al$_2$O$_3$

The plot shows a clear separation between the peridotitic and basaltic komatiites and almost all other rock types used in the comparison, including the meta-tholeiites from the lower formations. The peridotitic komatiites from the Komati formation, as well as a number of basaltic komatiites of the Geluk type, lie close to the aluminium-calcium tie line towards the calcium side of the diagram. The peridotitic komatiites from the Sandspruit formation, as well as all of the remaining basaltic komatiites, occur in the same region, but are farther removed from the tie line due to their higher alkali abundances (Fig. 4c). The average ankaramite picrite, basaltic kimberlite and omphacite-rich eclogites plot in the same region with respect to calcium and aluminium, but are generally distinguished by a considerably higher alkali content. The remaining rock types plot on the aluminium side of the tie line but are substantially removed from the latter owing to their higher alkali contents.

(iv) $Na_2O + K_2O$—SiO_2

In Fig. 4d the alkali abundances have been plotted against the silica abundances following a scheme of classification for the Hawaiian basalts proposed by Macdonald and Katsura (1964). Using this plot the latter authors were able to separate clearly the tholeiitic and alkalic basalts (indicated by means of a diagonal dashed line in Fig. 4d) on the basis of chemical composition. None of the Hawaiian basalts plot below the lowermost dashed line indicated in Fig. 4d. All classes of komatiite, except for the basaltic komatiites of the Barberton type, plot below and well out of the field of all known Hawaiian tholeiites given by Macdonald and Katsura (1964). The only other rocks that plot in this region are the varieties of peridotites quoted, the average achondrite, and the lunar sample. The basaltic komatiites of the Barberton type plot in the lower portion of the tholeiitic field as do the meta-tholeiites from the lower formations, which overlap the field of the basaltic komatiites of the Barberton type (Fig. 4d).

(v) Mgo—$FeO + Fe_2O_3$ (as FeO)

In the final plot (magnesium against total iron) a clear separation is again evident between the different types of komatiite, as well as between the latter and the majority of other rock classes quoted (Fig. 4e). Certain picritic rocks plot close to some of the komatiite fields and the lunar sample plots close to the field of the basaltic komatiite of the Barberton type.

VII. KOMATIITE — A PROPOSED NEW CLASS OF IGNEOUS ROCKS

From the above geochemical comparisons it is clear that the majority of mafic and ultramafic rocks of the lower formations of the Onverwacht Group are not chemically comparable to any well-established class of igneous rock.

In certain respects they bear some similarities to a variety of rocks including basaltic kimberlites, omphacite-rich eclogites, ankaramites, picrites, basaltic achondrites, and the lunar sample. Although some of these similarities may indicate some underlying genetic connection, the bulk chemical composition of the Barberton rocks is clearly different from all of these classes.

Based largely on geochemical data therefore, but also taking into account the petrology, geological setting, development, age and remarkable state of preservation and exposure of the Barberton rocks, it is considered that the main parameters of these are well enough defined to propose the presence of what we consider to be a new and hitherto unrecognised group of igneous rock. The name "komatiite" which has been used throughout this paper and which takes its name from the Komati river which traverses the Southern part of the Barberton Mountain Land, including the Komati formation where these distinctive rocks are best developed, is proposed. It is further suggested that the prefix peridotitic or basaltic be used, depending on the particular rock being described. Two varieties of peridotitic komatiite appear to be present, viz. those from the Sandspruit formation and those from the Komati and Theespruit formations (which are more ultrabasic). Three main types of basaltic komatiite, viz. the Barberton, Badplaas and Geluk types have been recognised and although there might be some overlap between their chemistry they appear for the most part to represent well-defined classes.

REFERENCES

[*Editors' Note:* Only the references cited in the preceding excerpt are reproduced here.]

COX, K. G., JOHNSON, R. L., MONKMAN, L. G., STILLMAN, C. J., VAIL, J. R. & WOOD, D. N., 1965. The Geology of the Nuanetsi Igneous Province, *Phil. Trans. Roy. Soc. Lond.*, A 257, 1078, 71–218.

DREVER, H. I. & JOHNSTON, R., 1967. (a), Picritic Minor Intrusions, In: *Ultramafic and Related Rocks*, Ed. Wyllie, P. J. Publ. John Wiley & Sons, 71–82.

DREVER, H. I. & JOHNSTON, R., 1967. The Ultramafic Facies in some Sills and Sheets, In: *Ultramafic and Related Rocks*, Ed. Wyllie, P. J., Publ. John Wiley and Sons.

ENGEL, A. E. J., ENGEL, C. & HAVENS, R. G., 1965. Chemical Characteristics of Oceanic basalts and the Upper Mantle, *Bull. Geol. Soc. Amer.* 76, 719–734.

GREEN, D. H., 1967. High Temperature Peridotite Instrusions, In: *Ultramafic and Related Rocks*, Ed: Wyllie, P. J., Publ. John Wiley & Sons, 212–222.

JOHANNSEN, A., 1937. A Descriptive Petrography of the Igneous Rocks, Vol. III, Univ. Chicago Press.

L. S. P. E. T. (Lunar Sample Preliminary Examination Team), 1969. Preliminary Examination of Lunar Samples from Apollo 11, *Science,* 165, 3899, 1211–1227.

MACDONALD, G. A., 1949, Hawaiian Petrographic Province, *Bull. Geol. Soc. Amer.* 60, 1541–1596.

MACDONALD, G. A. & KATSURA, T., 1964. Chemical Composition of Hawaiian Lavas, *J. Petrol.,* 5, 82–133.

MANSON, V., 1967. Geochemistry of Basaltic Rocks—major Elements, In: Hess and Poldervaart, (Eds.)—*Basalts* 1, 215–270, Interscience Publishers, John Wiley, New York-London-Sydney.

UREY, H. C. & CRAIG, H., 1953. The Composition of the Stone Meteorites and the Origin of Meteorites, *Geochim. Cosmochim. Acta* 4, 36.

WILLIAMS, A. F., 1932. The Genesis of Diamond (Two volumes), Publ. Ernest Benn Ltd., London.

WILSON, H. D. B., ANDREWS, P., MOXHAM, R. L. & RAMLAL, K., 1965. Archaen volcanism in the Canadian Shield, *Can. Journ. Earth Sci.* 2, 161–175.

[*Editors' Note:* Wagner, 1928, appears to be an unpublished reference.]

17

On the Significance of Komatiite

C. Brooks and S. R. Hart

ABSTRACT

The primary diagnostics for komatiite are chemical ($CaO/Al_2O_3 > 1$, $MgO > 9$ percent, $K_2O < 0.9$); however, komatiites can only be meaningfully isolated and compared with other rock classes if the chemical diagnostics are coupled with textural criteria (for example, quench or spinifex texture) that establish the occurrence of komatiites as silicate melts. Examination of analyses of more than 20,000 rocks, using both the textural and chemical criteria, reveals that peridotitic komatiite (PK) is at present unknown from modern environments, and basaltic komatiite (BK) is rare. While it can be demonstrated that neither PK nor BK are reliable tectonic province indicators, the consistent association of Archean BK and PK with felsic volcanics argues against a modern sea-floor analog for their origin.

INTRODUCTION

In a previous paper (Brooks and Hart, 1972), we described an occurrence of basaltic komatiite from an Archean terrain near Wawa, Ontario, and noted that its position in a continuous stratigraphic sequence of basalt-andesite-dacite-rhyolite was strongly indicative of an island-arc, rather than an oceanic-crust, tectonic origin. At that time, we considered komatiite to be a Precambrian phenomenon, because most descriptions of komatiites were from Archean terrains (see, for example, Viljoen and Viljoen, 1969a; Nesbitt, 1971; McCall and Leishman, 1971; Pyke and others, 1973).

Recently, however, Gale (1973) has reported several occurrences of basaltic komatiite in a lower Paleozoic volcanic sequence in Newfoundland. He concluded that the basaltic komatiites and associated low-K tholeiites were indicative of an oceanic-crust regime.

We examined more than 20,000 rock analyses, searching for rocks of komatiite-type chemistry, with a view toward further understanding of the tectonic environments in which such rock types may occur. In this paper, we present the results of our search, and show that komatiite-type rocks do not appear to provide definitive evidence for oceanic crust.

CHARACTERISTICS OF KOMATIITE

Viljoen and Viljoen (1969a) described a series of mafic and ultramafic rocks from the Swaziland Sequence and proposed the name komatiite for them. These rocks in general are characterized by high MgO contents (>9 percent), high CaO/Al_2O_3 ratios (>1) and very low alkali contents (<0.9 percent K_2O). They are divided into peridotitic komatiite or basaltic komatiite (hereafter PK and BK, respectively) based on their composition (SiO_2, MgO contents, and so forth). They may occur as pillow lavas and frequently exhibit remarkable quench textures (referred to as spinifex texture, a term describing skeletal crystal form [usually olivine] in which the crystals display parallel grouping; Nesbitt, 1971). As we will discuss later, it is important to recognize the distinction between the two types of komatiite and not to group them together in discussions of their genesis.

We have searched RKNFSYS, a computerized file of 11,500 Cenozoic volcanic rocks (compiled by F. Chayes, 1972, unpub. data) and GEOKEM, a file of 7,500 igneous rocks (compiled by B. Gunn, 1970, unpub. data) for komatiite-type rocks. Some 2,000 additional analyses in the literature were also examined. Analyses having MgO >9 percent, Al_2O_3 <10 percent, and SiO <54 percent were retrieved for further consideration. Numerous samples having the chemical character of peridotitic komatiite were isolated, and it is apparent that almost any ultramafic rock with significant clinopyroxene or calcic amphibole will fit the chemical description of PK

and will plot on a MgO-Al_2O_3-CaO diagram either within or close to the field observed by the Viljoens for Archean PK from Africa. Representative examples of these, plotted in Figure 1, include three lherzolites from the Indian Ocean floor (Vinogradov and others, 1969), pyroxene-bearing peridotite from the Union Bay complex, Alaska (Ruckmick and Noble, 1959) pyroxenite from the Tulameen complex, British Columbia (Findlay, 1969), pyroxenite and harzburgite from the Troodos complex (Moores and Vine, 1971), a mermechite from Siberia (Davidson, 1967), and an amphibole peridotite from the Lugar Sill, England (Drever and Johnston, 1967).

It is clear that rocks of PK chemistry occur in diverse tectonic regimes (for example, Alpine and high-temperature

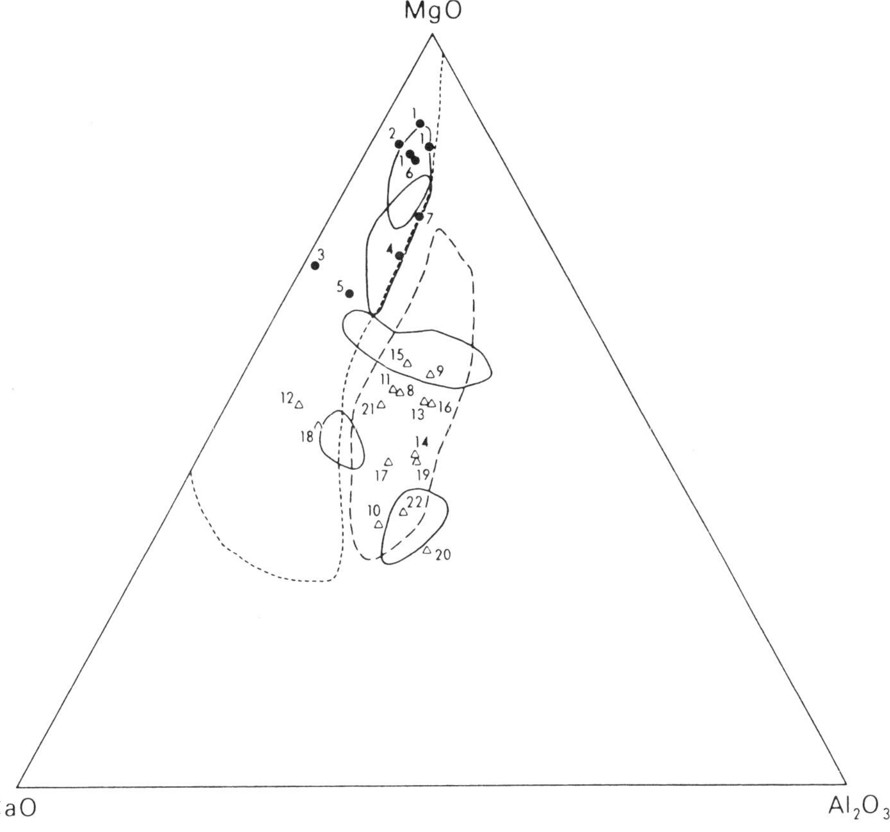

Figure 1. FMA plot showing relation between Viljoen and Viljoen (1969a) fields of PK and BK (solid lines) and fields of peridotite-pyroxenite (dotted line) and picrite-oceanite-ankaramite (dashed line). Numbers 1-7 are representative cumulate ultramafics with PK chemistry retrieved from data-files searched; 9, 13, 16, 20, 22 are young fine-grained rocks of BK chemistry; 12, 17, 18, 19, 21 are coarse-grained rocks of BK chemistry; 8, 10, 11, 14, 15 are picrite, ankaramite, or oceanite. (Numbers 8 to 22 correspond to those in Table 1.)

peridotites, peridotite-nodules, sea-floor peridotites, ophiolite complexes) throughout most of geologic time, and that the general chemical character is not helpful in specifying that tectonic regime. More subtle chemical indices may eventually be developed as trace element studies of ultramafic rocks are carried out.

In Figure 1, we have plotted a general field of pyroxenites and peridotites, using 65 analyses from the two data files and additional literature. The field clearly embraces the two types of PK discussed by the Viljoens (1969a). Apparently, the Viljoens did not recognize the common similarity of many ultramafic rock types with those of the PK class.

The critical feature of PK is that it was once a silicate melt, as evidenced by quench textures, pillowed structures, and other features characteristic of lava flows (Nesbitt, 1971; Pyke and others, 1973). To our knowledge, none of the post-Archean occurrences of rocks with PK chemistry exhibit spinifex texture or other unequivocal evidence of existence as "liquids" of the same composition, and it is obvious that differentiation effects that would concentrate clino-pyroxene could produce rocks of PK chemistry. Thus, at this stage, magmas of PK composition are apparently restricted to Archean occurrences, and consideration of mechanisms for their formation must include this observation.

Rocks of BK chemistry, on the other hand, are rare in the post-Archean record. The occurrences described by Gale (1973) are the only ones reported to date. In our data search, we found only five samples of BK chemistry that are probable liquid compositions (two picrites, two basalts, and one andesite; Table 1, analyses 9, 13, 16, 20, and 22) five others that have the right chemistry are coarse-grained rocks (one dolerite, two olivine gabbros, and two pyroxenites; Table 1, analyses 12, 17, 18, 19, 21), and their unusual chemistry may be due to accumulation of pyroxene. Some other young rocks of BK composition have been reported from Cyprus (Troodos), but their altered character (high CO_2 and H_2O) make original compositions somewhat uncertain (Searle and Vokes, 1969).

Numerous rocks classified as picrites, oceanites, and ankaramites (especially from oceanic islands) would be termed BK according to the criteria proposed by the Viljoens (1969a), but in all cases they have TiO_2 values well above the range reported by the Viljoens, and they are generally higher in K_2O. We have included five examples in Table 1 (analyses 8, 10, 11, 14, 15) and have plotted a general "oceanite-picrite-ankaramite" field in Figure 1, based on 77 analyses from the data files searched. It is clear that this field embraces the three BK fields presented by the Viljoens. However, these 77 analyses had an average TiO_2 of 2.2 percent; thus, we feel that a TiO_2 index should be added to the classification of BK. The following criteria seem sufficient to chemically distinguish BK: SiO_2, 46 to 53 percent; CaO/Al_2O_3 >1; TiO_2 <0.9; K_2O <0.9; MgO >9 percent. In general, BK also has K_2O <0.5 percent.

Four of the five analyses showing "liquid" BK chemistry are from rocks from island-arc environments (Solomons, New Hebrides, Aleutians, and Puerto Rico). Two of the rocks are called picrite but are clearly distinguished from picrites from oceanic islands by their TiO_2 contents (lower by a factor of three). Thus, TiO_2 may be diagnostic of the tectonic environment of BK. We have plotted the analyses of Bk from island arcs in Figure 1 for comparison with the various BK fields established by the Viljoens (1969a). Also plotted are the high TiO_2 analyses in Table 1, and the "plutonic" samples which are totally compatible with BK chemistry but may not represent original liquids. It should be noted that there may be a continuum of compositions between PK and BK (and tholeiites).

[*Editors' Note:* Material has been omitted at this point.]

REFERENCES CITED

[*Editors' Note:* Only the references cited in the preceding excerpt are reproduced here.]

Brooks, C., and Hart, S. R., 1972, An extrusive basaltic komatiite from a Canadian metavolcanic belt: Canadian Jour. Earth Sci., v. 9, p. 1250–1253.

Davidson, C. G., 1967, The kimberlites of the U.S.S.R., *in* Wyllie, P. J., ed., Ultramafic and related rocks: New York, John Wiley & Sons, Inc., p. 251–261.

Drever, H. I., and Johnston, R., 1967, The ultrabasic facies in some sills and sheets, *in* Wyllie, P. J., ed., Ultramafic and related rocks: New York, John Wiley & Sons, Inc. p. 51–63.

Findlay, D. C., 1969, Origin of the Tulameen ultramafic-gabbro complex, southern British Columbia: Canadian Jour. Earth Sci., v. 6, p. 410–411.

Gale, G. H., 1973, Palaeozoic komatiite and ocean-floor type basalts from northeastern Newfoundland: Earth and Planetary Sci. Letters, v. 18, p. 22–28.

McCall, G.J.H., and Leishman, J., 1971, Clues to the origin of Archean eugeosynclinal peridotites and the nature of serpentinization: Geol. Soc. Australia Spec. Pub. 3, p. 281–299.

Moores, E. M., and Vine, F. J., 1971, The Troodos massif, Cyprus and other ophiolites as oceanic crust: Evaluation and implications: Royal Soc. London Philos. Trans., v. A268, p. 443.

Nesbitt, R. W., 1971, Skeletal crystal forms in the ultramafic rocks of the Yilgarn block, Western Australia; Evidence for an Archaean ultramafic liquid: Geol. Soc. Australia Spec. Pub. 3, p. 331–350.

Pyke, D. R., Naldrett, A. J., and Eckstrand, O. R., 1973, Archean ultramafic flows in Munro Township, Ontario: Geol. Soc. America Bull., v. 84, p. 955–978.

Ruckmick, J. C., and Noble, J. A., 1959, Origin of the ultramafic complex at Union Bay, southeastern Alaska: Geol. Soc. America Bull., v. 70, p. 981–1018.

Vinogradov, A. P., Udintsev, G. B., Demitriev, L. V., Kanaev, V. F., Neprochnov, Y. P., Petrova, G. N., and Rikunov, L. M., 1969, The structure of the mid-ocean rift zone of the Indian Ocean and its place in the world rift system: Tectonophysics, v. 8, p. 377–401.

Viljoen, M. J., and Viljoen, R. P., 1969a, The geology and geochemistry of the lower ultramafic unit of the Onverwacht Group and a proposed new class of igneous rock: Geol. Soc. South Africa Spec. Pub. 2, p. 55–86.

18

Reprinted from page 320 of *Jour. Petrology* **18**(2):319-369 (1977)

KOMATIITIC AND IRON-RICH THOLEIITIC LAVAS OF MUNRO TOWNSHIP, NORTHEAST ONTARIO

N. T. Arndt, A. J. Naldrett, and D. R. Pyke

[*Editors' Note:* Only Table 1 is reproduced here.]

TABLE 1

Classification of komatiites in Munro Township

	Field characteristics	Petrologic characteristics	MgO *
Peridotitic komatiite	The most magnesian types (MgO > 30%) are massive dark gray to black, olivine-rich rocks, devoid of conspicuous volcanic features, and form the basal cumulate sections of spinifex-bearing flows, or the central parts of intrusions. Less magnesian types are paler gray-green rocks weathering chocolate-brown; they occur in the spinifex-textured sections of zoned flows, or in spinifex-free flows, or in the marginal parts of small intrusions. Spinifex-free lava is characterized by polyhedral jointing.	All types are composed of olivine grains and minor chrome spinel in a matrix of fine-grained clinopyroxene and devitrified glass. In cumulates olivine grains are close-packed, solid, roughly equant and comprise 60 to 80% of the rock; in spinifex texture olivine forms large skeletal platy grains (35 to 60%); and in spinifex-free, non-cumulate rock, olivine may be equant or skeletal (45 to 70%).	MgO > 20%
Pyroxenitic komatiite	Forms low, flat outcrops of soft, gray-green, crumbly lava that breaks with a hackly fracture; it may contain clino-pyroxene spinifex, polyhedral jointing, lava toes or varioles.	Equant solid, or platy skeletal grains of olivine (0 to 35%) in fine-grained matrix of clino-pyroxene and devitrified glass; or skeletal subcalcic clino-pyroxene needles in devitrified glass groundmass; or closely-packed equant grains of pyroxene and olivine. No plagioclase.	MgO between 12 and 20%
Basaltic komatiite	Forms massive outcrops of hard pale-gray-green lava, and may contain lava toes and hyaloclastite flowtops, or may be massive. Rare examples are clino-pyroxene spinifex-textured.	No olivine in the groundmass; plagioclase instead. Clino-pyroxene and plagioclase form spinifex texture, 'graphic' intergrowths, or normal sub-ophitic texture.	MgO < 12%

* Anhydrous values.

Part II

ORIGIN OF BASALTIC MAGMAS

Editors' Comments
on Papers 19 Through 33

19 BOWEN
Excerpt from *The Crystallization of Haplobasaltic, Haplodioritic, and Related Magmas*

20 FENNER
Excerpt from *The Crystallization of Basalts*

21 OSBORN
Excerpts from *Role of Oxygen Pressure in the Crystallization and Differentiation of Basaltic Magma*

22 YODER and TILLEY
Excerpts from *Origin of Basalt Magmas: An Experimental Study of Natural and Synthetic Rock Systems*

23 GREEN and RINGWOOD
Excerpts from *The Genesis of Basalt Magmas*

24 O'HARA and YODER
Excerpts from *Formation and Fractionation of Basic Magmas at High Pressures*

25 O'HARA
Excerpt from *The Bearing of Phase Equilibria Studies in Synthetic and Natural Systems on the Origin and Evolution of Basic and Ultrabasic Rocks*

26 WYLLIE
Excerpt from *The Dynamic Earth*

27 KUSHIRO
Excerpt from *The System Forsterite-Diopside-Silica with and without Water at High Pressures*

28 KUSHIRO
Abstract from *Effect of Water on the Composition of Magmas Formed at High Pressures*

29 EGGLER
Effect of CO_2 on the Melting of Peridotite

30 MYSEN and BOETTCHER
Excerpts from *Melting of a Hydrous Mantle: II. Geochemistry of Crystals and Liquids Formed by Anatexis of Mantle Peridotite at High Pressures and High Temperatures as a Function of Controlled Activities of Water, Hydrogen, and Carbon Dioxide*

31 PRESNALL et al.
Excerpts from *Generation of Mid-Ocean Ridge Tholeiites*

32 WALKER, SHIBATA, and DELONG
Abyssal Tholeiites from the Oceanographer Fracture Zone. II. Phase Equilibria and Mixing

33 STOLPER
A Phase Diagram for Mid-Ocean Ridge Basalts: Preliminary Results and Implications for Petrogenesis

Choosing a dozen papers from several hundred as the benchmarks in basalt petrogenesis is extremely difficult at best. The problem is compounded by the fact that some of the classic papers (e.g., Yoder and Tilley, 1962; Green and Ringwood, 1967; O'Hara, 1968) are extraordinarily long. Indeed, any one of these papers in its entirety would account for most of the space alloted for this part. Thus, we are forced not only to choose the most influential papers from a vast array but also to select only limited parts of many of these papers. We proceeded on the following bases: (1) the first paper to affect our thinking profoundly in a given area was chosen over later papers, even though the later papers were perhaps more sophisticated and ultimately more correct; and (2) the basic fabric of a work has remained influential although some aspects of the paper have subsequently been shown to be incorrect. Because of space constraints, many excellent papers and workers have necessarily been omitted; some, but by no means all, are mentioned in these introductory remarks.

Although some aspects of the origin and crystallization of basaltic magmas are still very much in doubt, there is a wide consensus that they form by partial melting of peridotite in the upper mantle. At the site of melting, either relatively high pressure, small degrees of partial melting, and/or volatiles rich in CO_2 apparently promote the generation of alkali basalts. Lower pressures, larger partial melt fractions, and/or H_2O-rich volatiles tend to produce tholeiitic basalts. The exact type of basalt formed depends on a myriad of factors including: depth at which partial melting occurs, magma is separated from its source, and crystal fractionation occurs; fugacities of O_2, H_2O, and CO_2; source composition; degree of melting; rate of ascent of the magma; other processes, the significance of which has not been completely evaluated, such as liquid immiscibility and the Soret effect (diffusion along a temperature gradient).

It is clear that N. L. Bowen's (1915) classic work was the beginning of the modern approach to basalt petrogenesis; part of his article is reproduced here as Paper 19. Bowen (1928) later concluded that the non-porphyritic central-type basalt at Mull (Bailey et al., 1924; later referred to as tholeiitic by Kennedy, 1933, and Tilley, 1950) could be derived from the plateau type (so-called alkali olivine basalt, Kennedy, 1933, and Tilley, 1950) by fractional crystallization of olivine, clinopyroxene, plagioclase, and magnetite. Tilley (1950) and Kuno et al. (1957), however, showed that tholeiitic magmas apparently cannot be derived from alkali basalts; thus, at least that conclusion of Bowen's was incorrect.

In the 1920s, a lively debate existed between Bowen, who felt that fractional crystallization of basaltic magma led to alkali and silica enrichment (the Bowen trend), and C. N. Fenner, who argued for iron enrichment (the Fenner trend) (Paper 20). Wager and Deer's (1939) study of the Skaergaard intrusion confirmed the Fenner trend in at least that type of differentiation. The first elegant resolution of the Fenner and Bowen argument was made by Osborn (1959, Paper 21), who demonstrated that variations in the fugacities of O_2 (and presumably H_2O), along with their effect on magnetite crystallization, could create at least two differentiation trends that were similar to the Bowen and Fenner trends. The application of Osborn's work has been questioned by Cawthorn and O'Hara (1976), as well as by Eggler and Burnham (1973), but Osborn's paper remains a major work.

If the first phase of the modern era of basalt petrology was ushered in by Bowen (1915, 1928), the second phase was certainly initiated by Yoder and Tilley (1962); a small part of their article is reproduced as Paper 22. Their monumental effort was the first to integrate a wide variety of experimental, field, and petrographic information into a coherent model for the genesis and crystallization of basaltic magma. We should also point out that theirs was the most comprehensive early effort to use melting experiments of natural rocks and minerals rather than synthetic systems that could be more rigorously treated by the phase rule. Since natural rocks are multicomponent systems, experimental work on them commonly can approach more closely the actual behavior of magmas in nature.

Yoder and Tilley's (1962) experimental work dealt with melting of natural basalts and eclogites, in addition to the behavior of synthetic systems, at a wide range of pressures and H_2O fugacities. Their principle conclusions were, first, if an eclogitic magma, formed by partial melting of a garnet peridotite below 60 km in the mantle, fractionated at relatively deep levels, then the omphacite component would be enriched in the residual liquids, leading to alkali basalts; second, if fractionation took place at shallower levels, the derivative magmas would be enriched in garnet components and thus form tholeiitic magmas; and third, most basalts have near-eutectic compositions. Although the exact role of eclogite fractionation remains uncertain today, Yoder and Tilley's work formed a major stepping stone for subsequent research.

Ito and Kennedy (1967), in addition to demonstrating experimentally the spinel peridotite–garnet peridotite transition with increasing depth in the mantle, concurred with Yoder and Tilley (1962) and O'Hara (1965) that basalt magmas seen at the earth's surface are derivative rather than primary magmas. Ito and Kennedy (1967), however, proposed a picritic basalt partial melt rather than an eclogitic melt at depths below 60 km. Other workers also prefer a picritic primary melt (for example, see Stolper, 1980). Schairer and Yoder (1964) expanded Yoder and Tilley's basalt tetrahedron to include compositions that allowed examination of crystallization paths for alkali basalts. This and subsequent work led Yoder (1976) to conclude that there are three *parental* basaltic magmas—olivine tholeiite, nepheline basanite, and olivine-melilite nephelinite.

The years 1967 and 1968 were among the most important for those interested in basalt petrogenesis. In addition to that of Ito and Kennedy (1967), three other major papers appeared—Green and Ringwood (1967, Paper 23), O'Hara and Yoder (1967, Paper 24), and O'Hara (1968, Paper 25). O'Hara's model of polybaric fractionation during rapid ascent of a basaltic magma is an elegant concept, and his proposal takes into account the metamorphic facies believed to exist in the upper mantle (i.e., from plagioclase- to spinel- to garnet-peridotite with increasing depth).

Green and Ringwood (1967) departed considerably from ideas and methodology of previous workers. Indeed, sharp disagreement between groups at the Australian National University and the Carnegie Geophysical Laboratory continued for the next decade. Wyllie (Paper 26) compared O'Hara's (1965, 1968) scheme with Green and Ringwood's (1967; see also Green, 1969). He (Wyllie, 1979) later attempted to summarize the salient points of the two schemes. Much of the disagreement revolved around, first, the role of aluminous orthopyroxene, garnet, and olivine during fractionation; second, the rate of ascent versus degree of melting as factors; third, the mineralogy of the mantle; and fourth, the use of a synthetic mixture of dunite and basalt (pyrolite) to approximate mantle peridotite (Ringwood, 1962). Since then, Ringwood (1975) has criticized O'Hara's scheme, and Yoder (1976) has criticized Green and Ringwood's.

The 1970s marked a new era in basalt petrogenesis, one that emphasized the effect of volatiles—specifically, H_2O and CO_2—but with some emphasis on O_2 or H_2. Earlier workers had treated this aspect only cursorily, but a major breakthrough was made with the work of Kushiro, culminating in major papers in 1969 and 1972. (See Papers 27 and 28.) Earlier workers had proposed schemes for basalt petrogenesis that relied upon partial melting and fractionation under primarily dry conditions (or under conditions in which fluid pressures could not be controlled accurately). Kushiro, however, demonstrated that a wide range in basalt compositions, from oversaturated to undersaturated in SiO_2, could be achieved depending on depth, degree of partial melting, and whether the system was open or closed with respect to H_2O. Somewhat similar conclusions were reported by Nicholls and Ringwood (1973).

The next major step dealt with the combined effect of CO_2 and H_2O. Although H_2O had been thought to play a major role by workers as far back as Bowen (1928), Roedder's (1965) discovery that fluid inclusions in peridotite xenoliths were primarily CO_2 sparked this new research. Subsequently, several excellent papers on this subject were published at about the same time; among these, Wyllie and Huang (1976) and Eggler (Paper 29, 1975, 1976, 1978) deserve special mention. In keeping with our first guideline of reproducing pioneering papers, Eggler (1974) is reproduced as Paper 29, even though we now know the phase diagrams to be preliminary and were revised in Eggler (1978). In addition to shedding light on the origin of kimberlitic and carbonatitic rocks, these studies demonstrated that high CO_2 fugacities increased the field of aluminous orthopyroxene at various pressures and thus provided an important mechanism for the origin of SiO_2-poor alkali basalts.

This type of research was continued and expanded by Mysen and Boettcher (1975a, 1975b), who simultaneously considered the roles of H_2O, CO_2, and H_2 (see Paper 30). They were able to show that partial melts of mantle peridotite become increasingly silica undersaturated, from andesites to olivine tholeiites to alkali basalts, with increasing CO_2/H_2O ratios in the fluid phase.

MORB (mid-ocean ridge basalts) are the most abundant volcanic rocks in the earth's crust. Literally hundreds of papers have been written about them, and choosing only a few for this part borders on the absurd, space constraints notwithstanding. Fortunately, the nonexperimental aspects of these rocks are treated in Part IV. At a time when emphasis had been on the role of volatiles in basalt magma generation, Presnall et al. (Paper 31) concluded that volatile-free phase relationships could be used to model the melting of peridotite to form MORB magmas. Moreover, they concluded that the least fractionated basalts were probably close in composition to primary basalts, in contrast to the beliefs of many workers cited in this part (Yoder and Tilley, 1962; Ito and Kennedy, 1967; O'Hara, 1968; Yoder, 1976) as well as more recent workers (e.g., Paper 33; O'Hara and Mathews, 1981). Presnall and his co-workers, however, do not stand alone. Wilkinson (1982, p. 2), in a review article, concluded that "at least a majority of mid-oceanic ridge basalts is intrinsically primary." Presnall et al. (Paper 31) found a cusp on the solidus curve, at the transition between simplified plagioclase and spinel lherzolite, which forms an invariant point at about 9 kb and 1,300° C. They proposed that the first liquids that melt at this cusp are quite close in composition to the least fractionated MORB.

Bowen (1928) clearly established the role of crystal fractionation as the principal means by which basaltic magmas are differentiated (for excellent reviews of this topic, see Kushiro, 1979, and Presnall, 1979). After 50 years of research, crystal fractionation remains the mechanism most often called upon. Three other processes, however, should at least be considered: (1) silicate liquid immiscibility, (2) the Soret effect, and (3) mixing of various magmatic products.

Roedder (1979) presented a comprehensive review of liquid immiscibility. An earlier paper by Irvine (1976) had shown that metastable liquid immiscibility in basaltic magmas possibly can lead to fractionation trends comparable to either the Bowen trend of silica enrichment or the Fenner trend of iron enrichment. The iron enrichment would be caused by a different mechanism than the magnetite fractionation proposed by Osborn (1959). The key to Irvine's scheme seems to be the presence of $KAlSi_3O_8$ in the melt; greater amounts lead to silica-rich compositions, whereas lesser amounts produce iron-rich melts.

The Soret effect, although mentioned for decades, was not examined seriously until very recently. A paper by Walker and DeLong (1982) presents evidence for a startling degree of chemical differentiation across a temperature gradient well above the liquidus of a MORB. Implications for other rock types are enormous.

With regard to mixing of magmatic products, we present a paper by Walker, Shibata, and Delong (Paper 32). They proposed that fractionation paths of MORB magmas produce minerals and melts that are later remixed to yield the more differentiated members of the MORB suite. This work supports an earlier proposal of O'Hara (1977) that magma chambers under oceanic ridges are periodically refilled with new magma batches, which then mix with earlier-crystallized materials. Similar ideas have been expressed by Rhodes and Dungan (1979), who particularly emphasize the imprtance of clinopyroxene equilibration.

Finally, we have included a paper by Stolper (Paper 33), who has attempted to prepare a phase diagram for MORB magmas at a variety of pressures. Stolper's data support early concepts to the effect that MORB magmas are not primary melts from the mantle but must have undergone fractional crystallization from some original melt. One possibility for the original melt is a picrite, but Stolper points out the complexity of the entire system.

REFERENCES

Bailey, E. B., C. T. Clough, W. B. Wright, J. E. Richey, and G. V. Wilson, 1924, *Tertiary and Post-Tertiary Geology of Mull, Loch Aline, and Oban,* Geological Survey of Scotland Memoir, Edinburgh,445p.

Bowen, N. L., 1915, The Crystallization of Haplobasaltic, Haplodioritic, and Rated Magmas, *Am. Jour. Sci.,* ser. 4, **40:**161–185.

Bowen, N. L., 1928, *The Evolution of the Igneous Rocks,* Princeton University Press, Princeton, N.J., 334p.

Cawthorn, R. G., and M. J. O'Hara, 1976, Amphibole Fractionation in Calc-alkaline Magma Genesis, *Am. Jour. Sci.* **276:**309–329.

Eggler, D. H., 1975, CO_2 as a Volatile Component of the Mantle: The System Mg_2SiO_4-SiO_2-H_2O-CO_2, *Physics and Chemistry of the Earth* **9:**869–881.

Eggler, D. H., 1976, Does CO_2 Cause Partial Melting in the Low-Velocity Layer of the Mantle?, *Geology* **4:**69–72.

Eggler, D. H., 1978, Effect of CO_2 upon Partial Melting of Peridotite in the System Na_2O-CaO-Al_2O_3-MgO-SiO_2-CO_2 to 35 kb, with an Analysis of Melting in a Peridotite-H_2O-CO_2 System, *Am. Jour. Sci.* **278:**305–343.

Eggler, D. H., and C. W. Burnham, 1973, Crystallization and Fractionation Trends in the System Andesite-H_2O-CO_2-O_2 at Pressures to 10 kb, *Geol. Soc. America Bull.* **84:**2517–2532.

Green, D. H., 1969, The Origin of Basaltic and Nephelinitic Magmas in the Earth's Mantle, *Tectonophysics* **7:**409–422.

Green, D. H., and A. E. Ringwood, 1967, The Genesis of Basalt Magmas, *Contr. Mineralogy and Petrology* **15:**103–190.

Irvine,T. N., 1976, Metastable Liquid Immiscibility and MgO-FeO-SiO_2 Fractionation Patterns in the System Mg_2SiO_4-Fe_2SiO_4-$CaAl_2Si_2O_8$-$KAlSi_3O_8$-SiO_2, *Carnegie Inst. Washington Year Book* **75:**597–611.

Ito, K., and G. C. Kennedy, 1967, Melting and Phase Relations in a Natural Peridotite to 40 kilobars, *Am. Jour. Sci.* **265:**519-538.

Kennedy, W. Q., 1933, Trends of Differentiation in Basaltic Magmas, *Am. Jour. Sci.,* 5th ser., **25:**239–256.

Kuno, H., K. Yamasaki, C. Lida, and K. Nagashima, 1957, Differentiation of Hawaiian Magmas, *Japanese Jour. Geology and Geography* **28:**179–218.

Kushiro, I., 1969, The System Forsterite-Diopside-Silica with and without Water at High Pressures, *Am. Jour. Sci.* **267A:**269-294.

Kushiro, I., 1972, Effect of Water on the Composition of Magmas Formed at High Pressures, *Jour. Petrology* **13:**311-334.

Kushiro, I., 1979, Fractional Crystallization of Basaltic Magma, in *The Evolution of the Igneous Rocks: Fiftieth Anniversary Perspectives,* H. S. Yoder, Jr., ed., Princeton University Press, Princeton, N.J., pp. 171-203.

Mysen, B. O., and A. L. Boettcher, 1975a, Melting of a Hydrous Mantle: I. Phase Relations of Natural Peridotite at High Pressures and Temperatures with Controlled Activities of Water, Carbon Dioxide and Hydrogen, *Jour. Petrology* **16:**520-548.

Mysen, B. O., and A. L. Boettcher, 1975b, Melting of a Hydrous Mantle: II. Geochemistry of Crystals and Liquids Formed by Anatexis of Mantle Peridotite at High Pressures and High Temperatures as a Function of Controlled Activities of Water, Hydrogen, and Carbon Dioxide, *Jour. Petrology* **16:**549-593.

Nicholls, I. A., and A. E. Ringwood, 1973, Effect of Water on Olivine Stability in Tholeiites and the Production of Silica-Saturated Magmas in the Island-Arc Environment, *Jour. Geology* **81:**285-300.

O'Hara, M. J., 1965, Primary Magmas and Origin of Basalts, *Scottish Jour. Geology* **1:**19-40.

O'Hara, M. J., 1968, The Bearing of Phase Equilibria Studies in Synthetic and Natural Systems on the Origin and Evolution of Basic and Ultrabasic Rocks, *Earth-Sci. Rev.* **4:**69-133.

O'Hara, M. J., 1977, Geochemical Evolution during Fractional Crystallization of a Periodically Refilled Magma Chamber, *Nature* **266:**503-507.

O'Hara, M. J., and R. E. Mathews, 1981, Geochemical Evolution in an Advancing, Periodically Replenished, Periodically Tapped, Continuously Fractionated Magma Chamber, *Geol. Soc. London Jour.* **138:**237-277.

O'Hara, M. J., and H. S. Yoder, Jr., 1967, Formation and Fractionation of Basic Magmas at High Pressures, *Scottish Jour. Geology* **3:**67-117.

Osborn, E. F., 1959, Role of Oxygen Pressure in the Crystallization and Differentiation of Basaltic Magmas, *Am. Jour. Sci.* **257:**609-647.

Presnall, D. C., 1979, Fractional Crystallization and Partial Fusion, in *The Evolution of the Igneous Rocks: Fiftieth Anniversary Perspectives,* H. S. Yoder, Jr., ed., Princeton University Press, Princeton, N.J., pp. 59-75.

Rhodes, J. M., and M. A. Dungan, 1979, The Evolution of Ocean-Floor Basaltic Magmas, in *Deep Drilling Results in the Atlantic Ocean: Ocean Crust,* M. Talwani, C. G. Harrison, and D. E. Hayes, eds., Maurice Ewing Series, vol. 2, American Geophysical Union, Washington, D.C., pp. 262-272.

Ringwood, A. E., 1962, A Model for the Upper Mantle, *Jour. Geophys. Research* **67:**857-867.

Ringwood, A. E., 1975, *Composition and Petrology of the Earth's Mantle,* McGraw-Hill, New York, 618p.

Roedder, E., 1965. Liquid CO_2 Inclusions in Olivine-Bearing Nodules and Phenocrysts in Basalts, *Am. Mineralogist* **50:**1746-1782.

Roedder, E., 1979, Silicate Liquid Immiscibility in Magmas, in *The Evolution of the Igneous Rocks: Fiftieth Anniversary Perspectives,* H. S. Yoder, Jr., ed., Princeton University Press, Princeton, N.J., pp. 15-57.

Schairer, J. F., and H. S. Yoder, Jr., 1964, Crystal and Liquid Trends in Simplified Alkali Basalts, *Carnegie Inst. Washington Year Book* **63:**65-74.

Tilley, C. E., 1950, Some Aspects of Magmatic Evolution, *Geol. Soc. London Quart. Jour.* **106:**37–61.

Wager, L. R., and W. A. Deer, 1939, Geological Investigations in East Greenland: Part III. The Petrology of the Skaergaard Intrusion, Kangerdlugssuaq, East Greenland, *Medd. Grönland 105,* 353p.

Walker, D., and S. E. DeLong, 1982, Soret Separation of Mid-Ocean Ridge Basalt Magma, *Contr. Mineralogy and Petrology* **79:**231–240.

Wilkinson, J. F. G., 1982, The Genesis of Mid-Ocean Ridge Basalt, *Earth Sci. Rev.* **18:**1–57.

Wyllie, P. J., 1979, Petrogenesis and the Physics of the Earth, in *The Evolution of the Igneous Rocks: Fiftieth Anniversary Perspectives,* H. S. Yoder, Jr., ed., Princeton University Press, Princeton, N.J., pp. 438–520.

Wyllie, P. J., and W. L. Huang, 1976, Carbonation and Melting Relations in the System $CaO–MgO–SiO_2–CO_2$ at Mantle Pressures with Geophysical and Petrological Applications, *Contr. Mineralogy and Petrology* **54:**79–107.

Yoder, H. S., Jr., 1976, *Generation of Basaltic Magma,* National Academy of Science, Washington, D.C., 265p.

Yoder, H. S., Jr., and C. E. Tilley, 1962, Origin of Basalt Magmas: An Experimental Study of Natural and Synthetic Rock Systems, *Jour. Petrology* **3:**342–532.

19

Reprinted from pages 174–185 *Am. Jour. Sci.*, ser. 4, **40**(236):161–185 (1915)

THE CRYSTALLIZATION OF HAPLOBASALTIC, HAPLODIORITIC, AND RELATED MAGMAS

N. L. Bowen

[*Editors' Note:* In the original, material precedes this excerpt.]

Fɪɢ. 10.

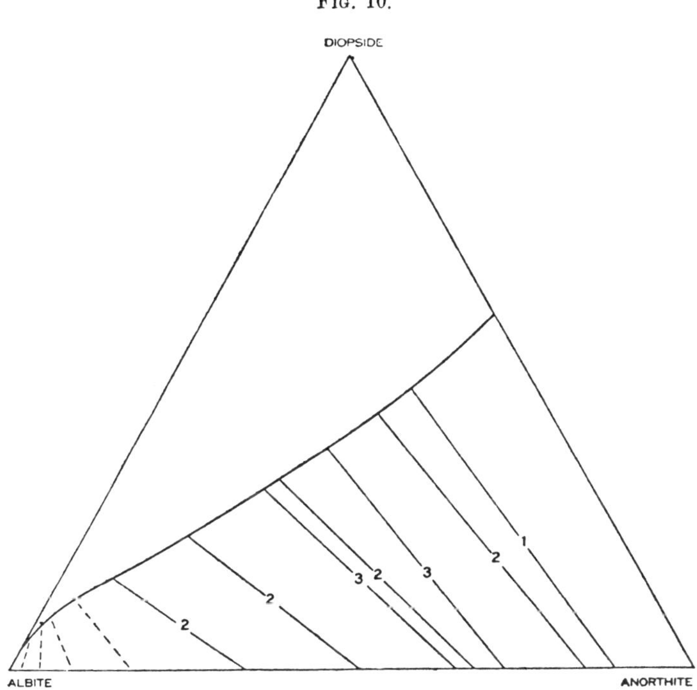

Fɪɢ. 10.   Three-phase-boundaries.

In fig. 10 the determined three-phase-boundaries are drawn in full lines and numbered according to the method of determination. In dotted lines the theoretical general direction of others is shown. A three-phase-boundary intermediate between two that are determined may be found by interpolation.

Crystallization of Mixtures in the Diopside Field.

Three-phase-boundaries being located, the course of crystallization (with perfect equilibrium) of any mixture whose composition is represented by a point in the diopside field can now be quantitatively described. Thus in fig. 11 the mixture F

73

(Ab$_1$An$_1$ 50 per cent–diopside 50 per cent) begins to crystallize at 1275°, diopside separating and the liquid changing along the straight line AFG towards G. At 1235°, when the liquid has the composition G, plagioclase of composition H (Ab$_1$An$_1$) begins to crystallize, the point H being determined by the three-phase-boundary (GH) through G. As the temperature is lowered the composition of the liquid follows the boundary

FIG. 11.

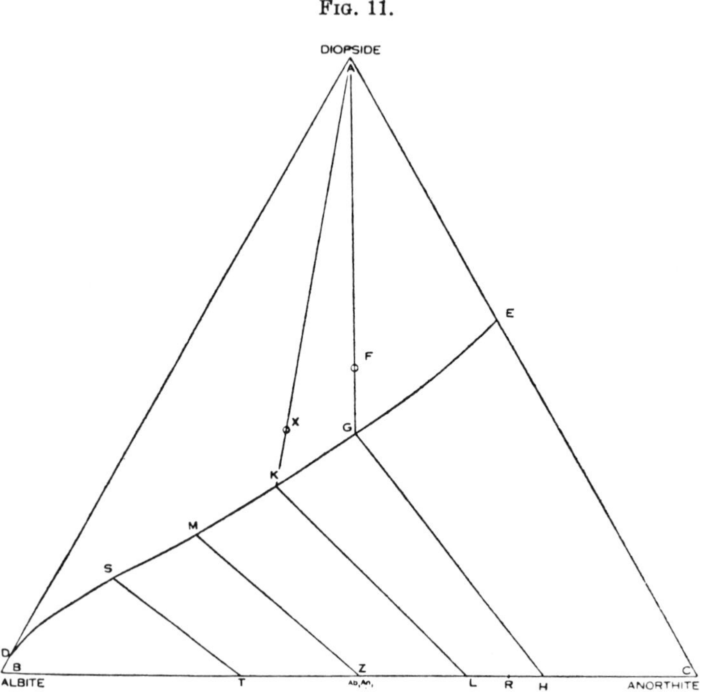

FIG. 11. Crystallization of mixtures in diopside field.

curve towards M. At 1218°, when the liquid has the composition K, the plagioclase has changed in composition from H(Ab$_1$An$_1$) to L(Ab$_1$An$_2$). Finally at 1200° the liquid is used up, the last minute quantity having the composition M, the plagioclase having now changed in composition to Ab$_1$An$_1$(Z). MZ is the three-phase-boundary through the composition Ab$_1$An$_1$ and fixes at the point M the temperature of final consolidation of F or of any other mixture of Ab$_1$An$_1$ and diopside.

The mixture X(Ab$_2$An$_1$ 60 per cent–diopside 40 per cent) begins to crystallize at 1252° with the separation of diopside,

74

and the liquid changes along the straight line AXK. At 1218°, when the liquid has the composition K, plagioclase of composition L(Ab₁An₂) begins to crystallize, KL being a three-phase-boundary. On further lowering of temperature the composition of the liquid moves along the boundary curve towards S, plagioclase increases in amount and changes continually in composition, diopside also increases in amount, until at 1176° the liquid is finally used up. The last of the liquid has the composition S and the feldspar has changed in composition to T, ST being a three-phase-boundary.

These examples make clear the necessity of determining three-phase-boundaries in order that the composition of plagioclase at any temperature may be known.

Crystallization of Mixtures in the Plagioclase Field.

For all mixtures in the diopside field the change of composition of the liquid until the boundary curve is attained is represented by a straight line. (Note AFG and AXK.)

For all mixtures in the plagioclase field, however, the liquid follows a curved course in reaching the boundary curve. The crystallization of any of these mixtures can not, therefore, be quantitatively described unless these crystallization curves in the plagioclase field are determined, and for this reason these curves were determined for two representative mixtures. Though applied to only two mixtures the method may perhaps prove useful in other more or less similar investigations and will therefore be described in full.

Determination of the Composition of Liquid and of Mix-crystals in a Two-Phase Mixture.—In order to find the composition of liquid in equilibrium with crystals at any temperature in a binary mixture, it is necessary only to hold a mixture at the desired temperature, quench it and determine the refractive index of the glass.* In a ternary mixture, however, the measurement of the refractive index of the glass is not sufficient to fix its composition. The composition can, nevertheless, be located as lying on the curve joining the composition of all glasses having that measured refractive index. Such curves will be referred to as isofracts.† But it is known also that the composition of the liquid must lie on the isotherm of the temperature at which the liquid was held. It must, therefore, lie at the point of intersection of the isofract and the

---

* N. L. Bowen : The Melting Phenomena of the Plagioclase Feldspars, this Journal (4), xxxv, p. 585, 1913.

† Objection to this term, based on its mixed derivation, seems to me to be outweighed by the fact that the prefix *iso* is that commonly accepted in this sense and therefore preferable to, say, *equi* while *fract* is mnemonic of refractive index.

isotherm. In order to apply this method, then, it is necessary to determine isotherms and isofracts, to hold the desired mixture at a measured temperature, quench and determine the refractive index of the glass. The point of intersection of the isotherm of the measured temperature and the isofract of the determined refractive index represents the composition of the liquid.

TABLE V.

Composition of glass		Refractive Index
Diopside	Plagioclase	
100	0	1·607
50	50 Ab	1·548
0	100 Ab	1·489
17·5	82·5 Ab_4An_1	1·523
60	40 Ab_2An_1	1·571
40	60 Ab_2An_1	1·553
25	75 Ab_2An_1	1·539
0	100 Ab_2An_1	1·517
50	50 Ab_1An_1	1·569
30	70 Ab_1An_1	1·553
0	100 Ab_1An_1	1·531
45	55 Ab_1An_2	1·573
25	75 Ab_1An_2	1·560
0	100 Ab_1An_2	1·545
60	40 An	1·594
0	100 An	1·575

The measurements of refractive indices, on which the location of isofracts is based, are given in Table V and the isofracts are drawn in fig. 12. The refractive indices were determined on glasses of known composition, the glass being compared with immersion liquids until a liquid whose index matched the glass was obtained. The index of the liquid was then determined on the refractometer. The probable error is usually not more than ·001 but an error of ·002 is possible in some cases.

If the mixture Ab_1An_1 85 per cent–diopside 15 per cent (D) is held at 1300° and quenched, the refractive index of the glass is found to be 1·539. Its composition is therefore fixed at the point P, fig. 12. The mixture, $Ab_{18}An_{32}$ 90 per cent–diopside 10 per cent (E), held at 1400° gives a glass of refractive index 1·561, the composition being, therefore, that of the point R, fig. 12.

It is important to note also that this determination of the composition of the liquid fixes the composition of the plagioclase crystals at the same time. Thus the composition of the crystals in the former case is given by joining PD and producing it to G which represents the composition of the plagioclase crystals. Similarly the point K on the straight line REK represents the composition of the plagioclase in the latter case. This method is the only one applicable to the determination of the composition of the plagioclase crystals in equilibrium with any liquid in the plagioclase field (i. e., not on the boundary

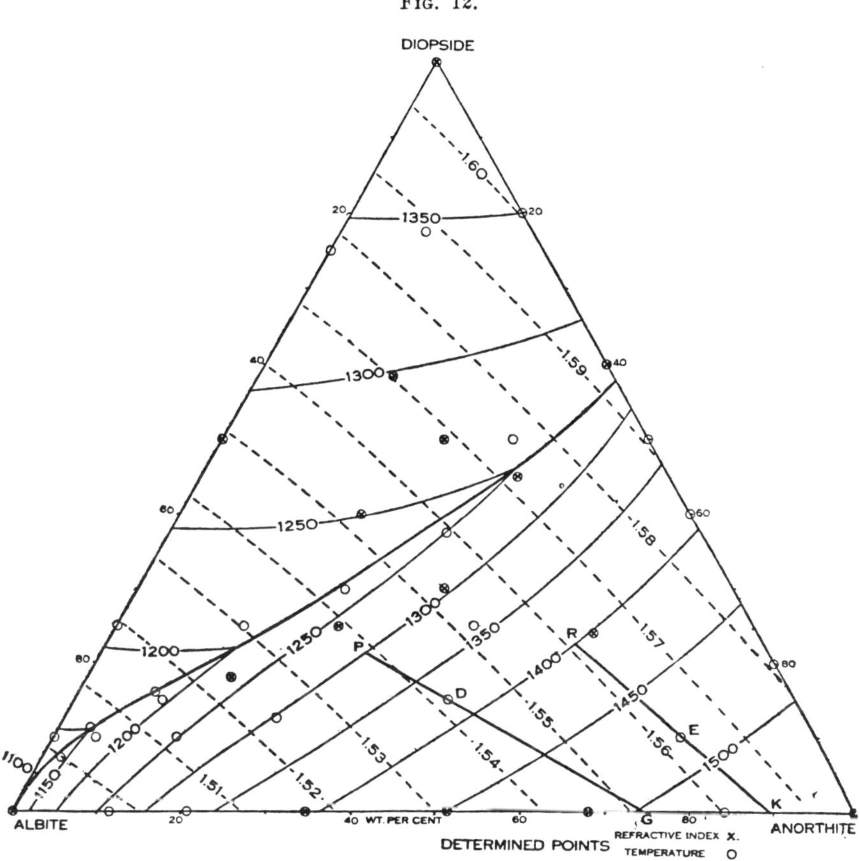

Fig. 12.

Fig. 12. Isotherms and isofracts.

curve, in which case the three-phase-boundaries fix the composition of the plagioclase) for the crystals themselves are too small for precise optical determination.

Crystallization Curves in the Plagioclase Field.—With the aid of the foregoing determinations of the composition of the liquid two representative crystallization curves in the plagioclase field can be drawn and the crystallization of the mixtures discussed. The mixture, Ab_1An_1 85 per cent–diopside 15 per cent (D, fig. 13), begins to crystallize at 1375° with the separation of plagioclase of composition Ab_1An_4. As the temperature falls the plagioclase increases in amount and changes in composition until at 1300° the liquid has the composition P and plagioclase the composition Ab_1An_3 (G of fig. 12). When the temperature has fallen to 1216° diopside begins to crystallize,

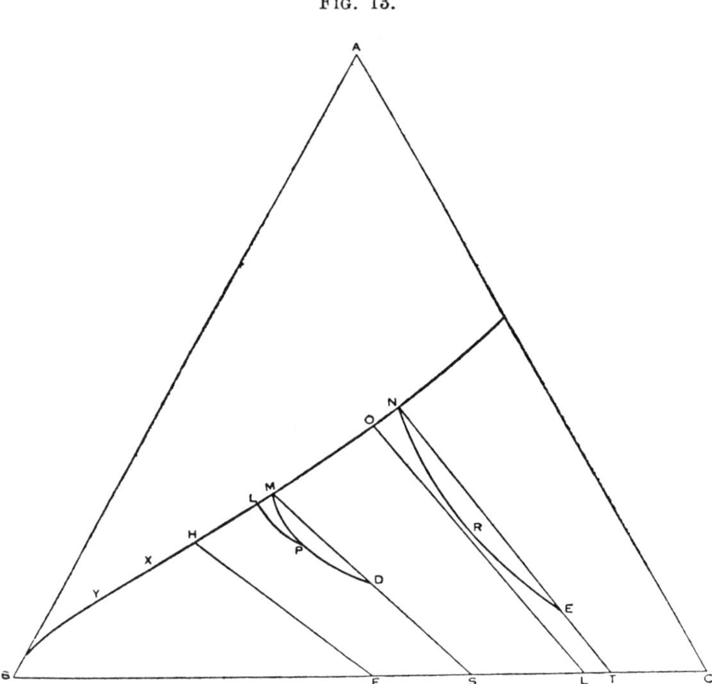

FIG. 13. Crystallization of mixtures in the plagioclase field.

the liquid then having the composition M (fig. 13) and plagioclase the composition $S (Ab_1An_2)$, SM being the three-phase-boundary through D. With further lowering of temperature the liquid follows the boundary curve, both diopside and plagioclase crystallizing, and at 1200° the liquid is all used up. The composition of the plagioclase is now Ab_1An_1 (F), FH being a three-phase-boundary.

In the case of the liquid E ($Ab_{18}An_{82}$ 90 per cent — diopside 10 per cent) crystallization begins at 1480° with the separation of Ab_5An_{95} and the composition of the liquid follows the curve ERN. At 1245° diopside begins to crystallize. The composition of the plagioclase is now T $(Ab_{15}An_{85})$, NT being a three-phase-boundary. As the temperature is lowered both plagioclase and diopside crystallize until at 1237° all the liquid is used up. The composition of the final liquid is O and the feldspar has attained the composition L $(Ab_{18}An_{82}.)$

It should be noted that the crystallization curves DPM and ERN apply to the liquids D and E respectively and to no other liquids. Thus the crystallization curve of the liquid P is not the curve PM but the new curve PL, i. e., if we start with a liquid P free from crystals the composition of the liquid follows the course PL. Only when the liquid P contains in it the crystals formed during the change from D to P does the further course of the liquid coincide with PM. Moreover, the liquid P when originally free from crystals becomes on cooling completely crystalline, not at 1200° (H), as before, but at a somewhat lower temperature.

Crystallization with Zoning.

Throughout the foregoing discussion of crystallization perfect equilibrium is assumed. The conditions are supposed to be such that crystals of plagioclase can change their composition through and through in response to the demands of equilibrium. It may be considered, however, that crystallization takes place in a quite different manner. When plagioclase of a certain composition has separated it may remain as such and become surrounded by layers of different composition deposited by the continually changing liquid. The liquid is in equilibrium at any instant only with the material crystallizing at that instant and not with crystals already formed.* A plagioclase crystal, once separated, does not participate further in the equilibria. As far as any effect on the course followed by the liquid is concerned the crystal may be considered absent. The course of crystallization of the liquid P, fig. 13, in the absence of crystals has been compared in the foregoing with that followed in the presence of crystals. If we examine also the liquid M, say, we find that if it crystallizes in the presence of crystals formed during the change in composition of the liquid from D to M, it then becomes completely crystalline at 1200° and the final liquid has the composition H. On the other hand, if the crystals referred to are separated from the liquid M complete crystallization does not take place until the temperature 1170° is attained and the final liquid has the composition X, i. e., is very much richer in albite. If in this latter case a second removal of crystals took place when the liquid

* This corresponds with "Erstarrung erster Art" of Schreinemakers, Zs. phys. Chemie, 1, p. 189, 1905.

had the composition H complete crystallization would not then take place until the temperature had fallen to 1125° and the final liquid has then the composition Y, exceedingly rich in albite.

If this separation of early crystals from liquid is a continuous process accomplished through zoning of the crystals, it is clear that this continual lowering of the temperature of final consolidation and offsetting in the composition of the liquid is limited only by the eutectic albite-diopside, 1085° and 97 per cent albite. There is a certain theoretical rate of cooling which will give maximal zoning, in which case this limiting temperature and composition of the liquid will be actually attained. An increased rate of cooling will bring about undercooling, and crystallization will be completed when the liquid has the composition Y or X or H, or with very rapid cooling, the liquid M may be cooled below 1170° before crystallization begins, in which case it will crystallize *in toto* without any offsetting of composition. A rate of cooling slower than that which gives maximal zoning will also limit the offsetting in the composition of the liquid on account of a certain amount of adjustment between liquid and crystals. The result is that the final liquid may be Y or X, or with exceedingly slow cooling (perfect equilibrium) H.

However, even when the cooling is slow enough to permit perfect equilibrium other factors may intervene to produce the same offsetting in the composition of the liquid as does zoning. These are the separation of crystals from the liquid, or from some of the liquid, by their sinking or by a squeezing-out or draining-off of residual liquid. Clearly the opportunity for both of these, especially the former, increases with slow cooling. The rate of cooling of any liquid of the present system is, therefore, of fundamental importance in determining the range of composition which will be covered by the liquid as it changes during crystallization, and likewise the range of composition of the crystalline products.

The Significance of the Results in Petrologic Problems.

The aim of experimental investigations of silicate melts is, of course, the explanation of some of the multitudinous, more or less disconnected facts concerning igneous rocks that have accumulated with the advance of descriptive petrography. It still is, and possibly always will be, a considerable extrapolation from the simple systems that can be investigated quantitatively to the more complex systems, or perhaps rather system, represented by magmas. Nevertheless, considerable progress is being made in the shortening of this extrapolation. It is not long since the discussion of the theory of the crystallization of igneous rocks was carried out by describing the crystallization of some simple binary eutectic mixture such as salt and water, and winding up with the statement that the crystallization of igneous rocks is in some measure analogous. A better grasp of the theoretical aspects of the problems and a considerable number of exact investigations of silicate melts themselves have now carried the matter well beyond this stage.

In the present investigation the mixtures treated are sufficiently close to certain natural magmas, in essential mineral composition, that I have, with more or less propriety, referred to them as haplobasaltic, haplodioritic and related magmas. (Note also analyses compared in Table VI.) The names seem further justified by the number of facts and principles governing the crystallization of their natural analogues that are brought out by the investigation of these artificial melts.

One fact stands out very clearly, viz., the great difference between the crystallization of the mixtures that have been described and crystallization in a system with a ternary eutectic. *There is no eutectic between diopside and any intermediate plagioclase, still less between plagioclase and a complex pyroxene solid solution such as augite.*

There is, then, nothing to be gained from the search for the gabbroic eutectic, the dioritic eutectic and so forth. Appreciation of the non-existence of such eutectics is an important matter. The existence of a gabbro eutectic would mean that there is a certain definite, lowest-melting mixture of calcic plagioclase and augite, with a little magnetite, towards which the liquid, during the crystallization of any gabbro, always proceeds and *beyond which it can not pass.* The sinking of crystals or the squeezing-out of residual liquid would avail nothing in enabling the liquid to pass the eutectic temperature and composition, the eutectic being a *necessary end-point.* A gabbro magma might give rise locally to anorthosite and pyroxenite by the sorting of crystals according to their densities, but, by crystallization-differentiation, ·it could never give diorite or syenite or granite.

On the other hand, if it is realized that the crystallization of gabbro is in large measure analogous to that of haplogabbro the possibility of the derivation of an igneous rock series becomes apparent. Haplogabbro magma of composition Ab_1An_2 50 per cent–diopside 50 per cent, may, if it is cooled very rapidly, give rise simply to haplogabbro with 50 per cent plagioclase crystals of the uniform composition Ab_1An_2 and 50 per cent diopside. Yet if the same original magma is cooled more slowly, different results may be obtained.* The plagioclase crystals may be zoned, with compositions ranging

* See discussion of figures 11 and 13.

from Ab₁An₃ to Ab₂An₁ and, if it had happened that the liquid was separated from the crystals before the later zones of plagioclase were formed, this liquid would, of course, have crystallized to a body of haplodiorite. Again, this separated haplodiorite magma might have crystallized under conditions which produced a zoning of its plagioclases, ranging, say, from Ab₁An₁ to Ab₄An₁, and if the liquid were separated at a late stage it could give a body of haplosyenite.

If sinking of crystals took place as they grew in the liquid it is plain that the effect on the uppermost liquid would be of the same kind, indeed only stirring of considerable vigor could prevent the formation of haplosyenite as the upper differentiate of a large, slowly-cooled mass of haplobasaltic magma. The only limit beyond which this differentiation could not pass is a mixture of 97 per cent albite and 3 per cent diopside.

Another feature of these differentiates is worthy of note, viz., the fact that as the plagioclase becomes more alkalic the percentage of diopside (colored constituent)* decreases rapidly. Thus in the original haplogabbro there was 50 per cent diopside, in haplodiorite derived in the manner outlined about 30 per cent, in haplosyenite 10–15 per cent. On the other hand, if it is imagined that these latter types were not derived by this process of development from more basic types but were, say, specially created, then there would be no necessary relation between the alkalinity of the plagioclase and the proportion of diopside.

The crystallization of the natural analogues of these melts, gabbro, diorite, etc., is beyond doubt a considerably more complex process, nevertheless the plagioclase mix-crystal series must exert a similar influence, and when we turn to the rocks this influence is plain in what is commonly termed the subalkaline series. The increasing alkalinity of the feldspar and the accompanying decreasing importance of the colored constituents in the later members (members of latest consolidation) of this series is a well-attested fact. One important difference between the artificial melts here described and the natural series is the prevalence of free silica as quartz in the later members. In another paper this matter will be discussed in some detail and it will be shown that the formation of olivine at an early stage and the later formation of hornblende and especially of biotite in the presence of water account satisfactorily for the development of quartz.

The natural series presents a further important difference from the 'haplo' series in that there is no special mixture, analogous to that containing 97 per cent albite and 3 per cent diopside, which can be referred to as a necessary end, unless

---

*I. e., corresponding to a colored constituent as these terms are ordinarily used though the pure, artificial diopside of these mixtures is not colored.

this is some frozen aqueous solution. The presence of volatile components and the existence of various equilibria between them and silicates removes the necessity for the existence of an end-point similar to that in the ' haplo ' series.

There appears then to be no room for reasonable doubt that the differentiation of the sub-alkaline series of igneous rocks is controlled entirely by crystallization. Moreover, the systematic decrease in the amount of colored constituent with increasing alkalinity of the feldspar indicates that the more acid types are not original magmas,* for in this case there would be no reason for such a balance in the proportion of these contrasted constituents, whereas this finds a natural explanation if the more acid magmas are regarded as derived from basic material, being, as it were, successive mother liquors from the crystallization of the basic magmas.

In this connection it seems worth while to point out the considerable degree of similarity in composition (when the absence of iron and potash and the fact that other constituents, principally lime, must make up this shortage are taken into account) between a haplodiorite magma from the investigated system and the average diorite as computed by Daly (Table VI). This haplodiorite is not any random one but is one which could be deprived as a mother liquor from the crystallization of haplobasaltic magma.† The similarity is suggestive in connection with the idea that diorites are related to basaltic magma in the same way.

J. H. L. Vogt believes that differentiation is accomplished in the liquid state,‡ but when he proceeds to discuss the problem he bases the discussion entirely on the progress of *crystallization*. Indeed, Vogt's paper, though professedly the opposite, is one of the best essays we have in favor of crystallization-differentiation.

TABLE VI.

	I	*mols.*	II	*mols.*
SiO_2	57·56		57·8	
TiO_2	·85		--	
Al_2O_3	16·90		17·0	
Fe_2O_3	3·20		--	
FeO	4·46		--	
MgO	4·23		6·5	
CaO	6·83		14·0	
Na_2O	3·44 }	·077	4·7	·076
K_2O	2·15 }		--	

I   Average diorite as calculated by Daly.
II   A haplodiorite represented by a point on the boundary curve.

*In the sense that they always existed as such.
† I. e., it lies on the boundary line.
‡ Uber anchi-monomineralische und anchi-eutectische Eruptivgesteine. Videnskabs-Selsk. Skr. I., Math. Naturv. Kl. 1908, No. 10.

Summary.

Mixtures of diopside with various members of the plagioclase series, referred to as haplobasaltic,* haplodioritic and so forth, according to the nature of the plagioclase, are studied by the quenching method of thermal analysis. Equilibrium, determined in this manner, is represented graphically in several diagrams. All the determinations which are necessary for the complete description of the crystallization of any mixture have been made and are presented.

The facts determined for haplodiorite and so forth are applied to their natural analogues and it is shown that there can be little reason to doubt that crystallization controls the differentiation of the subalkaline series of igneous rocks.

Geophysical Laboratory of the Carnegie Institution of Washington.

* From the Greek ἁπλόος = simple.

20

THE CRYSTALLIZATION OF BASALTS

C. N. Fenner

[*Editors' Note:* In the original, material precedes this excerpt.]

Let us review very briefly the course of the argument up to this point. From physico-chemical principles of general applicability it was deduced as probable that in the crystallization of the ferromagnesian silicates of igneous magmas, such as pyroxenes and olivines, an excess of magnesia should be concentrated in the crystals and an excess of iron in the residual liquid. Examination of the compositions of two sills of basic rock in which there had plainly been gravitative separation of crystals—one in Yellowstone Park and the other the Palisade sill in New Jersey—supported this inference. Further confirmation was given in a number of ways by comparing available analyses of early formed phenocrysts of augite and olivine in basalts and similar basic magmas with the analyses of the magmas in which they had been formed. In the final example cited a further step was taken, and confirmation was found for the logical deduction that when crystallization has become far advanced the residual liquid is rich in iron. We shall now endeavor to enforce this last conclusion by additional evidence. Before proceeding with this, however, a few words should be said regarding magnetite.

In the last example quoted it was evident that a large amount of the magnetite molecule was held in solution until late in the course of crystallization. No theoretical or experimental evidence of much weight can be adduced to explain this, but in the examples of basaltic crystallization that will be cited it will be found that this generally holds.

In a previous paper I have called attention to the frequency with which the final residue of basalts forms a mesostasis of brown glass, with magnetite crystals, and have pointed out the inconsistency of this with the supposition that crystallization of a basalt leads to a residue of granitic composition. Other writers also, even before the theory of crystallization differentiation was revived in its present form, have described the iron-rich residue of basalts. In answer it has been asserted that in such cases the basalt was probably undercooled before crystallization began, and the residue was simply undifferentiated basalt. It has also been said that there was no certainty that the brown glass was particularly rich in iron. Neither of these suggestions seems to be confirmed by critical examination of the evidence.

The Triassic Watchung basalts in New Jersey are generally regarded as the extrusive equivalents of the Palisade diabase sill. They were poured out as great surface flows over subaerial deposits of sandstones and shales. In the examination of a large number of thin sections from the first Watchung sheet certain general characteristics have been observed. The essential minerals are few in number, and consist of plagioclase feldspar, pyroxene, and magnetite. In an article published a number of years ago,[14] it was shown that when the magma arrived at the surface it was in an almost wholly molten condition, with very few phenocrysts, and that plagioclase and pyroxene appeared together at the beginning of crystallization, and continued their simultaneous growth to the end.[15]

The features of interest for the present purpose are illustrated by thin section No. 25. In this the lath-shaped plagioclases have usual dimensions of 0.02 to 0.04 mm. by 0.15 mm. The granular pyroxenes are of irregular outline, but an average diameter is 0.10 mm. A matter of great importance is that individual crystals of both minerals show variable composition as is indicated by wandering extinction, plainly visible when the stage is rotated between crossed nicols. They evidently made the normal response of solid solutions to crystallization

[14] Fenner, C. N., The crystallization of a basaltic magma from the standpoint of physical chemistry, this Journal, **29,** 217, 1910.

[15] Bowen, in his recent book, "The evolution of the igneous rocks" (p. 67), apparently believes that I have abandoned the view stated in the paper of 1910, with respect to the sequence of crystallization in basalts, and have reverted to the antiquated notion that all the feldspars crystallized first. I have not at any time entertained such a belief. In several places in his book Bowen apparently has misunderstood my views.

in a liquid of changing composition.   The evidence is distinctly against the idea of crystallization in a supercooled liquid. In the angular spaces between the crystals there are irregular patches of brown glass, 0.05 to 0.15 mm. across, scattered here and there over the whole section.   This is evidently a residue left after most of the minerals had crystallized.   In this glass magnetite is abundant, probably occupying nearly as much area as the undifferentiated glass.   A small part of this magnetite is in small grains, and might possibly be supposed to have been pushed into corners by the growth of larger crystals, but much the greater part is in connected branching growths, which frequently extend across the areas of glass, and evidently crystallized from the surrounding liquid.   In thin section No. 18 pyroxene and feldspar as well as magnetite can be seen to have crystallized in some of these areas, and in No. 47 most of the dark patches seem to be made up of a mixture of magnetite, pyroxene, and feldspar.   We deduce that these three minerals, with a proportion of magnetite much greater than that of the original magma, and with a large amount of pyroxene, represent the composition toward which these basalts tended through crystallization.   The pyroxene of this last crystallizing residue is probably richer in iron, and the feldspar richer in soda and potash, than are the corresponding minerals of early crystallization.   The place of titanium is not shown.

Similar features may be found in sections 64, 98, 31, 53, 137, and 97; in fact, they are characteristic of these rocks.

In a large collection of thin sections of Deccan traps, which Dr. Washington kindly lent me, analogous features to those described for the Watchung basalts are found.   In some of these the dark, magnetite-rich patches have become pretty well crystallized, though the grain is finer than in the general rock. In a section labeled Mainpat, Sarguja, the larger plagioclases show progressive zoning plainly, and in the pyroxenes wandering extinction is seen.   The residual liquid is represented by dark patches, consisting of arborescent magnetite and small but well individualized crystals of pyroxene and plagioclase. The presence of ilmenite is suggested by blade-like forms.

In the same collection we may pass to other sections which probably represent the final result of slow cooling.   No interstitial fine-grained material is found, as the pyroxenic and feldspathic material of the final residues has grown in continuous succession with previously formed crystals, but most of the magnetite is in a few large crystals, whose late deposition is indicated by an irregular straggling outline and by the manner in which its crystals have molded themselves around plagioclase and pyroxene.   In the Mainpat section just described a very pretty example is found, in which about 2/3 of a magnetite crystal consists of a stout, irregular form, lying adjaeent to large crystals of pyroxene and plagioclase, while the other 1/3 projects as a lattice growth into fine-grained pyroxene and plagioclase.   In a section from "Katem quarries, N. of Rajamundri," the molding of magnetite around silicate minerals in a uniformly coarse-grained rock is shown.

The last-named rock illustrates another matter to which it is desirable to call attention.   The rock is one which Dr. Washington has analyzed, and his calculation of the norm shows 8.16 per cent free quartz.   Now the section is so coarse-grained and shows such complete crystallization that if quartz were actually present in the mode it seems as if it should be visible, but careful examination with the microscope fails to reveal anything which suggests quartz.   This matter is brought up for the reason that those who believe that crystallization of a basalt should lead to a quartzose residual liquid have attached considerable importance to the fact that in many instances the norm of basalts shows free silica, and they deduce therefrom that quartz must finally appear if crystallization is complete. This is a conclusion of doubtful validity, as Dr. Washington, with whom I have discussed the subject, agrees.

The originators of the C. I. P. W. classification, in devising the system, found it necessary to make certain simplifying assumptions in calculating the norm.   It would be practically impossible to take account accurately of the great complexity of some of the natural minerals.   In the calculation, all of the Fe_2O_3 present is attributed to magnetite, and an equal molecular amount of FeO is assigned to it.   As a matter of fact, however, much of the Fe_2O_3 seems to be dissolved in pyroxene crystals as an independent substance, and the equivalent FeO is actually combined with SiO_2.   In this way free silica of the norm is lessened or disappears in the mode.   Pyroxenes also contain titanium, alkalies, and other substances, of which the actual form of combination is doubtful, so that the results arrived at by following the rules for calculation of the norm should be regarded only as a useful approximation to the modal composition, and not as a basis for theoretical conclusions unless supported by the evidence of the rocks themselves.

Out of a large number of sections of plateau basalts that I have studied, the great majority show features similar to those described, indicating that the residual liquid, up to complete crystallization, was highly ferriferous, and that it deposited magnetite, pyroxene, and feldspar. In a few sections, however, it was found that, for some reason, which can only be surmised, the interstitial glass was evidently of different composition. In a section of Deccan basalt from Panandrao, Kutch, there is the normal aggregate of plagioclase, pyroxene, and magnetite. All are rather coarse-grained, and the magnetite is of such form as to indicate that much of it is of late deposition. There is a little interstitial glass, most of which is fresh, and it is then of a clear brown color, is isotropic, and has an index of 1.510. In a few places incipient alteration has given it a bright yellow color. In the glass there are aggregates of minute pyroxenes and slender plagioclase laths, and many long, very thin needles, some branching, of a substance which is of high refraction but of such weak birefringence as to show no color even with the sensitive tint plate. It is probably apatite. The remarkable features are that the patches of glass contain no magnetite crystals, and that the index is so low. There is not much clue to the composition of the glass, but it may be suggested as a possibility that this little residue contains a considerable proportion of phosphates, borates, or silicofluorides. Phosphates are indicated by the apatite crystals, while the common occurrence of datolite and apophyllite as secondary minerals in trappean rocks suggests the other two. Possibly also the escape of easily volatile substances as the residual liquid became small in quantity carried away iron which would have appeared as magnetite.

The formation of microgranitic or myrmekitic intergrowths of quartz and feldspar in some doleritic or diabasic intrusions, such as the Palisade sill, may reasonably be ascribed to the action of late hydrothermal solutions containing chemically active substances of this nature. In the intrusive Triassic diabase at Goose Creek, Virginia, of which many of the features have been described by Shannon,[16] processes of this sort have been carried out on an unusually large scale. Hydrothermal solutions have formed in the diabase irregular masses of rock,

[16] Shannon, E. V., The mineralogy and petrology of intrusive Triassic diabase at Goose Creek, Loudoun County, Virginia, Proc. U. S. Nat. Museum, **66**, Art. 2, 1-86, 1924.

several feet in diameter, consisting mostly of albite and quartz, but with a large number of other minerals in small quantities. The albite and quartz are found both as separate individuals and as rather large graphic intergrowths, and in both forms they occur both in the mass of rock and in the numerous vugs. The nature of the solutions that effected these results is indicated by the presence of abundant datolite and apophyllite in associated veins, as described by Shannon, and of scapolite (chlorine-bearing mizzonite) which I have found. Boron, fluorine, and chlorine were therefore present, and by the action of these hydrothermal solutions the diabase was leached and recrystallized to a rock which has many of the characteristics of a granite.

Lacroix[17] suggests a secondary origin for the "micropegmatites of quartz and acid feldspars which fill intersertal cavities of so many diabases and exceptionally of certain doleritic basalts." His view is that they have been formed by pneumatolytic action. I should prefer to consider them as formed by hydrothermal solutions, largely of pneumatolytic derivation. The difference between the two conceptions is not great, and it is doubtful whether, in many cases, we have well established criteria by which to discriminate between them. Doubtless the action of vapors is often followed by the action of hydrothermal solutions, which continue the work of the vapors but give a somewhat different aspect to the final mineral assemblage.

These intersertal intergrowths of quartz and feldspar in diabases have been considered by the supporters of crystallization differentiation to represent the last crystallizing residues of diabase magmas, and much importance has been attached to them as indicators of the direction of differentiation. In the present paper, it has been pointed out that the phenomena of crystallization of basalts are opposed to the belief that a basaltic magma, when influenced by crystallization alone, differentiates toward such a residue, and the evidence given just above is believed to show that mineral aggregates closely allied to these intersertal quartz-feldspar intergrowths have rather plainly resulted from secondary reactions rather than from magmatic crystallization.

Similar features in basalts to those that have been described have been noted by other writers, and some have drawn especial

[17] Lacroix, A., Produits silicatés de l'éruption du Vésuve, Nouvelles archives du Muséum d'Histoire Naturelle, quatrième série, tome neuvième, p. 99, 1907.

attention to them. Iddings, writing of certain fine-grained recent basalts in Yellowstone Park,[18] speaks of "microlitic glass base, which is crowded with grains of magnetite, some forming skeleton aggregates of the usual stellate or cruciform shapes," thus recognizing its common occurrence in this form.

Washington, in his description of the Deccan traps, has referred to similar features to those described here, and speaks of "the late stage of crystallization of the magnetite and of much of the augite."[19]

Newton and Teall[20] have described basalts from Franz Josef Land, and state:

"We are therefore able to draw the important conclusion that in a magma of the type to which these basalts belong .... progressive crystallization leads to the formation of a mother-liquor poor in silica and alumina and rich in iron."

With reference to magnetite they state:

"The feldspar and augite are as a rule remarkably free from inclusions of this mineral; which certainly does not, in these rocks, belong to the earlier phases of consolidation, as it does in so many rocks of intermediate composition. In many cases it is found only in skeleton crystals in the interstitial matter .... and in some the iron oxides have remained wholly undifferentiated in a deep brown glass."

Tsuboi,[21] in his account of the rocks of the Volcano Oshima, gives careful descriptions of their megascopic and microscopic characteristics. Many are basaltic or close to basaltic in composition. Phenocrysts often form a large proportion of the rock, but however large in amount phenocrysts may be, the groundmass is generally composed of plagioclase, augite, and magnetite, the last often showing dendritic skeletal crystals. Brown glass is present in some. Tsuboi is evidently an adherent of the theory of crystallization differentiation, and seems somewhat puzzled over the mafic character of the groundmass. The description of one rock ends with the comment "It is a remarkable feature of the groundmass that the mafic com-

[18] Iddings, J. P., in Monograph 32, part 2, U. S. Geological Survey, p. 439, 1899.
[19] Washington, H. S., Deccan traps and other plateau basalts, Bull. Geol. Soc. Amer., **33**, 772, 1922.
[20] Newton, E. T., and Teall, J. J. H., Notes on a collection of rocks and fossils from Franz Josef Land, Quart. Jour. Geol. Soc., **53**, 477, 1897.
[21] Tsuboi, Seitarô, Volcano Oshima, Idzu, Jour. Col. Sci., Tokyo, **43**, Art. 6, 67-96, 1920.

ponents nearly equal, and at times even exceed the felsic, in quantity as well as in size."[22]

We may add here the observation of Washington[23] :

"Instances are accumulating of rocks in which much of the pyroxene belongs to the final stage of crystallization, or is occult in the glass, and this is true also of magnetite, contrary to the well-known dictum of Rosenbusch."[24]

In our examination of basalts we found that the last residuum from crystallization appears characteristically as small patches of brown glass and iron ore, or as a fine-grained aggregate of pyroxene, feldspar, and iron ore. From this, and for other reasons which have been presented, we deduce that the result of crystallization is to produce a final liquid rich in certain constituents, particularly iron and alkalies, and poor in silica, magnesia, and lime. This residual liquid would be likely to contain in addition a certain small amount of various volatile substances, such as water, borates, fluorides, and chlorides, especially if conditions were such that the escape of volatile compounds was somewhat impeded.

Substances of this last group would be likely to enter into evanescent combinations during the later stages of concentration. At one stage a certain amount of $FeCl_3$, for example, might be present. With increasing concentration this might escape as such, or it might react with water to form Fe_2O_3 and HCl, of which the HCl would be volatilized, and the Fe_2O_3 might temporarily form a borate, and finally enter into magnetite, ilmenite, or pyroxene. The direction of such reactions would depend upon the ease or difficulty of escape of volatiles, upon the temperature, and upon the concentration of each constituent.

In the basalts, evidences of the general nature of this crystallization residuum (especially of its high iron content) were plain. In the coarse-grained intrusive equivalents of the basalts—the diabases—we should be able to find evidences of a corresponding residuum. In searching for such evidence it

<hr>

[22] Tsuboi, S., op. cit., p. 78.

[23] Washington, H. S., Santorini eruption of 1925, Bull. Geol. Soc. Amer., 37, 383, 1926.

[24] For a number of years Dr. Washington has been working on the chemical composition of the pyroxenes, as related to the composition of the magmas in which they occur, and his analyses have been of great assistance in supplying data for the present paper.

is necessary, of course, to study fresh rocks, unaffected by metamorphic or weathering processes.[25]

In diabases of this kind we find, as we should expect, that most of the iron ore commonly occurs as a few large crystals, of which the late deposition is indicated by lack of idiomorphic outline, the crystals being of irregular or straggling form and molded upon adjacent crystals of silicate minerals. Of much significance is the fact that very frequently a little dark brown mica (probably biotite or lepidomelane) is closely associated with the ore, and is rare or absent elsewhere. So usual is this association of brown mica and iron ore in diabases that the need for some explanation has been recognized by petrologists, and it has been suggested that the mica is a secondary or reaction product between magnetite and adjacent feldspar, and was formed after the consolidation of the rock. In some instances this may be the correct explanation, but I should like to point out that these minerals are nearly equivalent in composition to what we deduced was the composition of the residuum in basalts, with a little water and fluorine added, and their mode of occurrence is in agreement with what might be expected as a final product of magmatic consolidation.

It has been said by some writers that the brown mica occurs by preference between ore and feldspar, but in many sections which I have examined the mica does not seem to be interposed between ore and feldspar oftener than between ore and pyroxene, and in very many instances it is not even immediately contiguous to ore grains, but is merely in close association. Furthermore, a reaction between magnetite and feldspar does not seem altogether satisfactory as an explanation for the formation of a mineral of the composition of biotite. I wish to suggest as a reasonable alternative that the process we observed in the basalts is supplemented in the diabases by a partial retention of water and fluorine to a late stage of crystallization, and that the magnetite-mica aggregates are the result.

If, in the consolidation of diabasic rocks, conditions should be such that the water and fluorine are able to escape almost completely before the stage for the formation of biotite is reached, plagioclase and pyroxene would continue to be deposited to the end, just as in some coarse-grained basalts, and

[25] In British petrographic literature the term diabase is generally understood to mean a somewhat altered rock. In this country and Germany it is commonly understood to mean the rock that the British term dolerite.

not much evidence of the course of differentiation would remain, except the segregation of iron ore, molded around the other minerals. These relations also are rather common.

The Triassic diabases and basalts of Northeastern United States show various phenomena indicative of the presence of a considerable quantity of volatiles. They also show in an interesting and instructive manner how the volatiles may escape, leaving hardly a trace, and a fallacious impression of their absence from the magma may be created. In most of the occurrences of the Watchung basalt flows, which are well exposed over wide areas, no evidence that volatiles were present can be found, but in certain localized areas quantities of datolite and apophyllite (associated with zeolites and other secondary minerals) occur. There is no reason to suppose that the original content of volatiles was greater in these areas than in the barren areas, but exceptional conditions (of which the nature is fairly well known) tended to restrict their escape.

No chlorine-bearing mineral has been recognized in these occurrences, but chlorine was probably present, for in the equivalent intrusive masses (the diabases) scapolite occurs, and some of the shales in the metamorphic contact zone have been changed to scapolite hornfels. Other minerals of interest which have been determined as fairly abundant in the hornfels are vesuvianite, tourmaline, apatite, biotite, augite, hornblende, garnet, spinel, cordierite, and zircon.[26] These and others are listed by Lewis and by Irving for New Jersey. In undescribed localities in Maryland and Virginia the results of contact metamorphism appear comparable.

Thorough metamorphism of the shales, with introduction of new constituents, is evident for several hundred feet above the Palisade intrusive and, in places, for a long distance below, but the coarser arkoses intercalated with the shales apparently allowed the gases or hydrothermal solutions to pass through without leaving much evidence, for they are usually little changed. In the intrusives themselves datolite and apophyllite have been found in abundance at numerous localities.

The view that basaltic and cognate magmas are almost lacking in volatiles occurs repeatedly in geologic literature, and has become almost an axiom. They probably contain much less than the siliceous magmas, but a little evidence such as this, that their content of volatiles is not insignificant, may not be amiss.

[26] Lewis, J. V., op. cit., p. 138.

In concluding this part of the discussion I can not refrain from referring to an instance illustrating the rather indefinite conception that many petrologists seem to have as to what crystallization differentiation is able to accomplish. This example can doubtless be matched by many others.

In the "Tertiary and post-Tertiary geology of Mull,"[27] the authors declare themselves to be in close agreement with the theory as an explanation of many of the phenomena they have observed, and their attitude is to regard the Plateau Magma Type of their area as the parental stock from which other types have been derived. Especially do they use this idea in Chapter 30 to explain, by gravitational separation, the passage of quartz gabbro into granophyre in a number of vertical dikes. In another section,[28] however, they describe certain phenomena, under the title of segregation veins, in which entirely different results are shown. These also are attributed to the separation of residual liquid from a mostly crystallized basaltic mass, and no explanation of the inconsistency is offered. With regard to these segregation veins I will quote a few extracts:

"A few of the Plateau Basalts show a somewhat remarkable tendency to segregate contemporaneous veins consisting mainly of augite, felspar, and analcite, without olivine. These veins differ in texture and composition from the parent-lava, and afford interesting evidence as to the manner of differentiation of the normal basalt magma. ....

"The more definite segregation veins traversing the lava are moderately coarse in texture. .... The augite builds elongated crystals that are highly colored and usually of a deeper tint towards their margins. The felspars are labradorite, and are often .... zoned with varieties that rapidly increase in alkalinity." (These phenocrysts are idiomorphic against the base, which consists of analcite, chlorite, a little alkali feldspar, pyroxene, and iron ore, presumably ilmenite, in thin plates and skeletal forms.) "The alkalinity of the base is emphasized by the frequent presence of a green augite. ....

"It will be seen from the above description that these segregation veins, which cannot be otherwise regarded than as normal differentiation products of a basalt magma, are in a general way mineralogically and structurally related to the lamprophyres."

Unless I have failed to understand the authors' views, they derive in one case a granophyre and in the other case veins

[27] Memoirs of the Geological Survey, Scotland, 1924.
[28] Op. cit., pp. 138 ff.

related to lamprophyres, as residual liquids from the crystallization of magmas of basaltic composition. Such inconsistencies, as these seem to be, illustrate the indefinite ideas that many petrologists have as to what may be attributed to crystallization differentiation. The theory appears to assume, at times, the character of a *deus ex machina,* to which any results may be ascribed.

It is believed that enough evidence has now been presented to show a notable inconsistency between the course of differentiation actually followed in the crystallization of plateau basalts and that which has been assumed by those who derive all other magmas from basaltic magma by crystallization. The necessity for an explanation of the discrepancies seems to be indicated.

Probably some will say that a part of the explanation is to be found in the fact that in magmas crystallizing at great depths the mineral assemblage will be somewhat different from that of surface basalts, and may be very different. This is an argument which the supporters of the theory should use with caution. The theory has received considerable support from laboratory experiments, in which the minerals produced are closely analogous to those of surface rocks, and in proportion as the mineral assemblage is conceived to differ will this support be withdrawn. In any event, however different the minerals may be supposed to be, so long as they are made up of solid solutions of various members there will always be a tendency for certain bases to be concentrated in the crystals and other bases in the residual liquid.

There may be others who will be content to point to the evidence that in many igneous rocks which have advanced in differentiation a certain distance away from basalt the later crystallizing portions are more siliceous than the early phenocrysts, are more albitic, and have a smaller content of ferromagnesian minerals, and they will ask whether this is not sufficient to establish their contention. I do not think it is. Here is a crucial point to which I wish to draw special attention. When a basaltic magma is extruded at the surface and there crystallizes we can be pretty sure that there is not much opportunity for anything but crystallization to affect the course of differentiation to any important degree. When the same kind of magma, deep in plutonic reservoirs or in the lower part of volcanic conduits, slowly crystallizes over a long period, we know that separation of crystals will change the composition of the residual liquid, but we have no knowledge whatever that other processes may not simultaneously affect it.

From the evidence cited in this paper it appears that a surface basalt, cooling with any degree of slowness required, leads by crystallization to a residual liquid rich in magnetite and pyroxene. If an abyssal magma differentiates in some other direction it seems to mean that, no matter how important crystallization may be quantitatively, other factors have intervened to change the course of differentiation.

My own inclination, as expressed in previous writings, is to look favorably upon several other processes as coöperating with crystallization in effecting differentiation. The relative importance of these I would hesitate to evaluate because of lack of information. Among them, however, I would place gaseous transfer, and since several writers have, for theoretical reasons, been inclined to minimize the importance of gases in volcanic phenomena, I may, perhaps, be permitted to digress for a moment.

Some of the objections raised to the presence of any notable quantity of gases in magmas, either as agents of transfer or as participating in any important manner in processes of any kind, seem to have disregarded to a considerable degree descriptions of volcanic phenomena to be found in the literature, I do not wish to take a captious attitude, but I believe that a great deal of irrelevant matter has been written, and general opinion unduly influenced by well-intentioned and sincere geologists and chemists who, through lack of opportunity to become acquainted with volcanoes of the gaseous type or with areas of recent explosive eruptions or with intense fumarolic activity, have formed an inadequate conception of the part played by gases in volcanism, and have devised a picture which has little relation to reality. Laboratory investigations and abstract reasoning are of demonstrated value as guides in explaining geologic phenomena, but they can not be used as substitutes for field observation and study, and they require to be checked at every point. Conclusions on the presence of an insignificant amount of gas in magmas are met in the field by disproofs which stare one in the face, and on scrutiny one finds that the conclusions set forth rest fundamentally on premises which represent little more than the authors' opinions. The ready acceptance of these conclusions by many petrographers causes one to wonder whether the outstanding works on volcanology, which most petrographers must have read, have failed to present a sufficiently realistic picture of the overwhelming effect of gases in explosive eruptions, which so impresses one in the field.

For those who believe that gases are unimportant or subsidiary elements in magmas, or that the little gas found in most solidified rocks is anything more than an insignificant residue of that present in magmas, it might be instructive to keep in mind that, according to well-authenticated information, explosions of Krakatoa during the eruption of 1883 were heard at many places 1,500 to 3,000 miles distant and that coarse particles were estimated to have been thrown to a height of 30 miles,[29] and try to conceive what sort of action this implies; or to picture the conditions under which the hundreds or thousands of cubic miles of fragmental tuffs and breccias were erupted in the Absaroka Range on the eastern border of Yellowstone Park, in the area of Tuscan tuffs in California, in the Cordillera Occidental of Peru and Chile, or in Porto Rico.[30]   Perret's narrative of gas discharge at Vesuvius during the eruption of 1906 might be read and pondered with profit.[31] Moreover, it would be well to remember that at the present time volcanic activity has diminished so greatly from that of the Tertiary as to seem almost insignificant, and that deductions from present phenomena are hardly an adequate basis for judging conditions during the greater and more violent activity in previous periods.

According to fundamental principles of the crystallization theory, rhyolite magma is the lowest melting residue of magmas, and is therefore incapable of fusing more basic rocks. As if unaware of this restriction, rhyolitic flows in a number of places in Yellowstone Park melted basaltic inclusions and formed a brown glass.   Iddings has described the phenomena in such a manner as to leave no doubt of the reality of the action.   He states: "It is evident, both from the occurrence of large fragments of basalt completely surrounded by rhyolite and from the microscopical character of the rocks, that the rhyolite fused the basalt and was the more recent eruption."[32]

[29] The eruption of Krakatoa, and subsequent phenomena: Report of the Krakatoa Committee of the Royal Society, pp. 84-87 and 379, 1888.

[30] Descriptions of these regions are to be found in the literature, but my impressions have been derived mainly from personal observation.

[31] Perret, F. A., The Vesuvius eruption of 1906, Published by the Carnegie Institution of Washington, pp. 43-46, 1924.

[32] Iddings, J. P., U. S. Geol. Surv., Monograph 32, pt. 2, 430-432, and 433, 1899.

It is especially striking that these surface flows, which presumably had cooled somewhat, should be capable of effecting this.

Harker, in his Natural History of Igneous Rocks,[33] has brought together a number of observations relating to the solvent action of acidic magmas upon basic rocks, with the production of hybrid material. In some instances solution apparently incorporated large quantities of basic rock. The effects do not seem to be explainable by the reaction principle, for that, in its strict application, would hardly permit minerals of earlier position in the reaction sequence to be dissolved by acid magmas in the definite and complete manner described.

Another good example is given by Tyrrell.[34] Between masses of gabbro and granite he found "a very heterogeneous region of gabbro-granite mixture-rocks, xenolithic and hybrid types, and also of quartz-diorite and diorite in which no traces of mixture or hybridism can be detected. . . . The petrographic evidence renders it probable that the pure, homogeneous diorite and quartz-diorite are the final results of the complete solution of the gabbroidal rocks within the granite magma. All stages in the process can be studied in the field and under the microscope."

At Novarupta Volcano, in the Valley of Ten Thousand Smokes, great quantities of basic rocks were melted down, apparently in a short time, by the newly erupted rhyolitic magma. Without reference to theory, there seem to be only three possible methods by which this could be accomplished: (1) The new magma was very greatly superheated; (2) Enormous quantities of hot gases were evolved from the new magma (although the body underlying the vent was probably a rather small, parasitic offshoot from the main mass); (3) Great heat was set free by exothermic reactions. Any one of these possibilities, and in fact the whole set of phenomena attending the outburst of Novarupta, seem to be absolutely and fundamentally irreconcilable with the theory of crystallization differentiation, with the reaction principle, and with ideas recently proposed in respect to gases in magmas, unless the possibility of modifications is recognized. No matter how attractive theories may appear in their abstract logic, until they are brought into closer accord with facts, one who has observed inconsistent phenomena must feel much hesitation in accepting

[33] Harker, A., op. cit., pp. 339 ff., 1909.
[34] Tyrrell, G. W., The geology of Arran, Memoirs of the Geological Survey, Scotland, 172, 1928.

the theories without modification and must seek an explanation more in harmony with what one has seen.

These comments are conceived to be applicable to certain tendencies in volcanology and petrogenesis; they are not intended as an introduction to the claims of gas transfer as an agent in differentiation. On that matter my views are the same as those that I have previously expressed: that this agency is certainly of considerable importance, but that data are not sufficient on which to base a quantitative estimate of its importance.

Moreover, it is not deemed desirable to present in this paper the evidences in favor of gas transfer, but to rest with the conclusion to which the data given seem to point, that processes of some kind, in addition to crystallization, play an essential part in differentiation. To those who wish to eliminate from consideration all other processes than crystallization, this is left as a problem for solution or as a conclusion for acceptance.

Geophysical Laboratory,
  Carnegie Institution of Washington.

21

ROLE OF OXYGEN PRESSURE
IN THE CRYSTALLIZATION AND DIFFERENTIATION
OF BASALTIC MAGMA*

E. F. OSBORN

College of Mineral Industries, Pennsylvania State University,
University Park, Pa.

ABSTRACT. Two types of crystallization paths in iron oxide systems are illustrated by describing equilibrium crystallization in the fayalite field of the system FeO–Fe_2O_3–SiO_2. In one type where total composition of the mixture remains constant, composition of the liquid changes along a path leading to a boundary curve and thence to an invariant point. In the other type oxygen partial pressure remains constant causing the liquid of the crystallizing mixture to follow an oxygen isobaric line on the liquidus surface. In this second type of crystallization the oxygen content of the mixture changes.

These principles are extended into an examination of crystallization phenomena in the quaternary system MgO–FeO–Fe_2O_3–SiO_2. In this system occurrence of extensive solid solution in olivine, pyroxene and magnetite structures along with incongruent melting of pyroxene necessitates consideration of fractional as well as equilibrium crystallization. It is shown that with fractional crystallization at *constant total composition* (Type I) the liquid moves to the magnetite (magnesioferrite) surface and thence down this surface continuously decreasing in amount and increasing in iron oxide content. With magnetite as one of the crystallizing phases, oxygen pressure decreases. Liquids take the same general direction of composition change as they do during fractional crystallization in the simpler system, MgO–FeO–SiO_2. Initial FeO/Fe_2O_3 ratio does not affect direction of movement of liquid after the magnetite surface is reached. With fractional crystallization at *constant oxygen partial pressure*, (Type II), the direction of movement of liquid is very different. Because the liquid must remain on an isobaric surface, change in composition of the liquid is in the direction of increasing silica and decreasing iron oxide content as soon as magnetite begins to crystallize. The liquid moves along the magnetite boundary terminating at the pyroxene-magnetite-silica univariant line, where the liquid completes its crystallization without further change of composition just as if it had reached a eutectic point. For mixtures analogous to basalt compositions, a small net increase in oxygen content is realized during the crystallization. If oxygen pressure increases during fractional crystallization, the liquid moves to even higher silica and lower iron oxide end points than under conditions of constant oxygen pressure. Thus the trend of change of liquid composition toward higher iron oxide contents during fractional crystallization of Type I shifts progressively to one of decreasing iron oxide content and increasing silica as the rate of decrease in oxygen pressure required in Type I crystallization becomes less and approaches the condition of constant or increasing oxygen pressure. The ratio of iron oxide to iron oxide plus magnesia in the fractionating liquid rises steeply with increasing silica content in Type I crystallization, and rises gently in Type II, for mixtures simulating basalt compositions.

Analyses applicable to igneous rocks are plotted for comparison with curves for fractionating liquids in the system MgO–FeO–Fe_2O_3–SiO_2. The iron oxide-magnesia trends for average tholeiitic basalts and Skaergaard liquids are of the same nature as those for liquids of Type I crystallization, whereas the basalt-andesite-dacite-rhyolite series exhibits trends of the nature of Type II crystallization. An example of the latter is the Cascade lava series. Basaltic magmas of the non-orogenic regions typically crystallize with total composition remaining essentially constant. Basaltic magmas entering the orogenic belts

* Contribution No. 58-78 from College of Mineral Industries, The Pennsylvania State University, University Park, Pennsylvania.

evidently crystallize with oxygen pressure remaining approximately constant. That this is true is further borne out by plotting subtraction diagrams. Where crystallization is of Type I, the FeO and Fe_2O_3 curves have a continuous slope, whereas with crystallization of Type II the curves for the two iron oxides have a discontinuity in slope because of addition of oxygen to the mixture during crystallization. The subtraction diagrams further suggest that high alumina basalts develop from tholeiitic olivine basalt by fractional crystallization under conditions such that plagioclase is delayed in separating from the liquid.

The tendency for oxygen pressure to decrease at a slower rate, or remain about constant or actually to increase during crystallization of basaltic magma entering orogenic belts instead of decreasing as in Type I crystallization may be the result of higher water contents of orogenic magmas coupled with their habit of accumulating in reservoirs where with slow cooling and fractional crystallization water pressure increases during crystallization. Liquids escaping from the reservoirs extrude as calc-alkali series of lavas and ash while those remaining below the surface cool as granodiorite and related plutonic rocks. The complementary crystal fraction is represented by periodotites and serpentinites.

[*Editors' Note:* Material has been omitted at this point.]

CRYSTALLIZATION OF BASALTIC MAGMA

The analysis just given of crystallization paths in the system MgO–FeO–Fe_2O_3–SiO_2 has shown that fractional crystallization can produce two distinctly different trends in change of liquid composition. One trend will be taken if total composition of the mixture remains constant, and in this case p_{O_2} decreases during crystallization. The other trend is followed if p_{O_2} remains about constant or increases during crystallization, and in this second case the total composition changes by an increase in oxygen content for the compositions of special interest in this study. These trends are a natural and necessary consequence of the relations which exist among the phases: Mg-Fe olivines, Mg-Fe pyroxenes, and magnetite.[6] Despite the greater complexity of relations during crystallization of basaltic magma, the same type of trends may be expected during the early stages of crystallization when olivine, pyroxene and mag-

---

[6] In the preceding discussion the spinel phase has been called magnesioferrite, for it is most simply viewed as a solid solution of magnetite and magnesioferrite. The composition of this phase in equilibrium with liquid c (figs. 5, 7 and 8) is approximately $13MgO$, $11FeO$, $76Fe_2O_3$. When referring to basaltic magma this spinel phase will be called magnetite, where titanium ions as well as magnesium occupy Fe^{2+} positions and other trivalent ions may substitute for part of the Fe^{3+}.

netite are principal phases, provided the two types of conditions are approached in nature: fractional crystallization at constant total composition of the mass, and fractional crystallization at constant or increasing p_{O_2}. Fractional crystallization undoubtedly occurs, but to a varying degree. The condition of constant total composition will be approached if the oxygen-containing gas phase is small compared to the mass of the condensed phases. The condition of constant or increasing p_{O_2} may be approached where sufficient water or carbon dioxide is present in the magma to give an appreciable p_{O_2} and to serve as a reservoir of oxygen to donate or accept oxygen from the liquid and crystalline phases without greatly changing its p_{O_2}. This will be discussed in a later section. Let us first see what trends we find in natural magmas, especially with respect to iron oxide, magnesia and silica. Do trends of change of liquid composition simulate those which occur in the system $MgO–FeO–Fe_2O_3–SiO_2$, and does it seem probable therefore that major lines of liquid descent can be explained as in this quaternary system by the manner of change of p_{O_2} during fractional crystallization?

Iron Oxide-Magnesia-Silica Trends in Basaltic Magma

Four series of chemical analyses have been plotted to ecompare trends of differentiating basaltic magmas with those established for mixtures in the $MgO–FeO–Fe_2O_3–SiO_2$ system. As representing tholeiitic basalts in general, Nockolds' (1954) averages have been used. Points I, II and III in figures 11a and 11b and table 2 are respectively: "tholeiitic olivine basalt", "normal tholeiitic basalt and dolerite," and "average tholeiitic andesite." Successive liquids of the Skaergaard intrusion according to Wager and Deer (1939), numbers Sk1, Sk2 and Sk3, are used as an example of a specific magma where differentiation by crystal fractionation is well established. Nockolds' (1954) three averages (points IV, V and VI) are used for the andesite-dacite-rhyodacite line of descent, while for a specific example of this type of magma series we have taken analyses of lavas from northwestern United States. The points labeled 2, 4, 5, 6 and 7 represent analyses of rocks selected by Turner and Verhoogen (1951) as representing the basalt-andesite association of the Cascade Province. The numbers for this last series correspond to those in Table XXI of these authors who list three olivine basalts. No. 2 was taken here as a representative olivine basalt from this province because it seems to be about average among those listed, not having the very large normative olivine content of their No. 1 nor the normative quartz of their No. 3. No. 4 is "basaltic andesite", No. 5 is "hpersthene andesite", No. 6 is "pyroxene andesite", and No. 7 is "porphyritic dacite." Analyses are listed in table 2.

If we compare figure 11a with figure 9a where the ratio of $FeO + Fe_2O_3/FeO + Fe_2O_3 + MgO$ is plotted against SiO_2, and figure 11b with figure 9b where total iron oxide is plotted against SiO_2, a similarity of trends of curves is evident. The tholeiitic basalts and the Skaergaard liquids have trends like the liquids formed from fractional crystallization in the $MgO–FeO–Fe_2O_3–SiO_2$ system at constant total composition. The trends of the andesite-dacite-rhyodacite series and of the basalt-andesite-dacite association of the Cascades are like those of liquids crystallizing at constant p_{O_2}.

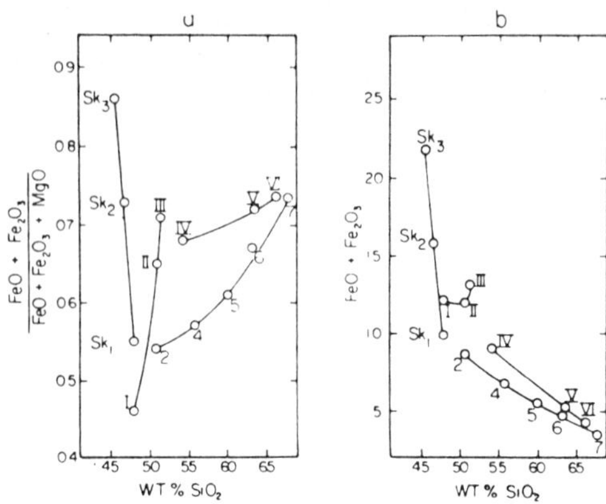

Fig. 11.   Curves for magma series. Compositions of points are listed in table 2.

The curves for the Skaergaard liquids in figures 11a and 11b have a negative slope like the section 132-40-24 of the curves in figure 9, suggesting not only that fractional crystallization at constant total composition was the process by which successive liquids developed but also that the olivine-pyroxene surface was approximately reached by the "first liquid" of Wager and Deer (1939), No. Sk1. In analogy with mixture No. 132 of figure 9a, liquid Sk1 may have been derived from a tholeiitic olivine basalt from which considerable olivine had precipitated with movement of the liquid to the vicinity of the pyroxene-olivine surface.

Explanation of Table 2.

I—Tholeiitic olivine basalt, average after Nockolds (1954)

II—Normal Tholeiitic Basalt and dolerite, average after Nockolds (1954)

III—Tholeiitic andesite, average after Nockolds (1954)

IV—Andesite, average after Nockolds (1954)

V—Dacite plus dacite-obsidian, average after Nockolds (1954)

VI—Rhyodacite plus rhyodacite-obsidian, average after Nockolds (1954)

Sk1—First liquid of Skaergaard intrusion according to Wager and Deer (1939)

Sk2—Second liquid of Skaergaard intrusion according to Wager and Deer (1939)

Sk3—Third liquid of Skaergaard intrusion according to Wager and Deer (1939)

2—Olivine basalt (late Pleistocene); Lava Top Butte, Newberry volcano. Williams (1935). From Table XXI of Turner and Verhoogen (1951)

4—Basaltic andesite (Pleistocene) with 6.3 percent normative quartz; near summit of Crater Peak, southern Oregon, Williams (1942). From Table XXI of Turner and Verhoogen (1951)

5—Hypersthene andesite (Pleistocene) with 11.7 percent normative quartz; Crater Lake, southern Oregon. Williams (1942). From Table XXI of Turner and Verhoogen (1951)

6—Pyroxene andesite (Pleistocene) with 18.7 percent normative quartz; Mount St. Helens, Washington. Verhoogen (1937). From Table XXI of Turner and Verhoogen (1951)

7—Porphyritic dacite (sub-Recent) with 21.8 percent normative quartz; Medicine Lake highland. Anderson (1941). From Table XXI of Turner and Verhoogen (1951)

PC—Porphyritic Central type basalt, Mull. Bailey et al (1924). Representative analysis as given by Tilley (1950).

104

Table 2

Rock Analyses

	I	II	III	IV	V	VI	Sk1	Sk2	Sk3	2	4	5	6	7	PC
SiO_2	47.9	50.8	51.4	54.2	63.6	66.3	47.9	46.7	45.7	50.7	55.8	60.1	63.2	67.7	50
Al_2O_3	11.8	14.1	13.0	17.2	16.7	15.4	18.9	15.3	12.7	18.0	18.0	17.8	18.2	16.3	18
Fe_2O_3	2.3	2.9	3.4	3.5	2.2	2.1	1.2	2.9	3.6	1.6	2.6	2.0	1.4	0.3	
															9
FeO	9.8	9.0	9.7	5.5	3.0	2.2	8.7	12.9	18.2	7.0	4.1	3.4	3.3	3.2	
MgO	14.1	6.3	5.3	4.4	2.1	1.6	7.8	5.9	3.5	7.6	5.1	3.5	2.3	1.3	5
CaO	9.3	10.4	8.8	7.9	5.5	3.7	10.5	9.9	8.3	9.7	7.4	6.3	5.2	3.3	10
Na_2O	1.7	2.2	3.2	3.7	4.0	4.1	2.4	2.8	3.2	2.7	3.6	4.2	4.1	3.9	2.5
K_2O	0.5	0.8	1.0	1.1	1.4	3.0	0.2	0.3	0.4	0.7	1.2	1.3	1.2	3.2	0.4
TiO_2	1.6	2.0	2.6	1.3	0.6	0.7	1.3	2.2	2.4	1.3	0.8	0.5	0.5	0.3	1.0

The curves in figures 11a and 11b joining the analyses I, II and III resemble those for the section n-132-40 of curves in figure 9. The tholeiitic series of magmas like the Skaergaard liquids develop trends resembling those to be expected where crystallization is at constant total composition. For tholeiitic liquid series I-II-III, however, the initial liquid is sufficiently rich in MgO that olivine precipitates first without pyroxene, accounting for the initial positive slope of the curves. Note that in figure 11b the curve I-II-III has the strong concavity to the left as shown by n-132-40 of figure 9b.

In contrast to these, the liquids representing basaltic magma series of the orogenic belts have curves resembling n-c-d. The analogy is so striking that certainly it is reasonable to view point c (figs. 5, 7, 8 and 9) as andesite-like liquid and point d as dacite-rhyolite-like liquid, and to surmize that andesites and dacite-rhyolite lavas form from basalts by fractional crystallization under conditions approaching constant p_{O_2}. In further analogy with points c and d, andesites of the orogens are liquids from which Mg-Fe pyroxenes crystallize as olivine tends to react with the liquid and disappear, and dacites are the stage in liquid descent at which tridymite and pyroxene coprecipitate with a complete absence of olivine.

The curve 2-4-5-6-7 (fig. 11a) if extended to the left intersects I-II-III between I (tholeiitic olivine basalt) and II (normal tholeiitic basalt), whereas curve IV-V-VI extended to the left intersects I-II-III at a higher point. It is possible to conclude from this arrangement of curves that the immediate parent magma for each of these two series, 2-4-5-6-7 and IV-V-VI is a composition on the tholeiitic trend (I-II-III), with the primary magma for all basalts being an olivine basalt of a composition similar to point I. Accordingly a primary basalt liquid during crystal fractionation takes off from I and starts to follow a total constant composition trend characteristic of the magmas we call tholeiitic. If in its subsequent history a gas phase is introduced or develops which is capable of maintaining oxygen pressure more or less constant, or of causing it actually to increase, then the trend of change in liquid composition shifts to a subhorizontal direction. This latter direction of change in liquid composition is that leading to the normal andesites and dacites and rhyolites.

Subtraction Diagrams

The liquid lines of descent for factionally crystallized laboratory mixtures (fig. 9) are derived by assuming continuous separation of crystals from the liquid. At any stage during this fractionation a subtraction diagram can be constructed showing the amount and composition of the two fractions. If the crystallization has been at constant total composition, the lines on the diagram for all oxides will be straight, whereas if p_{O_2} has remained constant, the oxygen content of the mixture will usually have changed appreciably causing the lines for FeO and for Fe_2O_3 to have a significant discontinuity in their slopes. To illustrate this, subtraction diagrams have been constructed for the primary liquid n of figure 9 fractionally crystallizing under the two different oxygen pressure conditions to give liquids No. 132 and c. Following a discussion of

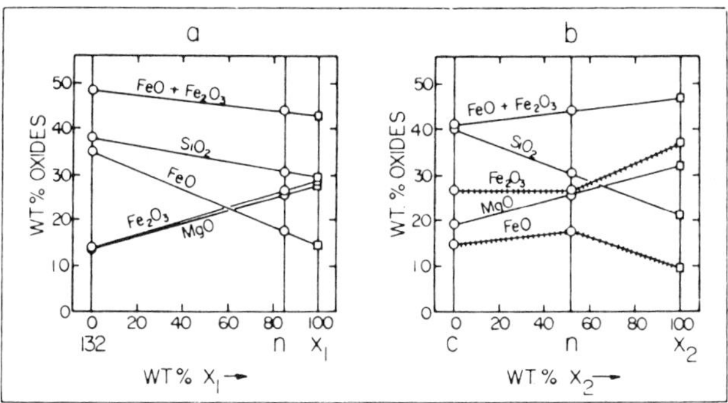

Fig. 12. Subtraction diagrams with mixture *n* of table 1 and figs. 8 and 9 as primary liquid. In (a) liquid 132 is derived by subtracting 85% as crystals of composition X₁. In (b) where constant p_{O_2} is maintained during fractional crystallization, liquid *c* is derived by subtracting 52% as crystals of composition X₂.

these cases, subtraction diagrams for basaltic magmas of figure 11 will be shown for comparison.

Mixture n in the system MgO–FeO–Fe₂O₃–SiO₂.—In figure 12a is shown the manner of derivation of liquid composition No. 132 (fig. 9 and table 1) from the primary liquid *n* by subtraction of the crystalline aggregate X₁. To construct this diagram it is necessary to make an estimate of the average composition of each crystalline phase. Then knowing the composition of liquids *n* and No. 132, the relative amounts of liquid and crystal fractions can be estimated. During fractional crystallization of *n* at constant total composition[7], the composition of the olivine changes from about 90M₂S, 10F₂S to about 50M₂S, 50F₂S at liquid No. 132. Taking into consideration the fact that early formed crystals will be in largest proportion in the crystalline aggregate, a reasonable estimate of the average composition of olivine during fractional crystallization of *n* as the liquid moves to point No. 132 is 75M₂S, 25F₂S. Pyroxene ranges in composition during this crystallization from about 88MS, 12FS to 55MS, 45FS, with an estimated average of 75MS, 25FS. The pyroxene coprecipitating with olivine at any temperature has a higher MS/FS ratio than the M₂S/F₂S ratio in olivine, but for mixture *n* much olivine has precipitated before pyroxene begins to crystallize with the result that the bulk composition of pyroxene has a ratio of FeO/MgO similar to that of olivine. The magnesioferrite precipitating has an average composition of about 55MF, 45FF. Using these compositions for the three crystalline phases separating while the liquid moves from *n* to No. 132, we find that approximately 85 percent of the mixture has separated as crystals. The chemical composition of the crystalline aggregate and the relative percentage of the phases is shown in table 3. It will be noted that the FeO and Fe₂O₃ lines, as well as others in figure 12a are straight as they must be under the assumption of constant total composition.

On fractionally crystallizing mixture *n* at constant p_{O_2}, the liquid moves

[7] Composition of olivine, pyroxene and magnesioferrite crystals appearing in the system MgO–FeO–Fe₂O₃–SiO₂ are given in terms of *weight percent* of the end members which are abbreviated as follows: M₂S = Mg₂SiO₄, F₂S =: Fe₂SiO₄, MS = MgSiO₃, FS = FeSiO₃, MF = MgFe₂O₄, FF = FeFe₂O₄.

first to point c of figures 6, 7, 8 and 9. In figure 12b is a subtraction diagram showing the crystal fraction (X_2) separated from n to give c. Only olivine and magnesioferrite precipitate in this case, making calculations simple. Assuming that the olivine averages $88M_2S$, $12F_2S$ and the spinel $65MF$, $35FF$, 52 percent of the mixture has been precipitated as crystals during movement of liquid from n to c with the composition of the crystalline aggregate as shown in table 3.

TABLE 3

Composition of Crystalline Aggregates X_1 and X_2 subtracted from mixture n
to give respectively liquids No. 132 and c (fig. 12)

	X_1		X_2	
SiO_2	29.5		21.0	
MgO	27.0		31.9	
FeO	15.0		9.7	
Fe_2O_3	28.5		37.4	
Olivine	30.9*	$75M_2S$, $25F_2S$	50.8	$88M_2S$, $12F_2S$
Magnesioferrite	38.2	$55MF$, $45FF$	49.2	$65MF$, $35FF$
Pyroxene	30.9*	$75MS$, $25FS$		

* It just happens that in using the figure of 85% crystals, olivine and pyroxene are in equal proportions. With a smaller percent of crystals olivine is in larger amount, and with a greater percent of crystals pyroxene is in larger amount. With 87% of crystals, which is nearly as good a solution as 85%, 28.7% is olivine and 33.7% pyroxene. Formula designations are explained in footnote 7.

In figure 12b the striking feature is the discontinuity in the slope of the FeO and Fe_2O_3 lines. At the composition of the primary liquid, n, the FeO line has its maximum and the Fe_2O_3 line its minimum value. This change in direction of these two lines at the composition of n is a requirement where fractional crystallization is at constant p_{O_2}, except for the unusual case of no net gain or loss of oxygen during crystallization. Inasmuch as the iron content of crystals plus liquid must remain constant, the line representing total iron oxide is approximately straight. The chemical composition of the crystalline fraction and the proportion of this fraction which must separate from n to give c is thus readily obtained except that the ratio of Fe_2O_3 to FeO is known only after an assumption of composition of the crystalline phases. With the crystalline phases having the compositions as given above, the Fe_2O_3 and FeO content of the crystalline fraction follows directly. The Fe_2O_3 and FeO contents of liquids n and c are known from the diagram (fig. 7) and are given in table 1. In the process of fractionally crystallizing liquid n to produce liquid c and crystals X_2, 0.6 percent oxygen has been added to the mixture.

Basaltic magmas.—Turning to basalts, let us now consider subtraction diagrams that may result as basaltic magma differentiates by fractional crystllization. The close analogy of the curves I-II-III and 2-4-5-6-7 of figure 11a to n-132-40 and n-c-d, respectively of figure 9a has been noted. Subtraction diagrams deriving magma II from I and deriving magma 4 from 2 (fig. 11) should therefore be of the same nature as diagrams 12a and 12b, respectively,

108

for the curve I-II has the trend of n-132 where total composition remained constant, and the curve 2-4 the trend of n-c where p_{O_2} remained constant. Furthermore, the magmas numbered Sk1 and 2 in figure 11a lie close enough

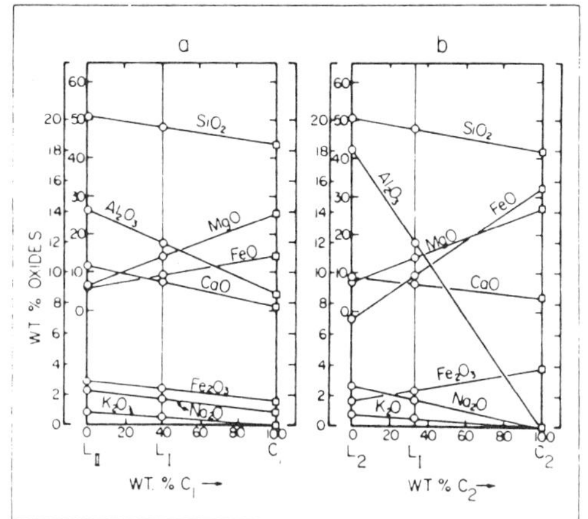

Fig. 13. Subtraction diagrams deriving by fractional crystallization liquids L_{II} and L_2 from primary tholeiitic olivine basalt, L_I. These liquids are respectively the composition II, 2 and I of table 2. The inside vertical scale applies to SiO_2 and MgO.

TABLE 4

Composition of Crystalline Aggregates C_1, C_2 and C_3*

	C_1	C_2	C_3
SiO_2	43.7	41.8	46.0
Al_2O_3	8.5	0.0	18.0
CaO	7.7	8.5	11.8
MgO	25.4	27.0	9.9
FeO	11.1	15.5	5.9
Fe_2O_3	1.5	3.8	4.4
Na_2O	0.85	0.0	2.2
K_2O	0.0	0.0	0.0
Plagioclase	27.2 Ab_1An_2	none	57.7 Ab_1An_2
Pyroxene	17.9 Di_4Fs_1	41.9 Di_4Fs_1	19.7 Di_2Fs_1
Olivine	51.7 Fo_4Fa_1	49.7 Fo_4Fa_1	14.4 Fo_9Fa_1
Magnetite	2.2	5.5	6.4

* C_1 and C_2 refer to crystalline aggregates subtracted from Tholeiitic Olivine Basalt, L_I to give respectively the Normal Tholeiitic Basalt liquid fraction (L_{II} of fig. 13) and the Orogenic Olivine Basalt fraction (L_2 of fig. 13). C_3 indicates crystalline aggregate subtracted from olivine basalt of Cascades to give Cascades basaltic andesite (fig. 14).

Summations are not 100 because TiO_2 and other minor constituents are omitted. Terms used in table 4 have the following compositions: Ab = $NaAlSi_3O_8$, An = $CaAl_2Si_2O_8$, Di = $CaMgSi_2O_6$, Fs = $FeSiO_3$, Fo = Mg_2SiO_4, Fa = Fe_2SiO_4, Magnetite = $FeO \cdot Fe_2O_3$. The mineralogical composition symbols are on a mol. percent basis.

109

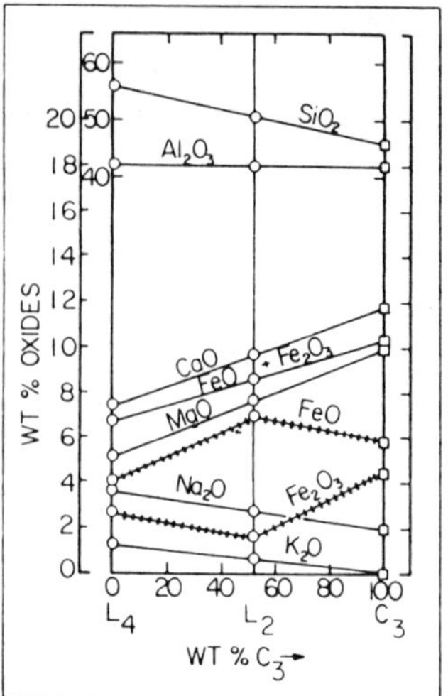

Fig. 14. Subtraction diagram deriving by fractional crystallization basaltic andesite, L₄, from olivine basalt L₂ (compositions 4 and 2 of fig. 11 and table 2), by subtracting 52% of crystalline aggregate C₃. The inside vertical scale applies only to SiO₂.

to the tholeiitic basalt line, I-II-III, that they also can be considered as derivatives of magma I by fractional crystallization at approximately constant total composition.

Subtraction diagrams are presented deriving magma II from I (fig. 13a), magma 2 from I (fig. 13b) and magma 4 from 2 (fig. 14). In each case the assumption has been made that the crystal fraction contains essentially no K_2O. Note that the FeO and Fe_2O_3 lines are straight in figures 13a and 13b, but contain a corner at the composition of the primary liquid in figure 14. The diagrams then have those features to be expected on the basis of comparing directions of curves in figure 11a with those of figure 9a; or in other words the sub-horizontal trend of curves in figure 11a develops as in figure 9a as a consequence of fractional crystallization at approximately constant p_{O_2} and the vertical trend as a consequence of fractional crystallization at approximately constant total composition. Details regarding these subtraction curves and their implications are developed in the two sections following.

(a) *Tholeiitic and high alumina basalts.* Normal tholeiitic basalts contain about 13 to 15 percent Al_2O_3 and occur chiefly in non-orogenic regions. Basalts and andesites of orogenic belts on the other hand are characteristically of a high-alumina variety, as recently emphasized by Tilley (1950). These latter contain of the order of 17 to 19 percent Al_2O_3, and although typical of the orogens are found in a few other interesting spots. It will be noted from table 2 that the olivine basalt, 2, and basaltic andesite, 4, of the Cascade orogenic belt are of the typical high-alumina variety, and that also the Skaergaard first liquid, Sk1, and Porphyritic Central type, PC, are high-alumina basalts. Subtraction diagrams deriving the normal tholeiitic basalt, II, and the high alumina basalt, 2, from the same primary tholeiitic olivine basalt, I (fig. 13), provide an interesting comparison, as well as a clue to the origin of the high-alumina basalts.

The diagrams of figures 13a and 13b develop readily on decreasing the K_2O content to zero at the vertical line repreesnting composition of the crystalline fraction. This fixes the percent of the crystalline and the derived liquid fractions and the chemical composition of the former. Thus in figure 13a, with 40 percent of crystalline fraction of composition C_1 subtracted from L_I, the normal tholeiitic basalt liquid II develops. The mineralogical composition of C_1 is estimated to be as shown in table 4.

In deriving liquid L_2 (olivine basalt 2) from L_I by separation of a crystalline fraction (fig. 13b), it is found, perhaps surprisingly, that Al_2O_3 and Na_2O decrease to zero in the crystalline fraction essentially simultaneously with K_2O. This occurs with 33 percent C_2, 67 percent L_2. The crystalline aggregate of composition C_2 can contain no feldspar. This suggests that the high-alumina basalts develop from tholeiitic olivine basalt by fractional crystallization under conditions such that plagioclase is delayed in separating.

The subtraction diagrams of figure 13 are significant in illustrating: (a) that a common parent is probable for the normal tholeiitic and the high alumina basalts, (b) that the high alumina content of the latter results from a lack of separation of plagioclase from the liquid at this differentiation stage, and (c) that the normal tholeiitic basalts and this particular high-alumina olivine basalt of the Cascade region were very likely derived by fractional crystallization under conditions of approximately constant total composition.

In the environments in which the high-alumina basalts typically develop apparently plagioclase crystals do not readily settle out of the basalt liquid. Possibly turbulence or convection currents characterize basalt and andesite liquids crystallizing in orogenic belts, and this movement within the liquid is a factor in keeping plagioclase crystals in suspension. A statement by Waters (1955) is of interest in this connection. In discussing the pyroxene andesites and basaltic andesites of the Cascade orogenic belt, he says, "Despite the great range in silica content, the rocks have certain features in common. They are characterized by high lime and alumina and are relatively low in iron. Most are crowded with phenocrysts of plagioclase that show a complex combination of oscillatory and progressive zoning." These observations are compatible with the idea that plagioclase crystallized but tended to remain in suspension. There is another possibility. Some of the high alumina magmas originating in orogenic belts may have had plagioclase crystallization supressed until a lower temperature point because of high water pressure. The experiments of Yoder (1954) indicate that the diopside-anorthite eutectic is shifted toward anorthite with increasing water pressure. Water pressure might therefore be a factor in causing delay in the appearance of plagioclase as a crystalline phase in certain of the high-alumina magmas.

Two other high alumina basalts, both of the Brito-Arctic region, are listed in table 2, labeled Sk1 and PC. Their compositions are very similar to that of the Cascade olivine basalt (no. 2, table 2), and hence can be derived from tholeiitic olivine basalt just as L_2 was derived from L_1, in figure 13. One is tempted to view both the "first liquid" of the Skaergaard and the Porphyritic Central type of Mull as magmas derived from a tholeiitic olivine basalt from which plagioclase was reluctant to separate. From petrologic evidence this would seem to be a not unreasonable conclusion with regard to the Porphyritic Central type. Wager and Deer (1939), however, believe that Sk1 is the primary liquid for the Skaergaard rocks and that is was "free from porphyritic plagioclase crystals when intruded".

(b) *Basalts and andesites of orogenic belts.* The series of Cascade magmas, 2-4-5-6-7 (table 2), is taken as representative of basalt-andesite suites of orogenic belts. As discussed earlier, the subhorizontal trend of the iron oxide-magnesia ratio line (fig. 11a) points to the formation of this series by fractional crystallization under conditions of approximately constant p_{O_2}. The subtraction diagram (fig. 14) where all of the major oxides are considered bears out this conclusion. In this diagram basaltic andesite, 4, of table 2 (L_4 of fig. 14) is derived from olivine basalt 2 (L_2) by subtraction of the crystalline fraction, C_3. The FeO and Fe_2O_3 lines in figure 14 have maxima and minima respectively at the vertical line representing the composition of the primary magma. Oxygen has been added to the total mass during fractionation, just as in the case of derivation of liquid, c, and crystalline fraction, X_2, of figure 12b. The fractions, L_4 and C_3 have a combined oxygen content 0.2 percent greater than the oxygen content of original magma, L_2.

The subtraction diagram of figure 14 is constructed by first finding the position for the vertical line representing the original magma, L_2, such that the K_2O line decreases to zero at composition of the crystalline aggregate, C_3. This occurs where the proportions of C_3 and L_4 are 52 and 48 percent respectively, by coincidence the same percentages found in deriving liquid c and X_2 from n (fig. 12b). Magnetite is assumed to have its stoichiometric composition, and plagioclase, pyroxene and olivine have compositions as given in table 4. Here the olivine has been assigned the mole composition, 90 Mg_2SiO_4, 10 Fe_2SiO_4, and plagioclase has the composition Ab_1An_2. Pyroxene then has the composition 40 $CaSiO_3$, 40 $MgSiO_3$, 20 $FeSiO_3$; and there is 6.4 percent magnetite. If olivine is assumed to have the composition 80 Mg_2SiO_4, 20 Fe_2SiO_4, as it was in C_1 and C_2, the pyroxene composition will have the higher Di:Fs ratio of 3:1, and the percent of magnetite will be reduced to 5.4. The mineralogical composition of C_3 (table 4) is thus by way of example. A reasonable olivine composition is assumed, and then this determines the MgO:FeO ratio in the pyroxene, the amount of magnetite, and the amount of O_2 which must be added to convert part of the FeO in L_2 to Fe_2O_3 in L_4 and C_3.

Conclusions Regarding Tholeiitic Basalt Derivative Magmas

From these considerations, and referring again to figure 11, we arrive at these following general conclusions. Tholeiitic basalt magmas of the orogens and of the non-orogens both have as a common parent a basalt of the general nature of Nockolds' average tholeiitic olivine basalt. With fractional crystallization one of two main trends will be followed. Where total composition remains about constant during crystallization, with a consequent decrease in p_{O_2}, a trend of the type exhibited by I-II-III and Sk1-Sk2-Sk3 curves in figure 11a is followed. This trend we note in the vast outpouring of lavas in the non-orogenic regions and in magmas of sheet intrusions, and apparently also in achondritic meteorites (Buddington, 1943). Where p_{O_2} has remained about constant or increased during fractional crystallization, curves for successive liquids follow trends such as 2-4-5-6-7 and IV-V-VI (fig. 11a). These trends are characteristic of the basalt-andesite series of orogenic belts.

These principal trends are a function of the manner of change of p_{O_2} during fractional crystallization. A secondary effect relating to the degree of separation of plagioclase from the liquid during the early stages of crystallization may be superimposed upon the main trends. If plagioclase fails to separate from the liquid in the proportions to be expected from fractional crystallization, high-alumina magmas form, characteristic of basalts and andesites of orogenic belts, but also found among some other magmas. This alumina effect is not noticed on the diagram of figure 11a because only MgO, iron oxide and SiO_2 are considered there, but is clearly evident in the subtraction diagrams (fig. 13).

[*Editors' Note:* Material has been omitted at this point.]

REFERENCES

Anderson, C. A., 1941, Volcanoes of the Medicine Lake highland, California: Calif. Univ., Dept. Geol. Sci. Bull. v. 25, p. 347-422.
Bailey, E. B., Thomas, H. H., and others, 1924, The Tertiary and post-Tertiary geology of Mull, Lock Aline and Oban: Scotland Geol. Survey Mem.
Buddington, A. F., 1943, Some petrologic concepts and the interior of the earth: Am. Mineralogist. v. 28, p. 119-140.
Nockolds, S. R., 1954, Average chemical compositions of some igneous rocks: Geol. Soc. America Bull., v. 65, p. 1007-1032.
Tilley, C. E., 1950, Some aspects of magmatic evolution: Geol. Soc. London Quart. Jour., v. 106, p. 37-61.
Turner, F. J., and Verhoogen, J., 1951, Igneous and metamorphic petrology: New York, McGraw-Hill Book Co.
Verhoogen, J., 1937, Mount St. Helens, a recent Cascade volcano: Calif. Univ., Dept. Geol. Sci., Bull., v. 24, p. 293.
Wager, L. R., and Deer, W. A., 1939, Geological investigations in East Greenland, part III. The petrology of the Skaergaard intrusion, Kangerdlugssuaq, East Greenland: Medd. om Grønland, v. 105, no. 4, p. 1-335.
Waters, A. C., 1955, Volcanic rocks and the tectonic cycle: Geol. Soc. America, Sp. Paper 62, Crust of the Earth, p. 703-722.
Williams, H., 1935, Newberry volcano of central Oregon: Geol. Soc. America Bull., v. 48, p. 295.
————, 1942, The geology of Crater Lake National Park, Oregon: Carnegie Inst. Washington, Pub. 540, p. 149.
Yoder, H. S., Jr., 1954, The system diopside-anorthite-water: Annual report of the director of the Geophysical Laboratory, Carnegie Inst. Wash. Year Book, No. 53, p. 106-107.

22

Reprinted from pages 342 and 504–509 of *Jour. Petrol.* **3**(3):342–532 (1962)

Origin of Basalt Magmas: An Experimental Study of Natural and Synthetic Rock Systems

by H. S. YODER, JR., *and* C. E. TILLEY

Geophysical Laboratory, Carnegie Institution of Washington, Washington, D.C., U.S.A.

ABSTRACT

Natural basalts and eclogites were investigated experimentally at a series of temperatures in the pressure range 1 atm to 40 kb and with water pressures 1 to 10 kb. Some runs were also made on related synthetic systems at 10 and 33 kb.

The two principal magma types recognized by field investigators—tholeiite and alkali basalt types—appear to be separated by equilibrium thermal divides at 1 atm. The principal divides were found by experiment at elevated pressures to give way to a new set of equilibrium thermal divides resulting from a new mineralogy. The change of the equilibrium thermal divides with pressure leads to the derivation of the two principal magma trends from the same bulk composition.

The melting behavior of basalts and eclogites indicates that both are the partial melting products of a more primitive rock (e.g. garnet peridotite). In the region of magma generation (below 60 km) the parental material, presumed to be garnet peridotite, yields an eclogitic magma and its fractionation depends on the garnet and omphacite of the eclogite, not on plagioclase and clinopyroxene of a basaltic magma. Increase of the garnet constituents in the magma at high pressure by effective removal of omphacite or shift of the garnet–omphacite boundary 'surface' will give rise to a tholeiite-type magma at low pressure. Similarly, increase of the omphacite constituents in the magma at high pressure by physical or physicochemical means will give rise to an alkali basalt-type magma at low pressure. In general, alkali basalt-type magmas are to be expected to be generated at greater depths than tholeiite-type magmas from the same primary source rock.

Establishment of the two major basalt series takes place in the region of generation; additional minor diversification of each series may come about after emplacement in or on the crust by crystal settling, oxidation or reduction, gas fluxing, contamination, and other processes. The derivative magmas are greatly restricted by the course of liquid thermal descent imposed at generation.

Pressure–temperature limits established experimentally suggest that the basalt–eclogite transformation may be responsible for the Mohorovičić discontinuity under the continents, but not under the oceans.

The field of stability of basalt is drastically reduced in the presence of water, and amphibolite is produced. The melting of amphibolite takes place over a much greater range of temperature than basalt. At 10 kb water pressure the beginning of melting of amphibolite closely approaches that of granite. Partial melting of amphibolite may yield anorthositic liquids having a relatively low anorthite content at exceptionally low temperatures. Eclogite itself is not stable in the presence of water and gives place to amphibolite or pyroxene hornblendite. Magmas which crystallize to basalt, gabbro, or eclogite must have had a low water-content at the time of crystallization.

Fifteen rock and twenty-three mineral analyses as well as numerous partial chemical analyses of experimental products were made by J. H. Scoon in the course of the investigation. These chemical analyses bear on many mineralogical and petrological problems.

[*Editors' Note:* Material has been omitted at this point.]

Application of results

Basalt–eclogite transformation under anhydrous conditions

A rough outline of the stability field of eclogite may now be made using the data in Tables 43 and 50 for the Glenelg eclogite and the data of G. C. Kennedy (1956 and 1959; verbal communication, 1961) between 10 and 14 kb at 500° C. These data are summarized in Fig. 43 along with runs on a natural basalt between 33 and 40 kb at 1,200° by Boyd & England (1959, p. 88). For convenient reference a depth scale is provided in Fig. 43 based on an average density of 2·73 (Washington, 1922, p. 392) for crustal rocks of the continent down to 35 km and an assumed average density of 3·33 for the upper mantle rocks in the region 35 to 140 km.[1] G. C. Kennedy (1959) reported the transformation of a natural basalt (not described), converted previously into a glass, into gabbro below 10 kb at 500° C in his squeezer apparatus. He described, in a pressure range above 10 kb at the same temperature, the gradual change of a gabbroic assemblage into a 'dominantly jadeitic pyroxene', identified by X-ray methods. No garnet was listed. The writers therefore have some reservations about these data demonstrating the transformation of gabbro to eclogite in which garnet is considered an essential phase. In later experiments, not as yet reported, Kennedy apparently obtained true eclogite rather than pyroxenite (personal communication, 1961). Using Kennedy's transformation zone, with reservations as noted, and the limitations imposed by the data at 10 and 20 kb, the slope of the transformation zone may be estimated. On these grounds, allowing for no over-pressure, it appears as though eclogite (without plagioclase) is not stable within either the oceanic or continental crust at any temperature. The transformation of basalt (or gabbro) to eclogite is relatively insensitive to temperature and is mainly dependent on the pressure. The pressure region of the transformation is adequately confirmed by the conversion of eclogite to basalt and, conversely, the conversion of basalt to eclogite (see Tables 43, 50C through 50G).

The nature of the transformation zone was described by Kennedy as a gradual decrease of feldspar with increasing production of pyroxene (and presumably, garnet). At some intermediate stage the assemblage Pl+Gr+Cpx should be found. The so-called plagioclase eclogites examined by the writers are con-

[1] Washington (1923e, p. 455) calculated the average specific gravity for the whole island of Hawaii and obtained 2·940. Goranson (1928) gave a density of 2·69 for the whole island core. These values have bearing on the pressure at the depth of magma generation on Hawaii.

sidered to be downgraded eclogites in various stages of transformation to gabbro and not equilibrium assemblages. In view of the production of a pyroxenite of basalt composition from both an eclogite (Table 44) and a basalt (Table 50), it is more likely that the transformation proceeds from basalt through pyroxenite to eclogite. A more detailed study in the 10 to 20 kb pressure range will no doubt resolve this question.

Melting relations of eclogite

In the pressure range 1 to 14 kb eclogite is transformed into basalt or gabbro before melting begins. An eclogite so transformed would, therefore, melt in the same fashion as a basalt or gabbro. In the pressure range 14 to 19 kb no data are available to describe the melting relations. When suitable equipment becomes available no doubt this region can be investigated by means of the quenching

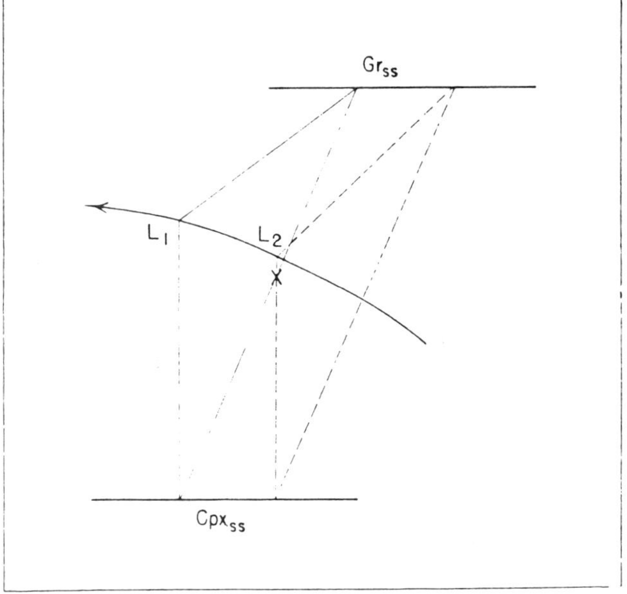

FIG. 47. Schematic presentation of the melting of two complex solid solutions, garnet and clinopyroxene. The assumed bulk composition is X, and L_1 and L_2 are projections of the liquid composition at temperatures T_1 and T_2 where $T_1 < T_2$. Pressure constant.

method. Above 19 kb pressure the two major minerals of eclogite, garnet and clinopyroxene, begin melting together and coexist with liquid over most of the narrow melting range of about 85° C. Clinopyroxene is the liquidus phase, and the liquidus rises at the rate of about 11° kb.

Because garnet and clinopyroxene melt together over such a narrow interval of temperature, the bulk compositions of eclogite must be close to or on the boundary 'surface' between garnet and clinopyroxene. The bulk composition of the Glenelg eclogite must lie on the clinopyroxene side of the boundary 'surface.' These observations may be illustrated by means of Fig. 47. The liquid L_1 is the first liquid to form from the bulk composition X; L_2 is the highest temperature and composition at which garnet and clinopyroxene coexist with liquid. The temperature increase represented by the change of liquid composition from L_1 to L_2 is probably of the order of $60°$ C. During this change of temperature the composition of clinopyroxene and garnet would change in the manner shown by the tie lines. At the temperature of L_2 the garnet is consumed and then clinopyroxene alone coexists with liquid up to the liquidus temperature. The Af–C–Fm plot given in Fig. 36 supports the notion that the compositions of eclogites and basalts lie along a narrow zone which one may presume outlines the projection of the boundary 'surface' between the two complex solid solutions.

The important conclusion to be gained from this interpretation of the melting behavior of eclogite is as follows. Because both major phases appear almost simultaneously on the liquidus and the melting interval is small, the eclogites themselves must be the partial melting product of an even more primitive rock. The writers suggest that the more primitive rock is a garnet peridotite. Eclogite is presumably related to garnet peridotite $(Ol + Cpx + Gr + Opx)$ at high pressure as granite is related to basalt at low pressure.[1] Neither basalt nor granite is stable in the region of the eclogites.

Differentiation of eclogites

An examination of Table 53 indicates that the omphacite contains most of the

TABLE 53

Norms of analyzed garnet and clinopyroxene from eclogites

	Clinopyroxene					Garnet				
	Glenelg 35090	Loch Duich 35083	Oahu 66118	Silberbach 958	'Weissen-stein'	Glenelg 35090	Loch Duich 35083	Oahu 66118	Silberbach 958	'Weissen-stein'
Or	0·11	0·06	—	—	0·56	—	—	—	—	—
Ab	23·68	14·46	12·84	14·34	14·47	—	—	—	—	—
Ne	6·42	2·10	1·56	7·28	5·50	—	—	—	—	—
An	3·70	—	11·95	5·00	8·90	45·23	33·67	26·41	38·64	39·75
Di	55·13	76·40	58·31	71·76	68·04	—	—	—	—	—
Hy	—	—	—	—	—	28·43	48·65	53·25	40·37	37·74
Ol	—	0·49	10·09	1·22	0·77	16·08	2·08	—	4·34	6·04
Ac	—	0·69	—	—	—	—	—	—	—	—
Wo	3·38	—	—	—	—	—	—	—	—	—
Sp	—	—	—	—	—	7·03	12·87	12·26	13·33	14·44
C	—	—	—	—	—	—	—	5·60	—	—
Il	0·82	0·62	1·52	0·15	1·06	0·55	0·35	0·61	0·30	0·46
Mt	6·50	5·01	3·94	0·23	0·70	2·69	2·85	2·32	2·09	1·62
TOTAL	99·74	99·83	100·21	99·98	100·00	100·01	100·47	100·45	99·07	100·05

normative diopside, albite, and nepheline and the garnet contains most of the normative anorthite, hypersthene, and olivine. Separation of either of these

[1] Harris & Rowell (1960) report in some preliminary experiments that at 1 *atm* the partial fusion of a synthetic peridotite yields basalt. The partial fusion product of basalt at the same pressure, however, is generally believed to be granite.

117

phases will have considerable influence on the composition of the residual liquid. The nature of the division of the normative minerals suggests an obvious method, either physical or physicochemical, for generation of the two principal magma types. Effective removal of garnet at high pressures will enrich the liquid in omphacite components and yield an alkali-type liquid at low pressure; effective removal of omphacite at high pressures will enrich the liquid in garnet components and yield a tholeiite-type liquid. Settling of garnet, such as that observed in some of the runs, would yield a liquid at high pressure which would produce an alkali-type magma at the surface.

More important, the boundary 'surface' between omphacite and garnet no doubt shifts with pressure as well. Such pressure changes would change the ratio of garnet and omphacite in the liquid. Presumably the boundary curve would move toward omphacite at the higher pressures and toward garnet at the lower pressures.[1] On these grounds the liquids which would produce alkali basalt at the surface would come from greater depths than those liquids which would produce tholeiite at the surface.

The distribution of the normative components of basalt between the garnet and omphacite of the eclogite has bearing on the olivine content of basalts. Bowen (1928, p. 164) had noted that basalts with olivine in excess of 12 to 15 per cent were usually porphyritic and were the product of crystal settling. The limitation of olivine content in basalts may be due to the restrictions imposed by the composition of garnet in its immediate source rock, eclogite, rather than the high liquidus temperature imposed at 1 atm by large amounts of olivine. The higher liquidus temperature of a garnet-enriched eclogite composition will not be as great, relatively speaking, as that of an olivine-enriched basalt composition. As already demonstrated, basalts with as much as 20 per cent normative olivine have reasonable liquidus temperatures.

Fractionation of eclogites

In spite of the small melting range, the fractionation during the melting of eclogite will be quite unique. It is to be noted that the fractionation depends *not* on plagioclase and clinopyroxene as in basalt, but on garnet and on omphacite. The fractionation scheme devised by Bowen (1928), which hinges in a large part on the plagioclases, obviously cannot be used to describe the changes in magmas at great depth. If the liquids which give rise to basalt originate at pressures in excess of 19 kb, the differences in basalt magmas cannot be expressed in terms of the minerals of basalt but in terms of the minerals of their equivalent eclogite. There appears to be no unambiguous way of computing an 'eclogite norm' from the analysis of a rock because of the extensive solid solution of each of the major phases. Although there are insufficient experimental data to outline the nature of the fractionation at pressures in excess of 19 kb, one may

[1] This argument is based on the assumption that the dT/dP of the garnet liquidus surface is greater than that of the omphacite liquidus surface.

deduce from the schematic diagram of Fig. 47 and the Af–C–Fm plot of Fig. 36 that with a given temperature change, for example, the clinopyroxene will become less hypersthenic and the garnet richer in almandite and grossularite. Further experimental work is required to ascertain whether such effects are achieved through rising or falling temperatures. It will be of considerable import to determine the relative compositions of the liquids schematically indicated as L_1 and L_2 (Fig. 47). Whereas both L_1 and L_2 are believed to be of basaltic composition (both involve melting of Gr and Cpx), designation as to which is related to the alkali basalt magma type or to the tholeiitic magma type cannot be made with the present data. It is clear that a new fractionation scheme is required to describe the behavior of magmas at high pressures.

Eclogites and the primary magma problem

It was demonstrated (Fig. 44 *a*, *c*) that the primary equilibrium thermal barriers, such as Fo–Ab and Fo–An which existed at 1 atm, were breached at

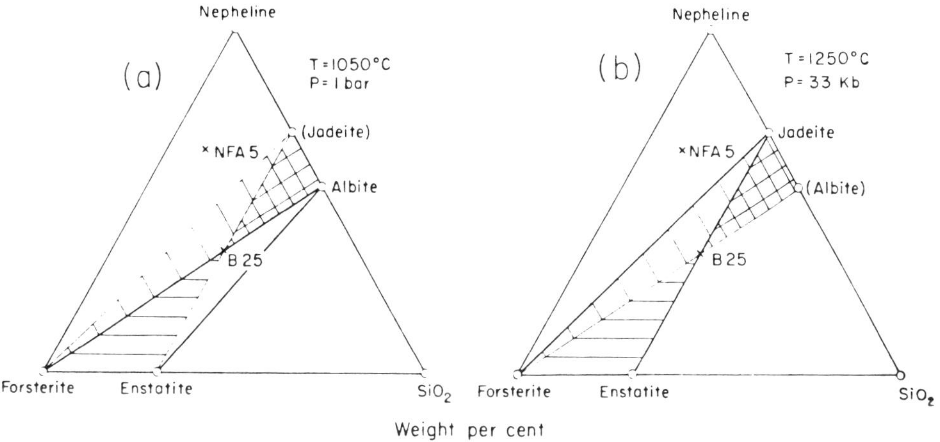

FIG. 48. The system Ne–Fo–SiO$_2$ at (*a*) 1,050° C and 1 bar and (*b*) 1,250° C and 33 kb. The heavy lines indicate the stable joins. Liquids in the horizontally ruled and cross-hatched areas of composition behave differently and in the opposite way at the two pressures. Liquids in the diagonally ruled area trend the same way at both pressures.

high pressures. The new equilibrium thermal barriers created impose a new set of restrictions on the courses of liquids. The Ne–Fo–SiO$_2$ diagram can be used in illustration of the argument (Fig. 48). Liquids having compositions in the horizontally ruled area in Fig. 48*a* will trend toward silica in the low-pressure region where the thermal barrier Fo–Ab is applicable. At high pressures liquids of the same compositions (Fig. 48*b*) will trend toward nepheline because the thermal barrier Jd–En is now in effect. In this way hypersthene-bearing nodules formed at high pressure may be inherited by nepheline–normative rocks crystallizing at low pressure. The opposite trends would be experienced, at the two pressure regions, by liquids having compositions in the cross-hatched areas. It is probable that nepheline-normative magmas may yield hypersthene- or quartz-bearing residua as a result of the pressure change during crystallization. The diagonally ruled area indicates those compositions which will produce at all pressures liquids which will have nepheline in the norm.

Extending this argument to more complex compositions, it may be concluded that there are wide ranges of bulk compositions capable of producing the two principal basalt magma trends by varying the pressure. A garnet peridotite,[1] for example, would yield nepheline-trending magmas at high pressures and silica-trending magmas at low pressures. On these grounds one can conclude that the alkali basalt magmas are derived at greater depths than tholeiite magmas. The writers believe that this argument constitutes strong support for the concept of a single primary source for all basalt magmas. That single primary source can obviously have a relatively wide range of bulk composition as illustrated in principle by the horizontally ruled and cross-hatched areas of Fig. 48 a, b. That single primary source would be of necessity eclogitic if the compositions of $Ne-Fo-SiO_2$ and $Ca-Tsch-Fo-SiO_2$ may be used as guides.

[*Editors' Note:* Material has been omitted at this point.]

REFERENCES

[*Editors' Note:* Only the references cited in the preceding excerpt are reproduced here.]

BOWEN, N. L., 1928. *The evolution of the igneous rocks.* Princeton: Princeton Univ. Press, 332 pp.

BOYD, F. R. & ENGLAND, J. L., 1959. Experimentation at high pressures and temperatures. *Carnegie Inst. Wash. Yearb.* **58**, 82–9.

GORANSON, R. W., 1928. The density of the Island of Hawaii, and density distribution in the earth's crust. *Amer. J. Sci.,* 5th ser., **16**, 89–120.

HARRIS, P. G., & ROWELL, J. A., 1960. Some geochemical aspects of the Mohorovicic discontinuity. *J. geophys. Res.* **65**, 2443–59.

KENNEDY, G. C., 1956. Polymorphism in the feldspars at high temperatures and pressures. *Bull. geol. Soc. Amer.* **67**, 1711–2 (Abstract).

——— 1959. The origin of continents, mountain ranges, and ocean basins. *Amer. Scientist* **47**, 491–504.

WASHINGTON, H. S., 1922. Isostasy and rock density. *Bull. geol. Soc. Amer.* **33**, 375–410.

——— 1923e. The density of the earth as calculated from the densities of Mauna Kea and Haleakala. *J. Wash. Acad. Sci.* **13**, 453–6.

---

[1] A 'high-pressure' peridotite is meant here; that is, one in which plagioclase is occult in garnet and pyroxene.

23

Reprinted from pages 104–105 and 159–167 of *Contr. Mineralogy and Petrology*
15:103–190 (1967)

THE GENESIS OF BASALT MAGMAS

D. H. Green and A. E. Ringwood

Abstract. This paper reports the results of a detailed experimental investigation of fractionation of natural basaltic compositions under conditions of high pressure and high temperature.
A single stage, piston-cylinder apparatus has been used in the pressure range up to 27 kb
and at temperatures up to 1500° C to study the melting behaviour of several basaltic compositions. The compositions chosen are olivine-rich (20% or more normative olivine) and include
olivine tholeiite (12% normative hypersthene), olivine basalt (1% normative hypersthene)
alkali olivine basalt (2% normative nepheline) and picrite (3% normative hypersthene). The
liquidus phases of the olivine tholeiite and olivine basalt are olivine at 1 Atmosphere, 4.5 kb
and 9 kb, orthopyroxene at 13.5 and 18 kb, clinopyroxene at 22.5 kb and garnet at 27 kb.
In the alkali olivine basalt composition, the liquidus phases are olivine at 1 Atmosphere and
9 kb, orthopyroxene with clinopyroxene at 13.5 kb, clinopyroxene at 18 kb and garnet at
27 kb. The sequence of appearance of phases below the liquidus has also been studied in detail.
The electron probe micro-analyser has been used to make partial quantitative analyses of
olivines, orthopyroxenes, clinopyroxenes and garnets which have crystallized at high pressure.

These experimental and analytical results are used to determine the directions of fractionation of basaltic magmas during crystallization over a wide range of pressures. At pressures
corresponding to depths of 35—70 km separation of aluminous enstatite from olivine tholeiite magma produces a direct fractionation trend from olivine tholeiites through olivine
basalts to alkali olivine basalts. Co-precipitation of sub-calcic, aluminous clinopyroxene with
the orthopyroxene in the more undersaturated compositions of this sequence produces derivative liquids of basanite type. Magmas of alkali olivine basalt and basanite type represent
the lower temperature liquids derived by approximately 30% crystallization of olivine-rich
tholeiite at 35—70 km depth. At depths of about 30 km, fractionation of olivine-rich tholeiite with separation of both olivine and low-alumina enstatite, joined at lower temperatures
by sub-calcic clinopyroxene, leads to derivative liquids with relatively constant SiO_2 (48 to
50%) increasingly high Al_2O_3 (15—17%) contents and retaining olivine + hypersthene
normative chemistry (5—15% normative olivine). These have the composition of typical
high-alumina olivine tholeiites. The effects of low pressure fractionation may be superimposed
on magma compositions derived from various depths within the mantle. These lead to diver-
gence of the alkali olivine basalt and tholeiitic series but convergence of both the low-alumina
and high-alumina tholeiites towards quartz tholeiite derivative liquids.

The general problem of derivation of basaltic magmas from a mantle of peridotitic composition is discussed in some detail. Magmas are considered to be a consequence of partial melting
but the composition of a magma is determined not by the depth of partial melting but by
the depth at which magma segregation from residual crystals occurs. Magma generation from
parental peridotite (pyrolite) at depths up to 100 km involves liquid-crystal equilibria between
basaltic liquids and olivine + aluminous pyroxenes and does not involve garnet. At 35—70 km
depth, basaltic liquids segregating from a pyrolite mantle will be of alkali olivine basalt
type with about 20% partial melting but with increasing degrees of partial melting, liquids
will change to olivine-rich tholeiite type with about 30% melting. If the depth of magma
segregation is about 30 km, then magmas produced by 20—25% partial melting will be of
high-alumina olivine tholeiite type, similar to the "oceanic tholeiites" occurring on the sea
floor along the mid-oceanic ridges.

Hypotheses of magma fractionation and generation by partial melting are considered in
relation to the abundances and ratios of trace elements and in relation to isotopic abundance
data on natural basalts. It is shown that there is a group of elements (including K, Ti, P, U,

Th, Ba, Rb, Sr, Cs, Zr, Hf and the rare-earth elements) which show enrichment factors in alkali olivine basalts and in some tholeiites, which are inconsistent with simple crystal fractionation relationships between the magma types. This group of elements has been called "incompatible elements" referring to their inability to substitute to any appreciable extent in the major minerals of the upper mantle (olivine, aluminous pyroxenes). Because of the lack of temperature contrast between magma and wall-rock for a body of magma near to its depth of segregation in the mantle, cooling of the magma involves complementary processes of reaction with the wall-rock, including selective melting and extraction of the lowest melting fraction. The "incompatible elements" are probably highly concentrated in the lowest melting fraction of the pyrolite. The production of large overall enrichments in "incompatible elements" in a magma by reaction with and highly selective sampling of large volumes of mantle wall-rock during slow ascent of a magma is considered to be a normal, complementary process to crystal fractionation in the mantle. This process has been called "wall-rock reaction". Magma generation in the mantle is rarely a simple, closed-system partial melting process and the isotopic abundances and "incompatible element" abundances of a basalt as observed at the earth's surface may be largely determined by the degree of reaction with the mantle or lower crustal wall-rocks and bear little relation to the abundances and ratios of the original parental mantle material (pyrolite).

Occurrences of cognate xenoliths and xenocrysts in basalts are considered in relation to the experimental data on liquid-crystal equilibria at high pressure. It is inferred that the lherzolite nodules largely represent residual material after extraction of alkali olivine basalt from mantle pyrolite or pyrolite which has been selectively depleted in "incompatible elements" by wall-rock reaction processes. Lherzolite nodules included in tholeiitic magmas would melt to a relatively large extent and disintegrate, but would have a largely refractory character if included in alkali olivine basalt magma. Other examples of xenocrystal material in basalts are shown to be probable liquidus crystals or accumulates at high pressure from basaltic magma and provide a useful link between the experimental study and natural processes.

[*Editors' Note:* Material has been omitted at this point.]

The Generation of Basaltic Magmas

In the previous sections we have discussed the fractionation of basaltic magmas at various depths in the mantle and demonstrated mechanisms by which a "primitive" olivine-rich tholeiite magma may produce derivative liquids of alkali olivine basalt, high-alumina basalt or quartz tholeiite type. An important alternative hypothesis (e.g. KUNO, 1960; KUSHIRO and KUNO, 1963) maintains that the compositions of the principal basaltic magmas are determined by the depth in the mantle at which partial melting occurs rather than by subsequent fractionation processes. The data obtained from our experiments are used in the following sections to evaluate this hypothesis in detail.

An essential prerequisite to a discussion of the generation of basaltic magmas in the mantle is a consideration of the mineralogical and chemical constitution of the upper mantle. We will take up this subject in the next section, proceed then to the physical conditions of magma formation and finally investigate the chemical and mineralogical equilibria involved and their effect upon the composition of the resultant magma.

a) Chemistry and Mineralogy of Parental Mantle

It can be argued plausibly on general petrological and geochemical grounds that the chemical composition of the primary undifferentiated upper mantle should be somewhere between those of typical basalt and typical alpine peridotite. A chemical and petrological model for the upper mantle based upon this postulate has been developed by RINGWOOD (1962a, b; 1966a, b), GREEN and RINGWOOD (1963) and GREEN (1966a). In this model, the primary undifferentiated composition of the upper mantle is assumed equal to approximately 1 part of basalt to 3 parts of peridotite. This primary composition is called pyrolite (pyroxene-olivine rock). It is emphasized that the 3:1 proportion is not regarded as critical or unique, and substantial variations in this ratio are possible. Nevertheless an approximate 3:1 ratio is suggested by certain geochemical considerations and is convenient for the formulation of a specific model. The composition of pyrolite as derived by RINGWOOD (1966a) is given in Table 20.

An important property of compositions close to pyrolite is the ability to crystallize in four distinct mineralogical assemblages over the range of P, T conditions existing in the upper mantle. These are

1. Olivine + amphibole ± enstatite ± spinel (Ampholite);
2. Olivine + pyroxenes + plagioclase + chromite (Plagioclase pyrolite);
3. Olivine + aluminous pyroxenes ± spinel (Pyroxene pyrolite);
4. Olivine + pyroxenes + garnet (Garnet pyrolite).

The stability fields of these mineral assemblages for the model pyrolite composition given in Table 20 are currently under experimental investigation. A preliminary outline of the pyrolite stability fields was given by RINGWOOD, MACGREGOR and BOYD (1964), and RINGWOOD (1966b). Our latest experimental results require some modifications of the earlier boundaries of pyrolite stability fields but these are not of a fundamental nature. Much of the revision is caused by the necessity to introduce a pressure correction to earlier results because of non-uniform distribution of pressure in the furnace assemblies (GREEN et al., 1966). A provisional outline of stability fields for the pyrolite composition (Table 20) according to our latest experimental results is given in Fig. 11.

Table 20

Model composition of pyrolite (RINGWOOD, 1966a)

SiO_2	45.16
TiO_2	0.71
Al_2O_3	3.54
Fe_2O_3	0.46
FeO	8.04
MnO	0.14
MgO	37.47
CaO	3.08
Na_2O	0.57
K_2O	0.13
Cr_2O_3	0.43
NiO	0.20
P_2O_5	0.06

For the present purposes, it is important to observe that the boundary between the stability fields of pyroxene pyrolite and garnet pyrolite intersect the pyrolite solidus at a depth of 100 km. *Thus, the formation of magmas by fractional melting of pyrolite at depths smaller than 100 km would occur in the stability field of olivine + aluminous pyroxenes. Garnet would not play a significant role in the genesis of magmas by fractional melting at depths smaller than 100 km.* These considerations constitute a serious objection to the views of YODER and TILLEY (1962), O'HARA (1965) and others, that the principal basalt magma types are formed by direct partial melting of "garnet peridotite" in the upper mantle. In order to stabilise garnet at shallower depths in the mantle, a pyrolite composition possessing a much higher ratio of R_2O_3 $(Al_2O_3 + Cr_2O_3 + Fe_2O_3)$ to total pyroxene than appears reasonable would need to be assumed. Furthermore, a model invoking such a composition would encounter further difficulties because of the tendency for garnet to melt incongruently to aluminous enstatite at an early stage of fractional melting in ultramafic compositions, so that magmas produced at relatively low pressures (15—30 kb) are in equilibrium with residual olivine and aluminous pyroxene for a very wide range of possible mantle compositions.

b) Physical Processes of Magma Generation

The formation of a magma in the mantle requires the supply of a large amount of thermal energy, in excess of 100 cals per gram of magma, to a localised region. Physical processes, e.g. thermal conduction, radioactive heat generation, mass transfer, which might be responsible for the supply of this energy operate on a comparatively long time scale. In contrast, the time scale required for separation of crystals from liquid within the mantle directly by gravity or indirectly by deformational processes ultimately of gravitational origin, is probably smaller by orders of magnitude. Because of these conditions, the formation of magmas in the mantle will almost always be the result of *partial* melting rather than of complete melting. Where a substantial degree of partial melting occurred throughout a large volume, the magma will tend to segregate from residual crystals into a self-contained magma body which thereafter evolves independently of the refractory residuum with which it was formerly associated. The degree of fractional melting which is required before the magma separates from residual crystals doubtless varies according to physical conditions, but perhaps ranges mostly between 20 and 40 percent (by volume).

Many processes of magma generation have been advocated in the past, e.g. melting by relief of pressure, localised melting caused by liberation of energy during earthquakes, melting caused by accumulation of heat in regions characterised by a' high concentration of radioactivity, and melting connected with rising "convection" cells or "advective movement" in the mantle. After an examination of possibilities, the authors are of the opinion that the only generally

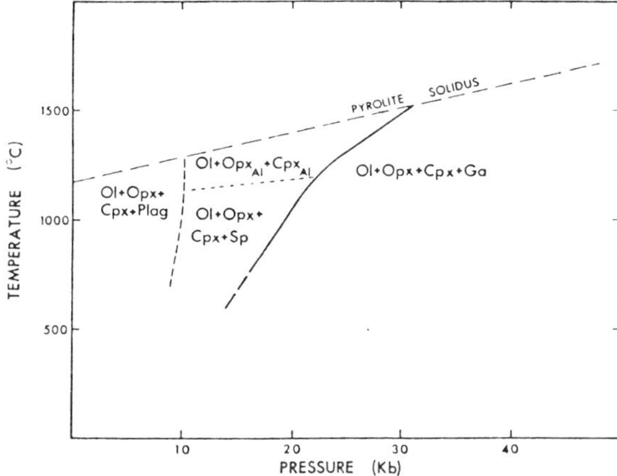

Fig. 11. Preliminary experimental determination of the stability fields of different mineral assemblages in the pyrolite composition of Table 20. The data are for an anhydrous composition and do not include the "ampholite" (olivine + amphibole ± enstatite) assemblage

satisfactory mechanism for producing basalt magmas and their observed distribution over the face of the earth and through geologic time, is some process connected with "convection" or "advection" in the mantle. Such processes have been advocated by many authors (e.g. HOLMES, 1926, 1927; VERHOOGEN, 1954). Accordingly we will base our discussion of magma generation on this type of process. Nevertheless it should be pointed out that much of the following discussion is of a general nature and could be applied to some other models of magma generation.

The condition for gravitational instability in the mantle is that the actual temperature gradient should exceed the adiabatic gradient. This condition is therefore most favourable in the upper few hundred kilometers of the mantle, where the geothermal gradient is greatly in excess of the adiabatic gradient. Indeed, as shown by CLARK and RINGWOOD (1964) it is probable that the density of the mantle actually decreases with depth down to 100 km or so in many regions. Although the outermost mantle many be potentially gravitationally unstable, the triggering-off of an actual instability leading to some form of mass-transfer requires other favourable conditions controlled particularly by rheological properties and by the presence of horizontal inhomogeneities in density, caused either by temperature or chemical differences. Because of the close approach of the actual temperature gradient to the melting point gradient in the uppermost mantle, this region will constitute a zone of low strength and high mobility as previously pointed out by numerous authors. Also, horizontal inhomogeneities are most pronounced in the upper mantle. All these factors contribute towards the occurrence of mass transfer processes in the upper mantle, and their essential restriction to this region.

Processes of mass transfer in the upper mantle are commonly referred to as "convection" a term which characteristically applies to quasiregular, thermally generated motions in a viscous fluid. The analogy has frequently been applied to the earth and it has been argued e.g. Vening Meinesz (1952, 1962), Runcorn (1962) that the mantle is characterised by regular arrays of convection cells, extending as deep as the core. Such schemes appear unrealistic and implausible for many reasons. Elsasser (1963) has provided a stimulating discussion of the subject, and argued convincingly for restriction of mass transport processes to the upper mantle. Furthermore he emphasizes the probable extreme irregularity both in time and configuration, of the processes to be expected in the upper mantle. An additional complication in the models which we shall discuss is that mass motions are accompanied by partial melting and chemical differentiation, and are hence irreversible. Clearly, "convection" in the conventional sense is not an ideal term[5] to apply to such complex processes.

The model for magma generation which we have in mind is given in Fig. 12. It is characterised by a highly specific relationship between the actual temperature distribution and the pyrolite solidus, as shown in the diagram. Gravitational instability in the upper mantle combined with a suitable combination of horizontal inhomogeneity and rheological properties causes a source-mass (S) of solid pyrolite to rise diapirically (in the manner of a salt dome) from the low-velocity zone. It is possible that the initial triggering-off was connected with stresses associated with seismic activity, i.e. the diapir may be derived from an earthquake source-region; however this is not essential. The rising diapir is sufficiently large and hence possesses sufficient thermal inertia in relation to its velocity, so that it cools adiabatically and does not interact by thermal conduction with the surrounding mantle. The adiabatic gradient, of the order of 0.3° C/km (Birch, 1952) is much smaller than the gradient of the pyrolite solidus. Accordingly partial melting in the rising diapir will occur as the temperature of the diapir, following the adiabat from S, intersects the solidus at F (Fig. 12). This causes an increase of the density contrast between rising diapir and surrounding mantle and accordingly, an increase in its rate of upward movement. We assume that the partially melted diapir remains adiabatic. As it rises, and the pressure decreases further, the degree of partial melting increases. The absorption of latent heat accordingly steepens the effective adiabatic gradient, and the temperature of

---

[5] In a previous paper (Ringwood and Green, 1966) we adopted the term "advection" for the process, following Elsasser (1963). Although this term adequately specifies the complexity of the mass transport envisaged, it also carries the implication that the horizontal dimensions of the motions greatly exceed the vertical dimensions. This may not necessarily apply in the upper mantle.

the rising diapir (now a crystal-liquid mush) behaves as shown. It is probable that at this stage, the velocity of upward movement is sufficiently slow to permit the liquid component of the mush to remain in chemical equilibrium with the residual unmelted crystals. Eventually, the degree of partial melting becomes sufficiently extensive (20—40%) so that the liquid segregates from residual refractory crystals, and forms an independent, homogeneous magma body. This may be termed the stage of *magma segregation* (M, Fig. 12). From this stage

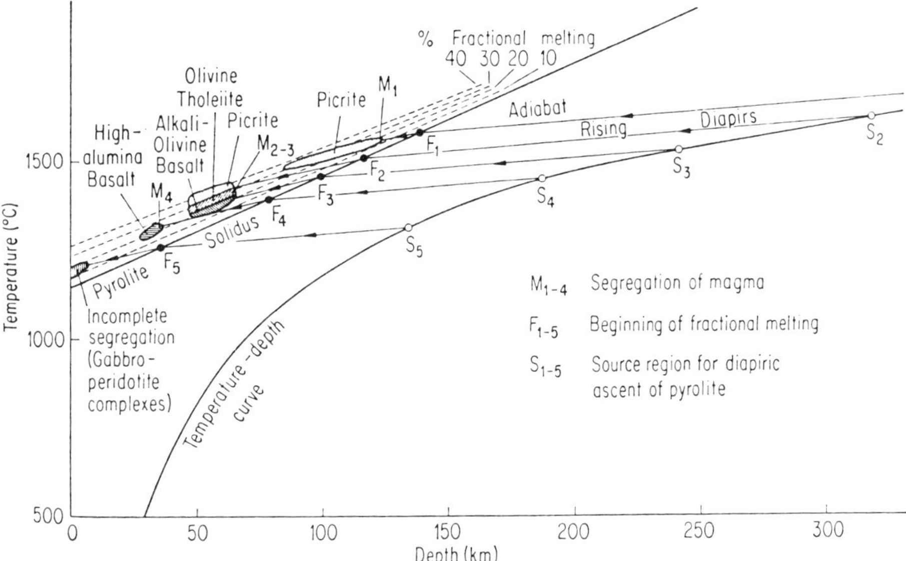

Fig. 12. A model for magma formation by fractional melting of mantle pyrolite. S_1—S_5 represent arbitrary source regions from which there is diapiric ascent of bodies of subsolidus pyrolite to intersect the pyrolite solidus at points F_1—F_5. Partial melting begins at these points and the rising crystal-liquid mushes follow the courses indicated until segregation of magma from residual crystals at M_1—M_4. The nature of the magma is determined by equilibria occurring at M_1—M_4 and not at F_1—F_4

onwards, the magma is no longer in equilibrium with the residual crystals with which it was originally associated. Instead, it may fractionate independently by cooling and crystal settling as it rises towards the surface.

The above outline suggests that the processes of magma generation in the mantle are more complex than sometimes assumed. Thus the initial source region (S) of the rising solid diapir, which may be associated with seismic activity, is probably much deeper than the point (F) at which the earliest liquid forms by fractional melting. As we shall see, the chemical composition of the liquid first formed at (F) is sensitively dependent upon depth. However, the magma which finally segregates at the much shallower level (M) may have an entirely different composition. As long as crystals and liquid in the rising crystal mush remain in chemical equilibrium, as appears likely, the nature of the liquid will change continuously with pressure, and will retain no "memory" of its earlier, deeper origin. The overall chemistry of the magma is in fact, not determined until it segregates from residual refractory crystals at M. Thus the depth of *magma segregation* M is decisive in determining the nature of the magma. This depth may not be related in any simple manner to the initial depth of partial melting (F). Accordingly, in subsequent sections, we will discuss the partial melting of pyrolite according to the depths at which magma may segregate from residual crystals.

c) Partial Melting of Pyrolite — General Discussion

The rationale of the pyrolite model was to select a composition which on fractional melting would yield a generally "basaltic" magma, leaving an unmelted residuum similar in composition to peridotite or dunite. In one sense, fractional melting may be regarded as the reverse of fractional crystallization, providing that the nature of the crystalline phases are similar in both cases. This relationship is independent of the actual proportions of phases which may be present. We have seen that at depths between about 30 kms and 100 kms, the mineral assemblage of pyrolite at the solidus consists essentially of olivine, aluminous enstatite and aluminous subcalcic clinopyroxene. The experiments upon the crystallization of basalt magmas at different pressures showed that the principal magma types could be derived by the separation of olivine and aluminous pyroxenes in various proportions according to P, T conditions and magma compositions. Moreover the Fe/Mg ratios of olivines and pyroxenes occurring near the liquidi of the experimental basaltic compositions are similar to the Fe/Mg ratios of olivines and pyroxenes of mantle-derived ultramafic material such as kimberlite inclusions, peridotite nodules and alpine peridotites. Because of these relationships, we are able to apply our results on fractional crystallization of basalt magmas to fractional melting of pyrolite within the above depth range. The inferred relationships are summarized in Fig. 13 and elaborated in the following sections.

d) Depth of Magma Segregation 0—15 km

It has long been known that at atmospheric pressure, the field of crystallization of olivine extends into quartz normative basaltic compositions. Accordingly, it is possible to derive quartz tholeiite from an olivine tholeiite magma by early crystallization of olivine, followed by segregation of crystals from magma, so that Bowen's reaction relationship is prevented. Conversely, segregation of small amounts of magma from fractional melting of pyrolite under restricted conditions may yield an oversaturated tholeiite magma together with residual dunite (Reay and Harris, 1964).

Boyd et al. (1964) showed that a pressure of a few kilobars was sufficient to prevent the incongruent melting of enstatite. T. H. Green is currently studying this relationship in natural basaltic magmas (personal communication). He finds that saturated tholeiites may form by fractional melting of pyrolite in the depth interval 0—15 km. Below this depth, the olivine reaction relationship does not occur, and the liquid fractions are always olivine-normative.

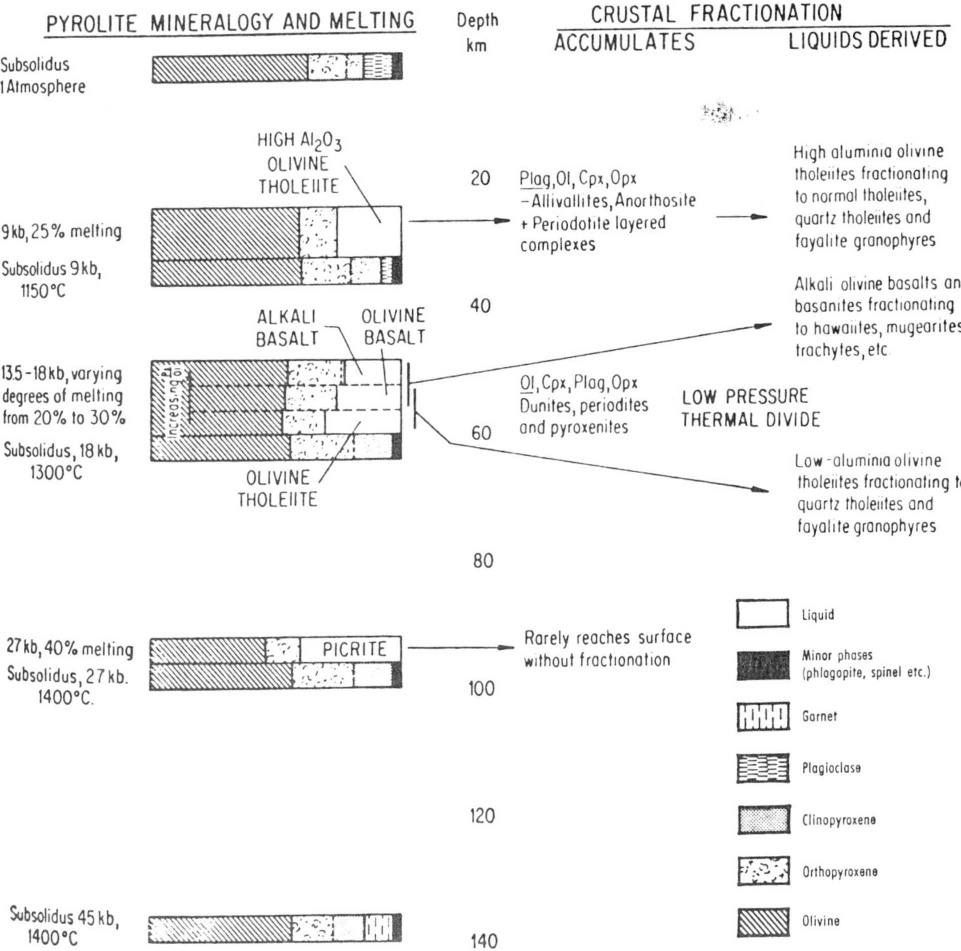

Fig. 13. Diagram illustrating the variation in the near-solidus mineralogy of pyrolite, the degree of partial melting and nature of both the liquid and the refractory residuum at various depths within the mantle

e) Depth of Magma Segregation 15—35 km

Results of experimental crystallization of the olivine tholeiite are given in Table 14. At 1,250° C, the separation of about 12 percent of olivine (Fo_{91-86}) and 3 percent of orthopyroxene (En_{87}, $Al_2O_3 = 5\%$) produced a residual liquid similar to high-alumina olivine tholeiite. Since the liquid is saturated with these phases under the stated conditions, it would be possible to increase these proportions to any desired amount without affecting liquid composition, e.g. 60 percent of olivine and 15 percent of enstatite ($5\% Al_2O_3$) could be added, giving a bulk composition essentially identical to pyrolite. It follows that fractional melting of 20—25% of pyrolite followed by magma segregation under these P, T conditions would yield a high alumina olivine tholeiitic liquid.

This liquid might thus be erupted directly to the surface after segregation or it might undergo crystal fractionation at a slightly higher level under closed system conditions leading to even greater enrichment in Al_2O_3 due to separation of olivine and pyroxenes as previously discussed.

f) Depth of Magma Segregation 35—70 km

The compositions of liquidus orthopyroxenes from the olivine tholeiite and olivine basalt are close to the compositions of natural orthopyroxenes of the olivine + aluminous orthopyroxenes + spinel mineral assemblage in peridotites. Previously GREEN and RINGWOOD (1964) have noted that near-liquidus olivine and orthopyroxene from the picrite composition at 13.5 kb closely match the compositions of olivine and orthopyroxene from the olivine + aluminous pyroxenes + spinel assemblage of lherzolite nodules. Olivine was not observed during the crystallization of olivine tholeiite or of the olivine basalt at 12—20 kb. However it was present on the liquidus of the picrite composition. It follows from a consideration of these three compositions that the olivine tholeiite and olivine basalt are almost saturated with olivine between 12—20 kb. We have already noted that they are saturated with highly aluminous orthopyroxene. Accordingly, a fairly large degree of fractional melting of pyrolite under these conditions would yield an olivine tholeiite magma closely resembling our chosen composition, but containing slightly more normative olivine. Such an olivine tholeiite would be in equilibrium with a refractory assemblage of olivine + aluminous orthopyroxene possessing Fe/Mg ratios identical to natural peridotite nodules.

The crystallization experiments upon picrite, olivine tholeiite and olivine basalt at 12—20 kb showed that these magmas would fractionate by means of the crystallization of substantial amounts of aluminous orthopyroxene ± aluminous clinopyroxene to the composition of alkali-olivine basalts. Thus we see that an alkali olivine basalt (with a slightly higher content of normative olivine than the one which we investigated), can be in equilibrium with olivine + aluminous enstatite ± aluminous clinopyroxene. This assemblage would be identical on cooling at pressure to the assemblage found in many lherzolite nodules which are commonly found in alkali olivine basalts and basanites (pages 181—185). These relationships show that olivine-rich alkali basalt may be formed by direct fractional melting of pyrolite.

Whether an olivine tholeiite or an alkali olivine basalt is formed by melting and magma segregation under these conditions depends simply upon the extent of partial melting. With a relatively small degree ($\sim20\%$) of partial melting of pyrolite, the residual crystals consist of olivine, abundant aluminous enstatite ± aluminous clinopyroxene, and the liquid has the composition of an olivine-rich alkali basalt. However, as the temperature is increased, the extent of partial melting increases, the clinopyroxene and then a large amount of aluminous orthopyroxene enter the liquid, changing its composition to an olivine tholeiite at about 30 percent partial melting. The situation is exactly the reverse of fractional crystallization of the olivine tholeiite.

The series of liquids developed by different degrees of partial melting at 35 to 70 km would be lower in SiO_2 content and in Al_2O_3 content than the series of liquids developed at 9 kb (30 km approx.) with similar degrees of melting of pyrolite. This is because the ratio of pyroxene/olivine, and particularly of ortho-pyroxene/olivine in the residual crystals is always higher at 13.5—18 kb than at 9 kb for a given degree of partial melting. In addition the residual pyroxene at 13.5—18 kb contains more Al_2O_3 in solid solution, resulting in lower Al_2O_3 content of the liquid fractions.

g) Depth of Magma Segregation Around 90 km

We have argued that the mineralogy of pyrolite at depths of around 90 km (P $\sim$ 27 kb) and temperatures near the basalt solidus or liquidus (1,420—1,520°C) will consist essentially of olivine + aluminous enstatite + aluminous clinopyro-xene. Garnet is absent from this assemblage but we have shown that garnet is a major phase in the subsolidus assemblages of the basaltic compositions at 27 kb and is the liquidus phase of at least two of the three basaltic compositions at 27 kb. These relationships demonstrate that the three basaltic compositions chosen are not compositions which may be derived by direct partial melting and magma segregation from pyrolite at 27 kb. Compositions which could be derived by partial melting of pyrolite at 27 kb should have olivine + orthopyroxene or olivine + clinopyroxene as liquidus phases or near-liquidus phases. The appearance of orthopyroxene near the liquidus of the picrite composition at 27 kb suggests that this composition is closer to being an appropriate partial melt at this pressure and re-inforces the previous arguments that liquids derived by partial melting and magma segregation at pressures near 30 kb (i.e. 100 km) would have the chemistry of picrites with over 30% normative olivine.

The deduced and observed nature and behaviour of the partial melt at 27 kb illustrates the important role which buffering by residual crystals may play in liquid fractionation. If a picritic liquid is segregated free from residual olivine + pyroxenes and fractionates essentially as a closed system then garnet and clino-pyroxene will be the major phases controlling fractionation. If, however, a picritic liquid does not segregate from residual crystals but begins to crystallize again then garnet and olivine are incompatible and will react principally to form aluminous enstatite + olivine. Garnet as a phase will have no role in either the fractionation sequence or in the solidus assemblage at 27 kb unless some segrega-tion of partial melt and residual crystals occurs.

[*Editors' Note:* Material has been omitted at this point.]

REFERENCES

[*Editors' Note:* Only the references cited in the preceding excerpts are reproduced here.]

BIRCH, F.: Elasticity and constitution of the earth's interior. J. Geophys. Research **57**, 227-286 (1952).

BOYD, F. R., J. L. ENGLAND, and B. T. C. DAVIS: Effects of pressure on the melting and polymorphism of enstatite, $MgSiO_3$. J. Geophys. Research **69**, 2101-2109 (1964).

CLARK, S. P., and A. E. RINGWOOD: Density distribution and constitution of the mantle. Rev. Geophys. **2**, 35–88 (1964).

ELSASSER, W. M.: Early history of the earth, chap. 1, p. 1–30. In: Earth science and meteoritics in honour of F. G. HOUTERMANS, edited by J. GEISS and E. GOLDBERG. Amsterdam: North Holland Publ. Co. 1963.

GREEN, D. H.: The origin of the "eclogites" from Salt Lake Crater, Hawaii. Earth and Planetary Sci. Letters **1**, 414–420 (1966a).

GREEN, D. H., J. EASTON, and A. E. RINGWOOD: Mineral assemblages in a model mantle composition: J. Geophys. Research **68**, 937–945 (1963).

GREEN, D. H., J. EASTON, and A. E. RINGWOOD: Fractionation of basalt magmas at high pressures. Nature **201**, 1276–1279 (1964)

GREEN, T. H., A. E. RINGWOOD, and A. MAJOR: Friction effects and pressure calibration in a pistoncylinder apparatus at high pressure and temperature. J. Geophys. Research **71**, 3589–3594 (1966).

HOLMES, A.: Contributions to the theory of magmatic cycles. Geol. Mag., **63**, 306–329.

KUNO, H.: High-alumina basalt. J. Petrology **1**, 121–145 (1960).

KUSHIRO, I., and H. KUNO: Origin of primary basalt magmas and classification of basaltic rocks. J. Petrology **4**, 75–89 (1963).

O'HARA, M. J.: Primary magmas and the origin of basalts. Scot. J. Geology **1**, 19–40 (1965).

REAY, A., and P. G. HARRIS: The partial fusion of peridotite. Bull. Volcanologique, **27**, 115–127 (1964).

RINGWOOD, A. E.: A model for the upper mantle. J. Geophys. Research **67**, 856–867 (1962a).

RINGWOOD, A. E.: A model for the upper mantle, 2. J. Geophys. Research **67**, 4473–4477 (1962b).

RINGWOOD, A. E.: The chemical composition and origin of the earth. In: Advances in earth science (P. M. HURLEY, ed.), p. 287–356. Boston, U.S.A.: M.I.T. Press 1966a.

RINGWOOD, A. E.: The mineralogy of the mantle. In: Advances in earth science (P. M. HURLEY, ed.), p. 357–417. Boston, U.S.A.: M.I.T. Press 1966b.

RINGWOOD, A. E., and D. H. GREEN: An experimental investigation of the gabbro-eclogite transformation and some geophysical implications. Tectonophysics **3**, 383–427 (1966).

RINGWOOD, A. E., I. D. MACGREGOR, and F. R. BOYD: Petrological constitution of the upper mantle. Carnegie Inst. Wash. Year Bk. **63**, 147–152 (1964).

RUNCORN, S. K.: Towards a theory of continental drift. Nature, **193**, 311–314 (1962).

VERHOOGEN, J.: Petrological evidence on temperature distribution in the mantle of the earth. Trans. Am. Geophys. Union **35**, 85–92 (1954).

VENING MEINESZ, F. A.: Convection currents in the earth and the origin of the continents. Koninkl. Ned. Akad. Wetensch., ser. B, **55**, 527–533 (1952).

VENING MEINESZ, F. A.: Thermal convection in the earth's mantle. International Geophysical Series, **3**, 145–176 (1962).

YODER, H. S., and C. E. TILLEY: Origin of basalts magmas: an experimental study of natural and synthetic rock systems. J. Petrology **3**, 342–532 (1962).

24

Reprinted from pages 67, 88, 99–115, 116, and 117 of *Scottish Jour. Geology*
3:67–117 (1967)

Formation and fractionation of basic magmas at high pressures

M. J. O'HARA and H. S. YODER, Jr.
Geophysical Laboratory, Carnegie Institution of Washington, D.C.

SYNOPSIS

Experiments were conducted at 30 kilobars on dry synthetic mixtures in the join diopside ($CaMgSi_2O_6$)-pyrope ($Mg_3Al_2Si_3O_{12}$), on a small number of compositions in adjacent parts of the plane $CaSiO_3$-$MgSiO_3$-Al_2O_3 and one composition in the plane diopside-pyrope-forsterite. In addition, experiments were made at the same pressure on the joins chrome diopside–enstatite, chrome diopside-pyrope, and chrome diopside-pyrope-olivine using the separated analysed minerals of a natural garnet-peridotite nodule in kimberlite, and the join garnet-omphacite using the separated analysed minerals of an eclogite nodule in kimberlite. A few experiments were made on the whole-rock sample of some eclogite nodules in kimberlite. The results have an important bearing on the range of composition of clinopyroxene solid solutions, the conditions of equilibration and original formation of eclogite and kimberlite, and the existence of an orthopyroxene-liquid reaction relationship in olivine and hypersthene-normative liquids at 30 kilobars pressure. They lead to the construction of a new fractionation scheme at high pressure in which a hypersthene-normative picrite or olivine tholeiite magma is the primary liquid, formed by partial melting of a garnet-peridotite mantle. Precipitation of garnets and clinopyroxenes (eclogite accumulates) leads to the development of alkaline and silica-poor mafic magmas from this primary liquid.

[*Editors' Note:* Material has been omitted at this point.]

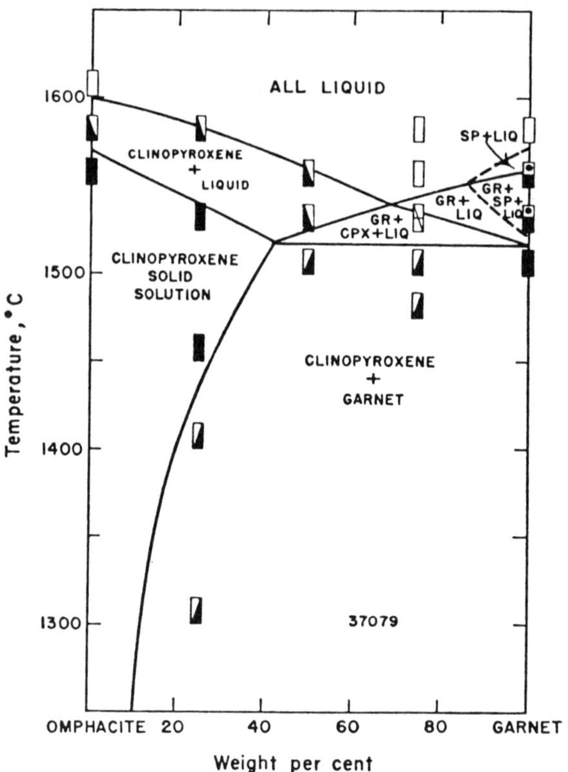

Fig. 6. Interpretation of results of experiments at 30 kilobars on mixtures of the omphacite and garnet from an eclogite, specimen 37079, from kimberlite.

EXPERIMENTS AT 30 KILOBARS ON NATURAL ECLOGITES
USING THE WHOLE-ROCK SAMPLE

A few runs have been made on the whole-rock sample of three eclogites from the Williams collection, made available by courtesy of W. Berg, University of Capetown. These samples include a kyanite-eclogite from the Dodoma mine, Tanzania, specimen No. TAN 503, originally described and figured by Williams (1932, pp. 333–5 and plates 84, 85 and 126) as a sillimanite eclogite. The aluminium-silicate phase in the original rock is kyanite as reported by Tilley (1937, p. 567) and kyanite is abundant in the sample supplied by W. Berg. The remarkable oriented intergrowth of kyanite in clinopyroxene in this rock suggests that much of the kyanite may have exsolved from a clinopyroxene solid solution stable at high pressures and high temperatures. This view is strengthened by the observation that nearly all the kyanite in this sample disappears, leaving garnet and clinopyroxene as the major crystalline phases at the beginning of melting (although it is conceivable, if unlikely, that the small amount of liquid present at 1525°C is excessively rich in Al_2O_3 and SiO_2). This sample passes from being predominantly crystalline at 1525°C to being virtually all liquid at 1550°C.

A second specimen from the Dodoma mine, Tanzania, TAN 501, is a very fresh bimineralic eclogite (Williams 1932, plates 82 and 83). It begins to melt at 1515°C±15°C and is all melted by 1540°C±15°C, the liquidus phase being clinopyroxene. Garnet persists as a separate phase at temperatures up to and presumably a little beyond the beginning of melting. Williams (1932, p. 333) gives analyses of both TAN 501 and TAN 503; the high Fe_2O_3, low FeO and absence of Na_2O reported for TAN 501 cast some doubts on this analysis. It is possible that TAN 501 and 37079 (above) are parts of the same specimen originally collected by Du Toit.

The third specimen is the corundum and spinel-bearing eclogite from the Roberts Victor mine, South Africa, specimen No. RV 372 (Williams 1932, plates 118 and 131). The proportion of alteration products in this sample is higher than in the two previous samples. Glass appears in this material in small amounts at temperatures as low as 1400°C but the charge is still predominantly crystalline at 1500°C. Garnet and abundant clinopyroxene are the only crystalline phases positively identified at this temperature. By 1550°C the charge is virtually all melted, clinopyroxene being the liquidus phase. It is not clear whether an unaltered rock of the same original composition begins to melt at a low temperature ($\sim$1400°C) or whether the glass observed in the runs is due largely to the presence of alteration products which may be rich in alkalis and water.

Yoder and Tilley (1962, pp. 500-1) have described experiments on natural eclogites from Glenelg and Loch Duich, Scotland. These eclogites contain some combination of quartz, amphibole, mica, rutile, iron ore, plagioclase, scapolite, sphene, epidote, pyrrhotite and apatite, all in small amounts, and the major minerals have a higher Fe/Mg ratio than is likely to be found in the minerals of most eclogite nodules from kimberlite. The run duration used in Yoder and Tilley's experiments was relatively long and may have produced a lowering of the apparent temperature at which beginning of melting occurs (Boyd and England 1963). A correction for friction was used in reporting the pressures of these runs. Although these results are not, therefore, directly comparable with those listed above, there seems little doubt that quartz-bearing eclogites will melt at temperatures of 1350–1450°C at 30 kilobars pressure.

Thermal divide at high pressure

Assemblages of garnet solid solution and clinopyroxene solid solution, with or without orthopyroxene, from the system $CaO-MgO-Al_2O_3-SiO_2$ simulate natural hypersthene eclogites and bimineralic eclogites respectively. These synthetic garnet-pyroxene assemblages define a plane of compositions within the tetrahedron $CaO-MgO-Al_2O_3-SiO_2$ (*see* Fig. 1). Forsterite-bearing assemblages simulating garnet-peridotites and spinel and corundum-bearing assemblages simulating spinel-corundum eclogite have bulk compositions on the SiO_2-poor side of this garnet-pyroxene plane. Quartz-bearing assemblages have bulk compositions on the SiO_2-rich side of this plane.

If the crystalline phases stable at the liquidus for compositions in the garnet-clinopyroxene-orthopyroxene plane are garnet and pyroxenes only (that is there is no field of forsterite, spinel or quartz on the liquidus in this composition plane) then:

(i) Partial melting of forsterite, spinel or quartz-bearing assemblages containing garnet and pyroxene must begin at temperatures below those at which their garnet-pyroxene sub-assemblage would begin to melt.

(ii) The liquid so produced by the partial melting of a forsterite or spinel-bearing assemblage must lie to the silica-poor side of the garnet-pyroxene plane. The liquid produced by the partial melting of a quartz-bearing assemblage must lie to the silica-rich side of the garnet-pyroxene plane.

(iii) Neither the fractional nor the equilibrium crystallisation of parental liquids lying on the silica-poor side of the plane can yield residual liquids lying on the silica-rich side of the plane, and vice versa. The plane is in fact a thermal divide.

Although no field of forsterite or quartz has been found on the liquidus in the synthetic system at 30 kilobars a field of spinel has. This potentially allows parental liquids on the silica-poor side of the composition plane to yield residual liquids on the silica-rich side during fractional crystallisation. However, the spinel field does not abut against a clinopyroxene or olivine field on the liquidus in this composition plane. It follows that the plane of garnet+clinopyroxene solid solution+orthopyroxene solid solution in the synthetic system at 30 kilobars acts as a thermal divide *provided that* the crystalline phases present include clinopyroxene (plus, for example, garnet). Provided clinopyroxene is present, the three conclusions listed above must hold. Penetration of the thermal divide at 30 kilobars is possible only for liquids which are crystallising spinel, and which cannot be crystallising clinopyroxene or forsterite. No such penetration is possible at 40 kilobars (Davis and Schairer 1965).

The above mentioned results of experiments on natural materials reveal an analogous situation to that inferred in the synthetic system. Olivine-bearing, spinel-bearing and quartz-bearing assemblages begin to melt at lower temperatures than the pure garnet-pyroxene assemblages.

In the more complex natural system a spinel appears in the presence of clino-pyroxene, garnet and liquid (but not olivine) in the experiments with mixtures of minerals from the specimen A.3/10596. The spinel is believed to be a chrome-spinel and its appearance is believed to be a consequence of the relatively high Cr_2O_3 content of the charges. Systems with lesser amounts of Cr_2O_3 such as the garnet-pyroxene pair from the eclogite 37079 do not develop spinel in the presence of clinopyroxene.

It is argued above that the liquid produced by partial melting of the natural garnet-lherzolite assemblage at 30 kilobars is silica-poor with respect to the garnet-pyroxene assemblages, despite the presence of a small amount of spinel. The fractionation of this liquid by the precipitation of clinopyroxene, garnet and spinel would undoubtedly lead to residual liquids lying to the silica-rich side of the garnet-pyroxene 'plane' because in the natural system there must be a piercing point in the clinopyroxene-garnet composition range for liquids in equilibrium with clinopyroxene, garnet and spinel.

However, spinel is unlikely to precipitate in this way and hence the residual liquids are unlikely to pass to the silica-rich side of the garnet-clinopyroxene 'plane' for the following reason. Spinel, orthopyroxene, and liquid appear together during the partial melting of the natural subsolidus assemblage olivine+clinopyroxene+garnet. Spinel as well as orthopyroxene must be in reaction relationship with this liquid, and therefore spinel, like orthopyroxene, will not precipitate from that liquid when it is subjected to fractional crystallisation.

FRACTIONATION AT HIGH PRESSURES

What then is the chemistry of this high pressure thermal divide? In an attempt to answer this question the norms of various garnet–clinopyroxene assemblages have been studied. The compositions of possible garnet–clinopyroxene mixtures have been calculated from the analysed minerals at intervals of 10 per cent by weight of the end members. The C.I.P.W. norms of these mixtures have been calculated using a computer programme written by F. Chayes of the Geophysical Laboratory. The results of one such calculation are presented in full (Table 3), this being the garnet-pyroxene assemblage from the eclogite nodule in kimberlite, specimen No. 37079. Partial results for the mineral pair from the garnet-lherzolite A.3/10596 are given in Table 5. The variation in amount of the C.I.P.W. norma-tive constituents with composition in six garnet-clinopyroxene pairs from eclogite facies rocks ranging from kyanite-eclogite to garnet-peridotite is represented in Fig. 9. Normative hypersthene, a tholeiitic characteristic, is associated with richness in garnet whereas normative nepheline, an alkali basalt characteristic, is associated with richness in clinopyroxene (Yoder and Tilley 1962, p. 506). Normative olivine culminates in the intermediate mixtures of garnet and pyroxene, and all of the mixtures contain normative olivine. Normative nepheline is restricted to the mixtures rich in clinopyroxene, except in those garnet-pyroxene

137

pairs such as Weissenstein and 37079 (eclogite in kimberlite) which are furthest removed in composition from the garnet-pyroxene pair which co-exists with orthopyroxene and olivine (*see* Fig. 2). Any mixture which is silica-poor with respect to the garnet–clinopyroxene assemblages must be relatively rich in normative olivine (~15–30 per cent by inspection of Fig. 9), but may contain either hypersthene or nepheline in the norm, depending upon the ratio of clinopyroxene to garnet molecules present in the liquid, and the extent of silica undersaturation with respect to the garnet–clinopyroxene assemblage.

Therefore the initial partial melting product of the garnet-two pyroxenes-olivine assemblage at high pressure is silica-undersaturated in the C.I.P.W. norm, and may be picritic in character. Bearing in mind the ratio of garnet to clinopyroxene molecules in the liquids formed by initial partial melting of garnet-pyroxene mixtures (compare Figs. 4, 6 and 7 with Figs. 5 and 9) and the appearance of some spinel in the systems constructed from natural minerals, it is probable that the initial partial melting product of garnet-lherzolite at 30 kilobars has the composition of a *hypersthene-normative picrite basalt*. Such a liquid could yield quartz-normative residual liquids (tholeiitic basalts) by fractionation of olivine at lower pressures. At high pressures the fractionation of this liquid will produce quite different geochemical changes, which will be discussed below.

It is suggested that the mineral assemblage of the garnet-peridotite and eclogite inclusions in kimberlite may represent equilibrium at high pressures and relatively low temperatures. Extensive solid solutions of potential kyanite molecule, garnet molecule and orthopyroxene molecule in clinopyroxene and of potential garnet in orthopyroxene occur as the temperature approaches that at which garnet-peridotite begins to melt at 30 kilobars, outlining a high-temperature subfacies of the eclogite mineral facies. Recent data (Boyd and England 1964; MacGregor and Ringwood 1964) indicate that the alumina-rich pyroxenes may, however, be restricted to pressures below 50 kilobars. The high-temperature eclogite assemblages only partly exsolved are found in nature among nodules in alkali basalt tuff from Hawaii, but textures implying the exsolution of garnet and kyanite from clinopyroxenes are also observed in eclogites from kimberlite, and indicate that these eclogites have been at high temperatures, perhaps high enough to have co-existed with a liquid in the dry system, i.e. partial melting and subsequent crystallisation of the liquid are processes which may have operated on some of the eclogite facies inclusions in kimberlite. The partial melting of garnet-peridotite at 30 kilobars occurs at a lower temperature than the partial melting of eclogite, and yields a liquid that is silica-poor with respect to the garnet-pyroxene assemblages, which constitute a thermal divide preventing the derivation of silica-enriched liquids by fractionation of garnet and pyroxene with or without olivine at this pressure. Partial melting of bimineralic eclogite is judged to be subordinate to the partial melting of garnet-peridotite as a source of basic magma. Furthermore, because of their higher melting temperatures the bimineralic eclogite inclusions in kimberlite manifestly cannot represent the compositions of the liquids formed by partial melting of garnet-peridotite nor can they represent the compositions of residual liquids formed by the fractional crystallisation of such liquids.

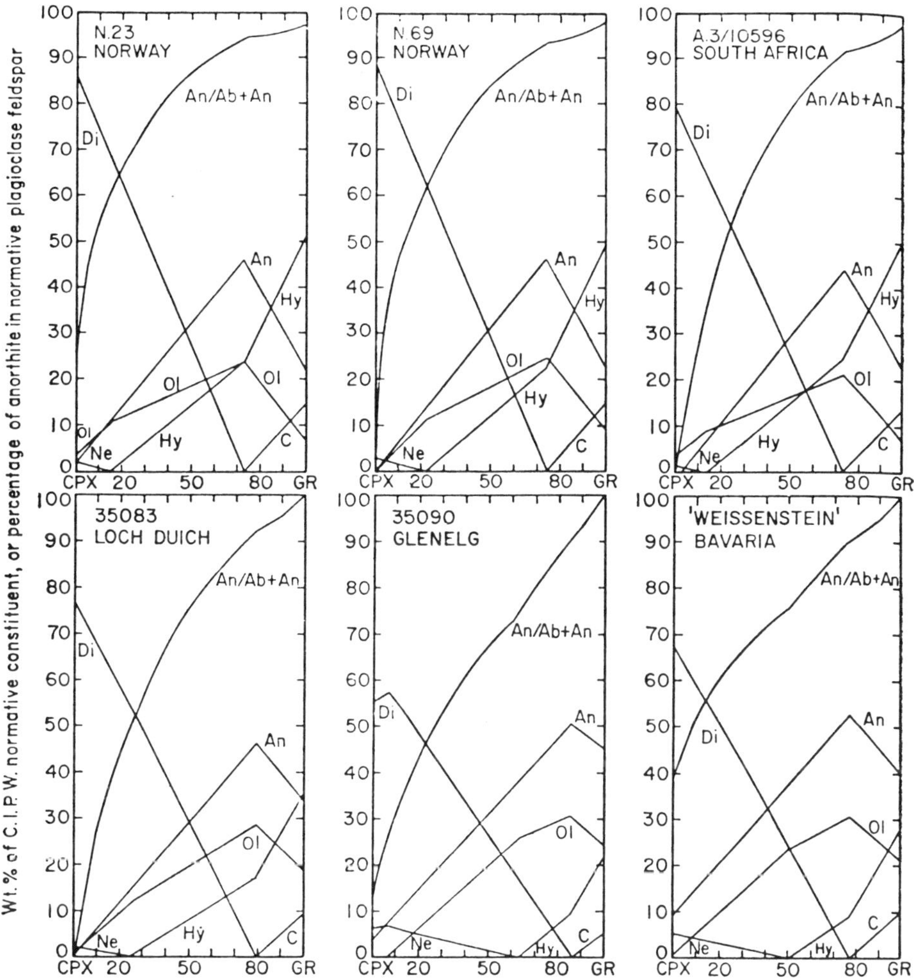

FIG. 9. C.I.P.W. norms calculated for mixtures of analysed co-existing garnets and clinopyroxenes from eclogite facies rocks. N.23, N.69, A.3/10596 are from orthopyroxene-bearing assemblages, the latter two containing olivine as well (O'Hara and Mercy 1963). The Glenelg and Loch Duich mineral pairs come from eclogites which contain a little quartz, but neither kyanite nor orthopyroxene. The 'Weissenstein' pair comes from a quartz-kyanite eclogite (Yoder and Tilley 1962).

The observation that the majority of the eclogites have higher FeO/MgO ratios and higher Na_2O/CaO ratios than the garnet–clinopyroxene pairs of the garnet-peridotite eliminates the possibility that the bimineralic eclogites represent crystalline residuals left behind during the partial melting of garnet-peridotite.

(The experimental evidence that garnet is the first and clinopyroxene probably the second crystalline phase to disappear during the partial melting of garnet-peridotite indicates the same conclusion.)   The same geochemical observation precludes garnet-peridotite as the partial melting product of eclogite, and again the experimental results support this conclusion—the melting products of eclogite are liquids which crystallise to further eclogites slightly richer in CaO and Al_2O_3 than the original. The bimodal character of the chemistry of the garnet-peridotites and eclogites and the scarcity of gradational types (O'Hara and Mercy 1963, Fig. 26) combined with the distinctive change in composition of clinopyroxene and garnet proves that the eclogites are not merely local segregations of the garnet and clinopyroxene present in the garnet-peridotite.

One physico–chemical mechanism appears to satisfy the necessary conditions of relationship between eclogite and garnet-peridotite. Bimineralic eclogite in kimberlite *is most likely a crystal accumulate formed during the high pressure fractionation of the liquid generated by partial melting of four-phase garnet-peridotite*. This hypothesis is consistent with the scarcity of eclogite relative to peridotite in the kimberlite pipes, absence of chemical gradation from peridotite to eclogite, and the higher beginning of melting temperature of eclogite relative to garnet-peridotite.

A four-phase rock such as garnet-peridotite should in general yield on partial melting a liquid which would at equilibrium crystallise completely to another four-phase rock. Only coincidence would place this liquid composition in the composition plane of the bimineralic eclogites themselves, and it was demonstrated that the liquid formed by partial melting of garnet-peridotite does not lie in this plane. Two-phase accumulates (bimineralic eclogites) might precipitate during the fractionation of a liquid formed by partial melting of a garnet-peridotite while the residual liquid composition is changing towards that at which additional crystalline phases might precipitate. This is the most likely interpretation consistent with the data which does not involve supposing either that there are two separate and wholly distinct origins for the eclogite and garnet-peridotite in kimberlite, or that there are physico–chemical processes operating which are outside present knowledge.

The experimental results also demonstrate that orthopyroxene has a reaction relationship to the liquid which forms by partial melting of garnet-peridotite. This is essential if orthopyroxene is not to precipitate during the subsequent fractionation of that liquid, and the proposed interpretation of bimineralic eclogite nodules in kimberlite as crystal accumulates from the partial melting product of garnet-peridotite requires the existence of this experimentally determined reaction relationship. This interpretation of the eclogites also requires that olivine should not precipitate from the partial melting product of garnet-peridotite, implying either that olivine also has a reaction relationship to the liquid at the beginning of melting, or that olivine is resorbed in the equilibrium olivine+clinopyroxene+ garnet+liquid on cooling. The location of the liquid relative to the tetrahedron of the four crystalline phases makes it unlikely that olivine is in reaction relationship with the liquid at the beginning of melting, but resorption of olivine in the equilibrium olivine + clinopyroxene + garnet + liquid may take place if the liquid composition lies on that side of the composition plane spinel–clinopyroxene–garnet which is remote from the olivine composition (*see* Fig. 1). In this case olivine must fail to precipitate during fractional crystallisation of such liquids.

The existence or non-existence of this required relationship between olivine and liquid is not elucidated by the above results. There is nevertheless some justification for an examination of the geological consequences of *assuming* that the eclogite nodules in kimberlites are crystal accumulates from the picritic liquids formed by the partial melting of garnet-peridotite.

Davis (1964) has suggested an alternative solution to the problem of obtaining garnet-clinopyroxene accumulates from the partial melting products of garnet-peridotite. Davis suggests that the locus of liquids which co-exist with forsterite, garnet and clinopyroxene lies within the composition volume which in the subsolidus is represented by garnet and a clinopyroxene which contains a small amount of forsterite in solid solution. Davis suggests that such a liquid will be of basaltic composition and will fractionate by eclogite precipitation. It must be noted, however, that such a liquid corresponds to no ordinary tholeiitic basalt because its C.I.P.W. norm (Fig. 5) contains at least 15 per cent olivine. Furthermore, the relationship suggested by Davis requires the liquid compositions in equilibrium with forsterite, garnet and clinopyroxene to lie in the garnet-pyroxene composition plane throughout. Liquids of the type which Davis suggests can only fail to precipitate forsterite during the fractional crystallisation if the liquids lie to the forsterite-poor, i.e. silica-rich, side of the garnet-pyroxene thermal divide plane with the residual liquids trending towards silica-rich compositions where quartz-eclogite would precipitate. The temperatures of beginning of melting of natural eclogites and synthetic garnet-clinopyroxene assemblages in the plane $CaSiO_3$-$MgSiO_3$-Al_2O_3 would then have to be lower than the beginning of melting of these same assemblages in the presence of olivine or forsterite. Results presented in this paper show the reverse to be true at 30 kilobars. Furthermore, such an interpretation is not in accordance with the geological circumstances found in kimberlite, where quartz-eclogites are unknown and there is nothing to suggest the development of silica-enriched residual liquids.

Alternatively the liquids postulated by Davis must fractionate by precipitation of garnet, clinopyroxene and small amounts of forsterite, in which case the liquids cannot lie within the composition volume which will be represented in the subsolidus by garnet and clinopyroxene solid solution only as required by Davis

but must lie to the forsterite-rich side of that composition volume, as postulated in this paper. Thus the situation postulated by Davis (1964) cannot explain the production of a suite of bimineralic olivine-free eclogites.

Even if all these difficulties could be circumvented, and the bimineralic eclogites in kimberlite be explained as crystal accumulates from a liquid, olivine being lost in the manner postulated by Davis (1964), the hypothesis would still fail to explain the observation that bimineralic eclogites from kimberlite begin to melt at significantly higher temperatures than those at which garnet-peridotites begin to melt.

GEOCHEMICAL EFFECTS OF ECLOGITE FRACTIONATION

Two aspects of the geochemistry of eclogite inclusions in kimberlite are examined in Fig. 10. A trend of compositions similar to that commented on by Yoder and Tilley (1962, p. 505 and Fig. 36) is revealed. The principal variation between eclogite samples is in the ratio $(MgO+FeO)/(CaO+Al_2O_3)$: neither MgO/FeO, nor CaO/Al_2O_3 (exclusive of Al_2O_3 in jadeite molecule) vary greatly. The amount of 'jadeite' molecule present in the eclogites of high $CaO+Al_2O_3$ content and of lowest MgO/FeO ratio is greater than in those with low $CaO+Al_2O_3$ content. It is suggested that those eclogites which are rich in $CaO+Al_2O_3$, rich in jadeite molecule and of lower MgO/FeO ratio have precipitated from the liquid at lower temperatures (i.e. at a later stage of fractionation) than those eclogites rich in $MgO+FeO$, this being the observed behaviour of these oxides in crystal-liquid relationships at low pressures. This is in agreement with the interpretation of experimental results for the system $CaO-MgO-Al_2O_3-SiO_2$ and the natural garnet-pyroxene pairs which also indicate that with falling temperature the garnet in garnet-pyroxene accumulates should become more calcic and the clinopyroxenes richer in CaO and Al_2O_3/MgO. As already noted, the trend shown in Fig. 10 may mark the change of bulk composition of the accumulates precipitating from a liquid lying on the garnet+ clinopyroxene+liquid ' surface ', but at some point on the silica-poor side of the garnet-pyroxene thermal divide and the olivine-poor side of the clinopyroxene-garnet-spinel plane. This trend of accumulate compositions lies very close in the projection to the trend of the minimum on the liquidus in the garnet-pyroxene plane (Fig. 4) but the two should not be confused. A comparable projection of the compositions of Hawaiian basalts (Fig. 11) reveals an almost identical trend. In contrast to the eclogites in kimberlite, the Hawaiian basalt compositions mostly represent liquids existing at low pressures, not crystal accumulates formed at high pressures. Furthermore, most of the basalts are richer in SiO_2 than the eclogites, a feature which is obscured by the nature of the projection, and their compositional variation is probably controlled largely by the fractionation of olivine at lower pressures and not of garnet and pyroxene at high pressures (Tilley *et al.* 1964, 1965).

The fractionation of tholeiitic magmas at low pressures is controlled by the early separation of olivine and a chrome-spinel phase followed in the middle stages by separation of clinopyroxene (and orthopyroxene), magnetite and plagioclase, and at a later stage by the separation of alkali feldspar. The geochemical features of the change in composition of the residual liquids at low pressure have been established by Nockolds and Allen (1954, 1956) and by Wager (1960) and by Wager and Mitchell (1956). These geochemical features are largely controlled by three factors: the original liquid composition, the nature and relative amounts of the solid phases crystallising, and the distribution ratios of various elements between the crystalline phases and the liquid. None of these factors will remain the same when considering fractionation at high pressures.

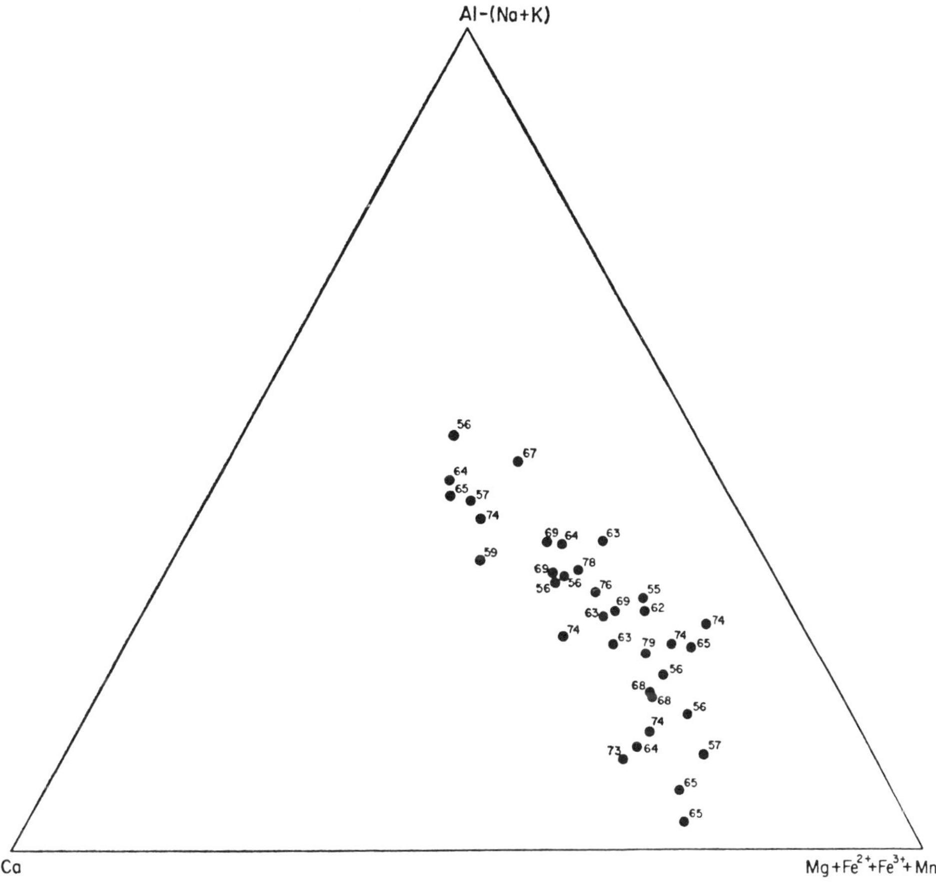

FIG. 10. Atomic per cent plot of the compositions of eclogites, kyanite eclogites, and grosspydite rocks from kimberlite pipes in South Africa (Williams 1932; Holmes 1936) and Siberia (Sobolev ed. 1959). The figures written beside the points indicate the ratio Mg/Mg + total Fe for the rocks.

The early decrease in ratios such as Cr/Al, Ni/Mg, Co/Mg and Mg/Fe observed in low-pressure basalt fractionation sequences is due to the early separation of olivine and chrome-spinel. Their continued but less dramatic decrease during the middle stages of fractionation, and the accompanying decrease in the ratio V/Al is largely due to the crystallisation of magnetite and to a lesser extent clino-pyroxene. The ratios Sr/Ca, Ba/Ca, Rb/Ca, Na/Ca, K/Ca all increase during the middle stages of low-pressure fractionation because of the preferential entry of

Ca into plagioclase feldspar, which does, however, accommodate small amounts of the elements Sr, Ba, Rb, K. All these ratios may fall in the late stage of fractionation where an independent alkali feldspar starts to crystallise.

As a result of these features, primitive or parental basalt liquids have associated with them relatively high contents of Mg, Cr, Ni, Co and V, while their residual liquids are poor in these elements, and richer in Na, K, Rb, Ba and Sr.

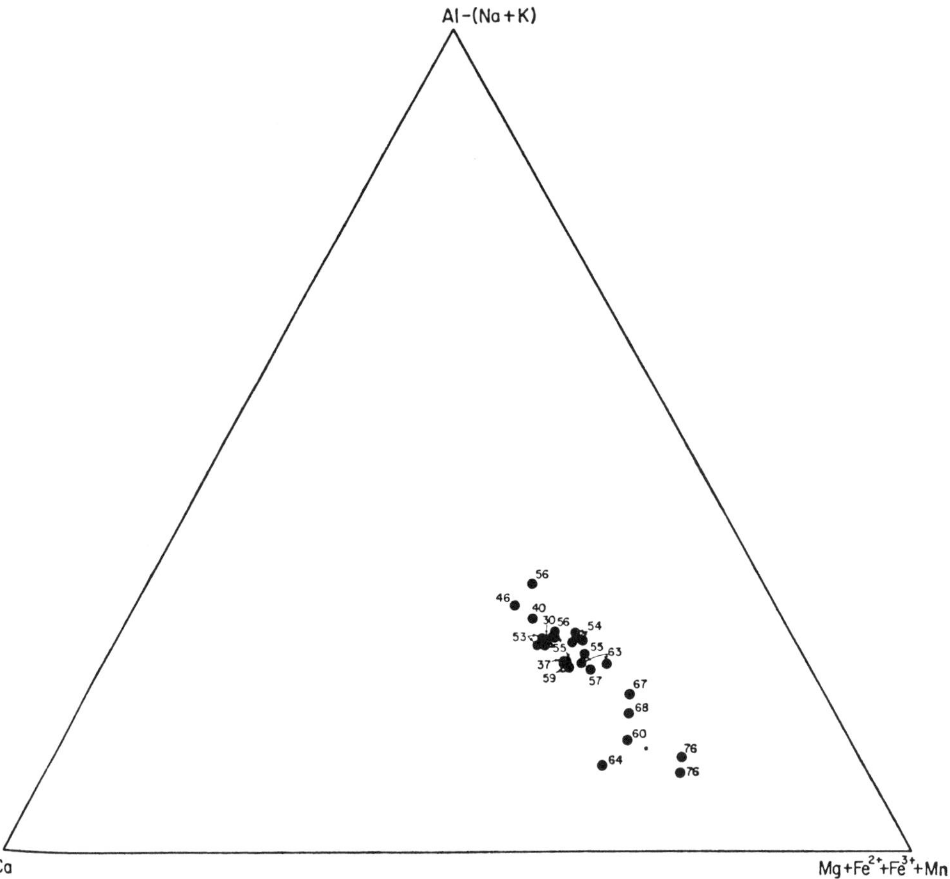

Fig. 11. Atomic per cent plot of the compositions of Hawaiian basalts (Yoder and Tilley 1962, Table 2; Muir and Tilley 1963, Table 10). The figures written beside the points indicate the ratio Mg/Mg+total Fe for the rocks.

The change in the SiO_2 content of the liquids at low pressure is again controlled largely by the early fractionation of two silica-poor minerals, olivine and spinel, which cause an enrichment of SiO_2 in the residual liquid. The low confining pressures accompanying this type of fractionation inhibit a strong concentration of volatile constituents in the residual liquids.

How different the influence of eclogite fractionation at high pressures will be can be illustrated by considering several specific aspects as affecting a hypersthene-normative picrite magma. These are its effects on the SiO_2 content, the volatile content, the Cr/Al, Ni/Mg, Co/Mg, V/Al and Mg/Fe ratios, the Sr/Ca, Ba/Ca, Rb/Ca, K/Ca and Na/Ca ratios and the Ti and P content.

The partial melting product of garnet-peridotite at high pressures (hypersthene-normative picrite) is already silica-poor in composition with respect to garnet-clinopyroxene mixtures and the fractionation of eclogite from this liquid can only decrease the proportion of SiO_2 to the sum of $CaO + Al_2O_3 + MgO + FeO +$ alkalis in the residual liquids. Abundant olivine, nepheline, dicalcium silicate, leucite and kalsilite might be expected to occur in the C.I.P.W. norms of these residual liquids.

Eclogite in kimberlite is devoid of primary hydrous phases. Any volatile constituents (H_2O, CO_2) present in the peridotite will be concentrated in the partial melt liquid when it is formed. The fractionation of dry eclogite from this liquid can only increase the amount of volatile constituents in the residual liquid. The residual liquids of high-pressure fractionation may come to contain large quantities of volatiles without the formation of a free vapour phase.

TABLE 9

Relative values of some geochemical ratios for co-existing minerals

	Olivine	Spinel	Magnetite	Clinopyroxene	Garnet
Cr/Al	—	90	16 000	6	1
V/Al	—	110	600 000	10	1
Ni/Mg	10	300	3 000	3	1
Ni/Fe^{2+}	17	20	8	13	1
Co/Mg	5	10	100	1	3
Co/Fe^{2+}	1	5	2	1	1

Figures given in Tables 9 and 10 were obtained from a consideration of co-existing clinopyroxene, spinel and magnetite (Howie 1955); clinopyroxene, garnet and plagioclase (O'Hara 1961); clinopyroxene, garnet and olivine (O'Hara and Mercy 1963); and clinopyroxene, plagioclase and potash feldspar (Howie 1955). The ratios cannot be determined in some cases because of the excessively small amounts of both elements present in the minerals.

In Table 9 the probable relative values of the ratios Cr/Al, V/Al, Ni/Mg, Co/Mg, Mg/Fe in co-existing olivines, spinels, garnets and pyroxenes are compared. No data exist for rocks containing all of these minerals, but it is possible to piece together the required information from various partial assemblages, e.g. olivine + garnet + pyroxene, olivine + pyroxene + spinel, olivine + pyroxene + magnetite. Although it is not known what are the appropriate figures for these ratios in a liquid co-existing with these solid phases, the differences are sufficiently large that it can be seen that the fractionation of garnet and clinopyroxene will not reduce the Cr/Al, V/Al, Ni/Mg, Co/Mg and Mg/Fe ratios to the extent produced by separation of an equivalent proportion of olivine + chrome spinel. Consequently relatively high concentrations of Cr, V, Ni, Co and Mg will not be restricted to the unfractionated or little fractionated liquids at high pressures, and the concentrations of these elements might even increase in the residual liquids during fractionation.

Table 10 contains figures for the relative values of the ratios Sr/Ca, Ba/Ca, Rb/Ca, K/Ca and Na/Ca in garnet, omphacite, pyroxene, augite and feldspar. (See subsequent remarks on the work of Harris and Rowell (1960) which is at variance with the result for Na/Ca presented here.) These were obtained in the same fashion as those of Table 9, and are used in the same way to indicate that the fractionation of eclogite will cause all of these ratios to increase by a larger factor in the residual liquids than would the separation of equivalent proportions of augite and feldspar. Consequently relatively high concentrations of Sr, Ba, Rb, K and Na will build up in the liquids fractionated at high pressure.

TABLE 10

Relative values of some geochemical ratios for co-existing minerals

	Plagioclase	Potash feldspar	Clinopyroxene	Garnet
Na/Ca	6–15	>50	~1	—
K/Ca	1	300	very small	—
Rb/Ca	4	50 000	<2	<1
Sr/Ca	1 000	20 000	1	1
Ba/Ca	40	250 000	10	<1

The crystallisation of apatite and ulvöspinel at an early stage from basalt magmas removes Ti and P in low-pressure fractionation. Neither of these phases are common as accessory constituents of the eclogites in kimberlite, and the garnet and clinopyroxene of the eclogites contain only small amounts of these elements. Both Ti and P would be expected to reach higher concentrations in liquids fractionated at high pressure than in low-pressure basalt fractionation sequences.

The residual liquid products of eclogite fractionation at high pressures might, therefore, be expected to be represented at the earth's surface as relatively scarce rock types poor or very poor in silica, and marked by either violently explosive eruption, or emplacement as magmas very rich in volatile constituents. These magmas should be characterised by the combination of high Cr, V, Ni, Co and Mg concentrations with high concentrations of Sr, Ba, Rb, K, Na, Ti and P, plus high K/Na ratios (because eclogite has a very high Na/K ratio).

There are three rock types which satisfy all the requirements of fluids which are the products of extensive fractionation at high pressures. These are the potassic mafic lavas (Higazy 1954), the groundmass of kimberlite (Dawson 1962) (that is kimberlite less its xenolithic and xenocrystal materials) and the carbonatites and alnöites with which must be included their fenitised country rocks. The intimate association of xenoliths of garnet-peridotite (mantle material and source rock) and eclogite (crystal accumulates) in kimberlite (residual liquid of eclogite fractionation plus xenolithic materials) in an environment which strongly suggests emplacement by the violent release of vapour from a volatile-rich magma at great depth, must be regarded as significant.

Harris and Rowell (1960, pp. 2454–5) proposed that the eclogite nodules in kimberlite are crystal accumulates from a *kimberlite* magma, and they suggested that this kimberlite magma formed originally at some greater depth than that at which it precipitated eclogite and incorporated garnet-peridotite xenoliths torn from the side wall or roof of the magma chamber. This hypothesis, like that of Holmes and Harwood (1932), differs from that now proposed in its choice of the initial liquid and it provides no origin for kimberlite magma itself.

TABLE 11

Minor element contents (as parts per million) of alleged parental magmas of tholeiitic and alkali basalt extrusive provinces, compared with average minor element contents from potassic mafic lavas of central Africa

	Tholeiitic basalt	Alkali basalt	Potassic mafic lava
Cr	400	400	380
Ni	170	190	110
Co	55	55	60
V	270	230	300
Na	15 000	28 000	16 280
K	4 000	9 000	28 500
Ti	9 000	18 000	23 620
P	900	3 170	3 290
Rb	10	70	190
Sr	230	1 300	5 690
Ba	100	740	2 990
Zr	115	170	825

Tholeiitic basalt figures are the average of Nockolds and Allen 1956, Table 32, columns 1, 2 and 4.

Alkali basalt figures are the average of Nockolds and Allen 1954, Table 11 (analyses), 2–4; Table 12, 4; Table 13, 1–7; and Table 14, 4 and 5.

Potassic mafic lava figures are the average of Higazy 1954, Table 2, column 7, and Table 4, column 7.

Harris and Rowell infer that the fractionation of eclogite from a kimberlite magma will result in an increase in the CaO/Na_2O ratio of the residual liquid and a reduction in the absolute amount of Na_2O in that residual liquid. These conclusions, which were fully justified by the data at their command, do not necessarily apply in the case of the hypothesis presented here. It is considered here that kimberlite may be a serpentinised mixture of xenocrysts, peridotite and a degassed fluid which is a residual liquid developed by the fractionation of eclogite from a liquid which is itself the partial melting product of garnet-peridotite.

If this view is correct no direct estimate of the Na_2O and CaO contents of the initial liquid formed at depth can be obtained from kimberlite. Moreover, Harris and Rowell based their conclusions on the average compositions of eclogites from the Roberts Victor Mine (Na_2O 2·15 per cent, CaO 9·8 per cent) and the

147

Jagersfontein Mine (Na_2O 1·33 per cent, CaO 10·7 per cent). Holmes (1936) published two analyses of eclogites from these mines, whose average yields Na_2O 1·32 per cent, CaO 9·88 per cent. It is difficult to find eclogite nodules in kimberlite which are free from alteration effects which may have involved the introduction of Na_2O, or CaO or both. Aggregates of the separated analysed garnet–clinopyroxene pairs used in this experimental study in the approximate proportions in which they appear to precipitate together from a liquid at 30 kilobars, yield figures of the order of Na_2O 0·7 per cent, CaO 12·0 per cent. Fractional crystallisation of sodium-poor eclogites of this type would be more likely to increase the Na_2O content and Na_2O/CaO ratios of the residual liquids.

It would thus appear that we are not yet in a position to deduce a unique effect of eclogite fractionation on the Na_2O, CaO or CaO/Na_2O values of the residual liquids, and more than one type of behaviour may be observed in practice. However, Harris and Rowell (1960, p. 2455) imply also that residual liquids in the system diopside-jadeite at high pressures will become impoverished in jadeite molecule, and that the initial partial melt of peridotite at high pressures will be impoverished in Na_2O relative to its parent rock. Bell and Davis (1965) have shown that jadeite will be enriched in residual liquids in the diopside–jadeite system, however.

The geochemical and geological links between kimberlite, alnöite and carbonatite have been recognised for some time (*see* Holmes and Harwood 1932; von Eckermann 1948) and it is tempting to think of carbonatites as potential kimberlite diatremes where the volatile pressures were not explosively released. The recent discovery of garnet-peridotite in carbonatite tuff at Lashaine volcano, Tanzania, and of diamonds in the nearby alluvium (Dawson 1964) and of pyrope crystals in alnöite breccia in the Solomon Islands (Allen and Deans 1965) strengthens this link.

Experimental data by Wyllie and Tuttle (1960) indicate that carbonate liquids can exist at such low temperatures that it is doubtful if they would ever crystallise at great depths because of the relatively high temperatures attained in the geothermal gradient. Carbonatite intrusions or kimberlite diatremes may be an inevitable end product of partial melting and fractionation processes at depths of 80-100 km and deeper, if the liquid so formed cannot escape to the surface at an early stage.

Many other hypotheses have been suggested to account for the origin of kimberlite, the potassic mafic lavas, and of the alkaline undersaturated magmas in general, and a review of these is given by Turner and Verhoogen (1960, pp. 235-50). The hypothesis that the combination of high Cr, Ni, Co, V and Mg with high Sr, Ba, Rb, K, Na, Ti and P in the potassic mafic lavas indicates assimilation of granitic material by basic magma does not explain the low silica content of these rocks relative to the other oxides, and paradoxically, the granitic xenoliths in such lavas have had just those elements added to them (Holmes 1945) which are alleged to be selectively extracted from granite by basalt to form the potassic mafic lava.

The resorption of biotite suggested by Bowen (1928) as a possible mechanism leading to the formation of silica-poor mafic lavas remains an interesting possibility, and the experimental results of Yoder and Tilley (1962, p. 453) indicate that biotite is a phenocryst phase in alkali basalt at 2-10 kilobars water vapour pressure. The selective refusion of biotite derived from crustal rocks would leave unexplained the absence of accompanying granitic liquid from the refusion of the quartzo-feldspathic materials in crustal rocks (which should occur at a lower temperature than the refusion of biotite).

Holmes and Harwood (1932, pp. 430-1) proposed an origin for the potassic mafic lavas whereby both eclogite and (garnetiferous) enstatite-peridotite were envisaged as crystal accumulates from a kimberlite-like peridotite magma. This hypothesis, which Holmes later abandoned, differs significantly from that proposed in this paper in the roles assigned to the various rock types, and in the choice of the original liquid. Nevertheless Holmes and Harwood observed the potential effect of eclogite fractionation as an agent in raising the K/Na ratio of the residual liquids.

Early removal of the partial melting product of garnet-peridotite (hypersthene-normative picrite in composition) to high levels in the mantle, or even in the crust followed by fractionation of olivine, obviously could lead to the formation of silica-saturated tholeiite magmas, while removal at an intermediate stage of high-pressure fractionation followed by further fractionation at low pressures might give rise to alkali basalt magmas.

SUMMARY OF CONCLUSIONS

1. The range of compositions which can yield a homogeneous clinopyroxene structure at high pressure and temperature is very much more extensive than at low pressures, with respect to the R_2O_3-bearing molecules. These substitutions can be expressed within the general formula $R^{2+}_{2-n} R^{3+}_n R^{3+}_n Si_{2-n}O_6$ but there are also indications of a substitution involving Al_2SiO_5 which would yield pyroxene with a deficit of divalent cations to fill the appropriate structural sites.

2. The present mineral paragenesis of eclogite and garnet-peridotite nodules in kimberlite reflects equilibration of the mineral assemblages at temperatures several hundred degrees lower than the beginning of melting in the dry state. There are, nevertheless, probable exsolution textures in such eclogites indicating that in some cases the present mineralogy has been derived by the recrystallisation of a higher temperature eclogite facies mineral assemblage.

3. The present mineral paragenesis of eclogite facies inclusions from an alkali basalt tuff in Hawaii indicates equilibration at high temperatures within the

eclogite facies, followed by partial exsolution of these minerals in response to changed conditions.

4. A reaction relationship at 30 kilobars between orthopyroxene and silica-poor liquids has been described from natural systems and from the quaternary system $CaO-MgO-Al_2O_3-SiO_2$. It is analogous to the well-known reaction relationship between olivine and silica-rich liquids observed at atmospheric pressure in natural and synthetic systems, and this orthopyroxene reaction relationship is likely to be equally important in its influence on the evolution of basic magmas at high pressures.

5. The partial melting product of a garnet-peridotite mantle at depths of 80-100 kilometres is believed to be a liquid with the composition of hypersthene-normative picrite basalt.

6. Garnet-peridotite begins to melt at high pressures at a temperature below that at which bimineralic eclogite begins to melt. Where both rock types are available to be melted under natural circumstances, liquid would be produced preferentially from the garnet-peridotite.

7. There is no apparent way of generating a basic liquid which is silica-saturated in the C.I.P.W. norm by the partial melting of garnet-peridotite at 30 kilobars pressure, nor of producing quartz-bearing eclogites from such a source rock while garnet and clinopyroxene are in equilibrium with the liquid. What happens at intermediate pressure is unknown, as yet, but it is well established that the low-pressure fractionation products of a hypersthene-normative picrite liquid (such as could be produced by the high-pressure partial melting of garnet-peridotite) include silica-saturated tholeiitic liquids. A garnet-peridotite mantle is, therefore, a possible source rock of parent liquids, which may yield contrasted extrusive products depending on the pressure at which fractionation occurs, in accordance with the conclusion expressed by Yoder and Tilley (1962, p. 518, conclusion 4).

8. Fractional crystallisation at high pressure of the liquid produced by partial melting of garnet-peridotite is believed to give rise to bimineralic eclogite accumulates and a series of silica-poor alkaline residual liquids which have the geochemical characteristics of the groundmass of kimberlite, or of the potassic mafic lavas. The association of nodules of garnet-peridotite (a possible source rock and mantle material) with nodules of eclogite (a possible accumulate) in kimberlite (a possible residual fluid of high-pressure fractionation) is regarded as significant.

REFERENCES

[*Editors' Note:* Only the references cited in the preceding excerpts are reproduced here.]

ALLEN, J. B. and DEANS, T. 1965. Ultrabasic eruptives with alnöitic-kimberlitic affinities from Malaita, Solomon Isles. *Mineralog. Mag.* (Tilley volume) **34,** 16–34.

ANDERSEN, O. 1915. The system anorthite-forsterite-silica. *Am. J. Sci.,* 4th ser. **39,** 407–454.

BELL, P. M. and DAVIS, B. T. C. 1965. Temperature composition section for jadeite-diopside. *Carnegie Instn. Wash. Yb.* **64,** 120–123.

BOWEN, N. L. 1928. *The evolution of the igneous rocks.* Princeton.

BOYD, F. R. and ENGLAND, J. L. 1962. Mantle minerals. *Carnegie Instn. Wash. Yb.* **61,** 107–112.

—— 1963. Effect of pressure on the melting of diopside, $CaMgSi_2O_6$, and albite, $NaAlSi_3O_8$, in the range up to 50 kilobars. *J. Geophys. Res.* **68,** 311–323.

—— 1964. The system enstatite-pyrope. *Carnegie Instn. Wash. Yb.* **63,** 157–161.

DAVIS, B. T. C. 1964. The system diopside-forsterite-pyrope at 40 kilobars. *Carnegie Instn. Wash. Yb.* **63,** 165–171.

DAVIS, B. T. C. and SCHAIRER, J. F. 1965. Melting relations in the join diopside-forsterite-pyrope at 40 kilobars and at one atmosphere. *Carnegie Instn. Wash. Yb.* **64,** 123–126.

DAWSON, J. B. 1962. Basutoland kimberlites. *Bull. geol. Soc. Am.* **73,** 545–560.

—— 1964. Carbonate tuff cones in Northern Tanganyika. *Geol. Mag.* **101,** 129–137.

GREEN, D. H. and RINGWOOD, A. E. 1963. Mineral assemblages in a model mantle composition. *J. geophys. Res.* **68,** 937–945.

HARRIS, P. G. and ROWELL, J. A. 1960. Some geochemical aspects of the Mohorovicic discontinuity. *J. geophys. Res.* **65,** 2443–2459.

HIGAZY, R. A. 1954. Trace elements of volcanic ultrabasic potassic rocks of south-western Uganda and adjoining part of the Belgian Congo. *Bull. geol. Soc. Am.* **65,** 39–70.

HOLMES, A. 1936. A contribution to the petrology of kimberlite and its inclusions. *Trans. geol. Soc. S. Africa,* **39,** 379–428.

—— 1945. Leucitised granite xenoliths from the potash-rich lavas of Bunyaruguru, South-West Uganda. *Am. J. Sci.* **243A,** 313–332.

HOLMES, A. and HARWOOD, H. F. 1932. Petrology of the volcanic fields east and south-east of Ruwenzori, Uganda. *Q. J. geol. Soc. Lond.* **88,** 370–442.

HOWIE, R. A. 1955. The geochemistry of the charnockite series of Madras, India. *Trans. R. Soc. Edinb.* **62,** 725–768.

MACGREGOR, I. D. and RINGWOOD, A. E. 1964. The natural system enstatite-pyrope. *Carnegie Instn. Wash. Yb.* **63,** 161–163.

NOCKOLDS, S. R. and ALLEN, R. 1954. The geochemistry of some igneous rock series: Part II. *Geochim. cosmochim. Acta,* **5,** 245–285.

—— 1956. The geochemistry of some igneous rock series—III. *Geochim. cosmochim. Acta,* **9,** 34–77.

O'HARA, M. J. 1960. A garnet-hornblende-pyroxene rock from Glenelg, Inverness-shire. *Geol. Mag.* **97,** 145–156.

—— 1961. Zoned ultrabasic and basic gneiss masses in the early Lewisian metamorphic complex at Scourie, Sutherland. *J. Petrology,* **2,** 248–276.

O'HARA, M. J. and MERCY, E. L. P. 1963. Petrology and petrogenesis of some garnetiferous peridotites. *Trans. R. Soc. Edinb.,* **65,** 251–314.

TILLEY, C. E. 1937. The paragenesis of kyanite-amphibolites. *Mineralog. Mag.* **24,** 555–568.

TILLEY, C. E., YODER, H. S. Jr. and SCHAIRER, J. F. 1964. New relations on melting of basalts. *Carnegie Instn. Wash. Yb.* **63,** 92–97.

—— 1965. Melting relations of volcanic tholeiite and alkali rock series. *Carnegie Instn. Wash. Yb.* **64,** 69–82.

TURNER, F. J. and VERHOOGEN, J. 1960. *Igneous and metamorphic petrology.* New York.

VON ECKERMANN, H. 1948. The alkaline district of Alnö Island. *Sver. geol. Unders., Ser. Ca.* **36,** 176 pp.

WILLIAMS, A. F. 1932. *The Genesis of the Diamond*. London.

WYLLIE, P. J. and TUTTLE, O. F. 1960. The system CaO-CO_2-H_2O and the origin of carbonatites. *J. Petrology*, **1**, 1–46.

YODER, H. S. Jr. 1964. Genesis of principal basalt magmas. *Carnegie Instn. Wash. Yb.* **63**, 97–101.

YODER, H. S. Jr., and TILLEY, C. E. 1962. Origin of basalt magmas: an experimental study of natural and synthetic rock systems. *J. Petrology*, **3**, 342–532.

25

Reprinted from pages 126–128 of *Earth-Sci. Rev.* **4**:69–133 (1968)

THE BEARING OF PHASE EQUILIBRIA STUDIES IN SYNTHETIC AND NATURAL SYSTEMS ON THE ORIGIN AND EVOLUTION OF BASIC AND ULTRABASIC ROCKS

M. J. O'Hara

[*Editors' Note:* In the original, material precedes this excerpt.]

CONCLUSIONS

(*1*) Garnet-lherzolite nodules in kimberlite yield an acceptable estimate of upper mantle composition and could yield ca. 15% or more of a picritic liquid on partial melting.

(*2*) Higher pressures increasingly favour the development of garnet-harzburgite rather than lherzolite residua during partial melting of garnet-lherzolites.

(*3*) The mantle is probably composed largely of garnet-lherzolite, and will probably yield garnet-harzburgite residua on partial melting at most, if not all, pressures if the compositions of garnet-lherzolite nodules in kimberlite are taken as representative of mantle chemistry.

(*4*) Partial melting of mantle peridotite is probably a constant total composition, self-buffering process with respect to partial pressure of oxygen in most instances. Variable partial pressure of oxygen is not, therefore, an important factor controlling the composition of the primary magma produced.

(*5*) The compositions of liquid in equilibrium with four phase lherzolite varies with increasing pressure, being quartz-normative at low pressure, olivine- and hypersthene-normative and of high-alumina character at ca. 5–8 kbar, nepheline-normative and increasingly picritic from 8 to ca. 20 kbar, then hypersthene-normative and picritic from 20 to above 40 kbar. More advanced partial melting yields liquids which tend to be hypersthene-normative at all pressures.

(*6*) The depth of origin of a primary magma exercises little direct control over the final erupted product but it does control the number of alternative evolutionary paths open to a particular magma batch.

(*7*) Erupted magmas are not *in general* primary liquids, but have undergone continuous fractionation *at least* of olivine during their ascent.

(*8*) Oceanic tholeiites like most other erupted basalts are probably not primary magmas.

(*9*) Wall-rock reaction between ascending magma batches and mantle peridotite is not an acceptable explanation of the control of major, minor or trace element contents of long sequences of basalts erupted through the same conduits.

153

(*10*) The existence of thermal divides not only controls fractionation trends at different pressures but also severely restricts what can be accomplished by such processes as hybridisation of magmas, or contamination by siliceous, calcareous or pelitic materials.

(*11*) Numerous reaction relationships exist between olivine-normative liquids and either olivine or orthopyroxene at elevated pressures. Such relationships are essential if liquids which fractionate without precipitating both olivine and orthopyroxene are to be derived from a mantle composition yielding harzburgite residua.

(*12*) Polybaric, polythermal crystallisation affects most natural magmas and introduces new complications in the interpretation of phase equilibria and accumulate sequences, particularly the lherzolite nodules in alkali olivine-basalts.

(*13*) The extent of fractionation of other phases in addition to olivine is controlled by the rate of movement of the magma batch.

(*14*) Tholeiitic magmas have probably been hypersthene-normative throughout their evolution.

(*15*) A pyroxenite stage in evolution of basic magmas is encountered in liquids already committed to the nepheline-normative volume at intermediate pressures.

(*16*) Fractionation schemes involving derivation of common basalt types by orthopyroxene fractionation at intermediate pressure from olivine-tholeiite liquids are impossible unless magma batches *descend* from higher levels, because there is no suitable reaction relationship by which olivine crystallisation can be suppressed in liquids formed by partial melting of peridotite.

(*17*) Neither orthopyroxene nor olivine fractionation alone can explain the contents of K, P, Ti in erupted basalts from any commonly observed peridotite composition. Eclogite fractionation offers a satisfactory mechanism, however, and enables observed K_2O contents of lavas to be obtained from a mantle source with less than 0.03% K_2O.

(*18*) Lherzolite nodules in basalts could be accumulates formed at intermediate pressures from ascending liquids originally formed at higher pressures in equilibrium with garnet-harzburgite residua.

(*19*) Fractionation of harzburgite and lherzolite nodules from magmas in the intermediate-pressure range probably is an essential part of the process whereby many alkali olivine-basalt magmas are generated, but it does not, by itself, explain their contents of K, P and Ti.

(*20*) Fractionation trends of iron-bearing liquids once they commence their ascent are very greatly influenced by the partial pressure of oxygen if the system is not closed with respect to volatiles.

(*21*) The most probable origin of the calc-alkali series magmas is by hydration, and fractionation under relatively high partial pressures of oxygen, of high-alumina basalt magmas passing from dry mantle into the wet environment of the orogen. There are, however, several possible alternative modes of origin.

(*22*) Many but not all erupted basalts display some evidence of being controlled in temperature and composition by low-pressure fractionation, although the extent may be small, and may partly arise by coincidence between the ratios of normative constituents *other than olivine* in liquids in high- and low-pressure cotectic equilibria.

26

THE DYNAMIC EARTH

P. J. Wyllie

[*Editors' Note:* Only Figures 8-15 and 8-16 are reproduced here.]

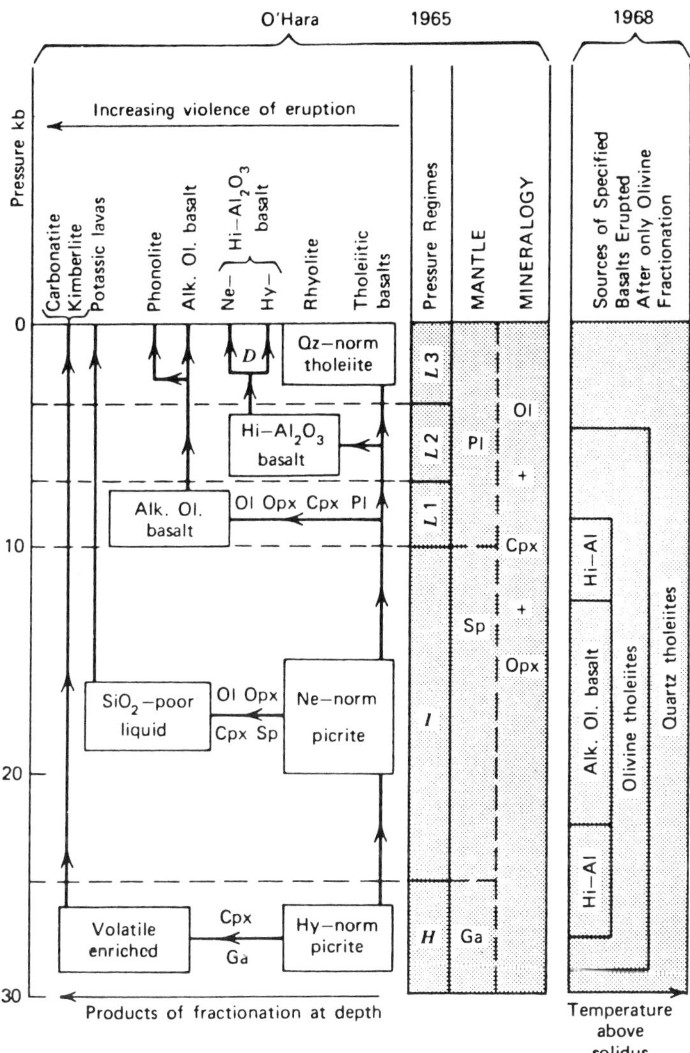

Figure 8-15. Summary of suggested basalt fractionation schemes dependent upon depth of origin and rate of migration to the surface according to O'Hara (1965, 1968). The main portion is based on O'Hara's Table 1 (1965). The main vertical sequence involves steady movement towards the surface through the successive pressure regimes (H = high pressure regime, I = intermediate, L = low, subdivisions 1, 2, and 3. Pressure scale is my estimate to permit comparison with Figure 8-16). Interrupted or delayed ascent at various depths yields the fractionation products shown. Minerals fractionating are listed near arrows. The right hand portion is a revised scheme (1968) showing the character of the erupted liquid when a magma batch formed and fractionated at various levels is taken from a particular pressure and temperature at such a speed that its composition remains near the boundary of, but always just within the olivine primary phase volume; hence it fractionates olivine only during its ascent. This is possible because of the increase in normative olivine content of liquids with pressure, as shown in Figure 8-14a. Olivine nephelinites and melilitites are erupted from fractionated materials at depth, with temperatures below that of the solidus depicted in the figure.

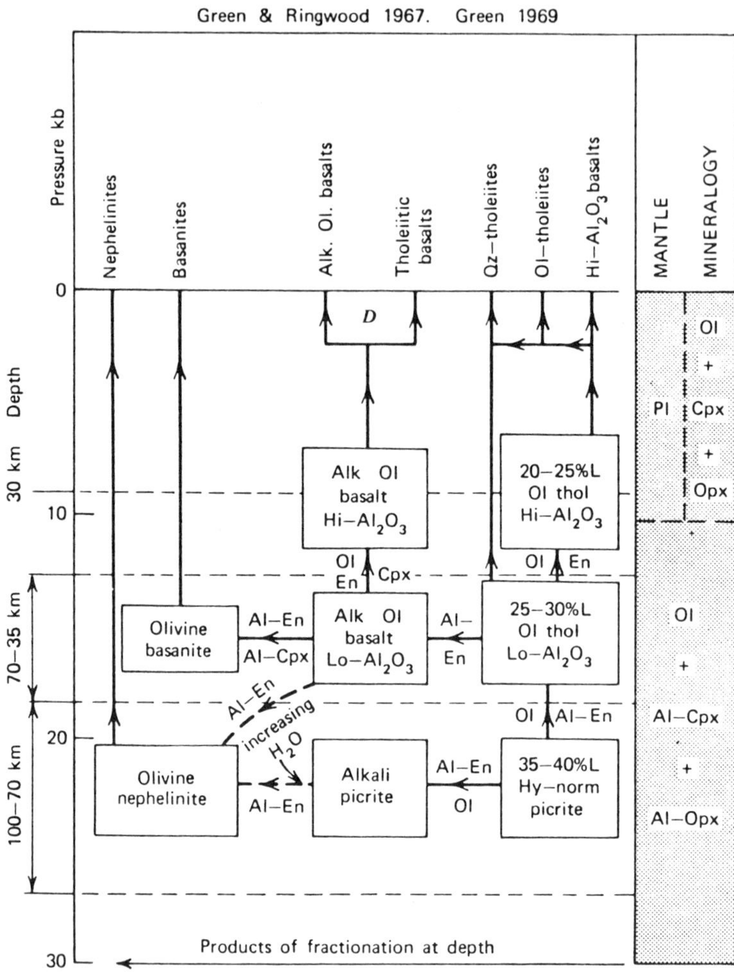

Figure 8-16. Simplified representation of Green and Ringwood's deduced crystal fractionation relationships among various basaltic magmas at moderate to high pressures; "closed system" fractionation. Based on Green and Ringwood (1967) and Green (1969). Contrast the mantle mineralogy between 10 and 30 kb (35 to 100 km depth) in this figure and Figure 8-15. Minerals fractionating are listed near arrows. Compare with Figure 8-17.

REFERENCES

[*Editors' Note:* Only the references cited in the preceding figure captions are reproduced here.]

Green, D. H., 1967, The stability fields of aluminous pyroxene peridotite and garnet peridotite and their relevance in upper mantle structure. *Earth Planet. Sci. Letters,* **3,** 151-160.

Green, D. H., 1969, The origin of basaltic and nephelinitic magmas in the earth's mantle. *Tectonophysics,* **7,** 409-422.

O'Hara, M. J., 1965, Primary magmas and the origin of basalts. *Scottish J. Geol.,* **1,** 19-40.

O'Hara, M. J., 1968, The bearing of phase equilibria studies in synthetic and natural systems on the origin and evolution of basic and ultrabasic rocks. *Earth Sci. Rev.,* **4,** 69-133.

27

THE SYSTEM FORSTERITE-DIOPSIDE-SILICA
WITH AND WITHOUT WATER AT HIGH PRESSURES

I. Kushiro

[*Editors' Note:* In the original, material precedes this excerpt.]

3. The system forsterite–diopside–enstatite at 20 kb.—The results of
the runs made on eight mixtures at 20 kb under anhydrous conditions
are shown in table 3. On the basis of these results and those obtained in
the Fo–Di system and the Di–En system, a portion of the liquidus diagram in the Fo–Di–En system at 20 kb under anhydrous conditions is
constructed (fig. 6). Forsterite crystallizing within the Fo–Di–En system at
20 kb may be a solid solution containing a small amount of the component $CaMgSiO_4$ as at 1 atm, although it was not confirmed in the present experiments.

The liquidus surface of diopside solid solution is very flat, the temperature of liquidus being within $1640 \pm 10°C$. As shown in table 1 the
melting point of pure diopside at 20 kb determined in the present experiments is $1645 \pm 10°C$. The solidus in this area is $1630 \pm 10°C$
which is close to the liquidus temperature. Enstatite liquidus surface is
steeper, the liquidus temperature increasing toward enstatite. As shown
in the Di–En join, enstatite is a primary phase in the compositions between enstatite and about $Di_{58}En_{42}$, and pigeonitic clinopyroxene is a
primary phase from about $Di_{58}En_{42}$ to $Di_{61}En_{39}$. The composition
$Fo_{30}Di_{60}Q_{10}$ (DFS-7) yields a pigeonitic clinopyroxene with quench crystals and glass, indicating that the liquidus field (primary phase field) of
pigeonitic clinopyroxene exists in the Fo–Di–En system. The boundary
between the primary phase fields of enstatite solid solution and pigeonitic clinopyroxene may be located at about $Di_{58}(En+Fo)_{42}$ as shown by a
dashed line in figure 6, although its position could not be determined
precisely because of insufficient data. The boundary between pigeonitic
clinopyroxene and diopside solid solution located at about $Di_{61}(En+Fo)_{39}$
as shown in figure 6.

On the Fo–pyroxene liquidus boundary, there should be two different isobaric invariant points. One is shown by point P, and the other
is shown by point P', although the position of P' is not determined di-

TABLE 3

Results of experiments in the system forsterite-diopside-

silica at 20 kb under anhydrous conditions

No.	Composition of starting material, weight percent			Temp., °C	Time, min.	Results
	Fo	Di	SiO_2			
DFS-4	20	75	5	1630	8	Di_{ss} + Fo*(s)
				1635	5	Di_{ss} + Fo(s)
				1635	7	Di_{ss} + q-Cryst(s)
				1640	7	Di_{ss}(s) + q-Cryst
DFS-5	23	70	7	1625	10	Di_{ss} + Fo(s)
				1635	10	Di_{ss} + q-Cryst(s) (Di_{ss}: Δ2θ= 0.355°+0.007, ≅29 wt % En)
				1640	7	Di_{ss} + q-Cryst + Gl(s)
				1650	5	q-Cryst + Gl
DFS-6	25	65	10	1550	20	Di_{ss} (Di_{ss}: Δ2θ=0.294°+0.007, ≅33 wt % En)
				1630	5	Di_{ss}
				1650	5	q-Cryst
DFS-7	30	60	10	1625	5	Di_{ss} + Fo?
				1630	5	Di_{ss} + Fo? + q-Cryst
				1635	5	Di_{ss} + Fo? + q-Cryst(s)
				1640	5	Cpx + q-Cryst(s) + Gl(s)
				1645	5	Cpx + q-Cryst
				1655	5	Cpx(s) + q-Cryst + Gl(s)
				1660	5	q-Cryst + Gl(s)
DFS-17	33	60	7	1630	7	Di_{ss} + Fo
				1650	7	Fo + q-Cryst
				1670	5	q-Cryst
				1690	5	q-Cryst
DFS-8	37	50	13	1630	5	Cpx + Fo(s)
				1635	30	Cpx + Fo(s)
				1650	6	Cpx + Fo?
				1670	5	Cpx + q-Cryst
				1680	5	q-Cryst + Gl(s)
DFS-9	50	30	20	1500	35	Cpx + Fo(s)
				1635	21	Cpx + En_{ss}? + Fo
				1650	5	Cpx + Fo + q-Cryst(s)
				1670	7	En_{ss} + Cpx + q-Cryst(s) + Gl(s)
DFS-10	65	10	25	1635	22	En_{ss} + Cpx + Fo
				1670	7	En_{ss} + Cpx + q-Cryst

Starting materials are glass, crystallized at 1 atm.

Abbreviations as in tables 1 and 2.

*Forsterite in this system may be a solid solution containing a small amount of Ca.

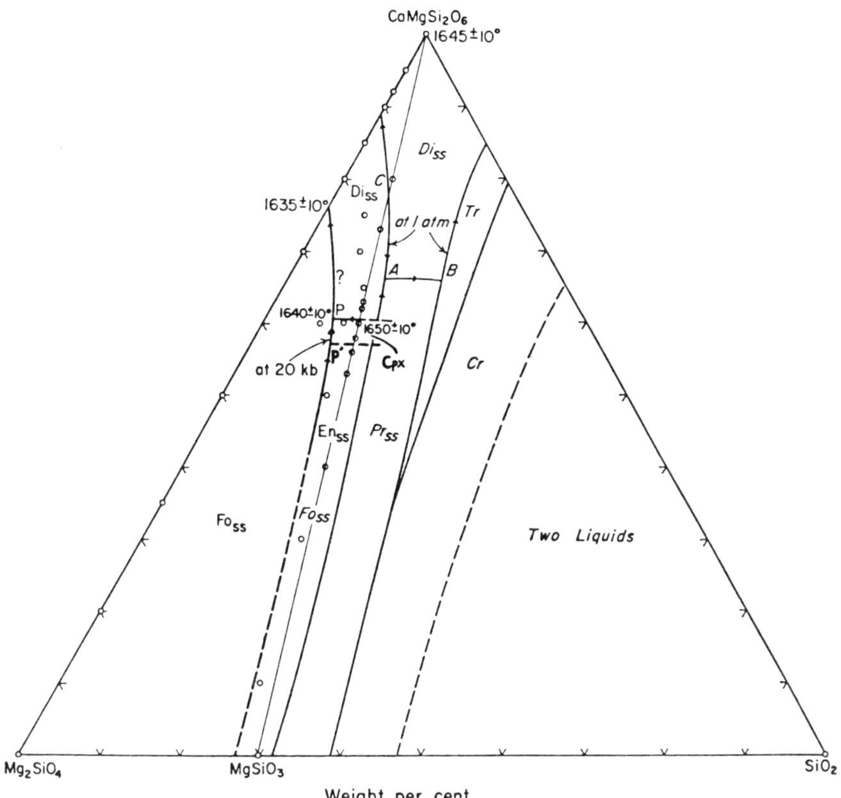

Weight per cent

Fig. 6. Liquidus diagram of the system forsterite (Mg_2SiO_4)–diopside ($CaMgSi_2O_6$)–enstatite ($MgSiO_3$) at 20 kb under anhydrous conditions. Liquidus diagram of the system forsterite–diopside–silica at 1 atm is from Bowen (1914), Schairer and Yoder (1962), and Kushiro and Schairer (1963). Italic letters are for diagram at 1 atm.

rectly. It is not certain whether the point P is a reaction point or a eutectic point. The nature of this point depends on the composition of the diopside solid solution in equilibrium with liquid at P. If the composition of the diopside solid solution is more enstatite rich than the intersection of the Di–En join and the extension of the Fo–P join ($\sim Di_{67}En_{33}$), the point P is a reaction point. In the present experiments, however, the composition of the diopside solid solution at P could not be determined because of the presence of a large amount of quench pyroxene. If the temperature of the Fo–Di_{ss} liquidus boundary drops continuously from the point P to the minimum on the Fo–Di join, the point P is a reaction point. However, the temperature difference between P and the minimum on the Fo–Di join is too small to determine with certainty the temperature profile along the Fo–Di_{ss} liquidus boundary. The point P' is most likely a reaction point where the reaction En_{ss} + $L \rightleftharpoons Fo + Cpx$ (pigeonitic clinopyroxene) takes place, although this reaction has not been observed in nature.

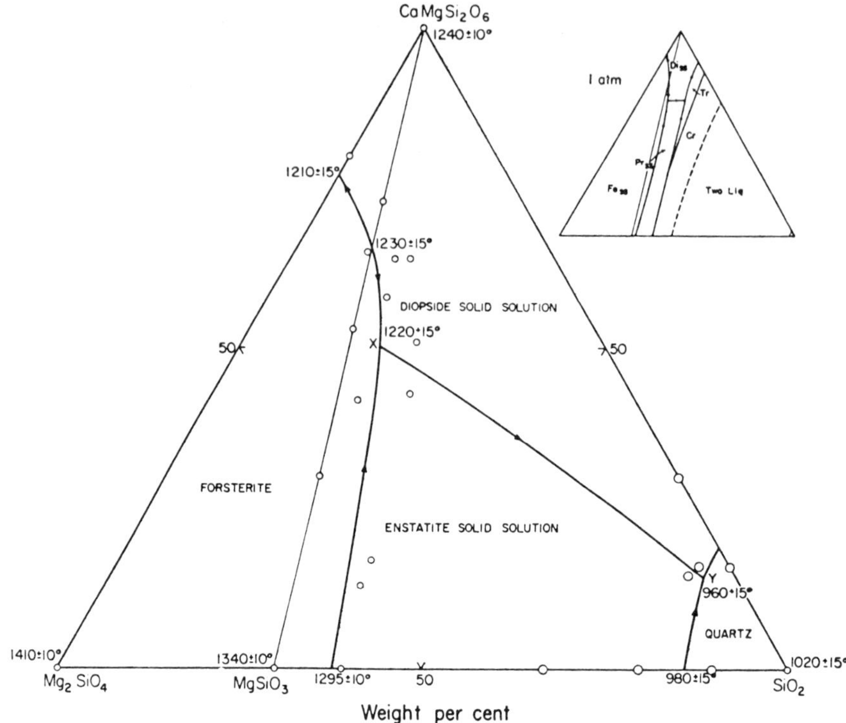

Fig. 7. The liquidus diagram of the system forsterite–diopside–silica at 20 kb P_{H_2O}. The boundaries are those projected onto this ternary system from the H_2O-apex in the quaternary system forsterite–diopside–silica–H_2O.

4. The system forsterite–diopside–silica at 20 kb under hydrous conditions.—In order to find out the effect of water pressure on the liquidus relations in the Fo–Di–Q system, experiments have been carried out on twenty-one selected compositions at 20 kb under hydrous conditions.[3] Seven of the compositions are mechanical mixtures as recorded in table 4. Results for $MgSiO_3$ at 20 kb P_{H_2O} (water pressure) are from Kushiro, Yoder, and Nishikawa (1968). Duplicate runs were made for compositions $Fo_{51.9} Di_{13} Q_{35.1}$, $Di_{53} En_{47}$, and $Fo_{37.8} Di_{42} Q_{20.2}$ from both glass and crystallized glass. The results are the same for different starting materials for $Di_{53} En_{47}$ and $Fo_{37.8} Di_{42} Q_{20.2}$ as shown in table 4. For the composition $Fo_{51.9} Di_{13} Q_{35.1}$, however the results are different for different starting materials. In this case, crystallized glass may be more reliable because there is a possibility that some silica might have been leached preferentially from glass by the gas phase. For all other compositions, therefore, crystalline materials were used as starting materials. The charges were wet, and free water existed after all the runs, suggesting that the water pressure was equal to load pressure during all the runs. However, there is a possi-

[3] "Hydrous conditions" used here means the conditions where water pressure is equal to total pressure, if not particularly mentioned.

Results of experiments in the system forsterite-diopside-silica at 20 kb under

hydrous conditions (20 kb water pressure)

Composition of starting material, weight percent			Starting material	Temp., °C	Time, min.	H_2O content, weight percent	Results
Fo	Di	SiO_2					
20	80	...	Crystal.	1200	50	25.0	Di_{ss} + q-Cryst
			Crystal.	1215	60	31.3	q-Cryst + Gl
19	73	8	Crystal.	1225	60	35.1	Di_{ss} + Gl + q-Cryst?
($Di_{73}En_{27}$)							
25	65	10	Glass	1175	40	42.2	Di_{ss}
			Glass	1225	30	39.0	Fo + Di_{ss} + q-Cryst(s)
			Crystal.	1235	40	53.7	Fo + Di_{ss} + q-Cryst
			Glass	1250	30	35.7	q-Cryst.
21.7	64	14.3	Crystal.	1220	60	26.8	Di_{ss}(s) + q-Cryst + Gl
19.6	64	16.4	Crystal.	1200	40	28.0	Di_{ss} + En_{ss}(s) + q-Cryst(s) + Gl(s)
			Crystal.	1225	60	41.0	q-Cryst + Gl(s)
25.9	58	16.1	Crystal.	1225	60	41.6	Di_{ss}(s) + q-Cryst + Gl(s)
			Crystal.	1235	60	30.6	q-Cryst + Gl(s)
33	53	14	Glass	1200	30	29.3	Fo + En_{ss} + Di_{ss}? + Gl(s) + q-Cryst(s)
($Di_{53}En_{47}$)							
			Crystal.	1225	40	29.7	Fo(s) + q-Cryst
			Glass	1250	30	34.6	Fo(s) + q-Cryst
			Crystal.	1250	40	38.4	Fo(s) + q-Cryst + Gl(s)
25.2	51	23.8	Crystal.	1215	60	29.8	Di_{ss}(s) + q-Cryst + Gl(s)
30.1	43	26.9	Crystal.	1215	60	31.8	En_{ss}(s) + q-Cryst + Gl(s)
37.8	42	20.2	Glass	1250	40	49.2	Fo + q-Cryst + Gl(s)
			Crystal.	1250	45	44.4	Fo + q-Cryst + Gl(s)
49	30	21	Glass	1150	60	24.3	En_{ss} + Di_{ss} + Fo(s) + Gl(s)
($Di_{30}En_{70}$)			Glass	1275	30	43.8	Fo + q-Cryst + Gl(s)
48.4	17	34.6	Glass	1200	40	31.9	En_{ss} + Di_{ss} + Gl?
			Crystal.	1250	40	36.1	En_{ss} + q-Cryst + Gl
51.9	13	35.1	Glass	1250	40	41.0	En_{ss} + Fo(s) + q-Cryst + Gl(s)
			Crystal.	1250	40	21.5	En_{ss} + q-Cryst + Gl(s)
61.2	...	38.8	Crystal.	1250	40	28.4	En + q-Cryst(s) + Gl
			Crystal.	1290	60	36.0	En + q-Cryst + Gl + Fo?
33.2	...	66.8	Crystal.	1100	80	28.2	En + q-Cryst + Gl
(Mech. mix.*)			Crystal.	1240	60	28.7	En(s) + q-Cryst + Gl
20.4	...	79.6	Crystal.	1100	70	26.9	En(s) + q-Cryst + Gl
(Mech. mix.*)			Crystal.	1125	60	26.1	En(s) + q-Cryst + Gl
			Crystal.	1150	90	37.7	q-Cryst + Gl
			Crystal.	{1250 ↓ / 1100}	{45 / 90}	62.4	En(s) + q-Cryst + Gl
10.2	...	89.8	Crystal.	1000	90	27.4	Q + Gl + En?
(Mech. mix.*)			Crystal.	1050	90	32.4	Gl
...	30	70	Crystal.	1100	100	30.2	Di(s) + Gl
(Mech. mix.**)			Crystal.	1150	100	35.0	Gl
...	16	84	Crystal.	975	120	41.8	Q(s) + Gl
(Mech. mix.**)							

TABLE 4 (continued)

Composition of starting material, weight percent			Starting material	Temp., °C	Time, min.	H_2O content, weight percent	Results
Fo	Di	SiO_2					
6.3	14.7	79	Crystal.	1000	120	45.5	$Di_{ss}(s) + En(s) + Gl$
(Mech. mix.***)			Crystal.	1025	120	42.9	$Di_{ss}(s) + En(s) + Gl$
			Crystal.	1050	130	36.4	$En_{ss}(s) + Gl + Di_{ss}?$
4.1	16	79.9	Crystal.	1035	172	42.5	$Di_{ss}(s) + Gl + En_{ss}?$
(Mech. mix.***)							
...	...	100	Cristo-balite	1005	150	24.4	Q + Gl
				1020	140	33.3	Q + Gl
				1050	120	40.3	Gl

Abbreviations: Crystal., glass crystallized at 1 atm; Q, quartz; others as in tables 1 and 2.

*Mechanical mixture of clinoenstatite and cristobalite.

**Mechanical mixture of diopside and cristobalite.

***Mechanical mixture of clinoenstatite and diopside solid solutions and cristobalite.

bility that free water observed after the run was completely dissolved in liquid at the conditions of the run and released during the quench. In this case water pressure could be lower than the load pressure during the runs even if free water existed after the run. However, the water content in the present experiments is higher than the maximum content of water that can dissolve in forsterite melt at 20 kb (about 20 wt percent) determined by Kushiro and Yoder (unpub. data), and the water pressure was probably equal to load pressure. Therefore, the water pressure of the experiments is considered to be 20 kb. Figure 7 shows the liquidus boundaries under water-saturated conditions. That is, these boundaries are between primary-phase volumes projected from the H_2O apex onto the Fo–Di–Q plane in the system Fo–Di–Q–H_2O. As shown in the figure, the Fo–En_{ss} liquidus boundary at 20 kb P_{H_2O} is located on the silica-rich side of that at 1 atm. The Fo–Di_{ss} liquidus boundary is located, however, on the forsterite side of that at 1 atm in the Fo–Di–En area. The most remarkable change is that the primary phase fields of diopside and enstatite solid solutions expand greatly toward silica, as shown in figure 7. This occurs because of large drop of melting temperature of quartz at high water pressures (Stewart, 1967) as compared with those of diopside, (Yoder, 1964) and enstatite (Kushiro, Yoder and Nishikawa, 1968). It may be possible, however, that metastable pyroxene persists even in the presence of liquid and vapor. To check this possibility, a run was made in which the mixture of composition $Fo_{20}Q_{80}$ was heated to a temperature well above the liquidus and then held at a temperature just below the liquidus. In this run a small amount of large euhedral enstatite was obtained with glass, indicating that enstatite is not metastable at least for composition $En_{20}Q_{80}$ at 20 kb P_{H_2O}.

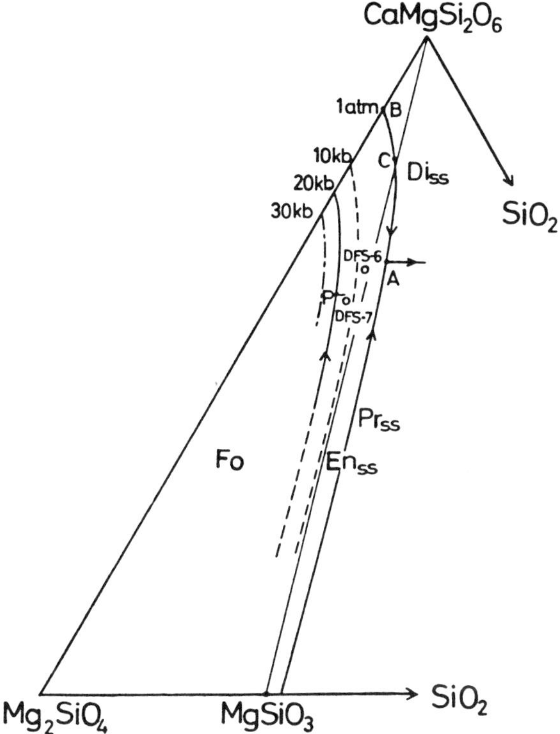

Fig. 8. Shift of the forsterite–pyroxene liquidus boundary with pressure under anhydrous conditions. The boundary at 30 kb is based on the liquidus relations on the join forsterite–diopside at 30 kb determined by Davis (1964). The boundary at 10 kb is very close to that predicted by Yoder and Tilley (1962, p. 412).

On the Fo-pyroxene liquidus boundary, there is a projection of the isobaric invariant point $Fo + En_{ss} + Di_{ss} + L + V$, which is located at about $Fo_{30}Di_{50}Q_{20}$ (X in fig. 7). This point is a reaction point where forsterite reacts with liquid to form diopside and enstatite solid solutions in the presence of vapor phase. The projection of another isobaric invariant point $Di_{ss} + En_{ss} + Q + L + V$ appears at about $Fo_5Di_{14}Q_{81}$ (Y in fig. 7). It is not certain whether this is a reaction point or a eutectic point.

On the liquidus of quartz (on the Di–Q and En–Q joins), rounded or square quartz crystals crystallize as a primary phase. They are probably high quartz, although double terminated pyramids are rarely seen. In the runs quenched from the temperatures above the liquidus, irregular-shaped glass including many bubbles and glass in the form of globules were obtained. However, it is not certain whether both vapor and liquid coexisted or only a super critical fluid existed during the runs made above the liquidus.

TABLE 5

Results of experiments on the compositions $Fo_{25}Di_{65}Silica_{10}$ (by weight)

(DFS-6) and $Fo_{30}Di_{60}Silica_{10}$ (DFS-7) under anhydrous conditions

Pressure, kb	Temperature, °C	Time, minutes	Results
		DFS-6	
7	1475	30	Di_{ss}
7	1490	20	q-Cryst
8	1495	20	Gl
9	1500	12	Di_{ss} + q-Cryst(s)
9	1520	15	q-Cryst
11	1550	15	q-Cryst
13	1550	20	Di_{ss} + Fo?
13	1570	20	q-Cryst + Gl
15	1520	20	Di_{ss} + q-Cryst
		DFS-7	
12	1540	7	Di_{ss} + Fo(s)
12	1550	15	Fo(s) + q-Cryst
12	1560	15	q-Cryst + Gl

Starting materials are glass, crystallized at 1 atm.

Abbreviations as in tables 1 and 2.

CHANGE OF LIQUIDUS RELATIONS WITH PRESSURE
UNDER ANHYDROUS CONDITIONS

The liquidus relations in the Fo–Di–En system change considerably with pressure. Figure 8 shows the Fo–pyroxene liquidus boundaries at 1 atm (Bowen, 1914; Schairer and Yoder, 1962; Kushiro and Schairer, 1963) and those at 20 kb under anhydrous conditions. As shown in the figure, the primary phase field of forsterite is much reduced at 20 kb as compared with that at 1 atm. The most significant difference between the liquidus relations at 1 atm and those at 20 kb is that at 1 atm there exists a reaction point (A) where forsterite reacts with liquid to form diopside and protoenstatite solid solutions, whereas at 20 kb forsterite does not react with liquid, but pigeonitic clinopyroxene may react with liquid to form forsterite and diopside solid solution at the invariant point (P).

In order to determine the pressure at which such a changeover takes place, additional runs have been made on two mixtures DFS-6 ($Fo_{25}Di_{65}Q_{10}$) and DFS-7 ($Fo_{30}Di_{60}Q_{10}$) at pressures between 7 and 15 kb (table 5). The composition DFS-6 does not lie in the primary phase field of forsterite at pressures at least higher than 7 kb, indicating that the primary phase field of forsterite is already on the forsterite-side of the composition DFS-6 at 7 kb. The composition DFS-7 lies in the primary phase field of forsterite at 12 kb. The pressure at which the changeover from a forsterite liquidus to a pyroxene liquidus takes place for this composition is estimated to be about 15 kb by comparing the results in table 5 with those at 1 atm and 20 kb. The Fo–pyroxene liquidus boundary at 10 kb, therefore, passes between the compositions DFS-6 and DFS-7 and is shown by a dashed line in figure 8. Although the data are insufficient, it is suggested that the reaction $Fo + L \rightleftharpoons Di_{ss} + Pr_{ss}$ or $Fo + L \rightleftharpoons Di_{ss} + En_{ss}$ takes place at pressures less than 6 or 7 kb under anhydrous conditions.

Because of the significant change of the liquidus relations in the Fo–Di–Q system with pressure under anhydrous conditions the trends of liquids by crystallization at 1 atm and at high pressures are significantly different from one another. As shown by Bowen (1914), Schairer and Yoder (1962), and Kushiro and Schairer (1963), the liquids from which forsterite crystallizes as a primary phase, except those in the area Fo–B–C in figure 8, change their compositions toward silica-saturated compositions across the Di–En join by fractional crystallization at 1 atm. On the other hand, at 20 kb all the liquids in the forsterite field never cross the Di–En join, and they change probably toward critically silica-under-saturated compositions across the Fo–Di join by fractional crystallization. Of course, if there is a maximum on the $Fo–Di_{ss}$ liquidus boundary, only liquids close to the Fo–Di join in the forsterite field can change their compositions across the Fo–Di join.

It should be noted that the primary phase field of forsterite is reduced continuously with increase of pressure under anhydrous conditions. In the $Fo–NaAlSiO_4–Q$, $Fo–CaAl_2SiO_6–Q$, and $Fo–MgAl_2O_4–Q$ systems, the primary phase field of forsterite is also reduced with increase of pressure (Kushiro, 1968). Therefore, if a magma is formed by the partial melting of olivine-rich lherzolite, which is probably an upper mantle material, at high pressures under anhydrous conditions, and crystallization of the magma takes place at low pressures, the first phase to crystallize from the magma is forsteritic olivine. The primary phase field of forsterite expands, and the composition of the magma enters into the forsterite field with decreasing pressure. In other words, crystallization of forsteritic olivine from a basaltic magma as a primary phase indicates that the magma has been formed at higher pressures than the pressure where crystallization takes place. This argument is based on the assumption that the partial melting of peridotite takes place at the invariant point where forsteritic olivine and orthopyroxene (or Ca-poor clinopyroxene) are involved or along the univariant or divariant surface, et cetera where forsteritic olivine and orthopyroxene (or pigeonite) are involved.[4] In some basalts, however, orthopyroxene is the primary phase to have crystallized. One example is a bronzite-bearing alkali basalt in Taka-sima (Kuno, 1964), although the bronzite is aluminous.

[4] This assumption is justified by the recent experiments on the melting of a natural lherzolite (Ito and Kennedy, 1967) and the "pyrolite" composition (Green and Ringwood, 1967b), in which forsteritic olivine and orthopyroxene persist through a large range of partial melting.

In experiments on some natural basalts, aluminous orthopyroxene also begins to crystallize as a primary phase (Green and Ringwood, 1967a). These results may be explained by the presence of water when the magma is formed, as discussed later.

EFFECT OF WATER PRESSURE ON THE LIQUIDUS RELATIONS IN
THE SYSTEM FORSTERITE–DIOPSIDE–SILICA

As shown in the experimental results at 20 kb under anhydrous and hydrous conditions, the presence of water significantly changes the equilibrium relations. Under anhydrous conditions, the primary phase field of forsterite becomes smaller with increase of pressure, and at 20 kb the Fo–En$_{ss}$ liquidus boundary is on the forsterite side of the Di–En join. Under hydrous conditions, however, the Fo–En$_{ss}$ liquidus boundary (and a part of the Fo–Di$_{ss}$ liquidus boundary) shifts toward silica with increase of water pressure, although the shift from that at 1 atm is not large. It is significant that the primary phase field of forsterite covers the Di–En join at high water pressures. This would suggest that the silica-saturated liquids can be formed from silica-undersaturated liquids (for example, olivine tholeiite magma) by fractional crystallization or from peridotites by partial melting at pressures at least up to 20 kb P_{H_2O}. Above 20 kb P_{H_2O} and up to at least 30 kb P_{H_2O} enstatite still melts incongruently to forsterite and liquid (Kushiro, Yoder, and Nishikawa, 1968), indicating that there is still a possibility of producing silica-saturated liquids from silica-undersaturated liquids or peridotites at 30 kb P_{H_2O}. This conclusion is important for the origin of silica-saturated magmas, such as quartz tholeiite magma, in the upper mantle. That is, if free water exists in some parts of the upper mantle (depths to at least 100 km) and if partial melting takes place in those parts, silica-saturated magmas can be generated. If the fractional crystallization of olivine tholeiitic magma containing water takes place in the upper mantle, silica-saturated magmas can also be formed. By further fractional crystallization, silica-rich magmas such as andesitic and dacitic magmas may be produced, as predicated from figure 7. It is noted that a considerable amount of water (more than 10 wt percent) would be necessary for the water pressure in the magma to become equal to total pressure at pressures higher than 20 kb because of the large solubility of water in the silicate melt at such high pressures. Even if water pressure is considerably less than the total pressure, however, the effect of the water pressure cannot be disregarded. That is, silica-saturated magmas can be formed at higher water pressures than under anhydrous conditions either by the fractional crystallization of olivine tholeiitic magma or by the partial melting of the upper mantle peridotite.

As has been discussed previously, under anhydrous conditions forsteritic olivine crystallizes as the earliest phase at lower pressures from the magmas formed by the partial melting of peridotite at higher pressures. If, however, a magma is formed in the upper mantle in the pres-

168

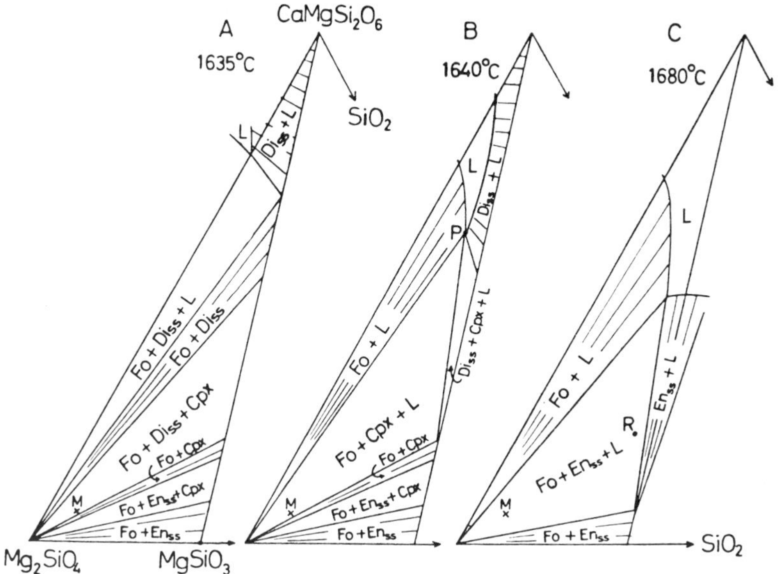

Fig. 9. Isobaric and isothermal sections in the system forsterite–diopside–enstatite at 1635°C (A), 1640°C (B), and 1680°C (C) at 20 kb under anhydrous conditions. Abbreviations as in tables 1 and 2. Ca in forsterite is not shown.

ence of water, and if water in the magma escapes during the ascent of the magma, it is possible that orthopyroxene will crystallize as a primary phase instead of forsteritic olivine. The primary phase field of forsterite relative to that of orthopyroxene is reduced with decrease of water content in the melt, and if the effect of escaping water is larger than the effect of lowering load pressure which expands the forsterite field, it is possible that the primary phase field of orthopyroxene relative to that of olivine field expands with ascending magma and orthopyroxene crystallizes as the earliest phase.

PARTIAL MELTING PROCESS IN THE SIMPLE PERIDOTITE AND PYROXENITE SYSTEM AT 20 Kb

The process of partial melting in the simple peridotite and pyroxenite system at 20 kb may contribute to an understanding of the process of partial melting of a peridotite upper mantle and also the process of generating magmas in the upper mantle. The mantle peridotite is considered to be a lherzolite which is rich in forsteritic olivine relative to enstatitic and diopsidic pyroxenes and has some normative plagioclase. Its composition (for example, average of four peridotite inclusions in basaltic rocks by Kushiro and Kuno, 1963; average lherzolite nodule in kimberlite by Ito and Kennedy, 1967) plots close to Fo in the Fo–Di–En system (M in fig. 9) after subtracting fayalite, ferrosilite, and feldspar normative components. The process of partial melting of this peridotite is considered in the isothermal sections at 20 kb (fig. 9). Fig. 9A shows the isothermal section at about 1635°C, in which 'mantle peridotite' is not included in the liquid-present region and, therefore, "mantle peridotite" is not melted at this temperature. Some of the simple wehrlites and pyroxenites are included in the three-phase region Fo + Di_{ss} + L and two-phase region Di_{ss} + L, and they are partially melted. At 1640°C, the reaction Fo + Di_{ss} $\rightleftharpoons$ Cpx (pigeonitic clinopyroxene) + L takes place and a new three-phase region Fo + Cpx + L is formed (fig. 9B). The "mantle peridotite" is included in this new three-phase region and begins to be melted. The temperature is constant until diopside solid solution in the "mantle peridotite" disappears. At this temperature, wehrlite is included in Fo + L, Di_{ss} + L, or L region and partially or completely melted depending upon the composition. On the other hand, simple saxonite (or harzburgite) is still not melted. When diopside solid solution disappears, the temperature begins to increase, and the composition of liquid begins to move toward the Fo–En join along the Fo–Cpx liquidus boundary, and the three-phase triangle also begins to move toward the Fo–En join. At about 1650°C, when the temperature of another invariant point (P′ in fig. 6) is attained, the reaction Fo + Cpx $\rightleftharpoons$ En_{ss} + L takes place, and new three-phase triangle Fo + En_{ss} + L is formed. When pigeonitic clinopyroxene disappears, the temperature begins to increase, and the liquid changes its composition toward the Fo–En join along the Fo–En_{ss} liquidus boundary. Part of the saxonite is now included in the three-phase region Fo + En_{ss} + L (fig. 9C). With further increase of temperature, the amount of liquid in the "mantle peridotite" increases, and composition of the enstatite solid solution coexisting with liquid becomes Ca-poor. When the liquid reaches the composition of the intersection of the Fo–En_{ss} liquidus boundary and the extension of the Fo–M join (R in fig. 9C), enstatite solid solution disappears, and with further increase of temperature the liquid moves toward M.

As shown above, the first liquid formed by the melting of the "mantle peridotite" at 20 kb has the composition P. If the liquid is removed from the system by a process such as filter pressing at the temperature of P, the material left consists of Fo, Cpx, and Di_{ss}, which may change to lherzolite by the inversion of Cpx to En_{ss} (having an exsolved Di_{ss}) with decreasing temperature. If the liquid is removed from the system at temperatures higher than that of the invariant point P but below that of P′, the material left consists of Fo and Cpx, which may change to saxonite with decreasing temperature. At temperatures higher than P′ but below R, the material left is saxonite, and at temperatures higher than that of R, the material left is dunite. However, wehrlite, clinopyroxenite, websterite, and orthopyroxenite cannot be residual materials left after the partial melting of the "mantle peridotite". These conclusions would still be valid even if small amounts of iron and alumina are added to this ternary system. If iron is added, for example, the invariant points P and P′ are replaced by univariant lines, and the univariant lines are replaced by divariant surfaces. In this case, the residual materials after the partial melting would become more magnesian. Lherzolite, saxonite (or harzburgite), and dunite can both be the residual materials after partial melting or crystal accumulates from basic magmas. These conclusions would also be applicable to the partial melting of the "mantle peridotite" which takes place in a wide pressure range from 1 atm to at least up to 30 kb under anhydrous conditions and also in a wide range of water pressure, although the compositions of liquids formed by the partial melting are different from those at 20 kb.

SUMMARY

1. The Fo–Di join is not binary and is not a thermal barrier at 20 kb under anhydrous conditions, diopside and forsterite crystallized from this join being solid solutions containing small amounts of the components $CaMgSiO_4$ and $MgSiO_3$, respectively. The maximum temperature of coprecipitation of forsterite and diopside solid solutions, which is a piercing point, is $1635\pm10°C$ at 20 kb.

2. The Di–En join is most likely binary and is a thermal barrier at 20 kb under anhydrous conditions. On this join there exist two peritectic points at temperatures near 1650°C. At one peritectic point enstatite solid solution reacts with liquid to form pigeonitic clinopyroxene, and at another peritectic point pigeonitic clinopyroxene reacts with liquid to form diopside solid solution of composition about $Di_{50}En_{50}$.

3. In the subsolidus region of the Di–En join there is a narrow field of pigeonitic clinopyroxene near the composition $Di_{20}En_{80}$. The Ca/(Ca+Mg) ratio of this composition is very close to the Ca/(Ca+Mg+Fe^{2+}) ratio of natural pigeonites. In this and adjoining fields, iron-free pigeonite has been synthesized.

4. In the Fo–Di–En system, there exist two isobaric invariant points on the Fo–pyroxene liquidus boundary. One is a reaction point where enstatite solid solution reacts with liquid to form pigeonitic clinopyroxene and forsterite, and the other may be a ternary eutectic point or a reaction point where pigeonitic clinopyroxene reacts with liquid to form diopside solid solution and forsterite. It is also not certain whether or not a temperature maximum exists on the Fo–Di$_{ss}$ liquidus boundary.

5. The Fo–pyroxene liquidus boundary shifts toward forsterite, and consequently, the primary phase field of forsterite is reduced with an increase of pressure under anhydrous conditions. The reaction Fo + L = Di$_{ss}$ + Pr$_{ss}$ or Fo + L = Di$_{ss}$ + En$_{ss}$ takes place, or the reaction relation between forsterite and pyroxene holds, at pressures less than 6 or 7 kb under anhydrous conditions.

6. The Fo–pyroxene liquidus boundary shifts toward silica with an increase of pressure under hydrous conditions ($P_{H_2O} \cong P_{load}$), and at 20 kb P_{H_2O} forsterite is on the liquidus over a wide compositional range in the Di–En join, and the join is not a thermal barrier.

7. The primary phase fields of diopside and enstatite solid solutions are greatly expanded toward silica at 20 kb P_{H_2O} as compared with those at 1 atm.

8. If partial melting of olivine-rich lherzolite, believed to be an upper mantle material, takes place under anhydrous conditions, silica-undersaturated liquids will be produced at pressures higher than 6 or 7 kb, whereas if partial melting takes place under hydrous conditions ($P_{H_2O} \cong P_{load}$), silica-saturated liquids will be produced at pressures from near 1 atm to at least 30 kb P_{H_2O}.

9. Silica-rich magmas, such as andesite and dacite magmas, may be produced from olivine tholeiitic magmas by fractional crystallization at high pressures under hydrous conditions.

10. The residual material after partial melting of the olivine-rich lherzolite are more magnesian lherzolite, saxonite, and dunite in a wide pressure range under both anhydrous and hydrous conditions. Wehrlite and pyroxenites cannot be residual materials of the olivine-rich lherzolite. They may be products of crystal accumulation from basaltic magmas.

ACKNOWLEDGMENTS

The author wishes to thank Dr. J. F. Schairer for the use of glass and crystallized glass in the system diopside–forsterite–silica and Drs. H. S. Yoder, Jr., D. H. Lindsley, and P. M. Bell for critical reading of the manuscript.

REFERENCES

[*Editors' Note:* Only the references cited in the preceding excerpt are reproduced here.]

Bowen, N. L., 1914, The ternary system diopside–forsterite–silica: Am. Jour. Sci., 4th ser., v. 38, p. 207-264.

Davis, B. T. C., 1964, The system diopside–forsterite–pyrope at 40 kilobars: Carnegie Inst. Washington Year Book 63, p. 165-171.

Green, D. H., and Ringwood, A. E., 1967a, The genesis of basaltic magma: Contr. Mineralogy Petrology, v. 15, p. 103-190.

Ito, K., and Kennedy, G. C., 1967, Melting and phase relations in a natural peridotite to 40 kilobars: Am. Jour. Sci., v. 265, p. 519-538.

Kuno, H., 1964, Aluminian augite and bronzite in alkali olivine basalt from Takasima, north Kyusyu, Japan: Adv. Front. Geology Geophysics, p. 205-220.

Kushiro, I., 1968, Compositions of magmas formed by partial zone melting of the earth's upper mantle: Jour. Geophys. Research, v. 73, p. 619-634.

Kushiro, I., and Kuno, H., 1963, Origin of primary basalt magmas and classification of basaltic rocks: Jour. Petrology, v. 4, p. 75-89.

Kushiro, I., and Schairer, J. F., 1963. New data on the system diopside–forsterite–silica: Carnegie Inst. Washington Year Book 62, p. 95-103.

Kushiro, I., Yoder, H. S., and Nishikawa, M., 1968, Effect of water on the melting of enstatite: Geol. Soc. America, Bull. (in press).

Schairer, J. F., and Yoder, H. S., 1962, The system diopside–enstatite–silica: Carnegie Inst. Washington Year Book 61, p. 75-82.

Stewart, D. B., 1967, Four-phase curve in the system $CaAl_2Si_2O_8$–SiO_2–H_2O between 1 and 10 kilobars: Schweizer. Min. pet. Mitt., v. 47, p. 35-59.

Yoder, H. S., 1964, Diopside–anorthite–water at five and ten kilobars and its bearing on explosive volcanism: Carnegie Inst. Washington Year Book 64, p. 82-89.

Yoder, H. S., and Tilley, C. E., 1962. Origin of basalt magmas: An experimental study of natural and synthetic rock systems: Jour. Petrology, v. 3, p. 342-532.

28

Reprinted from page 311 of *Jour. Petrology* **13**:311-334 (1972)

Effect of Water on the Composition of Magmas Formed at High Pressures

by IKUO KUSHIRO

Geophysical Laboratory, Carnegie Institution of Washington, Washington, D.C. 2008

(Received 14 May 1971; in revised form 17 August 1971)

ABSTRACT

Portions of the system $MgO-CaO-Na_2O-Al_2O_3-SiO_2-H_2O$ have been studied in the pressure range 13–35 kb at near-liquidus temperatures. The liquidus field of forsterite relative to that of orthopyroxene is considerably wider under hydrous than under anhydrous conditions and it covers part of the plane of silica-saturation in a wide pressure range. Partial melting of simple garnet lherzolite (= forsterite + orthopyroxene + clinopyroxene + garnet) with water produces quartz-normative liquids at pressures up to at least 25 kb regardless of water content. Hydrous minerals are not encountered at or near the solidus temperatures except in a Na-rich part of the system. Microprobe analysis of the run products in this synthetic system shows that the liquid (glass) in equilibrium with the lherzolite mineral assemblage is silica- and alumina-rich at 20 kb under vapor-present conditions. With increasing degree of partial melting, the liquid changes its composition, passing into a 'vapor-absent region' and becoming less silicic. Fractional crystallization of olivine tholeiitic magma under hydrous conditions also produces silica-rich magmas at high pressures. If the system is open to water, and water pressure is less than total pressure, the composition of the liquid varies from quartz-normative to olivine ($\pm$ nepheline)-normative depending on water pressure. It is suggested that in the presence of water, silica-rich magmas such as those of calc-alkalic andesite or dacite may be formed by direct partial melting of the peridotitic upper mantle at depths down to about 80 km. A large degree of partial melting of lherzolite under hydrous conditions would produce SiO_2- and MgO-rich magmas. The clinoenstatite rock from Cape Vogel, Papua, may have been formed by such a process. Peridotites with low $CaAl_2SiO_6$/jadeite ratios in the clinopyroxene could produce nepheline-normative magma by small degree of partial melting and tholeiitic magma by large degree of partial melting under hydrous conditions.

173

29

Reprinted from *Carnegie Inst. Washington Year Book 73,* pp. 215–224 (1974)

EFFECT OF CO_2 ON THE MELTING OF
PERIDOTITE

David H. Eggler

Carbon dioxide is the dominant volatile species of fluid inclusions in peridotite xenoliths (Roedder, 1965; H. W. Green, 1972). Because basalt is believed to originate in the upper mantle by the melting of peridotite, the role of CO_2 in peridotite melting is being investigated (Eggler, *Year Book 72*, pp. 457–467). These investigations of synthetic peridotite-CO_2 systems are complementary to the work of Boettcher, Mysen, and Modreski (1974) and Mysen and Boettcher (1974) on the melting of natural peridotite in the presence of H_2O and CO_2. The investigations are supplementary to phase equilibria studies modeling highly alkaline and silica-undersaturated magmas (Wyllie and Haas, 1965; Watkinson and Wyllie, 1965; Koster van Groos and Wyllie, 1968; Boettcher and Wyllie, 1969).

In investigations of both the system Mg_2SiO_4-SiO_2-H_2O-CO_2 (Eggler, *Year Book 72*, pp. 457–467) and natural peridotites (Boettcher, Mysen, and Modreski, 1974) it was found that the effect of CO_2 was opposite the effect of H_2O. Melting in the presence of both CO_2 and H_2O produces liquids that are less silica-saturated than those produced in the presence of H_2O alone. In both studies it is not clear whether changes in melt composition with change in CO_2/H_2O ratio are

due to variation in H_2O activity or to changes in both H_2O activity and CO_2 activity. In other words, is CO_2 essentially an inert component, or does its solubility in silicate melt (Eggler, *Year Book 72*, pp. 457–467; Yoder, *Year Book 72*, pp. 449–457; Eggler, Mysen, and Seitz, this Report) effect large changes in melt structure and melting behavior? The search for the answer led to an investigation of CO_2-saturated melting behavior on several joins in the system Na_2O-CaO-MgO-Al_2O_3-SiO_2-CO_2. The system contains the four phases that comprise peridotite. The compositions examined include several that were previously investigated by Kushiro (1968) under volatile-absent conditions and in the presence of excess H_2O (Kushiro, 1972). Comparison can thus be made of three limiting cases for melting: in the absence of volatiles, in the presence of CO_2, and in the presence of H_2O.

Inherent in these studies is the assumption that gases in experiments are oxidized; that is, that CO_2 and H_2O are dominant species rather than CO, CH_4, and H_2. Eggler, Mysen, and Hoering (this Report) show that this is the case in experiments such as those conducted for this study. Hydrogen fugacity in

174

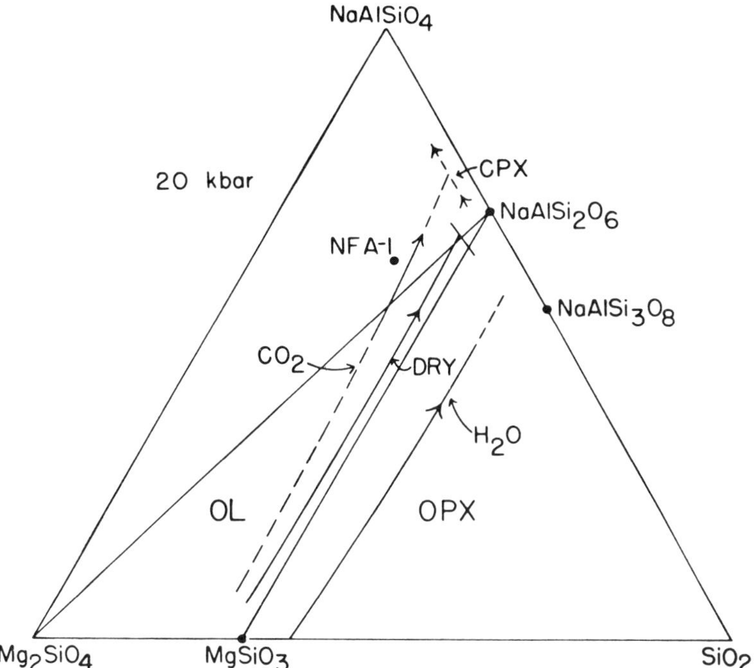

Fig. 1. Estimated position of phase field boundaries on the join Mg$_2$SiO$_4$-SiO$_2$-NaAlSiO$_4$ at 20 kbar pressure for excess-CO$_2$, excess-H$_2$O, and volatile-absent conditions.

capsules during runs is sufficiently low so that under run conditions vapors contain about 98 mole % CO$_2$, 1% CO, and 1% H$_2$O.

The join NaAlSiO$_4$ (nepheline)–Mg$_2$SiO$_4$ (forsterite)–SiO$_2$ is a critical join in the basalt tetrahedron (Yoder and Tilley, 1962) because it contains albite (NaAlSi$_3$O$_8$), and it is critical to peridotite petrogenesis because it contains jadeite (NaAlSi$_2$O$_6$). Composition NFA-1 (Ne$_{62}$Fo$_{18}$ silica$_{20}$, wt %) has been studied. The composition is of interest because under volatile-absent conditions (Fig. 1) NFA-1 lies across the olivine-orthopyroxene divariant boundary curve at 30 kbar and lies near the isobaric invariant point at which olivine, orthopyroxene, jadeitic clinopyroxene, and liquid coexist. (Clinopyroxene was found as a subliquidus run product by Kushiro, 1968. It did not crystallize in the temperature interval investigated in this study.)

The same NFA-1 glass, crystallized at 1 atm, that was used by Kushiro (1968) was loaded into Pt capsules with Ag$_2$C$_2$O$_4$ in amounts to yield about 10 mg total sample and about 20 wt % CO$_2$. The capsules were welded shut and run unbuffered in ½-inch solid-media, high-pressure apparatus.

Results are presented in Fig. 2. Melting is characterized by a change in liquidus phase from olivine to orthopyroxene at a pressure of about 22.5 kbar. The diagram is similar to that determined by Kushiro (1968) for volatile-absent conditions, except that the point at which the liquidus phase changes is displaced to lower pressure by about 7.5 kbar. The olivine liquidus temperature is also lower by about 90°C at 20 kbar, reflecting CO$_2$ solubility in the melt, analogous to CaMgSi$_2$O$_6$-CO$_2$ (Eggler, *Year Book 72*). These results are used in Fig. 1 to estimate the position of phase boundaries for

CO_2-excess conditions at 20 kbar pressure. The composition of the liquid at the invariant point is more silica-undersaturated than that at the volatile-absent invariant point and greatly different in composition from the more silica-saturated liquid produced by melting in the presence of H_2O.

Another join of interest is Mg_2SiO_4-$CaAl_2SiO_6$-SiO_2, as it contains the compositions $CaAl_2SiO_6$ (Tschermak's molecule) and $CaAl_2Si_2O_8$ (anorthite). Composition DP-65 lies near an invariant point at 26 kbar pressure where olivine, orthopyroxene, clinopyroxene, and liquid coexist and also lies near a high-pressure piercing point on the join diopside-pyrope ($Mg_3Al_2Si_3O_{12}$) (O'Hara and Yoder, 1967). The same material, crystallized at 1 atm pressure, that was used by Kushiro (1968) was loaded into Pt capsules with $Ag_2C_2O_4$ sufficient to yield 20 wt % CO_2 and was run as described above. Results are presented in Fig. 3 and contrasted with volatile-absent experiments (Kushiro, 1968). As with the NFA-1 composition, in the presence of CO_2 the pressure at which the liquidus phase changes from olivine to orthopyroxene is shifted to lower pressure relative to volatile-absent melting (26 kbar to 20 kbar). Liquidus temperatures are lowered about 75°C at 20 kbar pressure. With CO_2, as in the volatile-absent experiments, olivine or orthopyroxene is joined by clinopyroxene 10°–20°C below the liquidus. Composition DP-65 must lie very near the invariant point at 20 kbar (Fig. 4).

The third join studied was $CaMgSi_2O_6$-Mg_2SiO_4-SiO_2. Results are summarized in Fig. 5. Dramatic shifts of phase boundaries in the presence of CO_2 are evident. A critical join is $CaMgSi_2O_6$-Mg_2SiO_4;

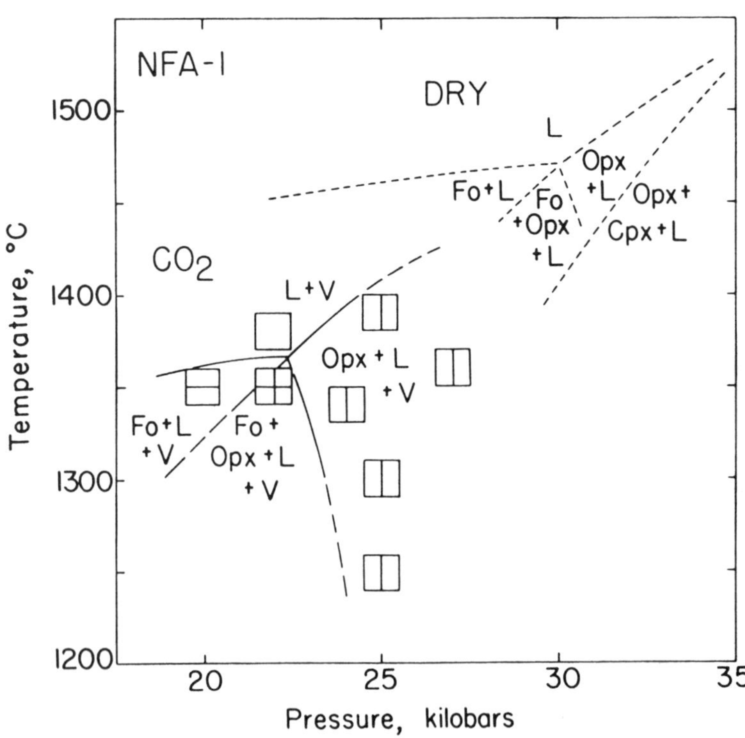

Fig. 2. Results of quenching experiments on NFA-1 crystallized glass with about 20 wt % CO_2. Volatile-absent phase equilibria are after Kushiro (1968).

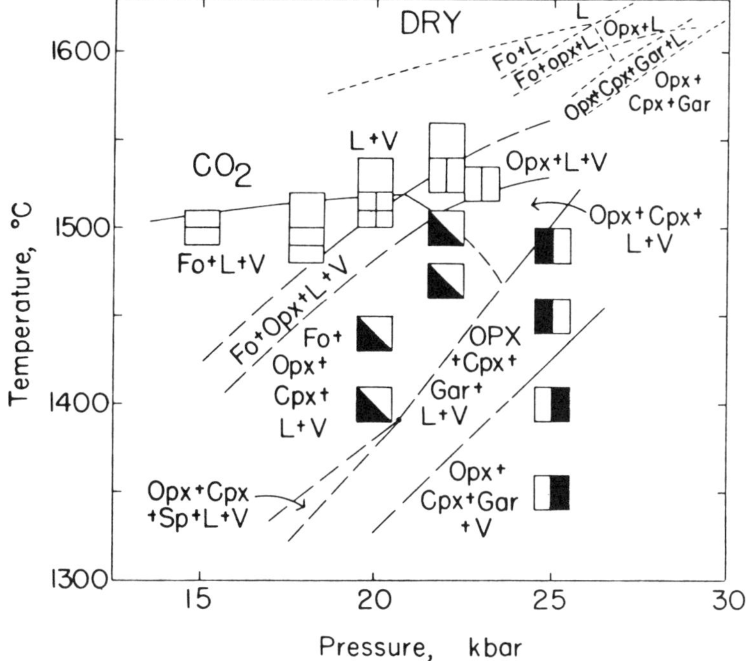

Fig. 3. Results of quenching experiments on DP-65 crystallized glass with about 20 wt % CO_2. Volatile-absent phase equilibria are after Kushiro (1968). The position of the spinel field is estimated from data of Kushiro and Yoder (1966).

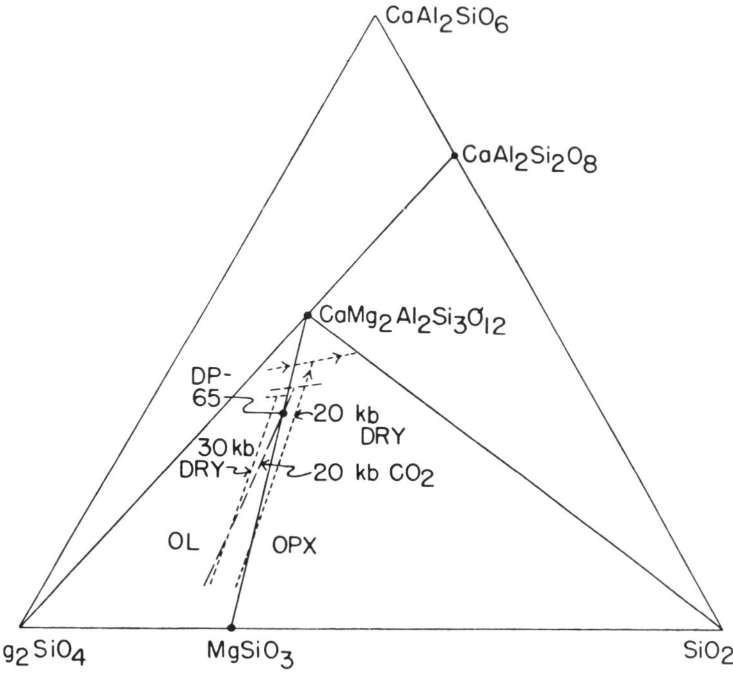

Fig. 4. Estimated position of phase field boundaries on the join Mg_2SiO_4-SiO_2-$CaAl_2SiO_6$ for excess-CO_2 and volatile-absent conditions.

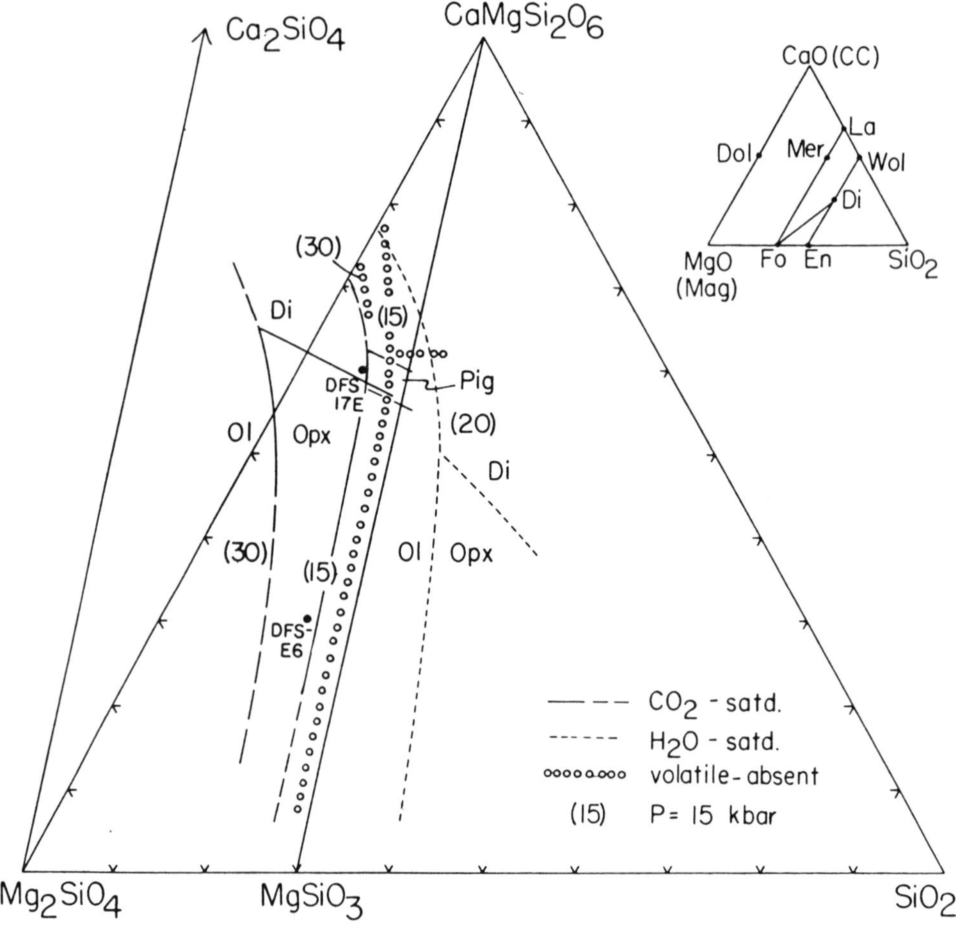

Fig. 5. Estimated position of phase field boundaries in a portion of the system CaO-MgO-SiO$_2$-CO$_2$. Boundaries with CO$_2$ are long-dashed where estimated.

phase relations on that join at 20 and 30 kbar pressure are shown in Figs. 6 and 7. Two types of starting materials were used. Several glasses or mixtures of glass and crystal synthesized by Kushiro and Schairer (*Year Book 62*, p. 96) were loaded with Ag$_2$C$_2$O$_4$. For other compositions, mechanical mixtures of CaCO$_3$, silica, and MgO were finely ground for 70 minutes and used as starting materials, either alone or with Ag$_2$C$_2$O$_4$ to supply CO$_2$ in addition to the amount generated by breakdown of CaCO$_3$.

Run products on the join were examined by a combination of optical petrography, x-ray diffractograms, and electron microprobe analyses. Microprobe analyses were particularly useful in determining the presence of orthopyroxene and pigeonite and in distinguishing stable clinopyroxene crystals from quench crystals (quenched liquid), which can attain large size. Quench crystals typically have compositions that lie within the orthopyroxene (or pigeonite)-diopside solvus (Kushiro, 1969) and are quite variable in composition. A useful method of detecting quench liquid was from the presence of quench carbonate among other quench crystals.

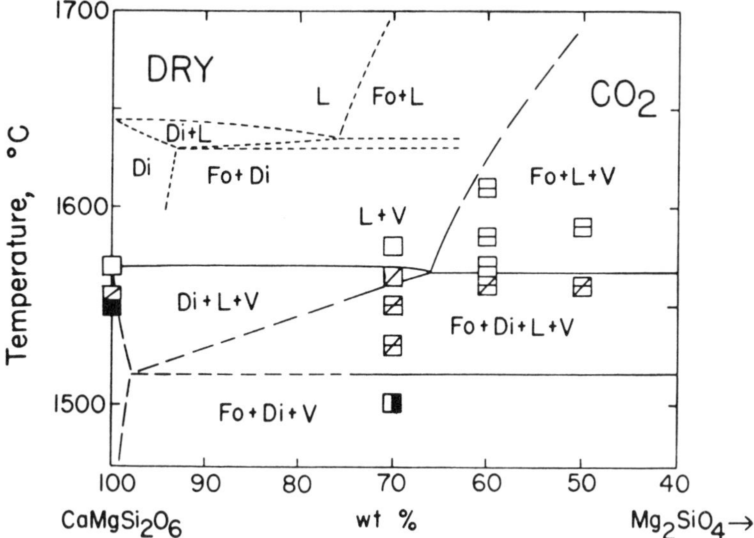

Fig. 6. Phase relations on the join CaMgSi₂O₆-Mg₂SiO₄ at 20 kbar pressure with 15–20 wt % CO₂ (CO₂-saturated).

The join diopside-forsterite at 20 kbar with excess CO_2 is compared with volatile-absent melting relations (Kushiro, 1969) in Fig. 6. The piercing point at which diopside, forsterite, CO_2-bearing liquid, and CO_2 vapor coexist lies at a composition of approximately $Fo_{35}Di_{65}$ and a temperature of about 1570°C, compared with $Fo_{25}Di_{75}$ and 1635°C under volatile-absent conditions. Another difference between CO_2-present and volatile-absent conditions is the significantly greater temperature range between liquidus and solidus with CO_2.

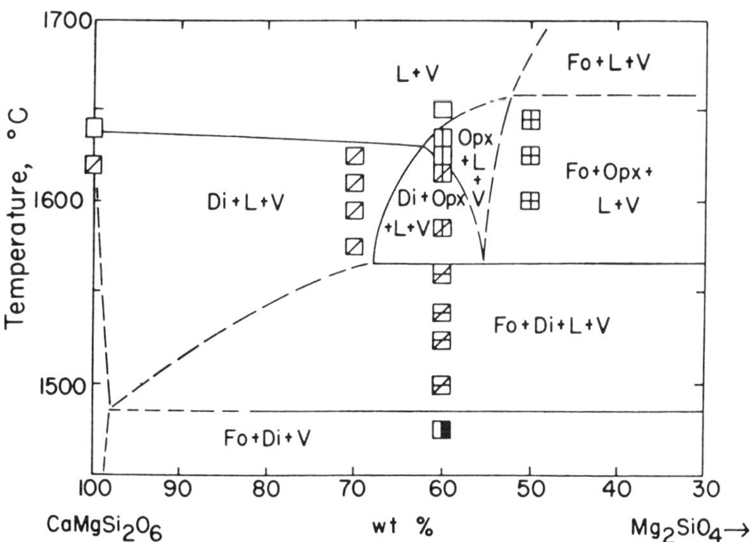

Fig. 7. Phase relations on the join CaMgSi₂O₆-Mg₂SiO₄ at 30 kbar pressure under CO₂-saturated conditions.

179

The join CaMgSi$_2$O$_6$-Mg$_2$SiO$_4$ at 30 kbar with excess CO$_2$ features the appearance of orthopyroxene on the liquidus (Fig. 7). Near-liquidus pyroxene for bulk composition Fo$_{40}$Di$_{60}$ has optical properties and composition (Ca$_{0.09}$Mg$_{1.01}$Si$_2$O$_6$) that indicate that it is orthopyroxene. Appearance of the orthopyroxene primary phase field indicates that the invariant point at which orthopyroxene, diopsidic clinopyroxene, olivine, and liquid coexist does not lie within the compositional triangle Fo-En-Di, as it does at 30 kbar dry, but rather lies to the left of the Di-Fo join (Fig. 5).

Two other compositions, prepared as CaCO$_3$-SiO$_2$-MgO mechanical mixtures, have been run to further define phase field boundaries within the compositional triangle Di-Fo-En. Composition DFS-17E has pigeonite on the liquidus at 17 kbar and olivine + pigeonite (Ca$_{0.18}$Mg$_{1.82}$Si$_2$O$_6$) on the liquidus at 15 kbar (Fig. 8). The olivine-pigeonite divariant boundary curve must pass through this composition at a pressure of 15 kbar (Fig. 5). These relations indicate that at 15 kbar, olivine, diopsidic clinopyroxene, and liquid coexist with pigeonite at an isobaric invariant point, as Kushiro (1969) found under volatile-absent conditions at 15 kbar. At 30 kbar with CO$_2$, olivine, diopsidic clinopyroxene, and liquid coexist with orthopyroxene at an isobaric invariant point (Figs. 5 and 7). The exact relations between the orthopyroxene and pigeonite primary phase fields at 15 and 30 kbar are not known, although phase relations of composition DFS-17E at 25 kbar suggest a reaction relationship between diopside, pigeonite, orthopyroxene, and liquid (Figs. 5 and 8).

Runs with composition DFS-E6 (Fig. 5) help define the olivine-orthopyroxene boundary curve with CO$_2$. The composition lies on that boundary at 30–40 kbar pressure under volatile-absent conditions (estimated from Kushiro, 1969). With CO$_2$ (Fig. 9), the liquidus phase changes from olivine (closely followed in crystallization by orthopyroxene) to orthopyroxene at a pressure of about 17 kbar.

These melting experiments in the pres-

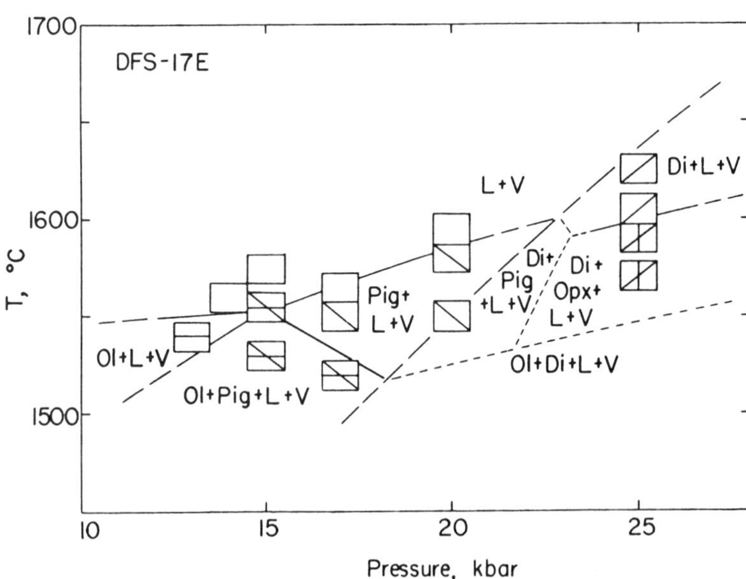

Fig. 8. Preliminary liquidus phase relations for composition DFS-17E under CO$_2$-saturated conditions.

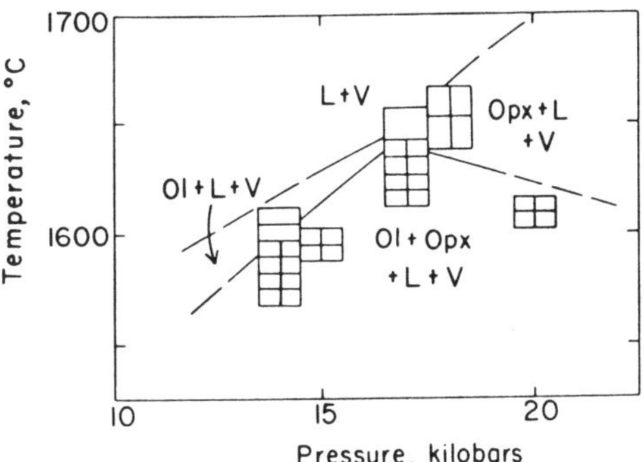

Fig. 9. Phase relations for composition DFS-E6 under CO_2-saturated conditions.

ence of CO_2 on joins of the peridotite system prove that CO_2 is not an inert component but rather that it dissolves in silicate melt and causes distinct changes in melt structure. In every join studied, isobaric invariant points are shifted to silica-undersaturated compositions relative to the position of these points in the absence of volatiles. The chief effect of CO_2 solution is to increase the size of the orthopyroxene primary phase field relative to its size when no volatiles are present. The magnitude of the effect increases with CO_2 pressure, as seen in the SiO_2-MgO-CaO-CO_2 system (Fig. 5). The primary phase field of clinopyroxene also increases in size relative to the olivine field (Figs. 4 and 5).

The only other peridotite melting studies with CO_2 at high pressures are those of Boettcher, Mysen, and Modreski (1974) and Mysen and Boettcher (1974). Glasses produced by melting natural peridotite in the pressure range 10–15 kbar using various mixtures of CO_2 and H_2O were analyzed with the electron microprobe. Glasses produced in the presence of vapor, varying in composition from 100 to 20 mole % H_2O, changed from quartz-normative through olivine-hypersthene normative and olivine-nepheline normative to nepheline-larnite normative. Experiments were not made on the same peridotite compositions with-

out volatiles. These results are broadly consistent with the pattern established from simple systems (Eggler, *Year Book 72*, pp. 457–467; this Report). They differ in detail, in particular in the pressure needed for liquids to become larnite-normative.

These new results have important implications for genesis of basalts. Models explaining the origin of basalt types are based principally on three factors—total pressure, H_2O pressure, and degree of partial melting. Another factor, CO_2 pressure, must now be added.

It is believed that CO_2 in the mantle is especially important to genesis of alkaline basalts. Fluid inclusions both in olivine of peridotite xenoliths in alkaline basalts (Roedder, 1965) and in phenocrysts (Sobolev, Bazarova, and Bakumenko, 1972) contain predominantly CO_2. Compositions of nepheline basanites, nephelinites, and melilite nephelinites do not have orthopyroxene on the liquidus in volatile-absent melting experiments at pressures to 40 kbar (Bultitude and D. H. Green, 1967; D. H. Green, 1973). These rocks cannot, therefore, be partial melting products of peridotite in the absence of volatiles. According to Bultitude and D. H. Green (1967) and D. H. Green (1973), small amounts of H_2O will stabilize orthopyroxene on the liquidus. Other petrologists have argued that H_2O decreases the primary phase field of orthopyroxene (e.g., Kushiro, 1972). Results presented here and those of Boettcher, Mysen, and Modreski (1974) indicate that in the presence of CO_2, not H_2O, the orthopyroxene field will expand, and these compositions then exhibit liquidus orthopyroxene. Therefore, if certain basanites, nephelinites, and melilite nephelinites are direct melting products of peridotite, they must be produced in regions of the mantle rich in CO_2. The absolute quantity of CO_2 need not be great. Silica-undersaturated basaltic melts are believed to be derived by only 3–7% melting (Gast, 1968). About 5 wt % CO_2 can dissolve in basalt melt at a pressure of 20–25 kbar (Eggler, Mysen, and Seitz, this Report). Thus in a mantle containing 0.15–0.35% CO_2 such melts would be CO_2-saturated. Other alkali basalts for which either volatile-absent or volatile-present melting origins are possible, on the basis of phase equilibria studies, can be generated at shallower depths in the mantle in the presence of CO_2 than is possible under volatile-absent conditions. Results of experiments on synthetic joins indicate that the minimum pressure for generation of silica-undersaturated melts will be decreased by an increment of 7 kbar (Mg_2SiO_4-SiO_2-$CaAl_2SiO_6$, Fig. 4) to 15 kbar (CaO-MgO-SiO_2-CO_2, Fig. 5).

The system CaO-MgO-SiO_2-CO_2 is a useful model with which to consider relations between peridotite, basaltic liquids, and carbonatite. Melting of peridotite (Fo-Di-En) under dry conditions to at least 40 kbar yields liquids within the subtriangle Fo-Di-En. With CO_2 at 30 kbar, the Fo-Di join is broken, and the invariant point lies in larnite (Ca_2SiO_4)-normative space, at a temperature of about 1560°C. Huang and Wyllie (1974) have shown that in the system CaO-SiO_2-CO_2 the join Ca_2SiO_4-CO_2 is not a thermal barrier; liquids in the composition range Ca_2SiO_4-$CaSiO_3$-CO_2 fractionate to yield calcite-rich liquid. Yoder (*Year Book 68*, pp. 449–457) has shown that at a pressure above 6 kbar, diopside + calcite (equivalent to akermanite [$Ca_2MgSi_2O_7$] + CO_2) melt to diopside, liquid, and vapor, implying that silicate melt richer in calcite than akermanite composition coexists with diopside. It thus appears in CaO-MgO-SiO_2-CO_2 that there is a high-temperature fractionation path from larnite-normative primary silicate melt (i.e., melilite-bearing) to calcite-rich liquid (carbonatite).

It is evident that a mantle rich in CO_2 in certain portions is an attractive model to explain the alkaline group of igneous rocks. If this is a reasonable hypothesis, it is also reasonable that there are variations in volatile content of the mantle, leading to variation in chemical character of partial melts. The volatile composition of the mantle when it was homogeneous is not known, but it probably consisted largely of CO_2 and H_2O if the outer part of the earth accreted from low-temperature, primitive, volatile-rich material (Turekian and Clark, 1969; Clark, Turekian, and Grossman, 1972). Such material might have been similar to type-I carbonaceous chondrites, which have an atom ratio of H/C of about 6 (Wiik, 1956). (Dominant species are *assumed* to be CO_2 and H_2O; however, Brett, 1971, has argued that H_2, CH_4, and CO are important species in the mantle.) Heterogeneity of mantle volatiles is believed to have been produced chiefly by magmatic processes. One process strongly fractionating volatiles would be formation of protocontinents by melting of the mantle and subsequent growth of those continents. Because granitic melts dissolve relatively little CO_2 but considerable H_2O (Eggler, *Year Book 72*, pp. 457–467), melting processes increase the CO_2 content of the residue (mantle) relative to the original material. Rubey (1951) has estimated that excess volatiles, the volatiles contained in the atmosphere, hydrosphere, and sedimentary rocks minus volatiles accounted for by rock weathering, are dominantly H_2O and CO_2. He believed that these volatiles have been added throughout geologic time. The ratio of H/C is 88 in the excess volatiles, suggesting strong fractionation of H_2O into the atmosphere and hydrosphere by additive processes. The mantle beneath continents would accordingly be CO_2-rich, unless that mantle were decoupled from the overlying lithosphere by plate motion.

Basaltic magmatic activity would also deplete source mantle in H_2O, but to a lesser extent, because basalt magmas dissolve more CO_2 than granite magmas (Eggler, Mysen, and Seitz, this Report). Volatile depletion may explain magma variation within volcanic centers, as in the Hawaiian Islands, where volcanoes built of tholeiite basalt flows are capped by alkaline basalts. Melting of mantle with CO_2-H_2O vapor containing approximately 40–60 mole % H_2O in the pressure range 15–20 kbar would produce tholeiite magma (Mysen and Boettcher, 1974; Eggler, *Year Book 72*, pp. 457–467). Source mantle would be depleted in H_2O relative to CO_2 and, provided all vapor were not dissolved in the melt and removed, would produce nepheline-normative basalt upon further melting.

Removal of all vapor would leave a volatile-absent mantle. Because there would be no volatile fractionation during later melting episodes, such a mantle might produce magma that is relatively homogeneous chemically. Gast (1968) has suggested that oceanic abyssal tholeiite basalts are derived from mantle previously depleted by a small degree of melting. If that small amount of melt were alkaline, it might have dissolved sufficient CO_2, as well as H_2O, to leave a mantle that was volatile-free.

In summary, evidence from volcanic gases and fluid inclusions in peridotite minerals and from cosmochemical considerations indicates that CO_2 is a volatile component of the mantle. Variation in the amount of H_2O and CO_2 can influence the derivation of magmas ranging from quartz-normative to olivine- nepheline normative by partial melting. Melting in mantle regions containing CO_2 but relatively little H_2O produces magmas more silica-undersaturated than can be produced by melting in the absence of volatiles. This process is believed to explain the origin of primary nephelinites and melilite nephelinites.

REFERENCES

Boettcher, A. L., B. O. Mysen, and P. J. Modreski, Melting in the Mantle: phase relationships in natural and synthetic peridotite-H_2O and peridotite-H_2O-CO_2 systems at high pressures, *Phys. Chem. Earth*, in press, 1974.

Boettcher, A. L., and P. J. Wyllie, The system CaO-SiO_2-CO_2-H_2O, III, Second critical end-point on the melting curve, *Geochim. Cosmochim. Acta, 33*, 611–632, 1969.

Brett, R., The earth's core: speculations on its chemical equilibrium with the mantle, *Geochim. Cosmochim. Acta, 35*, 203–221, 1971.

Bultitude, R. J., and D. H. Green, Experimental study at high pressures on the origin of olivine nephelinite and olivine melilite nephelinite magmas, *Earth Planet. Sci. Lett., 3*, 325–337, 1967.

Clark, S. P., Jr., K. K. Turkian, and L. Grossman, Model for the early history of the earth, in *The Nature of the Solid Earth*, E. C. Robertson, ed., McGraw-Hill Book Company, New York, pp. 3–18, 1972.

Gast, P. W., Trace element fractionation and the origin of tholeiitic and alkaline magma types, *Geochim. Cosmochim. Acta, 32, 1057–1086*, 1968.

Green, D. H., Experimental melting studies on a model upper mantle composition at high pressure under water-saturated and water-undersaturated conditions, *Earth Planet. Sci. Lett., 19*, 37–45, 1973.

Green, H. W., II, A CO_2 charged asthenosphere, *Nature, 238*, 2–5, 1972.

Huang, W. L., and P. J. Wyllie, Liquidus relationships between carbonates and silicates in parts of system CaO–MgO–SiO$_2$–CO$_2$–H$_2$O at 30 kb (abstract), *Eos, Trans. Amer. Geophys. Union, 55,* 479–480, 1974.

Koster van Groos, A. F., and P. J. Wyllie, Liquid immiscibility in the join NaAlSi$_3$O$_8$–Na$_2$CO$_3$–H$_2$O and its bearing on the genesis of carbonatites, *Amer. J. Sci., 266,* 932–967, 1968.

Kushiro, I., Compositions of magmas formed by partial zone melting of the earth's upper mantle, *J. Geophys. Res., 73,* 619–634, 1968.

Kushiro, I., The system forsterite-diopside-silica with and without water at high pressures, *Amer. J. Sci., Schairer Vol. 267A,* 269–294, 1969.

Kushiro, I., Effect of water on the composition of magmas formed at high pressures, *J. Petrology, 13,* 311–334, 1972.

Kushiro, I., and H. S. Yoder, Jr., Anorthite-forsterite and anorthite-enstatite reactions and their bearing on the basalt-eclogite transformation, *J. Petrology, 7,* 337–362, 1966.

Mysen, B. O., and A. L. Boettcher, Melting in a hydrous mantle, II, Geochemistry of crystals and liquids formed by anatexis of mantle peridotite with controlled activities of H$_2$O, CO$_2$, and O$_2$, *J. Petrology,* in press, 1974.

O'Hara, M. J., and H. S. Yoder, Jr., Formation and fractionation of basic magmas at high pressures, *Scot. J. Geol., 3,* 67–117, 1967.

Roedder, E., Liquid CO$_2$ inclusions in olivine-bearing nodules and phenocrysts from basalts, *Amer. Mineral., 50,* 1746–1782, 1965.

Sobolev, V. S., T. Y. Bazarova, and I. T. Bakumenko, Crystallization temperature and gas phase composition of alkaline effusives as indicated by primary melt inclusions in the phenocrysts, *Bull. Volcanol., 35,* 479–496, 1972.

Turekian, K. K., and S. P. Clark, Jr., Inhomogeneous accumulation of the earth from the primitive solar nebula, *Earth Planet. Sci. Lett., 6,* 346–348, 1969.

Watkinson, D. H., and P. J. Wyllie, Phase relationships on the join NaAlSiO$_4$ –CaCO$_3$–(25%) H$_2$O and the origin of some carbonatite-alkalic rock complexes, *Can. Mineral., 8,* 402–403, 1965.

Wiik, H. B., The chemical composition of some stoney meteorites, *Geochim. Cosmochim. Acta, 9,* 279–289, 1956.

Wyllie, P. J., and J. L. Haas, The system CaO–SiO$_2$–CO$_2$–H$_2$O: I, Melting relationships with excess vapor at 1 kilobar pressure, *Geochim. Cosmochim. Acta, 29,* 891–892, 1965.

Yoder, H. S., Jr., and C. E. Tilley, Origin of basalt magmas: an experimental study of natural and synthetic rock systems, *J. Petrology, 3,* 342–532, 1962.

[*Editors' Note:* The phase diagrams in this paper were preliminary and some appear in revised form in Eggler, D. A., 1978, Effect of CO$_2$ upon Partial Melting of Peridotite in the System Na$_2$O–CaO–Al$_2$O$_3$–MgO–SiO$_2$–CO$_2$ to 35 kb with an Analysis of Melting in a Peridotite–H$_2$O–CO$_2$ System, *Am. Jour. Sci.* **278**:305–343.]

Melting of a Hydrous Mantle :
II. Geochemistry of Crystals and Liquids Formed by Anatexis of Mantle Peridotite at High Pressures and High Temperatures as a Function of Controlled Activities of Water, Hydrogen, and Carbon Dioxide

by BJØRN O. MYSEN *and* A. L. BOETTCHER

ABSTRACT

The system peridotite–H_2O–CO_2 serves as a simplified model for the phase relations of mantle peridotite involving more than one volatile component. Run products obtained in a study of phase relations of four mantle peridotites in the presence of H_2O- and (H_2O+CO_2)-bearing vapors and with controlled hydrogen fugacity (f_{H_2}) at high pressures and temperatures have been subjected to a detailed chemical investigation, principally by the electron microprobe.

Mg/(Mg+ΣFe) of all phases generally increases with increasing temperature and with increasing Mg/(Mg+ΣFe) of the starting material. This ratio appears to decrease with increasing pressure for olivine, and for amphibole coexisting with garnet. Decreasing f_{H_2} from that of IW buffer to that of MH buffer decreases Mg/(Mg+ΣFe) of the partial melt from approximately 0·85 to approximately 0·50, whereas the Fo content of coexisting olivine increases slightly less than 3 per cent and the Mg/(Mg+ΣFe) of clinopyroxene increases about 4 per cent. However, the variations in Fo content of olivines are within those observed in olivines from natural mantle peridotite. The chemistry of other silicate minerals does not significantly reflect variations of f_{H_2}. Consequently, the peridotite mineralogy and/or chemistry is not a good indicator for the f_{H_2} conditions during crystallization.

All crystalline phases, except amphibole, and to some extent garnet, show increasing Cr content with increasing temperature and increasing Cr content of the starting material, resulting in a positive correlation with Mg/(Mg+ΣFe). Partial melts are depleted in Cr_2O_3 relative to the crystalline phases. High Mg/Mg+ΣFe) and Cr_2O_3 are thus expected in crystal residues after partial melting. The absolute values depend on degree of melting and the composition of the parent peridotite.

Liquids formed by anatexis of mantle peridotite are andesitic under conditions of $X^v_{H_2O} \gg$ 0·6 to at least 25 kb total pressure and to more than 200 °C above the peridotite solidus. This observation supports numerous suggestions that andesite genesis in island arcs may result from partial melting of underlying peridotite mantle. In contrast to basaltic rocks, the absence of amphibole (paragasitic hornblende) does not affect the silica-saturated nature of the liquids. Increasing K_2O content of the starting material (up to 1 wt. per cent K_2O) results in increasing potassium content of the amphibole ($\sim$1 wt. per cent K_2O) as well as the appearance of phlogopite. The liquid under these conditions is relatively K_2O-poor (less than 1 wt. per cent K_2O).

Partial melts are olivine normative with $X^v_{H_2O} \leqslant 0.5$, and initial liquids contain normative ol and ne at $X^v_{H_2O} \leqslant 0.4$. The alkalinity of these liquids increases with decreasing $X^v_{H_2O}$ below values of 0.5. The (ol+opx)-normative liquids resemble oceanic basalts whereas (ol+ne)-normative liquids resemble olivine nephelinite and melilite basalt. Low a_{H_2O} and high a_{CO_2} conditions may be those under which kimberlites and related rocks are formed in the mantle.

[*Editors' Note:* Material has been omitted at this point.]

186

With $X^v_{H_2O} < 0.5$, liquids formed by small degrees of melting are olivine- and often also nepheline-normative and extremely calcic, aluminous and alkaline, producing normative larnite at low pressures (Table 11). Ca and Al are likely to decrease with increasing pressure when garnet is stabilized, but analytical data are lacking because of overwhelming technical difficulties in analyzing hydrous liquids formed at $P > 20$ kb. With increasing temperature at $X^v_{H_2O} \leqslant 0.4$, the liquid compositions shift from nepheline-normative to hypersthene-normative at 10–15 kb; (ol+opx)-normative liquids at 1150 °C and 10 kb at $X^v_{H_2O} = 0.4$ shift to (ne+ol)-normative by decreasing $X^v_{H_2O}$ to 0.2 ($X^v_{CO_2} = 0.8$) (Table 11). The variations are shown in Figs. 22 and 23. This picture is not quite correct because $Fe^{3+}/(Fe^{3+}+Fe^{2+}) > 0$, and we have plotted in Fig. 23 one point with all iron as Fe_2O_3 to illustrate this phenomena. Fe^{3+}/Fe^{2+} determines whether a liquid is quartz- or olivine-normative for SiO_2 values close to the plane of silica

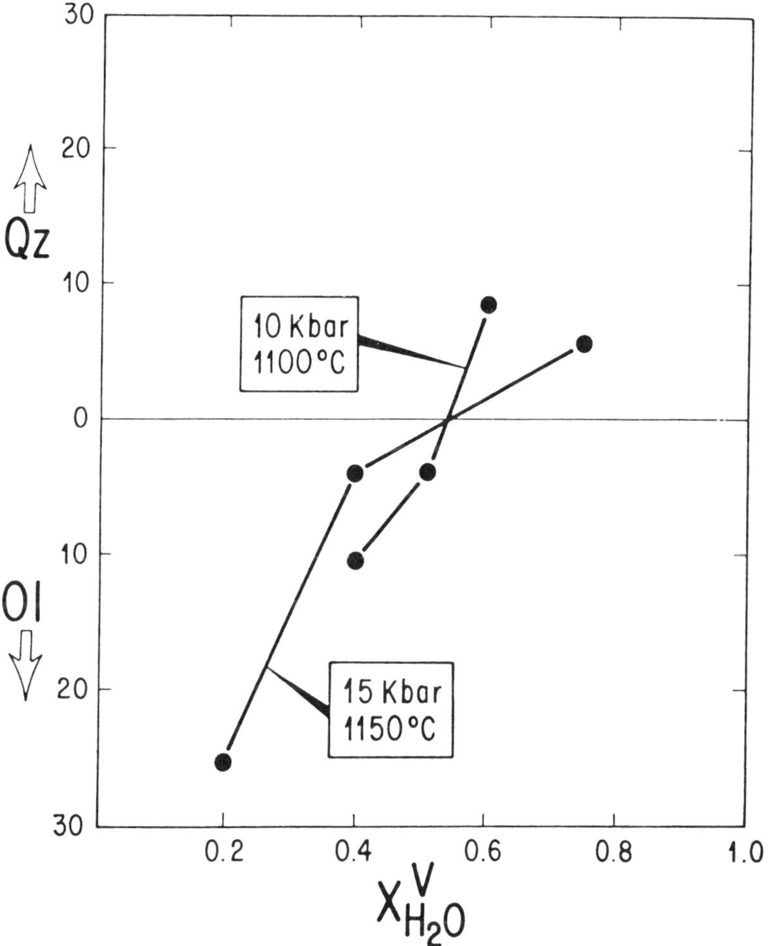

FIG. 22. Degree of silica saturation of partial melts as a function of P, T, and $X^v_{H_2O}$, with total iron as FeO.

saturation and whether it is (ne+ol)- or (ol+opx)-normative near the critical plane of silica undersaturation.

Changes in f_{H_2} may affect the liquid compositions, because the SiO_2 content of amphiboles formed with low f_{H_2} and $X^v_{H_2O} < 1·0$ is lower ($SiO_2 = 42$–43 wt. per cent) than that of the starting material, where the reverse is true when $X^v_{H_2O} \sim 1·0$ and f_{H_2} (NNO) (Table 4).

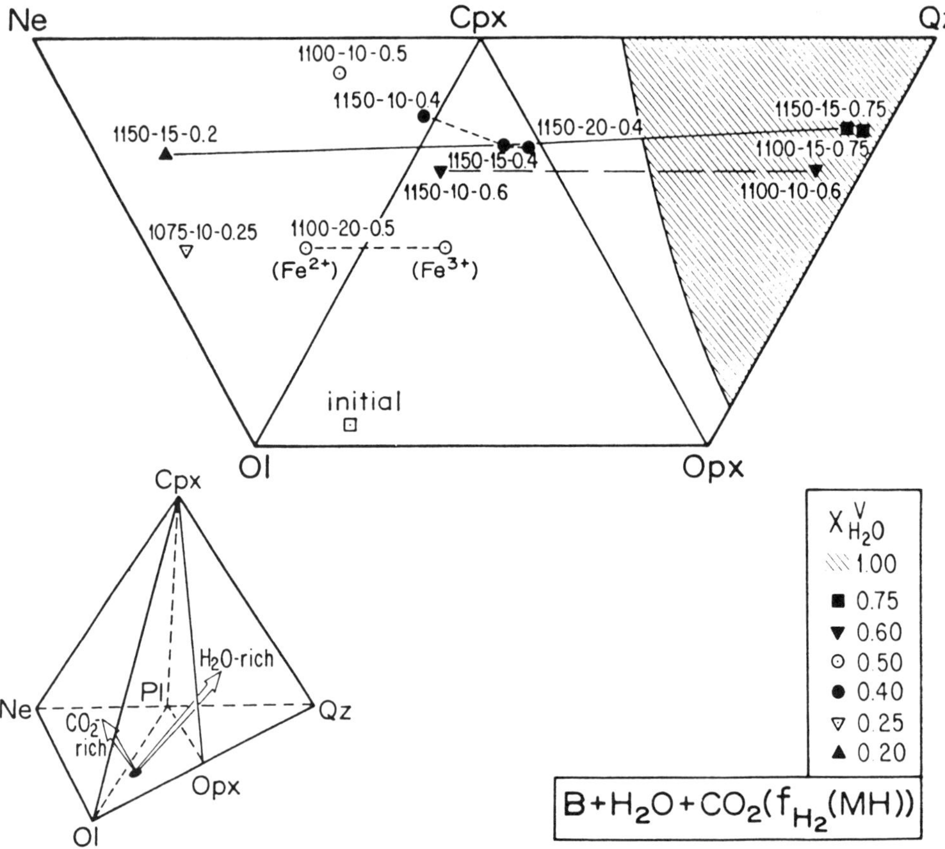

FIG. 23. Various liquid compositions in terms of normative ne, ol, opx, cpx, and qz with variable P, T, and $X^v_{H_2O}$. The sub-horizontal lines connect compositions at similar temperatures and/or $X^v_{H_2O}$.

The compositional variations outlined above can be understood by considering the phase relations in the systems Fo–An–Qz, Fo–Ne–Qz (Kushiro, 1972) and En–H₂O–CO₂ (Eggler, 1973) at high pressures. The data of Eggler (1973) show that at $X^v_{H_2O} < 0·58$, $En+V_{(H_2O+CO_2)}$ will melt according to the reaction:

$$En+V_1 \rightleftharpoons L+V_2 \tag{5}$$

whereas, at $X^v_{H_2O} > 0·58$, initial melting is described by:

$$En+V_1 \rightleftharpoons Fo+L+V_2 \tag{6}$$

Such melting reactions are divariant as H_2O is preferentially dissolved in the silicate liquids, and CO_2/H_2O increases over the melting interval. The same effect results in an interval, in Eggler's experiments, between $X^v_{H_2O} = 0.58$ and $X^v_{H_2O} = 0.75$, where reaction (6) is replaced by reaction (5). The width of this $X^v_{H_2O}$ interval depends on the solid/vapor ratio.

The liquid of Eqn.(6) is silica saturated whereas that of (5) is not. The transition occurs between $X^v_{H_2O} = 0.5$ and 0.6 in our experiments, which strongly suggests that the melting of orthopyroxene controls the liquid composition. The fact that, with increasing temperature, liquids with $X^v_{H_2O} = 0.6$ begin to resemble tholeiite before those with $X^v_{H_2O} = 0.75$ (Table 9) is probably a reflection of orthopyroxene shifting from melting according to (6) to (5) above the peridotite solidus. However, this change does not explain the appearance of Ca and Al-rich (ol+ne)-normative liquids with $X^v_{H_2O} < 0.4$ (Table 11). We may speculate that the reactions:

$$Jd+Fo+V_1 \rightleftharpoons En+V_2+L \tag{7}$$

and

$$Ca{-}Ts+Fo+V_1 \rightleftharpoons En+V_2+L \tag{8}$$

may occur at low $X^v_{H_2O}$. The high a_{CO_2} may also help promote the crystallization of orthopyroxene (Kushiro, 1973b; Eggler, 1974). Once in the (ne+ol)-normative field, Kushiro (1972) has shown that in the model mantle system Ne–Fo–CaTs–Qz$\pm H_2O$, increasing temperature shifts the liquid composition to (opx+ol)-normative. Kushiro's data also show that decreasing $X^v_{H_2O}$ will shift the liquid composition back towards the silica-poor, alkaline part of the system. There is thus agreement between the variations of liquid compositions with T and f_{H_2O} in the peridotite studied here and those in simplified model mantle systems.

In summary, it appears that we can produce almost any liquid composition from andesite to olivine nephelinite by partial melting of mantle peridotite provided we select the appropriate starting material composition, temperature, pressure, f_{H_2}, f_{H_2O}, and f_{CO_2}.

GEOLOGICAL APPLICATIONS

Our data show that andesitic liquids can form by anatexis of wet mantle peridotite over a broad range of temperatures and pressures and also within a limited range of f_{H_2O}. The Mg/(Mg+ΣFe) of such andesite is positively correlated with Mg/(Mg+ΣFe) of the starting material and with f_{H_2}.

The relation between andesitic and basaltic volcanism in island arcs has been a matter of intensive investigation (e.g. Kuno, 1966; Dickinson and Hatherton, 1967; McBirney, 1969; Gill, 1970; Boettcher, 1973). These rocks are closely related in space and time (e.g. Gill, 1970). They also exhibit similar Sr^{87}/Sr^{86} ratios (Hedge, 1966; Hedge, Hildreth, and Henderson, 1970; Peterman, Car-

189

michael, and Smith, 1970), and often have similar and low concentrations of trace elements (e.g. Taylor, 1969; Taylor, Ewart, and Kopp, 1968). Thus, although related genetically, fractional crystallization is hardly the mechanism that relates them. H_2O-bearing peridotite is probably present in or above the subducting slab, and initial melting of peridotite can yield andesitic magmas. Lowering of a_{H_2O}, for example by increasing degree of melting, results in basaltic melts. Such andesite and basalt will exhibit close similarity in the geochemical parameters listed above.

Dickinson (1970) suggested a positive correlation between depth of generation of island-arc andesites and K content. There is no pressure or temperature dependence of the K content of the liquids below at least 1150 °C at pressures below 15 kb because of the presence of phlogopite and/or potassic amphibole that buffer the K content of the liquid at a low level (<1 wt. per cent K_2O). The data of Modreski and Boettcher (1972; 1973) suggest that this situation is maintained to at least 35 kb pressure. Within the framework of the model for the formation of island-arc andesite suggested here, the variations of K content discussed by Dickinson (1970) must relate to a mechanism operative subsequent to the partial melting of peridotite.

The composition of liquids formed immediately above the peridotite solidus for any $X_{H_2O}^v < 0.5$ are strongly alkaline and resemble olivine melilites and olivine nephelinites (Table 11), which are spatially closely associated with kimberlite, and sometimes carbonatite (e.g. Ukhanov, 1965; Egorov, 1970). However, our liquids are too low in Na_2O, K_2O, TiO_2 and too high in CaO and Al_2O_3 for a perfect match. Our analyzed Na_2O is too low because of analytical error. The low K_2O and TiO_2 could probably be alleviated by melting phlogopite-bearing peridotite. However, the activity of H_2O must be sufficiently low for the breakdown of phlogopite and the solidus of peridotite to coincide. The Ca and Al contents will decrease with increasing pressure as we stabilize garnet and more aluminous pyroxene. Kimberlite does not contain amphibole but is commonly rich in phlogopite. Our data thus place a low pressure limit of approximately 20 kb for the formation of such rocks. The intercept between peridotite solidi at $X_{H_2O}^v < 0.5$ and the shield geotherm (see Fig. 5 of the preceding paper) occurs at a depth of 120 km or more. Our data suggest that phlogopite-bearing peridotite would melt to give melilitic and nephelinitic rocks at these pressures. Consequently, melting of peridotite relevant to kimberlite genesis probably occurs at $P_{total} \geqslant 35$ kb under conditions of relatively low f_{H_2O} and relatively high f_{CO_2}.

The intercept between the oceanic geotherm and peridotite solidi at $X_{H_2O}^v = 0.5$–0.25 occurs between 15 and 20 kb pressure (50–65 km depth; Fig. 5 of the preceding paper), which coincides with the depth corresponding to the low-velocity zone (LVZ) beneath abyssal plains (Anderson, 1962; Vinogradov, 1969). The low-velocity zone is often interpreted as a layer of incipient melting (e.g. Ringwood, 1969). Under these P, T, and f_{H_2O} conditions, the initial melt will be alkali basaltic or olivine tholeiitic with $Na \gg K$ (see Tables 9, 10, 11). A range in the composition of the peridotite itself may shift the upper lid of the $LVZ \pm 10$ km of depth. Such variations will probably not affect the overall melt compositions although we may expect minor variations of $Mg/(Mg+\Sigma Fe)$ and $Ca/(Ca+K+Na)$. This model is attractive for the following reasons: (1) The depth of magma generation is relatively shallow, and the depth corresponds to geophysical data for the depth to the oceanic LVZ. (2) Our model allows substantial amounts of melt (at least 5–10 per cent) to form without substantially changing alkalinity and silica content. Increasing temperature tends to decrease the alkalinity of the melt. However, f_{H_2O} also decreases with increasing degree of melting because of the high solubility of $_{H_2O}$ in the melt, tending to increase the alkalinity and decrease the silica content.

The chemistry of the residual phases of peridotite does not uniquely reflect f_{H_2O} and f_{H_2} conditions of equilibration. There is, however, a positive correlation between $Mg/(Mg+\Sigma Fe)$ and Cr contents of such crystalline phases that is useful in determining whether a peridotite nodule from basalt or kimberlite is of cumulate or residual origin. The former will be low in Cr, but may be Mg-rich whereas the latter is both Cr- and Mg-rich.

The orthopyroxene–clinopyroxene pair yields a pressure-independent Cr/Al^{VI} distribution that is a function of temperature. This geothermometer appears superior to that utilizing $Ca/(Ca+Mg)$ of opx coexisting with cpx (Boyd, 1973) that is both temperature- and pressure-dependent. The two may be combined to give a possible geothermometer and geobarometer, however. These data suggest that the Hawaiian nodules in these experiments are cumulative and have formed at depths of up to more than 100 km. However, their ranges of pressure (10->30 kb) and temperature (1000 °C->1200 °C) suggest no genetical relationship between them. Instead, the peridotite nodules were probably accidentally sampled by ascending magmas.

The experimental data thus show that the chemical differences between oceanic and island-arc volcanism as well as alkaline volcanism related to kimberlite genesis are best understood if melting of peridotite occurs with $0 < a_{H_2O} \leqslant 1$. High a_{H_2O} results in andesite liquids, intermediate a_{H_2O} yields oceanic tholeiites whereas low a_{H_2O} is required to obtain liquids of olivine nephelinite and melilite basalt compositions.

REFERENCES

[*Editors' Note:* Only the references cited in the preceding excerpts are reproduced here.]

ANDERSON, D. L., 1962. The plastic layer of the earth's mantle. *Sci. Am.* **205,** 2-9.

BOETTCHER, A. L., 1973. Volcanism and orogenic belts—the origin of andesite. *Tectonophysics* **17,** 231-44.

BOYD, F. R., 1973. The orthopyroxene geothermometer. *Geochim. cosmochim. Acta* **32,** 2533–46.

DICKINSON, W. R., 1970. Relation of andesites, granites, and derivative sandstones to arc-trench tectonics. *Rev. Geophys.* **8,** 813–60.

DICKINSON, W. R. & HATHERTON, T., 1967. Andesitic volcanism and seismicity around the Pacific. *Science,* **157,** 801–3.

EGGLER, D. H., 1973. Role of CO_2 in melting processes in the mantle. *Yb. Carnegie Instn. Wash.* **72,** 457–67.

EGOROV, L. S., 1970. Carbonatites and ultrabasic-alkaline rocks of the Maimecha-Kotui region. N. Siberia. *Lithos,* **3,** 341–59.

GILL, J. B., 1970. Geochemistry of Viti Levu, Fiji and its evolution as an island arc. *Contr. Miner. Petrol.* **27,** 179–203.

HEDGE, C. E., 1966. Variation in radiogenic strontium found in volcanic rocks. *J. geophys. Res.* **71,** 6119–26.

HEDGE, C. E., HILDRETH, R. A., & HENDERSON, W. T., 1970. Strontium isotopes in some Cenozoic lavas from Oregon and Washington. *Earth Planet. Sci. Lett.* **8,** 434–8.

KUNO, H., 1966. Lateral variation of basalt magma across continental margins and island arcs. *Can. geol. Surv. Prof. Pap.* **66-15,** 317–35.

KUSHIRO, I., 1972. Effect of water on the composition of magmas formed at high pressures. *J. Petrology,* **13,** 311–34.

McBIRNEY, A. R., 1969. Compositional Variations in Cenozoic calc-alkaline suites of Central America. In McBIRNEY, A. R. (ed.) *Proc. Andesite Conf.,* 185–9.

MODRESKI, P. J., & BOETTCHER, A. L., 1972. The stability of phlogopite+ enstatite at high pressures: a model for micas in the interior of the earth. *Am. J. Sci.* **272,** 852–69.

PETERMAN, Z. E., CARMICHAEL, I. S. E., & SMITH, A. L., 1970. Sr^{87}/Sr^{86} ratios of the Tala Sea series, New Britain, Territory of New Guinea. *Bull. geol. Soc. Am.* **81,** 39–40.

RINGWOOD, A. E., 1969. The composition of the crust and the upper mantle. *Am. Geophys. Un. Mon.* **13,** 1–17.

TAYLOR, S. R., 1969. Trace element chemistry of andesites and associated calc-alkaline rocks. In McBIRNEY, A. R. (ed.) *Proc. Andesite Conf.*

TAYLOR, S. R., EWART, A., & KOPP, A. C., 1968. Leucogranites and rhyolites: Trace element evidence for fractional crystallization and partial melting. *Lithos* **1,** 179–87.

UKHANOV, V. V., 1965. Olivine melilite from the diamond-bearing diatremes on Anabar. *Dokl. Akad. Sci. U.S.S.R. Earth Sci. Sec.* **153,** 176–8.

VINOGRADOV, A. P., 1969. Average contents of chemical elements in the principal types of igneous rocks of the earth's crust. *Geochemistry.* **7,** 641–64.

31

Reprinted from pages 3 and 20–30, 32, 33, 34, and 35 of *Jour. Petrology* **20**:3–35 (1979)

Generation of Mid-ocean Ridge Tholeiites*

by D. C. PRESNALL, J. R. DIXON, T. H. O'DONNELL†, *and*
S. A. DIXON‡

Department of Geosciences, The University of Texas at Dallas, P.O. Box 688, Richardson, Texas 75080

(*Received 14 October 1977; in revised form 27 June 1978*)

ABSTRACT

If a basaltic magma is separated from its mantle source region only after a large amount of fusion, as indicated for mid-ocean ridge tholeiites by large ion lithophile (LIL) element studies, volatile-free phase relationships may be used to model closely the fusion process at the site of origin. Volatile-free solidus curves in pressure–temperature space display low-temperature cusps where subsolidus phase transitions intersect the solidus, and these cusps are believed to be important in controlling the depth of magma generation and the compositions of primary magmas. In the system $CaO-MgO-Al_2O_3-SiO_2$, the solidus curve for simplified plagioclase and spinel lherzolite has been determined up to 20 kb, and a cusp has been found at 9 kb, 1300 °C, where the transition from simplified plagioclase to spinel lherzolite intersects the solidus and forms an invariant point. Simplified basaltic melts formed along the solidus curve change from quartz-normative compositions at low pressures to olivine-normative compositions at high pressures, and the composition of the first liquid produced at the 9 kb cusp resembles closely the composition of the least-fractionated mid-ocean ridge tholeiites. It is proposed that these least-fractionated basalts have been modified very little by fractional crystallization as they passed upward to the surface and are close to the composition of primary basalt generated at the mantle equivalent of the 9 kb cusp (30 km depth). Assuming a lherzolitic mantle source, the composition of this primary magma would be expected to remain fairly constant despite varying amounts of fusion as long as these amounts are between about 2 and 35 per cent, and despite variations in the proportions of minerals in the source region. For complex natural compositions in the mantle, the cusp would not be a point but would be expected to appear as a low temperature region on the solidus extending over a small pressure range at about 9 kb and 1200–1250 °C. It is suggested that magma generation at the cusp causes a sharp change in the slope of the geotherm beneath active ridges from approximately adiabatic at depths greater than about 30 km to strongly superadiabatic at depths less than 30 km. LIL element concentrations indicate that a close approach to fractional fusion is not applicable to the generation of primary magmas at spreading centers. Instead, a model involving varying amounts of equilibrium fusion is preferred, with fresh mantle being continually supplied at the 9 kb cusp as mantle material convects upward along the rising limb of a mantle convection cell. Primary magmas would thus be produced with fairly uniform major element compositions but widely different LIL element concentrations. Subsequent fractional crystallization of the primary magmas would produce most of the observed major element variations in the erupted basalts but would have only a second-order effect on the LIL element concentrations.

[*Editors' Note:* Material has been omitted at this point.]

---

* Department of Geosciences, The University of Texas at Dallas, Contribution No. 345.
† Present address: U.S. Geological Survey, 307 Federal Building, Ft. Myers, Florida 33901.
‡ Mobil Research and Development Corp., Field Research Laboratory, Dallas, Texas 75221.

PRODUCTION OF PRIMARY BASALT AT THE CUSP

In the basalt trends shown in Figs. 12 and 13, the atomic ratio $mg = \mathrm{Mg}/(\mathrm{Mg} + \mathrm{Fe}^{2+})$* decreases systematically from a maximum of 0·71† on the left to a minimum of 0·29 on the right. This type of variation occurs within groups of samples from the same locality as well as across the entire composite trend. Also, where data are available, the compositions of olivine phenocrysts show a similar trend of increasing fayalite content to the right. Thus, the least-fractionated basalts are concentrated to the left with the extension to the right being best explained by fractional crystallization involving the removal of olivine, plagioclase, and diopsidic pyroxene (Presnall & O'Donnell, 1976; Clague & Bunch, 1976; Bryan & Moore, 1977).

Assuming that (1) metamorphic peridotites at the base of ophiolite complexes are representative of the residual material remaining after removal of tholeiitic basalts from the mantle, and (2) olivine phenocrysts in mid-ocean ridge basalts are in equilibrium with their host liquid, comparison of olivine compositions in the peridotites with olivine compositions in the least-fractionated basalts should provide a test of the possibility that these basalts represent primary magmas unaffected by fractional crystallization as they moved upward to the earth's surface. The most magnesian olivines in the FAMOUS basalts have a composition of Fo_{87}, with mg values of the coexisting liquids being 0·70–0·72 (Hekinian *et al.*, 1976). Based on a $K_D = [(X^{Ol}_{FeO})(X^{liq}_{MgO})]/[(X^{liq}_{FeO})(X^{Ol}_{MgO})]$ of 0·3 for basalt, as recommended by Roeder & Emslie (1970), the olivine composition in equilibrium with liquids having these mg values would be Fo_{88-89}, which supports the assumption that the olivine and host liquid are in equilibrium. Frey *et al.* (1974) reported three analyses of oceanic tholeiites that they considered to be likely candidates for primary magmas. These basalts have compositions (mg values are 0·70–0·71) very similar to the least-fractionated FAMOUS basalts and contain olivine phenocrysts that are slightly more forsteritic (Fo_{90}), but still within 1–2 per cent Fo of the composition expected from a K_D of 0·3.

Except for one point at Fo_{84}, Coleman (1977, Fig. 5) shows the range of olivine compositions in metamorphic peridotites from ophiolites to range from Fo_{87} to Fo_{94}, with the main concentration falling in the range Fo_{90-92}. Thus, the most magnesian olivines from mid-ocean ridge tholeiites fall within the composition

---

* For mg values of basaltic glasses analyzed with an electron microprobe, it will be assumed throughout that $\mathrm{Fe}^{2+}/(\mathrm{Fe}^{2+} + \mathrm{Fe}^{3+}) = 0.86$.

† One partial analysis listed by Hekinian *et al.* (1976, table 4) has an mg value of 0·72.

range of olivines from metamorphic peridotites in ophiolites, but slightly on the iron-rich side of this range. This suggests that mid-ocean ridge tholeiites with *mg* values of 0·70–0·72 are either derived directly from their mantle source and erupted at the earth's surface in an essentially unfractionated condition or have fractionated only a very small amount corresponding to a change in olivine composition of about 2–3 mole per cent Fo.

The three analyses reported by Frey *et al.* (1974) plot, as would be expected from their *mg* values and magnesian olivines, in the left hand portion of the trend shown in Figs. 12 and 13 (points *f7, f9,* and *f10*) but it will be noted that they do not all plot in the *extreme* left-hand portion, a result that seems inconsistent with their apparently unfractionated character. It is possible to explain the apparent inconsistency as the result of surprisingly small variations in Na_2O and SiO_2, perhaps resulting from sample variation, analytical uncertainties, or both. To illustrate this effect, we have calculated the change in the norm of point *f10* (Figs. 12 and 13) caused by adding 0·4 wt. per cent NA_2O to the listed composition while fixing the amounts of all the other oxides. The resulting norm is shown as *f10'* in Figs. 12 and 13. Not only is the position of the point sensitive to small variations in Na_2O but the shift is approximately along the trend attributed above to fractional crystallization. The norm is much less sensitive to similar variations in the other oxides, but because SiO_2 is present in the largest concentration, it usually has the largest absolute uncertainty in a microprobe analysis*. Variations in SiO_2 have nearly the same effect on the norm as variations in Na_2O except that a change of one per cent SiO_2 (absolute) causes about half the shift produced by a change of 0·4 per cent Na_2O. Thus, part of the 'fractionation trend' may be due to small analytical and sampling variations in NA_2O. In fact, it is commonly observed that different samples from the same dredge haul (O'Donnell, 1974) or submarine sampling locality (Bryan & Moore, 1977) have nearly identical *mg* values but are dispersed along the 'fractionation trend'. Similarly, the three analyses reported by Frey *et al.* (1974) also have nearly identical *mg* values and are dispersed along the 'fractionation trend' (Figs. 12 and 13). Despite this problem, it is apparent that the *general* correlation of *mg* value and olivine composition with position along the trend requires that fractional crystallization be the main cause of the compositional spread.

If the fractionated basalts were deleted, there would remain a tight grouping of the least-fractionated basalts in the left-hand portion of the overall basalt trend shown in Figs. 12 and 13. The existence of such a parental basalt continuously produced in enormous volumes over long periods of time is a feature that appears to be a worldwide characteristic of volcanism at mid-ocean ridges. In the FAMOUS area, the average Mt. Pluto magma (Bryan & Moore, 1977) is representative of the least-fractionated tholeiites with high *mg* values erupted near the center of the rift valley, and it will be noted that this average analysis falls in the left-hand portion of the trend shown in Figs. 12 and 13.

* For basalt glasses from the mid-Atlantic ridge at 25°–29° N. O'Donnell (1974) found a standard deviation error (2 σ) of multiple spots analyzed by electron microprobe from a single polished section to be 0·2 wt. per cent for Na_2O and 0·6 wt. per cent for SiO_2 (absolute).

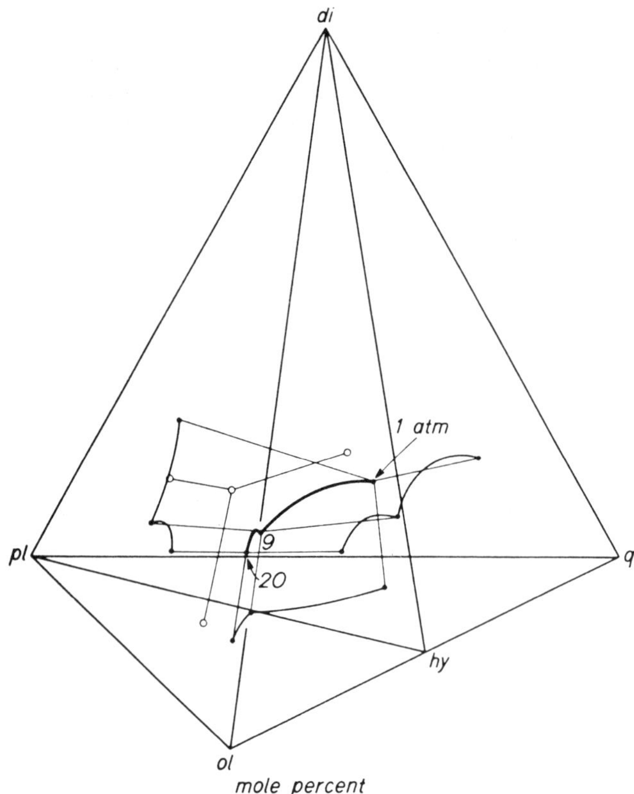

FIG. 15. Perspective drawing of liquid compositions along the solidus curve in Fig. 3 (heavy line with filled circles) and the average Mt. Pluto magma from the FAMOUS area (Bryan & Moore, 1977, open circle). Light solid lines are projections from *pl*, *di*, and *q* onto the faces of the tetrahedron. Numbers indicate pressures in kilobars.

Fig. 15 shows a perspective drawing of the Mt. Pluto average magma and the initial liquid compositions produced along the lherzolite solidus in the system $CaO-MgO-Al_2O_3-SiO_2$ at various pressures. It will be recalled that the addition of other components causes partial melts in equilibrium with lherzolite to shift to higher *di* and *pl* and to lower silica, changes that would shift the composition of the cusp at 9 kb toward the composition of the Mt. Pluto magma. If the magnitude of the effect of additional components is about the same as indicated by Kushiro's (1973) melting experiment on basalt T-87 at 7·5 kb discussed above and shown in Figs. 12, 13, and 14, the match of the 9 kb first liquid composition with the least-fractionated magma is very close. Based on this compositional correspondence, the close correlation of the compositions of olivine phenocrysts from mid-ocean ridge tholeiites and metamorphic peridotites from ophiolites, and the fact that the geotherm is most likely to intersect the solidus at the 9 kb cusp, it is suggested that the parental least-fractionated mid-ocean ridge basalts have been modified very little by fractional crystallization as they passed upward to the surface, and correspond closely to primary melts separated from the source region (rising limb of a convection cell) at about 9 kb.

Irvine (1977) calculated a primitive abyssal tholeiite based on the assumption of

(1) extensive olivine and chromite fractionation prior to the crystallization of other phases and (2) an assumed olivine composition of $Fo_{90.5}$ in equilibrium with this primitive tholeiite. Irvine's primitive liquid (point i in Figs. 12, 13, and 14) is picritic and lies toward olivine distinctly separated from the field of analyzed tholeiitic glasses. It will be recalled that the tholeiitic glasses reported by Frey *et al.* (1974) contain olivine phenocrysts with a composition of Fo_{90}, yet they are not picritic and plot along the trend of mid-ocean ridge tholeiites (points $f7$, $f9$, and $f10$ in Figs. 12 and 13). The direct evidence of Frey *et al.* (1974) on the composition of basalt liquid in equilibrium with olivine having a composition of Fo_{90} is preferred here to the calculation of Irvine (1977). Among abyssal tholeiites that have been reported so far, we are not aware of any *glass* analyses that show a picritic trend toward olivine.

O'Hara (1968b) suggested that mid-ocean ridge basalts are generated at 25–30 kb as picritic basalts with their observed compositions at the surface being the result of extensive olivine fractionation as they passed upward to the surface. However, as we have noted, no record of a fractionation trend extending away from the olivine apex is preserved in the glass compositions. Furthermore, none of O'Hara's arguments in favor of fractional crystallization require that it must take place at pressures between 9 and 30 kb. In the absence of any evidence for olivine fractionation at such high pressures, it is more reasonable to conclude that it never occurred, particularly since a mechanism now exists for producing the required parental magma at the 9 kb solidus cusp.

O'Hara (1968b, p. 685) assumed that if mid-ocean ridge basalts are derived as primary magmas at about 30 km, the geotherm would lie above the mantle solidus at greater depths near 100 km where picritic primary magmas would be produced. That is, if primary magmas are generated at low pressures, he argued that they must also be produced at higher pressures. However, if the mantle solidus has a cusp at about 9 kb, there is no necessity for the high-pressure portion of the geotherm to be at a higher temperature than the solidus. In fact, the reverse would be expected. In Fig. 16, the range of slopes for the geotherm at mid-ocean ridge crests calculated by various authors from heat flow data is shown as the shaded area. The exact shape of the geotherm at higher pressures is uncertain, but it is clear from the near-surface gradient based on heat-flow measurements that the geotherm must meet the solidus curve at a very shallow depth. The dashed portion of the geotherm is a hypothetical extrapolation, but illustrates the point that basalts could be generated at about 9 kb with no reason to expect, in addition, the generation of picritic basalts at 25–30 kb. The shape of the geotherm will be discussed further in a later section.

O'Hara's (1968b) main argument against deriving abyssal tholeiites as primary melts at shallow depths was his contention that compositions of these basalts do not touch the primary phase field of orthopyroxene at any pressure. Thus, he argued that these basalts could not represent unmodified primary magmas derived from a mantle containing olivine and orthopyroxene. It is clear from the above discussion that O'Hara's assumption about the location of the orthopyroxene field is incorrect. The available data indicate that the orthopyroxene field does indeed touch the compositions of the *least fractionated* mid-ocean ridge tholeiites over a narrow pressure range in the vicinity of 9 kb.

O'Hara (1973) cited the results of Kushiro & Thompson (1972) as experimental verification of his assertion about the location of the orthopyroxene field. Kushiro and Thompson reported a Ca-poor pyroxene at the liquidus of the mid-ocean ridge basalt T-87 over a narrow pressure range at about 7·5 kb, but O'Hara (1973) referred to this pyroxene as a pigeonite (Kushiro and Thompson stated that it was 'most probably orthopyroxene', although it contained 8·9 mole per cent $CaSiO_3$). O'Hara considered it to have a composition more iron rich than residual upper mantle enstatites, which is correct. The *mg* value for residual enstatites from mantle nodules ranges as high as 0·92 for highly depleted mantle and as low as 0·88 for undepleted mantle (Carter, 1970), whereas the pigeonite(?) in the 7·5 kb experiment of Kushiro and Thompson has an *mg* value of 0·84. However the *mg* value of the basaltic liquid in equilibrium with this pigeonite(?) is 0·61 compared to 0·70 for the Mt. Pluto magma of Bryan & Moore (1977), and it is reasonable to infer that if Kushiro and Thompson had carried out their study on a less fractionated basalt like the Mt. Pluto magma, the Ca-poor pyroxene would have also contained less Fe, comparable to the amount inferred by Carter for residual enstatites in nodules. Thus, the experiments of Kushiro and Thompson cannot be considered as verification of O'Hara's position.

A later experiment on another Fe-rich oceanic tholeiite (*mg* value = 0·61) by Fujii & Kushiro (1977) yielded enstatite at the liquidus at 8 kb with an *mg* value estimated by them to be 0·86. This further strengthens the argument that enstatite of an appropriate mantle composition can be in equilibrium with mid-ocean ridge basaltic compositions at about 9 kb.

AMOUNT OF FUSION AT THE CUSP AND EFFECTS OF MANTLE HETEROGENEITY

At constant pressure, fusion at the simplified lherzolite solidus (Fig. 3) occurs at a quaternary invariant point, which allows the amount of fusion to vary over wide limits without changing the composition of the melt (Presnall, 1969). One method of determining these limits would be to specify the starting composition. Then one could calculate the amount of melt produced just as the crystal path intersected a side, an edge, or a corner of the tetrahedron describing the four crystalline phases (*ol* + *en* + *di* + *sp* or *ol* + *en* + *di* + *an*) in equilibrium with the melt. At this point in the fusion process, one or more of the crystalline phases would be completely consumed and the composition of the liquid would leave the invariant point.

A different procedure will be followed here in order to avoid assumptions about the relative proportions of phases in the starting composition. The composition of the crystalline residue remaining just as fusion at the invariant point is complete will be assumed, and the starting composition will be allowed to slide along the line between the composition of this crystalline residue and the liquid composition at the invariant point. In the pressure range studied here, simplified plagioclase and spinel lherzolites always melt at a peritectic point. Thus, the line between the liquid composition and the bulk composition of the chosen residue would be only partially contained by the tetrahedron defined by the four crystalline phases in equilibrium with the liquid. Because the starting composition consists of all four crystalline phases, it could slide along the line between the residue and liquid compositions only as long as it remained within the tetrahedron defined by these four crystalline phases. The maximum amount of melt would be obtained for a starting composition situated on the crystalline residue-liquid line just at the point where this line passes out of the tetrahedron.

TABLE 3

*Compositions of phases in equilibrium at 9 kb, 1300 °C invariant point, in weight per cent**

	Diopside[†]	Enstatite[†]	Forsterite[†]	Liquid[‡]
SiO_2	51·86	55·19	42·53	49·65
Al_2O_3	7·70	7·50	—	20·68
MgO	20·35	35·25	57·07	14·05
CaO	20·08	2·06	0·40	15·61

   * Spinel and anorthite are assumed to be stoichiometric and are therefore not listed.
    † Based on unpublished data of J. R. Dixon and Presnall.
    ‡ Estimated from data in Table 2.

Based on the data of Carter (1970) for ultramafic nodules at Kilbourne's Hole, New Mexico, it will be assumed that fusion at the 9 kb cusp results in the simultaneous depletion of spinel and diopside and leaves a crystalline residue consisting of 87 per cent forsterite, 13 per cent enstatite by weight. Given the composition of the crystalline residue and the phases in equilibrium at the cusp (Table 3), the maximum proportion of liquid that can be extracted from a source capable of being expressed in both of the simplified lherzolite volumes *fo + en + di + an* and *fo + en + di + sp* is calculated by balancing the equation:

crystalline residue (*res*) + liquid (*liq*) − spinel (*sp*) + enstatite (*en*) + diopside (*di*)

This equation is most easily balanced by solving the following determinant (Korzhinskii, 1959, pp. 103–7) expressed in weight fractions:

$$\begin{vmatrix} & CaO & MgO & Al_2O_3 & SiO_2 \\ res & 0·0062 & 0·5423 & 0·0098 & 0·4417 \\ liq & 0·1561 & 0·1405 & 0·2068 & 0·4965 \\ sp & 0 & 0·2833 & 0·7167 & 0 \\ en & 0·0206 & 0·3525 & 0·0750 & 0·5519 \\ di & 0·2008 & 0·2035 & 0·0770 & 0·5186 \end{vmatrix} = 0$$

The solution, in weight proportions, is:

$$0·00954\ res + 0·01957\ liq = 0·00301\ sp + 0·01180\ en + 0·01430\ di \qquad (1)$$

from which the maximum proportion of liquid can be calculated, using the coefficients on the left-hand side of the equation, as 67 per cent. To produce this much liquid of a constant composition at the cusp, the starting composition would

199

consist of 41 per cent enstatite, 49 per cent diopside, and 10 per cent spinel (the proportion of phases indicated on the right-hand side of equation (1)), with forsterite being absent.

A more realistic approximation of a mantle source composition, this time containing forsterite, can be obtained by sliding the starting composition along the residue-liquid line back toward the crystalline residue composition. For example, if the upper limit on the amount of liquid obtainable at the cusp is reduced to 35 per cent, the proportions of phases in the source material, calculated from equation (1), are 42 per cent forsterite, 27 per cent enstatite, 26 per cent diopside, and 5 per cent spinel. These proportions are considered to be realistic for the mantle, inasmuch as they are essentially identical to those determined by Carter (1970, Table 2A) for an undepleted mantle under Kilbourne's Hole, New Mexico, containing olivine with a composition of Fo_{86}.

Other reasonable assumptions about the constitution of the residual material* would yield slightly different results but would not change the basic observation that for realistic model mantle compositions, large variations in the amount of fusion can occur without changing the composition of the initial melt. It should be noted also that considerable heterogeneity in the composition of the source can be tolerated without changing the composition of the initial melt (Yoder & Tilley, 1962, p. 519; Presnall, 1969, pp. 1190–2). It is necessary only that the source composition remain within both of the overlapping tetrahedra forsterite + enstatite + diopside + anorthite and forsterite + enstatite + diopside + spinel, the apices of these tetrahedra being defined by the compositions of the crystalline phases in equilibrium at the cusp.

For actual mantle compositions, fusion would not take place at an invariant point in a rigorous sense because of the presence of small amounts of additional components. Thus, the 9 kb cusp in the simplified system would be expected to correspond, in the mantle, to a low-temperature region spread over a small pressure interval, and some variation in the composition of primary magmas would be expected depending on the composition of the source and the amount of fusion. For example, the *mg* value of liquids would increase as the amount of fusion increases (Presnall, 1969 and in press). The amount of variation in the liquid composition is strongly dependent on the original composition of the source. There are many compositions in simplified systems that do not melt at an invariant point, yet produce liquids that vary in composition only a small amount for large variations in the amount of melting (Presnall, 1969 and in press). For amounts of fusion larger than about 2 per cent, Mysen & Kushiro (1977) have confirmed this type of melting behavior experimentally for natural mantle materials. They found that for lherzolites, an initial temperature interval of about 20–50 °C, in which the initial 1–2 per cent melting takes place, is followed by a second temperature interval of a similar magnitude over which a very large amount of fusion (up to 60 per cent)

---

* Harzburgite is the most commonly proposed material (for example, see Dickey, 1970; Menzies, 1973; Menzies & Allen, 1974) but the proportions of olivine and orthopyroxene vary somewhat. For example, in metamorphic harzburgites from ophiolites, the proportion of orthopyroxene varies from 5 to 30 volume per cent (Coleman, 1977, p. 25).

occurs without disappearance of any of the major mineral phases. Throughout this second interval, only small changes occur in most of the chemical constituents of the coexisting minerals and liquid. Thus, it would be expected that as mantle material rises in the ascending limb of a convection cell along a mid-ocean ridge, a small amount of fusion, perhaps 1–2 per cent, would occur at depths slightly greater than 30 km, and as the mantle material continues to rise, the amount of fusion would increase sharply at about 30 km, thus facilitating separation of the magma from the source while retaining a fairly constant melt composition.

If magmas at spreading centers are generated at about 30 km depth, the oceanic mantle at depths less than this would be predominantly depleted of its basaltic component whereas at greater depths the basaltic component would be retained (see also Kay *et al.*, 1970; Green, 1972; Oxburgh & Parmentier, 1977). The existence of such a depleted region would explain the rarity of plagioclase lherzolite nodules in oceanic volcanic rocks, for plagioclase lherzolite is stable only at shallow depths (Fig. 3) within the depleted region where the predominant rock type would be residual harzburgite.

A FIXED POINT ON THE MANTLE GEOTHERM BENEATH OCEANIC RIDGES

In one respect, the model proposed here for the generation of mid-ocean ridge tholeiites is similar to that presented by Green & Ringwood (1967*a*). That is, they suggested 30 km as the depth of magma separation from the source region, about the same depth as proposed here. They considered that fusion starts at a much greater depth*, however, and they attributed the characteristic composition of mid-ocean ridge tholeiites to very rapid upward movement of a mantle diapir such that separation of the melt fraction is delayed until the diapir reaches a depth of about 30 km. They proposed that if the rate of ascent decreases, the depth of magma separation would increase, thereby changing the composition of the separated magma to a low-alumina olivine tholeiite or an alkali olivine basalt. However, such basaltic *liquids* do not exist among any of the volcanic glasses analyzed to date from mid-ocean ridges (Figs. 12, 13, and 14). Schilling & Bonatti (1975) have reported four alkalic analyses of *whole rocks* from the east Pacific Rise, but this ridge is spreading rapidly and would be expected to be accompanied by rapidly rising mantle material. Thus, even if it is considered that these analyses represent liquids, their alkalic character is contrary to the prediction of the Green and Ringwood model.

In the model proposed here, ridges are considered to represent the surface expression of mantle material rising adiabatically over the entire length of the worldwide ridge system; that is, they represent the rising limbs of mantle convection cells. The cusp in the solidus at 9 kb is considered to be the key factor controlling the depth of magma generation and the composition of magma

---

* In the paper by Green & Ringwood (1967*a*, fig. 12), fusion of an anhydrous diapir rising along an adiabat was indicated as starting at 75–80 km. Later Ringwood (1975, pp. 150–7) modified this model by assuming a mantle with 0·1 wt. per cent water, initiating diapiric rise from about 150 km, and assuming the diapir to be partially liquid throughout its ascent.

produced. The rate of ascent is of subordinate importance. Faster rates of ascent of mantle material would simply result in the production of more lava of a similar composition, whereas a slower rate would decrease the thermal gradient causing a cessation of volcanic activity along the ridge. However, the lowered geothermal gradient would still be high enough to produce incipient fusion in the oceanic seismic low-velocity zone (Presnall, 1978).

It is suggested that the cusp on the mantle solidus acts as a thermal buffer that controls the shape of the geotherm at active mid-ocean ridges. The effect of the heat of fusion combined with separation and ascent of magma at 9 kb from upwardly convecting mantle material would depress the geotherm whereas the geotherm would rise during periods without magma production. The net result would be stabilization of the geotherm just at the cusp, as shown in Fig. 16.

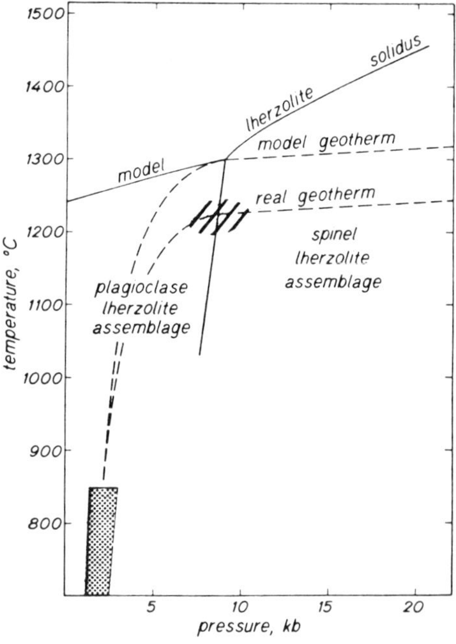

FIG. 16. Simplified lherzolite solidus curve from Fig. 3 and two dashed hypothetical extensions of the geotherm for the mantle beneath mid-ocean ridges, one for the model mantle system $CaO-MgO-Al_2O_3-SiO_2$ and one for the real mantle. The shaded area is the range of slopes calculated by Oxburgh & Turcotte (1968), Le Pichon & Langseth (1970), and Bottinga & Allègre (1973), and the hachured area is the estimated pressure and temperature at which primary tholeiites are generated in the real mantle. The geothermal gradient at pressures above 9 kb is adiabatic (0·5 °C/km).

The model geotherm drawn is for a mature ridge and is adiabatic (0·5 °C/km) at pressures greater than 9 kb. Such a gradient is consistent with the concept of rapid, and therefore essentially adiabatic, convective upwelling of mantle material along ridges, and with the gradient of approximately 1 °C/km determined from magneto-telluric data by Hermance & Grillot (1974) beneath Iceland. For newly-formed ridges, mantle material rising adiabatically might intersect the solidus curve at some pressure higher than 9 kb, but as the ridge matures, the intersection would be expected to migrate to lower pressures and finally stabilize at the cusp. In this case, the early volcanics would be generated at higher pressures and would tend to be more alkalic in composition. It is interesting to note that this trend is exactly what is observed in the Ethiopian Rift (Gass, 1970; Mohr, 1971), a feature that is presumably an example of a spreading center in its initial stages of formation.

Assuming the cusp corresponds to a fixed point on the geotherm beneath active oceanic ridges, an estimate of the effect of additional components on the temperature and pressure at the cusp would be useful in constructing thermal models for the mantle. For pyrolite, Green & Ringwood (1967b) found the solidus temperatures at 9 and 20 kb to be lower than our solidus curve (Fig. 2) by 75 and 65 °C, respectively. Mysen & Kushiro (1977) took the lherzolite 66SAL-1 as representative of the least chemically depleted oceanic mantle. They found that at 20 kb, large amounts of liquid (1·5 to 60 per cent) were produced at 1350–1375 °C, a temperature range 70–95 °C lower than our solidus at 20 kb. From these studies, it appears that our solidus curve is about 50–100 °C higher than temperatures for the occurrence of large amounts of melting in mantle compositions containing additional components. Thus, the mantle equivalent of the cusp in the simplified system would probably occur at about 1200–1250 °C. Because of the large amount of fusion indicated at the cusp, the presence of small amounts of CO_2 and H_2O in the mantle would not alter this estimate appreciably.

The effect of additional components on the plagioclase to spinel lherzolite transition, which at high temperatures is essentially the pressure of the cusp, has been studied by Green & Hibberson (1970) and Emslie (1971). Green and Hibberson found the maximum stability of plagioclase in a pyrolite bulk composition to be 11·2 kb. As would be expected in a complex composition, plagioclase and spinel were found to coexist over a range of pressure, but the lower limit of this range was not defined. They listed one run containing both plagioclase and spinel at 4·5 kb and 1200 °C, but it is important to note that their runs in the absence of a liquid phase were a maximum of 4 hours long, far shorter than we have found to be necessary for runs containing a liquid phase. Emslie, using a mixture of natural olivine and plagioclase, found the maximum stability of plagioclase to be 10 kb. No minimum pressure for the coexistence of plagioclase and spinel was determined. Thus, the available data are inconclusive, but suggest that the transition between plagioclase and spinel lherzolite occurs over a range of pressure that is not appreciably displaced from the 9 kb transition found in the system $CaO–MgO–Al_2O_3–SiO_2$.

From the above discussion, it is estimated that the region of magma generation for the real mantle beneath active ridges would be centered at about 9 kb and 1200–1250 °C. Fig. 16 shows this region as the hachured area together with the estimated real geotherm.

FRACTIONAL VERSUS EQUILIBRIUM FUSION

The constraint imposed by LIL element studies that mid-ocean ridge tholeiites are the result of a large amount of partial fusion (Gast, 1968; Kay *et al.*, 1970; Schilling, 1971, 1975) requires that equilibrium fusion rather than fractional fusion (Presnall, 1969) applies. Langmuir *et al.* (1977) have also shown that a close approach to fractional fusion, that is, stepwise equilibrium fusion (Presnall, in press) and extraction of melt from the same source in small increments, would produce highly variable La/Yb ratios, which are not observed. However, Schilling (1975) has shown that the observed rare earth element (REE) patterns can be explained by variations in the amount of equilibrium fusion between about 10 and 30 per cent if each basalt sampled is extracted from a fresh source. In agreement with Schilling's conclusion, White & Bryan (1977) calculated for the FAMOUS basalts that the major element differences between the most primitive and most fractionated basalts can be explained by about 23 per cent crystallization, whereas LIL elements (K, Rb, Ba, and REE) are enriched by factors of 2–4. Because 23 per cent crystallization is inadequate to explain these enrichment factors, they invoked varying degrees of equilibrium fusion. Hart *et al.* (1973) have also discussed the inadequacy of fractional crystallization as a mechanism for explaining the LIL element concentrations.

In the FAMOUS area, Langmuir *et al.* (1977) found the additional complication of crossing REE distribution patterns and concluded that one way to explain these results would be to postulate a source with heterogeneous REE distribution patterns. However, they preferred to explain the crossing patterns by invoking a homogeneous source and postulating a combination of at least (1) varying amounts of single-step equilibrium fusion, with each melt extracted from a fresh portion of the source, and (2) stepwise equilibrium fusion in moderately large steps from the same source with retention in the residue of some of the melt produced at each step. In either case, the arguments against a close approach to fractional fusion would hold.

Thus a satisfactory explanation of both the major and trace element data appears to require variations in the amount of equilibrium fusion as well as variations in the amount of subsequent fractional crystallization. Varying amounts of equilibrium fusion (up to a maximum of 35 per cent in our model calculation) would generate primary magmas with fairly uniform major element compositions but widely different LIL element concentrations, whereas subsequent fractional crystallization of these primary magmas would produce most of the observed differences in the major element compositions of the erupted lavas, but would have only a second-order influence on the concentrations of the LIL elements. It would be expected that the resulting partially fused uppermost portion of the oceanic mantle (about 8–30 km depth) would display varying degrees of depletion, as is observed for alpine peridotites believed to be samples of this portion of the mantle (Dick, 1977, p. 827; Richard *et al.*, 1976, p. 272). Wide variåtions in the amount of equilibrium fusion at the mantle equivalent of the 9 kb cusp would require temperature differences of only 20–50 °C (Mysen & Kushiro, 1977).

ERRATUM

Page 21, line 25 should read: "analytical and sampling variations in NaO_2 and SIO_2."

REFERENCES

[*Editors' Note:* Only the references cited in the preceding excerpts are reproduced here.]

BOTTINGA, Y., & ALLÈGRE, C. J., 1973. Thermal aspects of sea-floor spreading and the nature of the oceanic crust. *Tectonophysics,* **18,** 1-17.

BRYAN, W. B., & MOORE, J. G., 1977. Compositional variations of young basalts in the Mid-Atlantic Ridge rift valley near lat. 36° 49' N. *Bull. geol. Soc. Am.* **88,** 556-70.

CARTER, J. L., 1970. Mineralogy and chemistry of the earth's upper mantle based on the partial fusion-partial crystallization model. *Ibid.* **81,** 2021-34.

CLAGUE, D. A., & BUNCH, T. E., 1976. Formation of ferrobasalt at east Pacific midocean spreading centers. *J. geophys. Res.* **81,** 4247-56.

COLEMAN, R. G., 1977. *Ophiolites,* Springer-Verlag.

DICK, H. J. B., 1977. Partial melting in the Josephine peridotite I, the effect on mineral composition and its consequence for geobarometry and geothermometry. *Am. J. Sci.* **277,** 801-32.

DICKEY, J. S., JR., 1970. Partial fusion products in alpine-type peridotites: Serrania de la Ronda and other examples. *Spec. Pap. Mineralog. Soc. Am.* **3,** 33-49.

EMSLIE, R. F., 1971. Liquidus relations and subsolidus reactions in some plagioclase-bearing systems. *Yb.Carnegie Instn. Wash.* **69,** 148-55.

FREY, F. A., BRYAN, W. B., & THOMPSON, G., 1974. Atlantic Ocean floor: geochemistry and petrology of basalts from legs 2 and 3 of the Deep Sea Drilling Project. *J. geophys. Res.* **79,** 5507-27.

FUJII, T., & KUSHIRO, I., 1977. Melting relations and viscosity of an abyssal tholeiite. *Yb. Carnegie Instn. Wash.* **76,** 461-5.

GASS, I. G., 1970. The evolution of volcanism in the junction area of the Red Sea, Gulf of Aden and Ethiopian rifts. *Phil. Trans. R. Soc. Lond.* **A 267,** 369-81.

GAST, P. W., 1968. Trace element fractionation and the origin of tholeiitic and alkaline magma types. *Geochim. cosmochim. Acta,* **32,** 1057-86.

GREEN, D. H., 1972. Composition of basaltic magmas as indicators of conditions of origin: application to oceanic volcanism. *Phil. Trans. R. Soc. Lond.* **A 268,** 707-25.

GREEN, D. H., & HIBBERSON, W., 1970. The instability of plagioclase in peridotite at high pressures. *Lithos,* **3,** 209-21.

GREEN, D. H., & RINGWOOD, A. E., 1967*a.* The genesis of basaltic magmas. *Contr. Miner. Petrol.* **15,** 103-90.

GREEN, D. H., & RINGWOOD, A. E., 1967*b.* The stability fields of aluminous pyroxene peridotite and garnet peridotite and their relevance in upper mantle structure. *Earth planet Sci. Lett.* **3,** 151-60.

HART, S. R., SCHILLING, J. -G., & POWELL, J. L., 1973. Basalts from Iceland and along the Reykjanes Ridge: Sr isotope geochemistry. *Nature Phys. Sci.* **246,** 104-7.

HEKINIAN, R., MOORE, J. G., & BRYAN, W. B., 1976. Volcanic rocks and processes of the Mid-Atlantic Ridge rift valley near 36° 49' N. *Contr. Miner. Petrol.* **58,** 83-110.

HERMANCE, J. F., & GRILLOT, L. R., 1974. Constraints on temperature beneath Iceland from magnetotelluric data. *Phys. Earth planet. Interiors* **8,** 1–12.

IRVINE, T. N., 1977. Definition of primitive liquid compositions for basic magmas. *Ibid.* **76,** 454–61.

KAY, R., HUBBARD, N. J., & GAST, P. W., 1970. Chemical characteristics and origin of oceanic ridge volcanic rocks. *J. geophys. Res.* **75,** 1585–1613.

KORZHINSKII, D. S., 1959. *Physiochemical basis of the analysis of the paragenesis of minerals.* New York: Consultants Bureau.

KUSHIRO, I., 1973. Origin of some magmas in oceanic and circum-oceanic regions. *Tectonophysics,* **17,** 211–22.

KUSHIRO, I. & THOMPSON, R. N., 1972. Origin of some abyssal tholeiites from the Mid-Atlantic Ridge. *Yb. Carnegie Instn. Wash.* **71,** 403–6.

LANGMUIR, C. H., BENDER, J. F., BENCE, A. E, HANSON, G. N., & TAYLOR, S. R., 1977. Petrogenesis of basalts from the FAMOUS area: Mid-Atlantic Ridge. *Earth planet. Sci. Lett.* **36,** 133–56.

LEPICHON, X., & LANGSETH, M. G., JR., 1970. Heat flow from the mid-ocean ridges and sea-floor spreading. *Tectonophysics.* **8,** 319–44.

MENZIES, M. A., 1973. Mineralogy and partial melt textures within an ultra-mafic body, Greece. *Contr. Miner. Petrol.* **42,** 273–85.

MENZIES, M. A., & ALLEN, C., 1974. Plagioclase lherzolite-residual mantle relationships within two eastern Mediterranean ophiolites. *Ibid.* **45,** 197–213.

MOHR, P. A., 1971. Ethiopian Rift and plateaus: some volcanic petro-chemical differences. *J. geophys. Res.* **76,** 1967–84.

MYSEN, B. O., & KUSHIRO, I., 1977. Compositional variations of coexisting phases with degree of melting of peridotite in the upper mantle. *Am. Miner.* **62,** 843–65.

O'HARA, M. J., 1968*b*. Are ocean floor basalts primary magma? *Nature,* **220,** 683–6.

O'HARA, M. J., 1973. Non-primary magmas and dubious mantle plume beneath Iceland. *Ibid.* **243,** 507–8.

OXBURGH, E. R., & PARMENTIER, E. M., 1977. Compositional and density stratification in oceanic lithosphere—causes and consequences. *J. geol. Soc. Lond.* **133,** 343–55.

OXBURGH, E. R., & TURCOTTE, D. L., 1968. Mid-ocean ridges and geotherm distribution during mantle convection. *J. geophys. Res.* **73,** 2643–61.

PRESNALL, D. C., 1969. The geometrical analysis of partial fusion. *Am. J. Sci.* **267,** 1178–94.

PRESNALL, D. C., 1978. A double seismic low-velocity zone beneath mid-oceanic ridges. *Trans. Am. geophys. Un.* **59,** 369–70 (abstr.)

PRESNALL, D. C., (in press.) Fractional crystallization and partial fusion. In YODER, H. S., JR. (ed.), *The Evolution of the Igneous Rocks: A Fiftieth Anniversary Perspective.* Princeton: Princeton University Press.

PRESNALL, D. C., & O'DONNELL, T. H., 1976. Origin of basalts from 25°–29° N on the Mid-Atlantic ridge. *Trans. Am. geophys. Un.* **57,** 341 (abstr.).

RICHARD, P., SHIMIZU, N., & ALLÈGRE, C. J., 1976. 143Nd/146Nd, a natural tracer: an application to oceanic basalts. *Earth planet Sci. Lett.* **31,** 269–78.

RINGWOOD, A. E., 1975. *Composition and petrology of the earth's mantle.* New York: McGraw-Hill.

ROEDER, P. L., & EMSLIE, R. F., 1970. Olivine-liquid equilibrium. *Contr. Miner. Petrol.* **29,** 275–89.

SCHILLING, J. -G., 1971. Sea-floor evolution: rare-earth evidence. *Phil. Trans. R. Soc. Lond.* **A 268,** 663–706.

SHILLING, J. -G., 1975. Rare-earth variations across 'normal segments' of the Reykjanes Ridge, 60°–53° N, Mid-Atlantic Ridge, 29°S, and East Pacific Rise, 2°–19° S, and evidence on the composition of the underlying low-velocity layer. *J. geophys. Res.* **80,** 1459–73.

SCHILLING, J. -G., & BONATTI, E., 1975. East Pacific Ridge (2° S–19° S) versus Nazca intraplate volcanism: rare-earth evidence. *Earth planet Sci. Lett.* **25,** 93–102.

WHITE, W. M., & BRYAN, W. B., 1977. Sr-isotope, K, Rb, Cs, Sr, Ba and rare-earth geochemistry of basalts from the FAMOUS area. *Bull. geol. Soc. Am.* **88,** 571–6.

YODER, H. S., JR., & TILLEY, C. E., 1962. Origin of basalt magmas: an experimental study of natural and synthetic rock systems. *J. Petrology,* **3,** 342–532.

32

Abyssal Tholeiites From the Oceanographer Fracture Zone

II. Phase Equilibria and Mixing

David Walker[1], Tsugio Shibata[2], and Stephen E. DeLong[3]

[1] Hoffman Laboratory, Harvard University, Cambridge, Massachusetts 02138, USA
[2] Department of Earth Sciences, Okayama University, Tsushima, Okayama 700, Japan
[3] Department of Geological Sciences, State University of New York at Albany, Albany, New York 12222, USA

Abstract. Unusual effects have been discovered in the major element phase relations of basalts from the Oceanographer Fracture Zone (OFZ) which suggest that magma mixing of primitive basalt with the differentiation residue of a previous batch of primitive magma has occurred. These effects include a reversal in mineral crystallization sequence which cannot happen during normal differentiation processes or be explained by any plausible change in physical conditions. This unusual effect is encountered as a result of curvature in composition space of the liquid-line-of-descent equilibria involving olivine, plagioclase, and high-calcium clinopyroxene. Mixing of magmas at different stages of their evolution produces mixtures that do not lie on the curved liquid-line-of-descent. Observation of such anomalous compositions in the OFZ suite supplements accumulating petrographic and trace-element geochemistry evidence that magma mixing is an important petrogenetic process.

Mixing of fractionated residual liquids can produce mixtures which are either superheated or supercooled depending on the sense of 'thermal curvature' of the liquid-line-of-descent. Both senses are encountered in the tholeiitic system, and this effect may exert a qualitative control on the crystallization texture of the mixture.

A comparison of approximately 2,000 abyssal tholeiite compositions to the experimental liquid-line-of-descent reveals that erupted differentiates which would be expected from advanced fractionation are scarce. Just this sort of phenomenon (the 'perched' steady state) was proposed by O'Hara (1977) as an earmark of the operation of a continuously fractionating magma chamber into which fresh magma is periodically remixed.

Introduction

From time to time the mixing of magmas has been advocated as an important process responsible for compositional variation in igneous rocks. Indeed mixing was once thought to be responsible for rocks transitional between acid and basic extremes although little attempt was made to account for the origin of the separation of the end members. This idea was discarded when it became apparent that compositional and mineralogical variation along such semi-continuous series as basalt-rhyolite was too complex to be explained in terms of mixtures of reasonable numbers of end members. The failure of mixing to explain the complexities of magmatic evolution led to the ascendancy of crystal fractionation as a more popular mechanism. It is a curious irony of recent developments in basalt petrology that mixing should now be reconsidered because of the failure of crystal fractionation to cope adequately with observed complexities – a failure which originally led to the demise of mixing as a serious hypothesis.

A number of investigations of continental and oceanic island volcanism have pointed out the incongruence between phenocryst compositions observed in rocks and the compositions that might be expected to crystallize (Kuno, 1936; 1968; Eichelberger, 1975; Anderson, 1976). Coupled with these obvious indications of disequilibrium between phenocryst and host lava is the occurrence of melt inclusions within the anomalous phenocrysts with compositions more appropriate to growth of those anomalous phenocrysts. These observations have fostered the interpretation that mixing of magmas with their various characteristic phenocrysts and melt inclusions is responsible. Mixing has also been proposed from bulk chemical evidence (Wright, 1973).

208

It has been pointed out that the magmas which are mixing need not be unrelated. In fact, one of the principal attractions of the new magma mixing is that the end members may be different consanguineous compositions on a line of descent so that a peculiar or unexplained range of primary compositions is unnecessary. O'Hara (1977) has developed the theme that remixing of primitive material with differentiation products of previous batches of primitive material is a natural consequence of reasonable volcanic plumbing systems. Stagnation of ascending magma in shallow chambers promotes normal crystal differentiation processes which are periodically interrupted by the introduction of fresh pulses of primitive magma into the chamber. O'Hara outlined a number of geochemical consequences of this mixing model which should be observable and incompatible with normal crystal fractionation. The anomalous phenocryst evidence is clearly consistent with the existence of such a process.

The conjunction of mixing and fractionation is implicit in several of the models of ocean-ridge magma chamber operations which have been proposed from inspection of ophiolite complexes and erupted basalts (Greenbaum, 1972; Cann, 1974; Dewey and Kidd, 1977; and Bryan and Moore, 1977). These models provide for a continuously operating magma chamber beneath the ocean ridges. Mixing is a natural consequence of the entry of fresh magma into this fractionating body. The zoning which may become established in such a chamber is graphically illustrated by the work of Bryan and Moore (1977) in the FAMOUS area.

Perhaps the most striking confirmation of the importance of mixing in conjunction with fractionation and the most detailed account of the supporting evidence comes from work by Rhodes et al. (1979) and Dungan und Rhodes (1978) on DSDP basalts from the Atlantic and by Bryan and Moore (1977) and Bryan et al. (1979) from work on the FAMOUS area. These workers have made a very strong case for mixing based not only on the presence of anomalous phenocrysts of olivine and plagioclase and their associated melt inclusions but also on the basis of an apparent excess of incompatible elements in moderately evolved compositions over what could be produced during normal crystal fractionation. Furthermore, Dungan and Rhodes (1978) showed that the paucity of clinopyroxene phenocrysts in many DSDP basalt suites, which is so curious when considered in terms of the substantial amounts of clinopyroxene fractionation required to balance calculations of crystal fractionation in these suites, may be explained in terms of popular reconstructions of a ridge-crest magma chamber. Gabbro cumulates formed in

the more differentiated chamber margins are the sink for the cryptic clinopyroxene.

In this context, the results of our melting experiments on basalts from the Oceanographer Fracture Zone are of interest because they provide evidence in the major element phase relations for the importance of magma mixing.

Oceanographer Fracture Zone Basalts

Fresh, unaltered basalt samples were recovered from dredge hauls in the Oceanographer Fracture Zone which offsets the Mid-Atlantic Ridge about 130 km at 35° N, 35° W (Fox et al., 1976). Shibata and Fox (1975) reported the petrography and major element chemistry of 48 of these basalts, which are characterized by phenocrysts of plagioclase, olivine and/or clinopyroxene and are tholeiitic with olivine and hypersthene in the CIPW norm. The 48 samples define systematic, continuous AFM and SI variation trends of iron enrichment without SiO_2 increase which can be reproduced by least squares fractionation calculations with very small residuals using as crystal extracts the phenocryst phase compositions observed (Shibata et al., 1979). This suggests that the basalts are related by crystal fractionation along a continuous liquid line of descent. The effects of fractionation observed in this one geographically-restricted area appear to produce variation of some chemical parameters that is virtually as large as that shown on a global scale by abyssal tholeiites.

Experimental melting studies were undertaken to investigate whether the proposed fractionation was possible in terms of the major element phase relations. For this purpose, the most primitive (highest Mg/Mg + Fe, V30-RD8-P12, henceforth P12) and the most differentiated (lowest Mg/Mg + Fe, V30-RD8-P22, henceforth P22) were selected to see whether fractionation of the primitive composition would, in fact, yield the differentiated composition as a residual liquid.

Experimental Procedures

Approximately 20 grams of each rock were ground to powder and homogenized in a Spex Shatterbox. Platinum loops of 0.005 in. diameter wire were presaturated with Fe in a melt of P12 at 1250°C for 60 h in a CO/CO_2 gas stream with a pO_2 N 'QFM'. The melt was later dissolved away in warm HF. These presaturated Pt–Fe loops were then used to suspend 10–15 mg aliquots of the samples in a quenching furnace with an atmosphere of CO/CO_2 mixed to give a pO_2 approximately at 'QFM' at 1,200° C. This mixing ratio ($CO = 4.75\%$; $CO_2 = 95.25\%$) was used for all experiments. Furnace temperature was calibrated against the melting point of Au (1,064° C). Pellets of rock powder with polyvinyl alcohol (Dupont Evanol) as a binder were sintered to the Pt–Fe loops. After equilibration at one atmosphere total pressure for times from 18–95 h at various temperatures (Table 1), the charges were quenched in less than 3 s in water introduced at the base of the gas stream. The products observed in the charges are recorded in Table 1 and Fig. 1. No quench crystals were observed. We reused the loops from previous experiments by cleaning them between experiments in warm HF. A single loop was used for all experiments on P12, but a second loop was required for experiments 21–32 on P22. The procedure of reusing presoaked loops is found to cause minimal iron loss in the relatively long runs made here.

Electron microprobe analyses of the products showed that the glasses and olivines were generally homogeneous, whereas the clinopyroxenes and plagioclases below the liquidus retain detectable inhomogeneity, presumably from the variable mineral composi-

Table 1. Experimental Results

Phases present

$T°C$ / hours	V30-RD8-P12 Primitive P12	V30-RD8-P22 Differentiated P22	Experiment #
1,239 / 18.5	glass		20
1,225 / 18	glass, plag, sp		19
1,199 / 17.9	glass, plag, ol, sp	glass	18
1,185 / 21.3	glass, plag, ol, px, sp	glass, plag	17
1,170 / 23	glass, plag, ol, px, sp	glass, plag, sp	16
1,160 / 22.9	glass, plag, ol, px, sp	glass, plag, px, sp	14
1,155 / 20.2	glass, plag, ol, px, sp	glass, plag, px, sp	23
1,154 / 40	glass, plag, ol, px, sp	glass, plag, px, sp	21
1,152 / 39.1	glass, plag, ol, px, sp	glass, plag, px, ol, sp	12
1,133 / 88	glass, plag, ol, px, sp	glass, plag, px, ol, sp	13
1,106 / 65	glass, plag, ol, px, sp	glass, plag, px, ol, sp	22
1,085 / 95		glass, plag, px, ol, sp	30
1,078 / 93		glass, plag, (ol), sp, ilm, low-Ca + high-Ca px	32

(All experiments with rock powder starting material)

Liquid compositions from additional experiments (wt. %)

#	P12–33	P12–35	P12–28	P22–30	P22–32	P22–26	P22–31
SiO_2	51.24	50.84	50.89	50.42	53.86	54.95	69.53
TiO_2	0.55	0.90	1.32	4.14	3.54	3.57	1.26
Al_2O_3	15.89	15.68	15.34	11.38	11.76	11.36	10.17
Cr_2O_3	0.06	0.04	–	0.02	0.01	0.02	–
FeO	4.07	6.47	8.21	16.04	15.14	14.28	7.22
MnO	0.07	0.13	0.16	0.28	0.27	0.24	0.12
MgO	9.82	9.72	8.54	3.33	2.53	2.31	1.49
CaO	16.60	15.77	13.78	7.71	7.49	6.55	3.24
K_2O	0.18	0.29	0.39	1.07	1.49	1.39	2.30
Na_2O	2.03	1.91	2.35	2.43	2.42	2.34	1.41
P_2O_5	–	–	–	0.46	–	0.53	–
Sum	100.51	101.75	100.98	97.28	98.51	97.54	96.74
$T°C$	1,235	1,222	1,202	1,085	1,078	1,131/1,070	1,051
Hours	17[a,b,c]	22[a,b,c]	12[a]	95	93	15/76[e]	110[d]
Phases present	plag cpx ol	plag cpx ol	plag cpx ol Sp	plag cpx ol sp	plag pxs (ol) ilm, sp	plag pxs (ol) ilm, sp	plag pxs (ol?) ilm, silica

[a] Crucible Tamela diopside
[b] An_{70} admixed in starting rock powder
[c] San Carlos olivine as cap
[d] Devitrified silica crucible
[e] Dropped from 1,131 to 1,070 in 4 h

tions present in the rock powder starting material (20 μ max. grain-size, with most < 5 μ) even for runs of 1 3 days duration. Selected analyses are given in Table 1. Crystal phase compositions are discussed in the companion paper (Shibata et al., 1979).

Additional experiments were conducted in order to extend the temperature range over which three-phase-saturated liquids could be recovered for microprobe analysis. At temperatures above 1,200° C the 3 phases olivine, clinopyroxene, and plagioclase no longer coexist with liquid in the P12 bulk composition. Crystals of diopside from Tamela, Finland (Harvard U. #29751) with dimensions on the order of a cm were used as crucibles by preparing cavities with a WC drill. P12 powder was loaded into these diopside crucibles and equilibrated in the same furnace as before at temperatures above cpx disappearance in P12. The crucible imposes cpx saturation on the liquid generated as it effectively changes the bulk composition by partly dissolving. Tamela diopside is not the cpx in equilibrium with P12 liquids so the cpx added to the liquid by dissolution is somewhat too Ca- and Mg-rich. There is also a tendency for ion exchange between crucible and liquid to occur once there has been enough dissolution to saturate the liquid with cpx. The ion exchange process appears to be limited by slow diffusivity in the cpx crucible. Composition gradients of Fe, for

Table 1 (Continued)

Phase compositions from P12 experiments (wt. %)

#	20	19	18	17	16	14	23	12	
SiO_2	48.73	49.85	49.39	50.14	50.71	50.06	50.13	50.50	
TiO_2	1.21	1.14	1.25	1.55	1.80	2.16	2.27	2.35	
Al_2O_3	16.19	16.29	15.70	14.53	13.97	13.49	13.52	13.46	
Cr_2O_3	0.08	0.06	0.08	0.09	0.05	0.02	0.07	0.02	Liquid
FeO	9.29	9.32	9.89	10.96	12.25	12.87	12.06	13.49	
MgO	8.05	8.00	8.13	7.36	6.66	5.86	6.09	5.76	
MnO	0.18	0.16	0.16	0.20	0.20	0.22	0.23	0.22	
CaO	12.43	12.40	12.28	12.10	11.19	10.54	10.91	10.24	
K_2O	0.35	0.33	0.34	0.43	0.55	0.71	0.65	0.72	
Na_2O	2.24	2.38	2.41	2.57	2.60	2.77	2.57	2.68	
Sum	98.91	99.93	99.65	99.93	99.98	98.70	98.49	99.45	
SiO_2				48.57	48.31	49.57	50.60	49.50	
TiO_2				1.41	1.36	1.06	1.01	1.50	
Al_2O_3				5.43	5.63	4.89	4.78	5.30	
Cr_2O_3				0.41	0.39	0.43	0.61	0.36	
FeO				10.14	8.99	10.05	8.10	11.23	Clinopyroxene
MgO				13.56	14.13	14.86	14.58	12.77	
MnO				0.22	0.21	0.24	0.21	0.27	
CaO				19.90	19.71	19.67	18.84	19.45	
K_2O				0.04	0.03	0.06	0.06	0.05	
Na_2O				0.46	0.31	0.29	0.38	0.40	
Sum				100.13	99.09	101.13	99.15	100.83	
SiO_2		45.74	46.82	49.11	50.81	49.12	49.56	51.78	
Al_2O_3		33.70	32.85	30.86	29.86	31.83	30.80	29.94	
FeO		0.68	0.71	1.07	1.14	1.14	0.80	1.21	
MgO		0.26	0.24	0.46	0.70	0.71	0.27	0.35	Plagioclase
CaO		17.14	16.18	15.26	13.94	15.64	14.72	13.64	
K_2O		0.04	0.03	0.05	0.08	0.14	0.07	0.11	
Na_2O		1.33	2.06	2.44	3.21	2.33	2.84	3.51	
Sum		98.89	98.90	99.25	99.72	100.92	99.07	100.56	
SiO_2			38.70	38.79	37.78	38.08	37.46	37.81	
Al_2O_3			—	0.11	0.32	0.15	0.52	—	
Cr_2O_3			0.05	0.05	—	0.02	—	0.02	
FeO			15.11	16.71	20.50	22.68	22.15	23.68	Olivine
MgO			46.39	43.76	40.71	38.85	37.70	36.82	
MnO			0.25	0.32	0.31	0.35	0.35	0.33	
CaO			0.41	0.36	0.48	0.45	0.55	0.50	
Sum			100.90	100.09	100.10	100.58	98.73	99.16	
$T°C$	1.239	1.225	1.199	1.185	1.170	1.160	1.155	1.152	

example, extend less than 5 μ into diopside. While interaction of the charge with the diopside crucible changes the details of the chemistry of the liquid, no important change in the normative proportions of olivine, pyroxene, and plagioclase in the liquid are introduced by this procedure. This was determined by running an experiment below the 3-phase-saturation temperature in a diopside crucible and comparing the liquid with runs made on Pt loops. Parameters such as $Mg/Mg + Fe$ change, but the liquid still plots within the band of experiments from Pt loops on a diagram such as Fig. 2.

To maintain olivine and clinopyroxene and plagioclase saturation at temperatures above 1.220° C, it was necessary to add crystalline olivine and plagioclase to the contents of the diopside crucible. San Carlos olivine and plagioclase (Harvard U. #10084, locality unknown, ~ An_{70}) were used for this purpose. Plagioclase was admixed and ground into P12 powder before insertion in the diopside crucible. Olivine was added as a single crystal on top of the cavity filled with powder. The partly molten basalt has a strong tendency to wet the solids. Creep of the liquid out of the cavity and into fractures and crevices in the diopside becomes a severe

Table 1 (Continued)

Phase compositions from P22 experiments (wt. %)

#	18	17	16	14	23	12	13	22	
SiO_2	51.61	50.82	52.02	50.95	50.90	50.50	50.66	50.55	Liquid
TiO_2	2.05	2.03	2.16	2.45	2.40	2.68	3.00	3.45	
Al_2O_3	15.00	14.54	14.05	13.61	13.58	13.23	12.71	11.95	
Cr_2O_3	0.06	0.06	0.06	0.05	0.05	0.02	0.02	0.04	
FeO	11.66	11.94	12.27	12.36	11.90	13.70	14.67	16.30	
MgO	5.30	5.54	5.67	5.73	5.67	5.58	5.12	4.06	
MnO	0.22	0.25	0.23	0.20	0.25	0.27	0.28	0.30	
CaO	10.76	10.59	10.67	10.44	10.42	9.80	8.86	8.12	
K_2O	0.52	0.50	0.52	0.57	0.58	0.60	0.69	0.96	
Na_2O	3.02	2.93	3.01	3.07	2.94	2.96	2.98	2.94	
Sum	100.20	99.20	100.66	99.43	98.69	99.34	98.99	98.68	
SiO_2				50.18	50.85	49.92	49.73	49.80	Clinopyroxene
TiO_2				0.89	0.84	1.34	1.19	1.06	
Al_2O_3				3.04	3.36	3.93	3.53	2.98	
Cr_2O_3				0.25	0.38	0.17	0.13	0.23	
FeO				9.54	8.58	10.84	11.26	12.05	
MgO				15.71	15.80	15.13	15.90	15.72	
MnO				0.26	0.23	0.25	0.32	0.34	
CaO				19.42	19.30	18.54	17.29	17.23	
K_2O				0.06	0.06	0.02		0.04	
Na_2O				0.29	0.32	0.33	0.30	0.30	
Sum				99.65	99.73	100.46	99.66	99.76	
SiO_2		52.22	53.20	53.29	53.11	53.68	53.03	53.89	Plagioclase
Al_2O_3		28.80	28.76	28.52	28.72	28.79	28.97	28.02	
FeO		1.16	1.14	1.07	1.07	1.12	1.18	1.23	
MgO		0.29	0.21	0.28	0.25	0.21	0.29	0.16	
CaO		13.15	12.05	12.65	12.60	12.53	12.16	11.23	
K_2O		0.11	0.11	0.16	0.10	0.09	0.18	0.16	
Na_2O		3.76	4.33	4.26	3.98	4.39	4.16	4.68	
Sum		99.49	99.82	100.23	99.84	100.81	99.97	99.37	
SiO_2							36.57	35.86	Olivine
Al_2O_3							0.14	—	
Cr_2O_3							0.02	0.09	
FeO							27.58	34.77	
MgO							34.21	26.59	
MnO							0.51	0.62	
CaO							0.36	0.59	
Sum							99.40	98.52	
$T°C$	1,199	1,185	1,170	1,160	1,155	1,152	1,133	1,106	

problem at these higher temperatures. Using a single crystal of olivine as a cap to the cavity slows this escape, apparently since the space between the olivine and diopside forms a surface tension trap. The crystalline olivine and plagioclase added are not too far from the compositions which precipitate from P12, so that their use introduces no important dislocation in the saturation curves.

To extend the coverage of 3-phase-saturated liquids to low temperatures, two approaches were used. The first was to produce fractionated residual liquids not in equilibrium with the bulk composition. P22 was held at a temperature in its melting interval and then dropped to a temperature below the equilibrium solidus. Low-Ca pyroxene coexisted with liquid, plagioclase, and high-Ca clinopyroxene. Olivine was also present but was armoured with pyroxene. The reaction relation was thus exploited to derive residual liquids beyond olivine saturation below the solidus. The second approach was to produce silica-saturated liquids at low temperature by using devitrified silica glass as a crucible. Powder or chunks of P22 could not successfully be run directly at low temperature because P22 was below its solidus and no interaction with

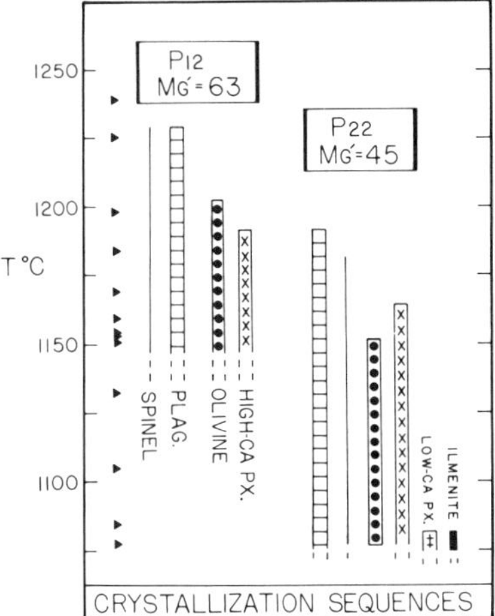

Fig. 1. Experimental crystallization results for two OFZ basalts. *Mg'* is molar $Mg/Mg + Fe$. *Triangles* near left side indicate the temperature of experiments. Note that primitive composition P12 crystallizes olivine before clinopyroxene whereas the differentiated composition P22 crystallizes clinopyroxene before olivine. Experiments at 1 bar with pO_2 near QFM in Fe-soaked Pt loops

the silica occurred. Instead, a puddle of partially molten P22 was generated in the silica crucible so that a chemical bond of liquid was established with the crucible surface. The liquid remaining after equilibration at low temperature was at the crucible/charge interface. Probe analysis of this material required considerable care since it collected in very small pools. It appeared not to be uniform in composition, perhaps as an artifact of the difficulty in finding a 'clean' probe spot or perhaps as a result of a real lack of homogenization.

Discussion of Results

Plagioclase is the liquidus phase at 1 atm (approximately 'QFM') for both primitive (P12) and differentiated (P22) members of the Oceanographer Fracture Zone suite. This is in accord with the petrography of both samples, which bear phenocrysts of plagioclase. Spinel is also a near-liquidus phase in both samples. The first ferromagnesian silicate to crystallize from P12 is olivine followed by pyroxene, whereas the reverse is true for P22 – pyroxene crystallizes before olivine. Although the temperature difference between the appearance of olivine and pyroxene is relatively small in each case, the amount of the first ferromagnesian silicate to crystallize before the second in each case is significant. P12 crystallized plagioclase and approximately 2% olivine compared with P22 which crystallizes plagioclase and approximately 5% clinopyroxene before plagioclase + olivine + clinopyroxene saturation is reached. (Amounts calculated

by mass balance from microprobe analysis of experimental charges and confirmed by visual estimate.) Furthermore, this reversal of olivine and clinopyroxene entry between P12 and P22 appears to be recorded in the samples' petrography. P12 has approximately 1.7% olivine phenocrysts and none of clinopyroxene, whereas P22 has noticeably more clinopyroxene (approximately 0.5%) than olivine (approximately 0.3%) phenocrysts, in addition to the ubiquitous plagioclase phenocrysts. It should be noted that the presence of clinopyroxene phenocrysts cannot explain the departure of P22 from olivine-clinopyroxene cosaturation since P22 has approximately 5% of excess clinopyroxene over the cosaturated composition, of which only approximately 0.5% can be accounted for as phenocrysts. In contrast P12 has roughly enough olivine phenocrysts (1.7%) to account for the departure of P12 from olivine-clinopyroxene cosaturation ($\sim 2\%$ needed).

The crystallization of all phases occurs at significantly lower temperatures in P22 than in P12. These relations are broadly consistent with the idea that P22 might be a residual liquid from a magma similar in composition to P12. However the details of the experimental results clearly indicate that if such a fractional crystallization process is responsible for the chemical variation in the OFZ suite, it cannot have occurred at the physical conditions of the experiments. This conclusion is the result of three observations on the experimental crystallization sequences. First, plagioclase is the liquidus phase in both compositions and is separated by a significant temperature and composition interval (5–10% plagioclase crystallization) from the appearance of the second silicate phase. If the differentiated composition were the daughter of the primitive composition by fractional crystallization operating at conditions near the ocean floor, it should be multiply saturated with plagioclase, olivine, and pyroxene at its liquidus since the primitive composition has reached multiple saturation at the liquidus temperature of the differentiated composition. This is not observed. Second, the fact that olivine and pyroxene appear in reversed order in the experiments on the two compositions shows that the two cannot be simply related, because they approach multiple saturation from different directions, not from the same direction as required in a single fractionation sequence. Third, microprobe analyses of experimental residual liquids from P12 show that the composition of P22 is not produced. This should be a logical expectation from the first two observations.

Primitive ocean floor tholeiites of high $Mg/Mg + Fe$ generally have olivine and/or plagioclase phenocrysts and have clinopyroxene late in their crystallization sequence as is the case for the primitive

OFZ sample P12. Clearly the anomaly in the OFZ suite is the experimental observation (consistent with sample petrography) that some of the evolved basalts crystallize clinopyroxene before olivine. Perhaps this anomaly could be removed if the fractionation responsible for the chemical variation of the basalts did not happen at the physical conditions of the experiments, even though these conditions appear to be relevant to the natural crystallization of the samples upon eruption. To explain the anomaly successfully, the phase relations must change so that the appearance temperature of clinopyroxene decreases relative to olivine (and plagioclase) or at least does not increase as fast. Said another way, the liquidus volume in composition space of clinopyroxene should shrink with the change of physical conditions. Increase of load pressure, for example in a buried magma chamber, would be an obvious way to change the physical conditions at the site of differentiation. Studies of basalt melting as a function of pressure have shown that clinopyroxene appearance temperatures have steeper dT/dP slopes than either olivine or plagioclase. As a result, fractionation of 3-phase-saturated liquids at depth produces liquids of clinopyroxene-undersaturated character upon eruption. In such a scenario, the OFZ anomaly would be magnified rather than reduced. So clearly increase of load presure is not a satisfactory explanation of the anomaly. Load pressure coupled with volatile components might be considered. Moderate pCO_2 probably operates in the same direction as load pressure whereas increasing pH_2O may produce the desired shrinking of the clinopyroxene volume. Unfortunately ocean ridge basalts in general appear to be conspicuously dry (Delaney et al., 1977). In consequence then, the OFZ clinopyroxene appearance anomaly is not easily dismissed as an artifact of inappropriate experimental conditions. Therefore, we must conclude that it is not possible in detail to regard the differentiated composition, P22, as a residual liquid from fractionation of the primitive composition, P12, along a liquid line of descent.

The clinopyroxene anomaly renders an apparently simple fractionation scheme into a complex one. We must keep in mind that the very low residuals on the fractionation mass-balance calculations strongly suggest that the resolution of this anomaly must not involve exotic or cryptic substances. The actors have been identified but the script evidently has not yet been found, because the obvious plot fails in detail to relate properly the actors. Furthermore, because plausible changes of physical conditions magnify the problems with the obvious script, we must seek a resolution which does not involve fiddling with the set and props. The plot must thicken.

Liquidus Phase Boundaries

Compositional Curvature

The major normative constituents of the OFZ basalts are plagioclase, diopside, olivine, and hypersthene. A convenient way to treat compositional variation in this system is to use a modification of O'Hara's CMAS (1968 a) combinations to construct the tetrahedron plagioclase, olivine, diopside, and silica. Formulae for plotting compositions in this way are given in the caption to Fig. 2. Because the OFZ basalts have plagioclase on the liquidus, a feature shared with many other abyssal tholeiites, a plagioclase-saturated liquidus projection on the olivine-diopside-silica plane represents much of the interesting compositional variation in the system. The phase boundaries dividing the fields of olivine (+ plagioclase) crystallization from high-Ca clinopyroxene (+ plagioclase), etc. can be constructed from electron microprobe analyses of

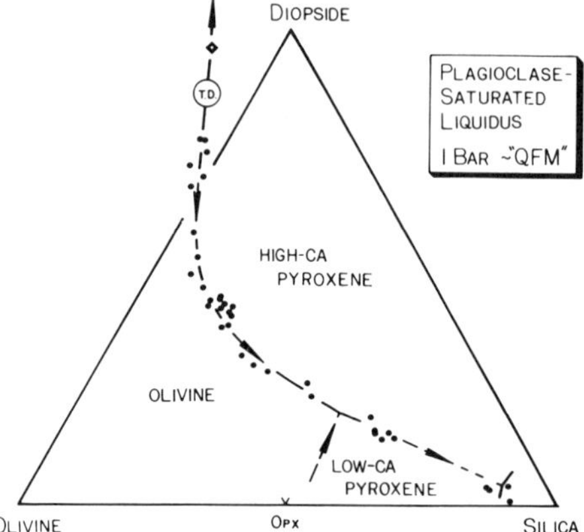

Fig. 2. Projected plagioclase-saturated liquidus on diopside-olivine-silica. Points used to construct boundary curves are microprobe analyses of 3-phase-saturated liquids (+/− spinels and ilmenite). All Fe is as FeO. *Circled T.D.* is thermal divide. *Diamond symbol* across the thermal divide is point on the diopside-olivine-plagioclase curve from Presnall et al. (1978). Note strong concavity of this curve towards diopside.

Compositions may be plotted in this projection by dividing weight percent of oxides by molecular weight to obtain molecular proportions. These are combined according to the following:
Plagioclase = $Al_2O_3 + Na_2O + K_2O$.
Diopside   = $CaO − Al_2O_3 + Na_2O + K_2O$.
Olivine   = $(FeO + MgO + MnO + 2Fe_2O_3 + Al_2O_3 − CaO − Na_2O − K_2O)/2$.
Silica   = $SiO_2 − (Al_2O_3 + FeO + MgO + MnO + 3CaO + 11Na_2O + 11K_2O + 2Fe_2O_3)/2$.
Proportions of diopside, olivine and silica are then normalized to the sum of diopside + olivine + silica and plotted on a molecular basis. The plot of Fig. 8 is obtained by normalizing the proportions of diopside, olivine and plagioclase to the sum of diopside + olivine + plagioclase

experimental glasses equilibrated with the three silicate phases. Such a liquidus diagram is presented in Fig. 2 for the experiments described above. Topologically this diagram is similar to the diagram for the end-member system diopside-forsterite-silica without plagioclase saturation summarized by Kushiro (1972). One difference, besides the cosmetic change from weight to mole units, is the broadening of the high-Ca clinopyroxene field in the presence of plagioclase, and the increase of normative silica in silica-saturated liquids. Another difference is that the thermal divide on the 3-phase curve olivine-plagioclase-cpx projects on the undersaturated side of the diopside-olivine join. This is partly an artifact of the procedure we have used in the projection since all Fe is counted as Fe^{2+}. This arbitrary procedure is adopted for convenience in making all projections in this paper, in the absence of Fe^{2+}/Fe^{3+} determinations. Including some Fe^{3+} in the total Fe would shift the projected position of the thermal divide towards the olivine-diopside join. For the purpose of this discussion, the important topological difference between this diagram and the end-member system without plagioclase is the sense of curvature of the olivine-plagioclase-cpx boundary on the tholeiitic side of the thermal divide. This curve is strongly concave towards diopside rather than towards olivine. The sense of curvature determined here for this boundary by microprobe analysis of three-phase-saturated glasses agrees with the sense of curvature shown schematically by Presnall et al. (1978) in the anorthite-diopside-olivine-silica tetrahedron. The importance of this curvature will be developed below.

These phase boundaries can be used to illustrate the crystallization of the OFZ basalts. The compositions of the 6 OFZ basalts with complete chemical analyses (Shibata and Fox, 1975) are plotted in this projection in Fig. 3. The 'primitive' sample P12 plots in the olivine field as it should because we determined that olivine crystallized before cpx in this sample. Also the differentiated sample P22 plots in the cpx field as it should. The composition of residual liquids produced during fractionation can be followed graphically. For example, P12 crystallizes olivine (and plagioclase) and the liquid evolves towards the olivine and cpx (+plag) boundary with falling temperature. Once this three-phase boundary is reached the residual liquid will follow this boundary as it precipitates all three phases. During equilibrium crystallization the residual liquids would evolve as far as the olivine +low-Ca pyroxene+cpx+plag+liquid reaction point which is reached at the solidus of this olivine-normative composition. During fractional crystallization the residual liquids could evolve further along the low-Ca pyroxene+cpx+plag curve. At no point

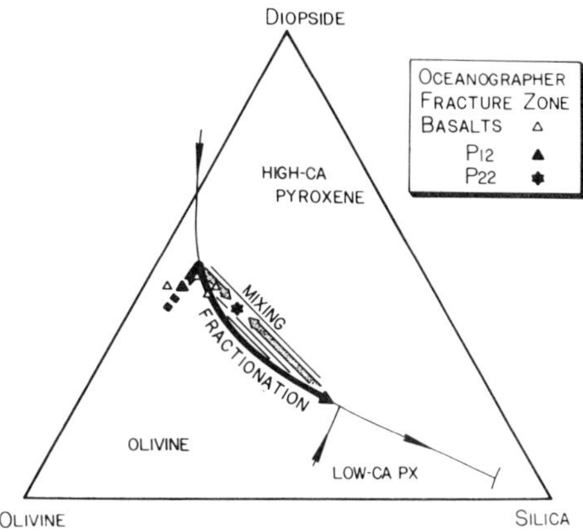

Fig. 3. OFZ basalts on liquidus of Fig. 2. Fractionation of primitive sample P12 produces residual liquids along the heavy arrow. The primitive sample P12 does not produce the differentiated sample P22 as a residual liquid at 1 bar (~QFM). Increase of pressure moves fractionation curve away from P22. Mixing of various liquids on the fractionation curve can give mixtures crystallizing clinopyroxene before olivine

in a crystallization sequence for P12 (either equilibrium or fractional) is P22 encountered as a residual liquid. When we reached this conclusion above we noted that no plausible change in physical conditions would accomplish this goal. In terms of Fig. 3, this is equivalent to saying we cannot get the 3-phase boundary to move towards P22.

How then can we produce P22 as a differentiation product of P12 without introducing additional primitive magmas in the cpx (+plag) field which could differentiate to P22? The curvature of the 3-phase boundary now becomes important. Residual liquids from P12 can be produced along this boundary as shown by the heavy FRACTIONATION arrow. If batches of such residual liquid are mixed with one another as indicated by the mixing vectors, P22 can be produced. It is the concavity of the 3-phase curve towards diopside which allows mixtures to be produced within the cpx field. Thus a mechanism exists for producing compositions that crystallize cpx before olivine even though the parent composition crystallizes olivine before cpx. In this manner the OFZ cpx-appearance anomaly can be resolved.

The concept of compositional curvature of liquidus saturation boundaries is not new. Irvine (1970) has used the principle to good advantage in explaining the variable proportions of cumulus mineral phases observed with stratigraphic height in differentiated layered intrusions. Analogous applications are made in the companion paper (Shibata et al., 1979) to explain the variable proportions of phenocrysts observed in the OFZ fractionation sequence.

It should be noted that the proposed mixing model does not violate the requirement that cryptic or exotic substances be excluded, a restriction imposed by the low residuals in the least-squares calculations of fractionation. The residual liquids mixed are generated by different amounts of fractionation of the phenocryst phases and so the resultant linear mixture also gives the appearance of being able to be produced directly by some combination of extracted crystals, even though it cannot be. The process we advocate relies heavily on normal crystal fractionation to produce a range of residual liquids. However, mixing of some of these liquids is required as a supplementary process to account for the cpx-appearance anomaly which is not satisfactorily explained by a simple fractionation process.

In our companion paper we present independent petrographic and trace element evidence for the operation of mixing. Briefly the presence of anomalous plagioclase phenocrysts and melt inclusions is recorded. Ni abundances are not consistent with simple fractionation but suggest the operation of mixing. Also there are a number of curious circular to subcircular structures which may be generated in a mixing event. We shall now explore some thermal consequences of mixing which should influence crystallization textures and will show that many crystallization textures observed are consistent with the operation of mixing.

Liquidus Phase Boundaries

Thermal Curvature

The explanation of the clinopyroxene appearance anomaly arises through compositional curvature of the liquid-line-of-descent equilibria involving olivine, cpx, and plagioclase. Additional effects occur as a result of what we shall call, for lack of a better term, 'thermal curvature'. These effects may be observable in a sample's crystallization texture. Figure 4 shows the composition of the 3-phase-saturated liquids in our experimental charges as a function of temperature in terms of the proportion of normative diopside and olivine. The upper extreme of the curve is the thermal divide. Compositions dashed downward to the right from the thermal divide become increasingly nepheline-normative whereas the experimental liquids plotted to the left of the thermal divide become more hypersthene- and then quartz-normative with falling temperature. The change from olivine to low-Ca pyroxene as the third saturating phase produces a no-

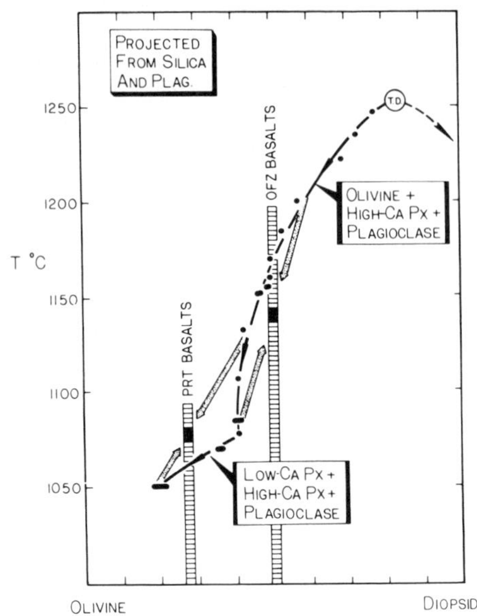

Fig. 4. Temperature vs. composition of 3-phase-saturated liquids (+/− spinels and ilmenite). Portion of olivine + cpx + plagioclase curve to right of the thermal divide descends into nepheline-normative compositions approaching melilite saturation. Composition range of Oceanographer Fracture Zone (OFZ) and Puerto Rico Trench (PRT) basalts indicated by the width of striped vertical bars. The upper termination of these striped vertical bars has no temperature significance. Mixing arrows for OFZ basalts indicate that the mixture (solid part of OFZ bar) will be supercooled below the 3-phase curve. Hence such a mixture will preserve suspended crystalline debris and undergo extra crystallization as a result of mixing. By contrast a mixture produced as indicated by the PRT arrows will be superheated

ticeable inflection in the curve. The lower termination of the curve corresponds to silica saturation. 'Thermal curvature' results from composition not changing uniformly as a function of temperature and shows up as a deviation from a straight line on a plot such as Fig. 4.

Two senses of 'thermal curvature' are evident in Fig. 4 as they pertain to a mixing process. The right set of speckled mixing arrows converges on the diopside/olivine ratio of OFZ basalts and shows that the mixture, indicated by the solid portion of the vertical bar of OFZ basalts, has a lower temperature than the 3-phase-saturated liquid of that olivine/diopside ratio. Two liquids mixed in this way crystallize spontaneously because the 3-phase curve is the lowest temperature at which a particular composition can be all liquid. We assumed in drawing straight mixing lines that the heat capacities of the liquids are the

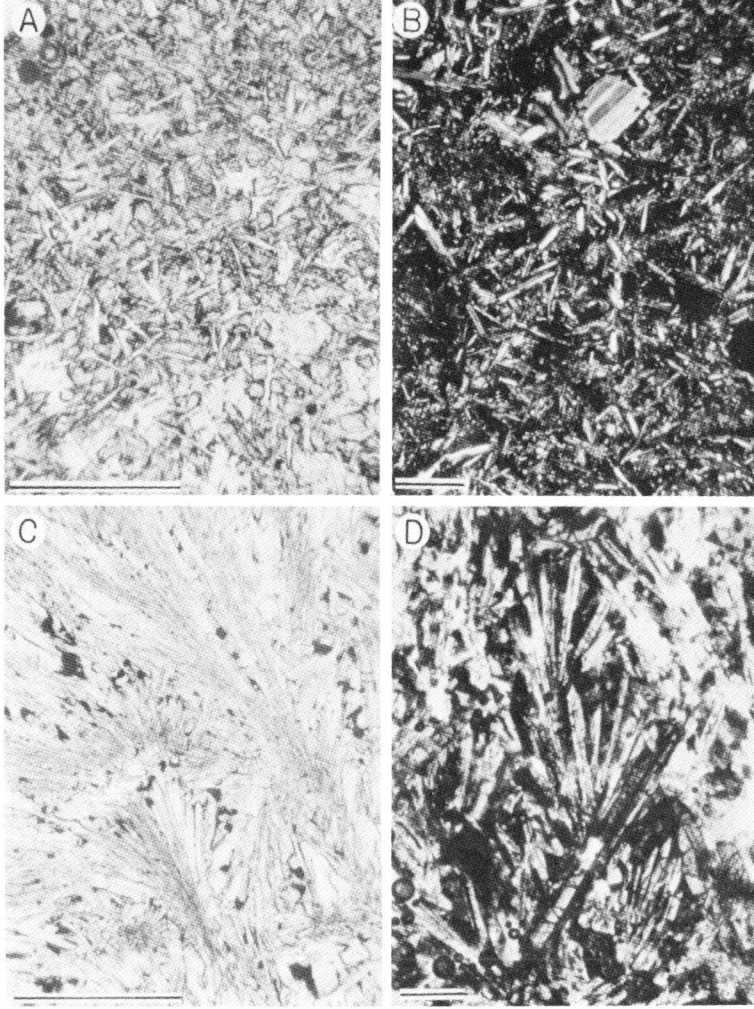

Fig. 5A–D. Crystallization textures. Scale bars are $^1/_2$ mm. in each case. **A** Experimental mesh to sub-ophitic texture produced when cooling begins from below 3-phase-saturated temperature. Experiment on meteoritic basalt Stannern (Walker et al., 1978). **B** Mesh to sub-ophitic texture commonly observed in evolved OFZ basalts. Note 'anomalous' plagioclase which was not dissolved. **C** Experimental fasciculate texture produced when cooling begins from mildly superheated condition. Experiment on meteoritic basalt Stannern (Walker et al., 1978). **D** Fasculate texture commonly observed in PRT basalts

same and that there is no significant enthalpy of mixing (Carmichael et al., 1977). Inspection of Fig. 3 shows that the supercooled mixture should precipitate clinopyroxene to regain thermal equilibrium. A further consequence of this sense of 'thermal curvature' is that the mixtures produced have no tendency to dissolve any crystals which may be suspended. Hence any evidence of mixing which may be present in anomalous phenocrysts and their melt inclusions will be preserved in the mixing process and not erased. As we report in the companion paper, evidence of this sort has been preserved in the OFZ samples.

In contrast, the left arrows converge on the olivine/diopside ratio of Puerto Rico Trench (PRT) basalts (Shido et al., 1974) and suggest that mixtures produced in this way, indicated by the solid portion of the vertical bar for PRT basalts, may be superheated above the 3-phase-saturation curve. Such mixtures would tend to dissolve suspended crystals, although the capacity for such action probably cannot exceed $\sim 10\%$ of the mixture.

Recent studies of the crystallization textures produced from compositions close to 3-phase-saturation showed that the textures were strongly controlled by the temperature at the start of cooling (Walker et al., 1978). If cooling begins from below the 3-phase temperature, nucleation is controlled by the distribution of crystallites already present and mesh to sub-ophitic textures develop. However, if cooling begins from above the 3-phase temperatures, nucleation centers are limited and fasciculate textures develop. If these results are general, and if the OFZ and PRT basalts were mixtures before eruption as suggested in Fig. 4, then we might expect mesh to sub-ophitic and fasciculate textures, respectively, to be the rule in these different suites. This expectation is realized as may be seen in Fig. 5.

This may seem, at first, to be a striking independent confirmation of the operation of mixing and the importance of thermal curvature, and it may be so. However, it is ambiguous because the textures observed could be produced in other ways – unlike the cpx-appearance anomaly which admits no simpler explanation. Nevertheless the textures observed do constitute permissive circumstantial evidence which is entirely consistent with the combined fractionation and mixing model developed here.

TiO₂ Abundances

Plots of TiO_2 versus $Mg/Mg+Fe$ have seen much service in the interpretation of chemical variation in various basaltic suites terrestrial, lunar, and meteoritic (Bender et al., 1978; Rhodes et al., 1979; Papike and Vaniman, 1978; Stolper, 1977). Ti is to a first approximation an incompatible element whereas Fe and Mg are compatible and show a strong relative fractionation in crystal-liquid equilibria. Inspection of such a plot may be useful in determining the processes responsible for the variation. Incipient melting should produce variation of TiO_2 while $Mg/Mg+Fe$ changes little. Crystal fractionation should increase TiO_2 and decrease $Mg/Mg+Fe$ as the proportion of liquid remaining decreases. The trend need not be linear. Mixing of two end-members should produce a linear variation.

Figure 6 shows such a plot for 2,100 oceanic basalts from the PETROS data bank (Mutschler et al., 1976). Only rocks and glasses too hydrated ($H_2O > 4\%$), oxidized ($Fe^{3+} > Fe^{2+}$), or obviously petrologically degraded to be of interest were excluded from this tabulation. TiO_2 and $Mg/Mg+Fe$ show antipathetic variation, so incipient melting is probably not the major control on the composition of the world's abyssal tholeiite or mid-ocean ridge basalt (MORB) population. This population resembles a linear smear which could be consistent with either crystal fractionation or mixing (or both!).

Also plotted on Fig. 6 are the liquids from the OFZ experiments. Open circles are for temperatures above 1,080° C when no ilmenite crystallizes. TiO_2 increases with falling temperature until precipitation

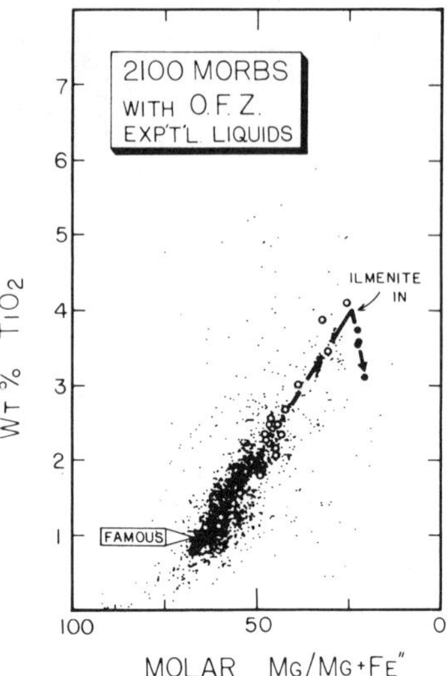

Fig. 6. TiO_2 vs. $Mg/Mg+Fe$ for 2,100 mid-ocean-ridge basalts and OFZ experimental liquids. Solid blob for FAMOUS glasses. Ilmenite crystallization terminates the build up of TiO_2 in residual liquid at ~4% for 1 bar QFM fractionation. Natural examples of liquids with $TiO_2 > 2.5$ are rare relative to the heavy cluster between 1–2%. Data base is the PETROS file of Mutschler et al. (1976)

of ilmentie below 1,080° C drives TiO_2 rapidly down as shown by filled circles. It appears that the MORB smear and the OFZ experimental liquids show a similar trend. Previous attempts to relate MORB trends in general (Rhodes et al., 1979) or the FAMOUS glass trend in particular (Bender et al., 1978) to crystal fractionation have concluded that there is 'too much' Ti in liquids of evolved $Mg/Mg+Fe$. These conclusions were based on calculated fractionation patterns and are supported only marginally by the crystal fractionation sequence shown in Fig. 6. This may be because the calculations did not include the effect of concurrent spinel and clinopyroxene crystallization. Inspection of Fig. 6 shows that crystal fractionation is an adequate mechanism for understanding the broad features of the MORB distribution,. This conclusion by no means negates the possibility that mixing of the sort discussed above is an important process. It is merely not strongly required by an apparent TiO_2 excess. Likewise complex melting may occur but it is not a requirement of these data.

Shibata et al. (1979) show, however, that there is a large excess of Ni in at least one derivative composition at the same time Rb shows negligible anomalous enrichment. The contrast in anomalous enrichment behavior of compatible and incompatible elements is expected in a fractionation and mixing

218

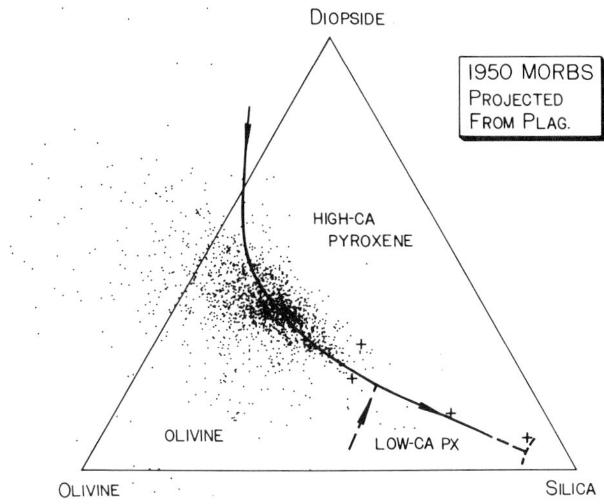

Fig. 7. MORBS on liquidus of Fig. 2. Densest cluster of MORB analyses conforms to the 1 bar olivine + clinopyroxene (+ plagioclase) saturation boundary. This implies that the major control on MORB chemistry is at low pressure. Note the scarcity of MORB compositions approaching low-Ca pyroxene saturation on this curve. These compositions, however, are common in OFZ basalt mesostasis and might be expected in the fractionation process that produces evolved MORBS from primitive ones. Crosses are OFZ mesostasis compositions reported by Shibata et al. (1979). The MORB population is "perched" with respect to these potential differentiates, a phenomenon suggesting the operation of O'Hara's (1977) periodically recharged, continuously fractionating magma chamber

process which starts its cycles with fresh magma. The negligible anomalous enrichments of TiO_2 and Rb compared with large excess of Ni do seem to be stronger evidence for the combined fractionation and mixing process than the small TiO_2 anomaly itself.

One feature of note in Fig. 6 is the sparseness of the MORB population with TiO_2 greater than 2.5%. Residual liquids in this system may have up to 4% TiO_2 and such liquids are common as glass and devitrified glass in OFZ basalt mesostasis. Evidently such liquids can be generated either in the laboratory or in a hand specimen of basalt during the time scale of eruption. Surely such liquids should be produced in natural magma chambers also. (Indeed such liquids are required in the mixing resolution to the cpx-appearance problem posed above.) The scarcity of such liquids in the world MORB population is a curious observation which will be developed and explained in the next section.

Other Oceanic Basalts

The OFZ basalts are not a particularly distinctive suite of abyssal tholeiites in terms of their major element chemistry. It might be reasonable to expect, therefore, that the liquidus phase relations determined from OFZ samples have some measure of general applicability to the problems of fractionation of oceanic tholeiite. Figure 7 shows the composition of 1950 mid-ocean ridge basalts on the plagioclase-saturated liquidus diagram. It can be seen that the densest cluster overlaps the OFZ basalts, so that the OFZ basalts are representative of MORBS in the PETROS file (Mutschler et al., 1976) which served as the data base for Fig. 7.

Figure 8 is a projection from silica onto the plane olivine-diopside-plagioclase. Because silica is not a saturating phase of the rocks and equilibria plotted in this diagram, this plot is not useful as a liquidus diagram, as is Fig. 7 for plagioclase-saturated equilibria. Nevertheless multiply-saturated experimental glass compositions and the PETROS rocks may be plotted together. The point of doing this is to check whether the impression that the densest cluster of MORB is controlled by the olivine-clinopyroxene-plagioclase curve is valid. It can be seen that this impression gained in Fig. 7 is substantiated by Fig. 8. The densest MORB cluster indeed conforms to the ol-plag-cpx curve.

The combined impact of Figs. 7 and 8 is that MORB seem to have their major element chemistry controlled by these equilibria. It should be emphasized that these curves were generated at 1 atm, and because we know that these curves move appreciably with pressure (O'Hara, 1968a and b; Shibata, 1976; Presnall et al., 1978), we must conclude that low pressure crystal-liquid fractionation is a major control on MORB chemistry. This conclusion was reached 10 years ago by O'Hara (1968a) but seems to have been eclipsed in subsequent work on 'primary' MORB magmas. We reaffirm O'Hara's conclusion as a result of detailed crystallization work on a MORB and an inspection of a great number of MORB analyses.

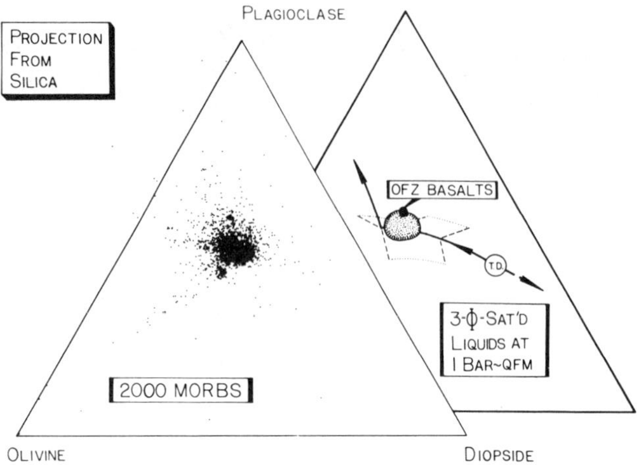

Fig. 8. MORBS plotted in terms of their normative plagioclase, olivine, and diopside contents. Right-hand triangle shows the path of 3-phase-saturated liquids used to construct Figs. 2 and 4. Inflection in this curve corresponds to the change from olivine to low-Ca pyroxene as the third saturating phase. Speckled blob in right-hand triangle is the densest MORB cluster from left hand triangle. This figure shows that the use of a plagioclase saturated liquidus for many MORBS, as in Fig. 7, is justified

One curious feature of the distribution of MORB compositions in Fig. 7 is the paucity of the differentiates that should be produced at lower temperature. The heavy cluster is centered along the olivine-plag-cpx curve in a region where the temperature is approximately 1,125–1,175° C. The scarcity of erupted lower temperature fractionates with higher silica contents approaching low-Ca pyroxene saturation is all the more puzzling in the context of the previous discussion. Such liquids are an integral requirement of our resolution of the cpx-appearance anomaly. These liquids are indeed found in OFZ basalt mesostasis. Some examples of mesostasis compositions reported by Shibata et al. (1979) are shown as crosses in Fig. 7. Clearly such liquids can be and are produced on the rapid time scale of eruption so it is unreasonable to expect that they are not produced in a fractionating magma chamber as well. Why do we not find more examples among the world population of MORB?

O'Hara (1977) has developed a model of a fractionating magma chamber which is periodically recharged with fresh magma. A number of curious effects may result from this style of volcanic plumbing. One of the effects is the possibility that a steady-state composition of erupted magma could be produced which is neither the primitive magma nor any composition obviously related to a chemical or thermal buffer. Magma mixing of the primitive material and its own differentiates is an integral part of this process. Fresh pulses of hot primitive magma keep the residual liquids in the fractionation process from sliding downhill in temperature and composition as far as they might. Extreme fractionates are reincorporated into the mixture before eruption. The steady state composition may then appear to be 'perched' above alternate, lower temperature, more differentiated compositions which might naturally be expected to develop.

We appear to have a good example of a 'perched' steady state in the global MORB population. As noted above, magmas of 2.5–4% TiO_2, approaching low-Ca pyroxene saturation, should not be as rare as they apparently are; and the world MORB population is 'perched' with respect to them. This situation is difficult to explain in simpler fractionation models but can be a natural result of O'Hara's; consequently we must seriously consider his model..

An important feature of the O'Hara model which can produce 'perched' steady states is that the magma chamber must be in continous operation. This is required to suppress the successful eruption of late differentiates by continuing to remix them into primitive material. If the chamber shuts down often, 'unperched' derivative compositions should result. Their dearth among the world's MORB suggests that the magma chamber erupting MORB is in at least semicontinuous operation. A corollary is that most of the material erupted has been through the process to some degree. The widespread occurrence of anomalous phenocrysts and their melt inclusions in MORB certainly encourages this view.

It is interesting to note that Schilling (1975) has proposed that mixing of magmas should have played some role in the petrogenesis of the OFZ basalts because they are geographically and chemically intermediate between the 'enriched' basalts of the Azores platform and more 'normal' ridge segments to the south. Schilling mixes magmas with different REE signatures to provide for the transition observed between the 'enriched' rocks of the Azores and the background of 'normal' ridge basalts. Primary magmas of diffferent REE signatures are required in Schilling's model, and the mixing can occur anywhere from the source regions to the sub-ridge magma chamber. It is important to distinguish mixing of this sort from the mixing of the present discussion which requires only a single primary magma and must occur in the sub-ridge chamber or other low pressure environment. Schilling's mixing addresses the question of what is input into the magma chamber; the mixing of the present discussion addresses the question of what happens to magma once it has arrived in the chamber. Although both are mixing processes, they are clearly distinguishable, produce different results, and may operate independently from one another. Batches of magmas entering the chamber may well differ from one another at the same time local fractionation and mixing produce the cpx anomaly, the phenocyst anomalies, and anomalous trace element enrichments from either batch or their mixture. Neither mixing process precludes the operation of the other.

The magma mixing of the 19th century was a mechanism for extending the range of compositions observed. Multiple primary magmas produced an even greater number of magma types through the operation of mixing. However the magma mixing of the present discussion is a mechanism *reducing* the range of compositions derived from a *single* sort of primary magma. It is ironic that a process should be resurrected in order to produce results opposite to those originally supposed. But this irony is probably no more curious than the one noted in the introduction – mixing is being reconsidered for just those reasons which originally led to its dismissal.

Conclusions

The operation of a process in which primitive magma is mixed with its own differentiation products can explain the anomalous appearance of cpx before olivine in the crystallization sequences of some OFZ basalts. This anomaly cannot be explained by simpler fractionation models.

The importance of this sort of differentiation process may be appreciated by realizing that the world population of MORBS is a 'perched' steady state-compositional distribution. 'Perched' steady states may be generated in a magma chamber where complex fractionation and mixing occur; this effect was

specifically anticipated by O'Hara (1977). The implication is that the MORB magmas come from continously operating chambers which employ normal fractionation and mixing.

Acknowledgments. We thank E.M. Stolper for valuable suggestions during the course of this work and continuing discussion of the results. We also thank G.M. Biggar, M.J. O'Hara, and D.C. Presnall for their comments at various stages. This work was supported by NASA grant NGL 22-007-247 (J.F. Hays, Principle Investigator), NSF grants OCE 75-21753 and OCE 76-21886, and the Commitee on Experimental Geology and Geophysics of Harvard University. Part of this work was undertaken while D.W. was a Visiting Scientist at the Lunar and Planetary Institute, Houston, and is Basaltic Volcanism Study Project contribution number 37. (Basaltic Volcanism Study Project is organized and administered by the Lunar and Planetary Institute/Universities Space Research Association under NASA contract NSR 09-051-001.)

References

Anderson, A.T.: Magma mixing: petrological process and volcanological tool. J. Vol. Geoth. Res. **1**, 3–33 (1976)

Bender, J.F., Hodges, F.N., Bence, A.E.: Petrogenesis of basalts from the Project FAMOUS area: experimental study from 0 to 15 kb. Earth Planet. Sci. Lett. **41**, 277–302 (1978)

Bryan, W.B., Moore, J.G.: Compositional variations of young basalts in the Mid-Atlantic ridge rift valley near lat 36° 49′ N. Geol. Soc. Am. Bull. **88**, 556–570 (1977)

Bryan, W.B., Thompson, G., Michael, P.G.: Compositional variation in a steady-state zoned magma chamber: Mid-Atlantic ridge at 36° 50′ N. Tectonophysics (in press, 1979)

Cann, J.R.: A model for oceanic crustal structure developed. Roy. Astron. Soc. Geophys. J. **39**, 169–187 (1974)

Carmichael, I.S.E., Nicholls, J., Spera, F.J., Wood, B.J., Nelson, S.A.: High temperature properties of silicate liquids: applications to the equilibration and ascent of basic magma. Philos. Trans. Roy. Soc. Lond. [A.] **286**, 373–431 (1977)

Delaney, J.R., Muenow, D.W., Ganguly, J., Royce, D.: Anhydrous glass-vapor inclusions from phenocrysts in oceanic tholeiite pillow basalts. EOS **58**, 530 (1977)

Dewey, J.F., Kidd, W.S.F.: Geometry of plate accretion. Geol. Soc. Am. Bull. **88**, 960–968 (1977)

Dungan, M.A., Rhodes, J.M.: Residual glasses and melt inclusions in basalts from DSDP legs 45 and 46: Evidence for magma mixing. Contrib. Mineral. Petrol. **67**, 417–431 (1978)

Eichelberger, J.C.: Origin of andesite and dacite: evidence of mixing at Glass Mountain in California and at other circum-Pacific volanoes. Geol. Soc. Am. Bull. **86**, 1381–1391 (1975)

Fox, P.J., Schreiber, E., Rowlett, H., McCamy, K.: The geology of the Oceanographer Fracture Zone: a model for fracture zones. J. Geophys. Res. **81**, 4117–4128 (1976)

Greenbaum, D.: Magmatic processes at ocean ridges: evidence from the Troodos massif, Cyprus. Nature **238**, 18–21 (1972)

Irvine, T.N.: Crystallization sequences in the Muskox intrusion and other layered intrusions: 1. Olivine-pyroxene-plagioclase relations. Geol. Soc. S. Afr. Spec. Publ. **1**,, 441–476 (1970)

Kuno, H.: Petrological notes on some pyroxene andesites from Hakone volcano, with special reference to some types with pigeonite phenocrysts. Jap. J. Geol. Geogr. **13**, 107–140 (1936)

Kuno, H.: Origin of andesite and its bearing on the island arc structure. Bull. Volcanol. **32**, 141–176 (1968)

Kushiro, I.: Determination of liquidus relations on synthetic silicate systems with electron probe analysis: the system forsterite-diopside-silica at 1 atm. Am. Min. **57**, 1260–1271 (1972)

Mutschler, F.E., Rougon, D.J., Lavin, O.P.: PETROS – a data bank of major-element chemical analyses of igneous rocks for research and teaching. Comput. Geosci. **2**, 51–57 (1976)

O'Hara, M.J.: The bearing of phase equilibria studies in synthetic and natural systems on the origin and evolution of basic and ultrabasic rocks. Earth Sci. Rev. **4**, 69–133 (1968a)

O'Hara, M.J.: Are ocean floor basalts primary magma? Nature **220**, 683–686 (1968b)

O'Hara, M.J.: Geochemical evolution during fractional crystallization of a periodically refilled magma chamber. Nature **266**, 503–507 (1977)

Papike, J.J., Vaniman, D.T.: The lunar mare basalt suite. Geophys. Res. Lett. **5**, 433–436 (1978)

Presnall, D.C., Dixon, S.A., Dixon, J.R., O'Donell, T.H., Brenner, N.L., Schrock, R.L.: Liquidus phase relations on the join diopside-forsterite-anorthite from 1 atm to 20 kbar: their bearing on the generation and crystallization of basaltic magma. Contrib. Mineral. Petrol. **66**, 203–220 (1978)

Rhodes, J.M., Dungan, M.A., Blanchard, D.P., Long, P.E.: Magma mixing at mid-ocean ridges: evidence from basalts drilled near 22° N on the Mid-Atlantic ridge. Tectonophysics (in press, 1979)

Schilling, J.-G.: Azores mantle blob: rare-earth evidence. Earth Planet. Sci. Lett. **25**, 103–115 (1975)

Shibata, T.: Phenocryst – bulk rock composition relations of abyssalt tholeiites and their petrogenetic significance. Geochim. Cosmochim. Acta **40**, 1407–1417 (1976)

Shibata, T., Fox, P.J.: Fractionation of abyssal tholeiites from the Oceanographer Fracture Zone (35° N, 35° W) Earth and Planet. Sci. Lett. **27**, 62–72 (1975)

Shibata, T., DeLong, S.E., Walker, D.: Abyssal tholeiites from the Oceanographer Fracture Zone: I. Petrology and fractionation. Contrib. Mineral. Petrol. **70**, 89–102 (1979)

Shido, F., Miyashiro, A., Ewing, M.: Basalts and serpentinite from the Puerto Rico Trench, 1. Petrology. Marine Geology **16**, 191–203 (1974)

Stolper, E.M.: Experimental petrology of eucritic meteorites. Geochim. Cosmochim. Acta **41**, 587–611 (1977)

Walker, D., Powell, M.A., Lofgren, G.E., Hays, J.F.: Dynamic crystallization of a eucrite basalt. Proc. 9th Lunar and Planet. Sci. Conf. pp. 1369–1391 (1978)

Wright, T.L.: Magma mixing as illustrated by the 1959 eruption, Kilauea volcano Hawaii. Geol. Soc. Am. Bull. **84**, 849–858 (1973)

Received January 18, 1979; Accepted May 2, 1979

A Phase Diagram for Mid-Ocean Ridge Basalts:
Preliminary Results and Implications for Petrogenesis

Edward Stolper

Division of Geological and Planetary Sciences, California Institute of Technology, Pasadena, CA 91125, USA

Abstract. Samples of a primitive mid-ocean ridge basalt (MORB) glass were encapsulated in a mixture of *ol* (Fo90) and *opx* (En90) and melted at 10, 15, and 20 kbar. After quenching, the basaltic glass was present as a pool within the *ol+opx* capsule, but its composition had changed so that it was saturated with *ol* and *opx* at the conditions of the experiment. By analyzing the quenched liquid, the location of the *ol+opx* cotectic in the complex, multicomponent system relevant to MORB genesis was determined.

As pressure increases from 1 atm to 10 kbar, the dry *ol+opx* cotectic moves from quartz tholeiitic to olivine tholeiitic compositions. With further increases in pressure, the cotectic continues to move toward the *ol-di-plag* join (i.e., toward alkalic compositions). Between 15 and 20 kbar, *ol+opx+di*-saturated liquids change from tholeiitic to alkalic in character, although part of the *ol+opx* cotectic is still in the tholeiitic (i.e. hy-normative) part of composition space. At pressures of 10–15 kbar, tholeiitic liquids may be able to fractionate to alkalic liquids on the *ol+di* cotectic.

Primitive MORB compositions come close to but do not actually lie on the *ol+opx* cotectic under any conditions studied. This suggests that not even the most primitive of known MORBs are primary melts of the mantle. The correspondence of most MORBs to the 1 atm *ol+di+plag* cotectic suggests that low pressure fractionation was involved in their genesis from parent liquids. Picritic liquids that have been proposed as parents to the MORB suite could equilibrate with harzburgite (or lherzolite) at 15–20 kbar and thus could be primary. Fractionation of *ol* from these liquids could yield primitive MORB liquids, but other primary liquids or more complex fractionation paths involving others phases in addition to *ol* cannot be ruled out. The possibility that these picritic liquids could equilibrate with *ol+opx* at 25–30 kbar cannot be ruled out.

Introduction

Models of mid-ocean ridge basalt (MORB) petrogenesis generally fall into one of two categories. In the first, most recently championed by Presnall et al. (1979), the most primitive of known MORB samples are considered to represent "primary" magmas; that is, their compositions are believed to be similar to those of liquids generated by partial melting of the mantle beneath the mid-ocean ridges, little modified by crystal fractionation or other processes since they segregated from their source regions. Presnall and his coworkers have suggested, based principally on studies of crystal-liquid equilibria in the simplified basalt-peridotite system $CaO-MgO-Al_2O_3-SiO_2$, that primitive MORB samples were generated by partial melting of an olivine(*ol*)–orthopyroxene(*opx*)–diopsidic clinopyroxene(*cpx*)–plagioclase(*plag*)–spinel(*sp*) assemblage at about 9 kbar.

In the second category of petrogenetic model, it is maintained that none of the currently sampled MORBs are primary; that is, all have developed from primary magmas by substantial amounts of fractionation of principally olivine, but perhaps of other phases as well, at pressures less than those at which the primary magmas segregated from their mantle source regions. Primary magmas, according to this class of models, would be rich in an olivine component (i.e., "picritic"), and could contain as much as 20 wt.% MgO. Three distinct suggestions have been made as to the conditions under which the proposed picritic primary magmas could have been generated. O'Hara (1968a), who originally proposed the concept of picritic primary magmas for MORBs, favored generation by partial melting at 25–30 kbar, leaving *ol+opx+cpx+garnet* in the mantle as residual phases. Green et al. (1979) favor partial melting at 20 kbar, leaving only *ol+opx±sp* in the residue.

Elthon (1979) suggests that the high-MgO primary liquids last equilibrated with an $ol+opx$ residue at pressures between 5–10 kbar.

Although there is probably a range of conditions under which primary liquids are generated beneath mid-ocean ridges and a spectrum of primary magma compositions produced in these environments, it is unlikely that all of the above models of MORB petrogenesis can be correct. In this paper, I present new experimental data relevant to phase equilibria in MORB systems. These data were obtained with a new technique in which mantle minerals are used as capsules for basalt melting experiments. This technique has been used to determine the characteristics of liquids saturated with mantle minerals under a range of conditions. The technique and the results of the work were first reported by Stolper (1980). Although still preliminary, my results provide constraints on permissible petrogenetic models for MORB systems. These results suggest that the petrogenetic models of Presnall et al. (1979) and of Elthon (1979) are probably not valid. The model of Green et al. (1979) is consistent with available data, but may not be favored over other permissible models. Picritic compositions that have been proposed as representative of primary MORB magmas could equilibrate with a harzburgite or lherzolite residue at pressures of 15–20 kbar. My data do not permit evaluation of the hypothesis of O'Hara (1968a) that these picritic compositions could represent partial melts of garnet lherzolite at 25–30 kbar.

Experimental Approach

One ground rule has been fundamental to nearly all considerations of MORB petrogenesis, or indeed of all basalt petrogenesis: olivine and orthopyroxene are present in the source regions of MORBs, and are left behind in these source regions when the primary magmas segregate from them. Thus, identifying basalts that can be, under some conditions of P, T, fO_2, fH_2O, ..., in equilibrium with both olivine and orthopyroxene or determining the characteristics of such liquids has been and remains a major preoccupation with experimental petrologists (in the case of MORB petrogenesis, dry phase equilibria appear to be most applicable; Presnall et al. 1979). It was based on his conclusion that MORBs would not be saturated with orthopyroxene at any pressure that O'Hara (1968a) rejected the idea that MORBs were primary. In contrast, Presnall et al. (1979) concluded that the least fractionated MORBs would be saturated with olivine and orthopyroxene at about 9 kbar and proposed that they are either primary or have only been fractionated a small amount.

How does one go about characterizing liquids that could equilibrate with mantle minerals under different conditions of P and T? Presnall et al. (1979) have studied the compositions of liquids in the system $CaO - MgO - Al_2O_3 - SiO_2$ that are saturated with ol, opx, cpx, and $plag$ and/or sp at pressures between 1 atm and 20 kbar. Insofar as these results can be generalized to more complex natural systems, they provide a guide to the nature of primary magmas in the mantle. Despite the importance of these studies to petrogenetic theory, their applicability to determining the details of the petrogenesis of a particular suite of rocks is limited by the often substantial effects on phase equilibria of the other components found in natural systems.

A complementary approach has been to partially melt natural peridotite samples at various pressures and then to analyse the liquid compositions. Ideally, this can yield insights into the compositions of liquids produced by melting of peridotites under a range of conditions in the mantle. Although such experiments are valuable, they have, to date, been disappointing in terms of their ability to provide the kind of information needed to evaluate the details of the evolution of a particular group of rocks. This has been partly due to experimental difficulties such as failure to reach equilibrium, problems with maintaining bulk composition, changes of liquid composition during quenching, and alkali mobility during microprobe analysis (e.g., Cawthorn et al. 1973; Mysen and Kushiro 1977). The general conclusions of these studies remain valid: olivine tholeiites versus alkali olivine basalts or quartz versus olivine tholeiites as primary liquids under different conditions. However, in order to constrain the petrogenetic history of a specific suite of rocks, greater precision than these broad labels is required. In addition, the nature of melts produced from peridotites depends on the peridotite compositions, and it is difficult to choose the specific peridotite whose melting behavior most closely matches that of MORB source regions.

A direct approach to the problem of determining the characteristics of liquids saturated with olivine and orthopyroxene at high pressure would be to take several MORB samples, study their crystallization behaviors at high pressures, and compare the characteristics of samples that crystallize $ol+opx$ at their liquidi at different pressures with those that do not. In this way, the types of liquids that could be primary under different conditions could be identified. This approach has not been very useful when applied to MORB samples because few appear to crystallize both orthopyroxene and olivine at their liquidi at any pressures. Kushiro (1973) and Fujii and Kushiro (1977) have identified two compositions that appear to have both olivine and low-Ca pyroxene on their liquidi at 7.5–8 kbar, but none of the other MORB samples studied show this behavior (e.g., Bender et al. 1978, Green et al. 1979) and no MORB samples show this multiple saturation at higher pressure. Another potential difficulty with this experimental approach is that due to the various reaction relationships involving olivine, orthopyroxene, and liquid at high pressures (e.g., O'Hara 1968b), even a sample saturated with $ol + opx$ at its liquidus at a given P and T might not be recognized as such since these phases might not crystallize from it under these conditions.

Green et al. (1979) studied a primitive MORB sample and found that it did not crystallize olivine and orthopyroxene simultaneously at its liquidus at any pressure. However, by adding 17% olivine to their MORB sample and studying the high pressure liquidus of the olivine-enriched sample, they showed that this composition has olivine and orthopyroxene at its liquidus at 20 kbar. This is the only information available on the characteristics of liquids in MORB systems that are saturated with both olivine and orthopyroxene at pressures higher than 8 kbar.

The approach that I have taken in this study to determine the characteristics of $ol + opx$-saturated liquids in MORB systems can be summarized as follows: A sample of MORB glass is placed in a capsule made of olivine and orthopyroxene. This configuration is then brought to the desired conditions of pressure and temperature above the solidus of the basalt; the basalt melt exchanges with the "harzburgite" capsule until it approaches equilibrium with it. The sample is then quenched and the glass, with a composition typically different from that of the starting glass but now at or near equilibrium with olivine and orthopyroxene at the conditions of the experiment, is analysed with the electron microprobe. In this way, the characteristics of liquids that are in equilibrium with olivine and orthopyroxene in the complex, multicomponent system relevant to MORB genesis can be determined as functions of P and T.

How does the technique employed in this study compare with the alternatives? In some ways, it is most like a "pyrolite" melting experiment and thus is analogous to peridotite melting experiments. Pyrolite is a model mantle composition derived by adding basalt to a residual peridotite; in my experiments, a basalt is equilibrated with a harzburgite, equivalent to a residual peridotite, and thus the results are equivalent to melting a pyrolite composition constructed by adding the basalt to the enclosing olivine + orthopyroxene mixture. My experimental approach does, however, differ somewhat from a traditional peridotite or pyrolite melting experiment. The glass in my experiments is present as a large, homogeneous pool

and thus has fewer problems with quench growth and the electron beam can be defocussed to minimize alkali loss during probe analysis; these are problems encountered in peridotite melting experiments. In addition, the petrogenesis of a particular suite of rocks can be addressed since the "pyrolite" composition effectively studied is derived from a specific basalt in the suite. In contrast to direct melting experiments on specific basalts, which rarely show both olivine and orthopyroxene at the liquidus under any conditions, this approach permits the systematic examination of the olivine + orthopyroxene cotectic as a function of P and T for a particular region of composition space relevant to a specific basalt type and thus provides a framework for exploring the petrogenesis of a whole suite of rocks. In this respect, it differs from the simplified, model systems approach of Presnall et al. (1978, 1979) since the results are directly applicable without the need for extrapolation to the complex, natural system. By comparison of the locations in composition space of the olivine + orthopyroxene cotectics determined for different starting basalt compositions, the effects of variable composition on this cotectic can be assessed. The approach taken by Green et al. (1979) yields results similar to those obtainable through my experiments. However, whereas many experiments were needed by Green et al. (1979) to determine the location of one composition on the olivine + orthopyroxene cotectic, *each* experiment carried out with the approach used in this study can yield an olivine + orthopyroxene − saturated composition.

Experimental Techniques

In each experiment, a chunk of mid-ocean ridge basalt glass was embedded in a powdered mixture of olivine and orthopyroxene inside of a graphite capsule. This capsule was then loaded into a 1/2″ piston-cylinder apparatus and held at run conditions for several hours. After quenching, the glassy portion of the charge was analyzed with an electron microprobe. In most runs, the MORB sample retained its original shape and did not flow out into the enclosing olivine + orthopyroxene mixture, providing a large, homogeneous pool of glass, free of quench crystals, for microprobe analysis. In this way, a series of $ol + opx$ − saturated liquid compositions, differing from the starting MORB composition primarily in the amounts of olivine and orthopyroxene they dissolved or crystallized, were generated under a variety of conditions of P and T.

Walker et al. (1979) employed a similar technique to determine cotectics relevant to MORB systems at 1 atm. The success of this approach as applied by Walker et al. (1979) was the stimulus to try to apply it at higher pressures. Watson (1980ab) has used capsules made of single crystal apatites both at 1 atm and higher pressures to determine the P_2O_5 contents of apatite-saturated silicate melts. Takahaski (1980) reports the use of powdered olivine capsules at high pressures to determine partition coefficients for Ni between olivine and silicate melt.

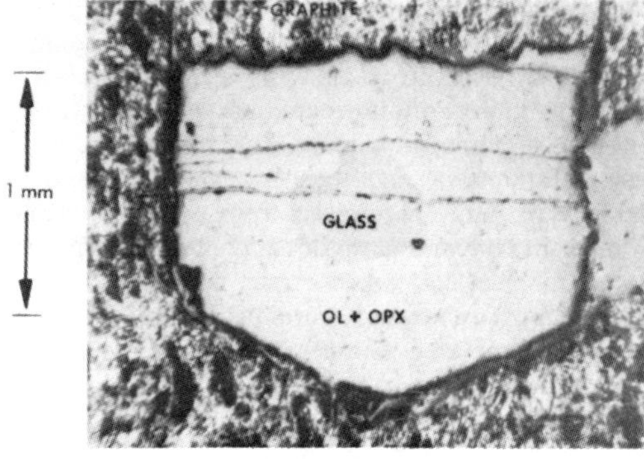

Fig. 1. Photograph of a polished cross section of an experimental product (Run No. 519-14) showing the basaltic glass enclosed within an olivine + orthopyroxene mixture, enclosed within a graphite capsule. The intermediate reflectivity material at the right of the photograph is epoxy

Table 1. Starting materials

	ALV-519-4-1	Olivine	Orthopyrox-ene	Spinel
SiO_2	49.37	40.47	54.31	0.05
TiO_2	0.77	0.01	0.17	0.12
Al_2O_3	15.95	–	5.09	60.01
Cr_2O_3	0.04	0.02	0.34	8.70
FeO	9.00	10.39	6.58	10.50
MnO	0.20	0.15	0.15	0.13
MgO	9.61	49.90	32.59	21.74
CaO	12.50	0.07	0.84	–
Na_2O	2.07	–	0.14	–
K_2O	0.09	–	–	–
NiO	–	0.31	–	–
Total	99.60	101.32	100.21	101.25

A basaltic glass from the FAMOUS region of the Mid-Atlantic Ridge, sample ALV-519-4-1, was used as the starting material. The sample was provided by W.B. Bryan. This sample was included in the study of Langmuir et al. (1977) and is one of the most primitive of known MORB glasses. A microprobe analysis of the glass is given in Table 1. A chunk of the glass was cut into approximately cubic shapes about 1 mm in each dimension. These cubes were then rounded in a crystal rounder (Bond 1951).

The approximately spherical pieces of glass were then embedded in a powdered mixture of olivine and orthopyroxene within a graphite capsule. Olivine and orthopyroxene crystals were handpicked from a spinel lherzolite from Kilbourne's Hole, provided by A.J. Irving (sample KH77−6). The olivine is approximately Fo89.5; the orthopyroxene is En89.8. Full analyses are given in Table 1. The orthopyroxene separate contained several percent spinel; an analysis is given in Table 1. Each mineral separate was ground in an agate mortar under ethyl alcohol for several hours then passed through a 325 mesh (44 µm) nylon sieve. A mixture of the olivine and orthopyroxene powders, 50–50 by weight, was then made and reground for 1 h in an agate mortar.

Experiments were conducted in a 1/2″ piston-cylinder apparatus similar to that of Boyd and England (1960). The sample assemb-

Table 2. Run conditions

Run #	P (kbar)	T (°C)	Dura-tion (hours)	Comments
519-45	10	1,350	2.3	liquid beginning to escape into ol + opx capsule
519-12	10	1,350	2.0	liquid beginning to escape into ol + opx capsule
519-7	10	1,300	2.0	
519-10	10	1,300	18.8	no recognizable pool of glass remaining in center of ol + opx capsule
519-11	10	1,250	3.0	ol + cpx + plag ± sp in contact with glass, no opx
519-52	15	1,400	2.0	small rims of quench cpx on some crystals
519-51	15	1,350	2.0	inner edge of ol + opx capsule rimmed with cpx
519-16	20	1,450	2.0	rim of quench px at edge of pool of glass
519-14	20	1,400	2.0	rim of quench px at edge of pool of glass, primary cpx?

All experiments conducted with ALV-519-4-1 glass embedded in ol + opx mixture

lies and run procedures were those described by Chipman and Hays in Johannes et al. (1971), with the following exceptions: W3Re W26R thermocouples were used (no correction for the effects of pressure on EMF were applied) and a crushable alumina spacer was used in place of the fired pyrophyllite. Piston-in runs were used and reported pressures include an 8% friction correction.

After quenching, a polished mount of a vertical cross section of the graphite capsule was prepared for examination in reflected light and for electron microprobe analysis. Analyses were made with a MAC − 5 − SA3 electron microprobe interfaced to a PDP-8/L computer for control of the stage and the crystal spectrometers and for on-line data processing. Operating conditions were 15 kV accelerating voltage and 0.02 µA sample current on brass. Analytical procedures were essentially those described by Chodos et al. (1973). Most analyses were done with a focussed beam, but each sample was also analyzed with a beam defocussed to 15 20 µm to demonstrate that alkali loss had not occurred in the focussed beam analysis.

A photomicrograph of a completed experiment is shown in Fig. 1.

Experimental Results

The run conditions of the experiments are listed in Table 2 and microprobe analyses of the glasses and selected crystalline phases are given in Table 3. For those experiments for which no mineral compositions are given, the analyses were essentially identical to those of the starting ol and opx and hence are not reported. The possibility that small rims of modified material are present on these mineral grains cannot be ruled out. Projections of the glass compositions in the system OLIVINE-PLAGIOCLASE-DIOPSIDE-SILICA are shown in Figs. 2 and 3. Also shown in these

Table 3. Selected phase compositions from experiments

	519-45 Glass	519-12 Glass	519-7 Glass	519-10 Glass	519-11 Glass	519-52 Glass	519-51 Glass	519-16 Glass	519-14 Glass
SiO_2	50.12	49.90	49.58	48.34	45.71	48.17	47.39	46.86	47.50
TiO_2	0.68	0.73	0.73	0.80	2.43	0.70	0.78	0.72	0.86
Al_2O_3	15.95	17.57	16.01	18.53	12.81	15.13	16.03	14.59	16.67
Cr_2O_3	0.14	0.11	0.14	0.06	0.00	0.14	0.04	0.14	0.00
FeO^*	8.81	8.28	8.23	8.40	18.17	9.71	9.38	10.30	10.06
MnO	0.23	0.17	0.21	0.21	0.23	0.21	0.20	0.12	0.17
MgO	13.29	11.21	11.37	10.10	5.92	15.07	11.93	15.98	10.18
CaO	9.38	10.77	10.9	10.91	9.60	9.35	11.26	9.29	10.74
Na_2O	1.74	1.98	1.96	1.88	2.51	1.80	2.38	1.94	2.50
K_2O	0.09	0.09	0.08	0.12	0.29	0.09	0.11	0.10	0.13
Total	100.43	100.81	99.21	99.35	97.67	100.37	99.50	100.04	98.81

	519-12 Ol	519-12 Opx	519-7 Ol	519-10 Ol	519-10 Opx	519-11 Ol	519-11 Cpx	519-11 Plag	519-16 Opx
SiO_2	41.06	55.02	40.98	41.08	54.97	39.80	50.29	54.13	54.04
TiO_2	–	0.14	–	–	0.15	–	0.92	0.06	0.09
Al_2O_3	–	4.76	–	–	4.96	–	6.23	29.14	5.41
Cr_2O_3	0.16	0.25	0.08	0.20	0.34	0.15	0.17	0.01	0.35
FeO^*	10.87	6.63	10.98	10.92	6.63	16.72	9.97	0.83	6.22
MnO	0.10	0.16	0.13	0.18	0.15	0.30	0.21	–	0.13
MgO	49.55	33.12	48.59	49.13	32.79	42.99	15.45	0.13	32.39
CaO	0.21	0.82	0.22	0.23	0.80	0.30	16.64	12.30	1.74
Na_2O	–	0.15	–	–	0.11	–	0.51	4.30	0.11
NiO	0.26	–	–	0.29	–	–	–	–	–
Total	102.21	101.05	100.98	102.03	100.90	100.26	100.39	100.90	100.48

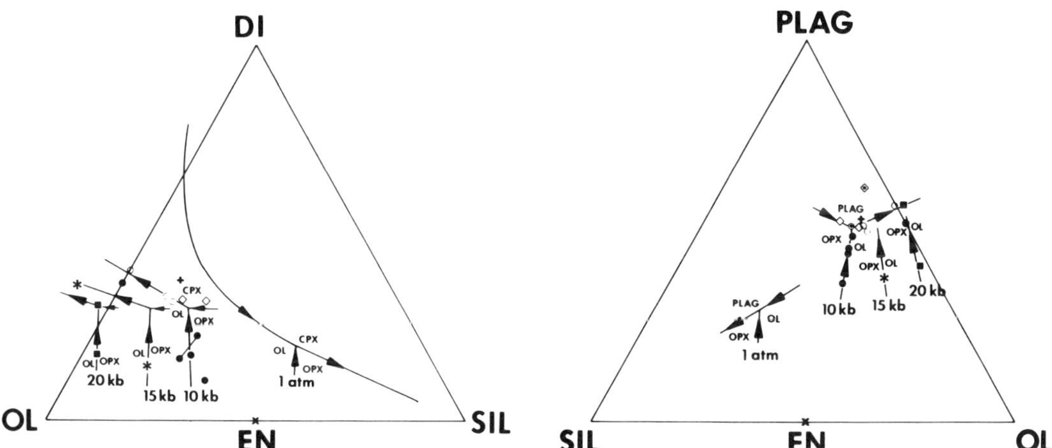

Fig. 2. Projection from PLAGIOCLASE onto the plane OLIVINE-SILICA-DIOPSIDE of experimental results. The algorithm used to calculate the projection is given in Walker et al. (1979; caption to Fig. 2). The *cross* is the starting basalt composition used in this study. *Solid dots* are glass compositions from 10 kbar experiments. The *two dots* joined by a line are glasses from 519-7 and 519-10 discussed in the text. *Asterisks* are glass compositions from 15 kbar experiments. *Solid squares* are glass compositions from 20 kbar experiments. Glass compositions are given in Table 3. *Open circles* are *ol + cpx*-saturated compositions (10 kbar) from Bender et al. (1978). *Open diamonds* are *ol + cpx + plag + low-Ca pyroxene*-saturated compositions from Kushiro (1973) and Fujii and Kushiro (1977).

One atm cotectic phase boundaries are from Walker et al. (1979); 10–20 kbar cotectics are based on data points in the figure. The "opx" field includes pigeonite-saturated liquids. *Arrows* indicate direction of falling temperature on the cotecties

Fig. 3. Projection from DIOPSIDE onto the plane OLIVINE-SILICA-PLAGIOCLASE of experimental results. Symbols and comments on Fig. 2 apply, except: *Circle enclosing star* is composition of liquid saturated with *ol + cpx + plag* at 8 kbar (Bender et al. 1978); *diamond enclosing star* is the composition of T-89 (Kushiro and Thompson 1972), which has plagioclase on its liquidus at 10 kbar

figures are compositions from Bender et al. (1978) that are *ol+ cpx ± plag* – saturated at 8 and 10 kbar, compositions from Kushiro (1973) and Fujii and Kushiro (1977) that are *ol + cpx + low-Ca pyroxene + plag*-saturated at 7.5–8 kbar, and a composition from Kushiro (1973) that is *cpx + plag*-saturated at 10 kbar. These compositions, along with those from this study, were used to draw the phase boundaries at 10, 15, and 20 kbar pressure shown in Figs. 2 and 3. The 1 atm phase boundaries are from Walker et al. (1979).

Although the phase boundaries shown in Figs. 2 and 3 and reproduced in subsequent figures are shown as curves and their intersections as points, they should more appropriately be shown as intersecting bands. The widths of these bands would reflect, in part, the uncertainties in the positions of the data points used to constrain the locations of the phase boundaries. In addition, the locations of the phase boundaries are dependent on aspects of liquid composition not represented in these projections (e.g., Fe/Mg, TiO$_2$ content). However, based on the data shown in Fig. 2 and 3 and presented in Walker et al. (1979), the phase boundaries for a restricted range of compositions such as the MORBs can be reasonably well approximated by narrow bands whose widths are small relative to their movement with changing pressure.

The greatest difficulty in experiments of this sort is demonstration of an approach to an equilibrium state. There is no question that true equilibrium was not achieved in these experiments. The maintenance of a pool of liquid within the olivine + orthopyroxene capsule indicates that textural equilibrium was not achieved. Due to the lack of reversals of the liquid compositions and of the solid compositions, the results presented in this study must be regarded as preliminary. Nevertheless, several observations suggest that the compositions listed in Table 3 are not far removed from equilibrium values and that the phase boundaries constructed from these compositions are meaningful.

1. Runs 519-7 and 519-10 were conducted under the same conditions, but 519-7 was 2 h in duration while 519-10 was 19 h in duration. In 519-7, the glass was still present as a spherical pool in the center of the capsule; in 519-10, however, the liquid had escaped into the surrounding olivine + orthopyroxene "capsule", and was only present in small pools near the junctions of individual crystals and along crystal faces. Nevertheless, both glass compositions are significantly different from the starting material composition and, in terms of the components portrayed in Figs. 2 and 3, are very similar. In other words, considerable change in liquid composition occurred in 2 h, but little further change occurred in the next 17 h.

2. In those runs in which a pool of glass remained in the capsule center, no zoning was observed in the glass from the center to the edge of the pool, a distance of up to 300 μm, yet in all cases, the glass composition changed significantly from the composition of the starting glass.

3. The Mg-number (100 Mg/(Mg + Fe) molar) of the basalt starting material is about 65. Except in those experiments in which extensive crystallization of the basalt prevented equilibration between the sample and the enclosing *ol + opx* (see below), the Mg-numbers of the glasses in the completed experiments are between 68 and 73. The Fe/Mg K_D values for measured *ol-liq* pairs in the experiments are between 0.27 and 0.33; for *opx-liq* parts they are between 0.24 and 0.30. The changes in the Mg-number of the basalt and the observed K_D values indicate that at least with respect to Fe/Mg partitioning, equilibrium between the basalt and the enclosing *ol + opx* was approached (Roeder 1974; Longhi et al. 1978; Cawthorn et al. 1973).

The Al$_2$O$_3$ contents of some of the glasses given in Table 3 are higher than expected based on the Al$_2$O$_3$ content of the 519-4-1 starting material. I attribute this to dissolution of the aluminous spinel in the olivine + orthopyroxene mixture. Although spinel was present in the starting mixture, none was observed near the edges of the glass pool, supporting this hypothesis for the high Al$_2$O$_3$ contents. Note that the principal difference between 519-7 and 519-10 is the much higher Al$_2$O$_3$ content of 519-10 (Table 3); this is consistent with the idea that interaction with the spinel in the surrounding crystalline powder is responsible for the high Al$_2$O$_3$ contents. In 519-10, the liquid escaped from the center of the pool, and thus contacted and dissolved more spinel than the liquid in those experiments (e.g., 519-7) in which the glass remained in a pool in the center of the capsule and had no interaction with spinels far from the liquid crystal interface.

Two experiments, 519-11 and 519-51, were conducted at temperatures below the beginning of clinopyroxene crystallization in the glass starting material (based on the liquidus of DSDP3-18-7-1 given by Green et al. (1979), which is similar in composition to 519-4-1). What appears to have occurred in these experiments is that crystallization of the sample at the interface between the olivine + orthopyroxene crystals and the MORB sample resulted in a barrier that prevented equilibrium between the liquid and the surrounding *ol + opx* crystals. Only olivine and clinopyroxene were found in contact with the quenched glass; thus, these points cannot be used to constrain the characteristics of *ol + opx*-saturated liquids, though they can be used to locate the *ol + cpx* cotectics. Run 519-14 was conducted just below the liquidus of the starting basalt and may have primary clinopyroxene at the "capsule" rim.

Comparison With Previous Work

Green et al. (1979) presented the results of a study that can be compared with this one. Their experimental results, and my interpretation of a pseudo-liquidus phase diagram consistent with these data, are shown in Fig. 4. The movements of the *ol + opx + cpx* multiple saturation "point" with pressure deduced from the results of Green et al. (1979) are compared with my results in Fig. 5. The paths followed by this cotectic in my experiments and in those of Green et al. (1979) are virtually identical, except for an apparent discrepancy in our pressure calibrations. My 10 kbar *ol + opx + cpx* cotectic coincides with their 12 kbar cotectic; my 15 kbar cotectic would correspond to their 19 kbar cotectic. Discrepancies of this magnitude have been observed in studies reported in the lunar literature (e.g., compare the results of Kesson 1975 and Walker et al. 1977, on sample 15555).

Also shown in Fig. 5 is the movement of the *ol + opx + cpx + (plag* and/or *sp)* isobaric invariant point in the simplified system CaO − MgO − Al$_2$O$_3$ − SiO$_2$ (Presnall et al. 1979). The directions of movement of this point in the simplified system and of the *ol + opx + cpx* cotectic for natural MORB composition are similar. At pressures above 10 kbar, my results and those of Presnall et al. (1979) are not strictly comparable since my *ol + opx + cpx* cotectics at 15 and 20 kbar are not saturated with spinel. Based on the results of Presnall et al. (1978), the spinel-saturated *ol + opx + cpx* cotectics would probably have higher OL/ DI ratios than those shown in Fig. 5. The differences between my results (and those of Green et al. 1979)

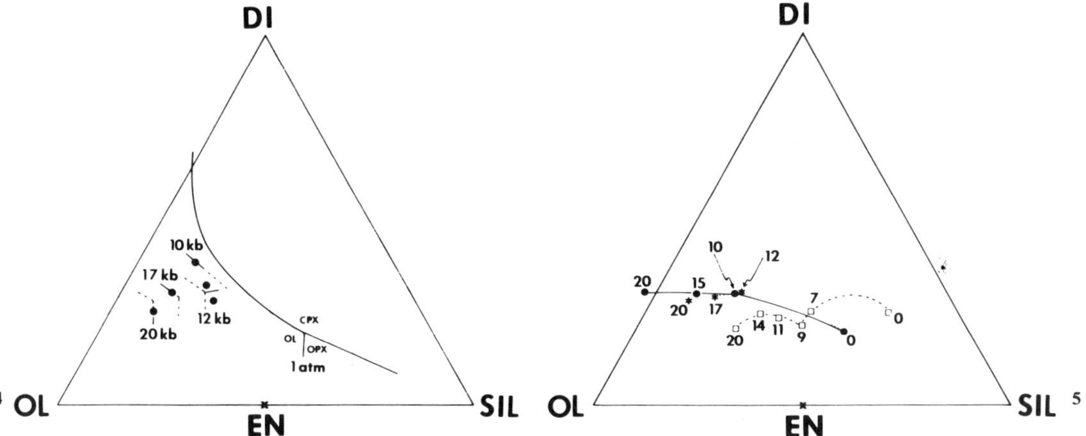

Fig. 4. Summary of the results of Green et al. (1979). *Closed circles* are compositions studied by Green et al. (1979); *dashed* and *solid lines* are my interpretation of cotectic curves consistent with their experimental results. One atmosphere cotectics from Walker et al. (1979)

Fig. 5. Comparison of the changes in composition of *ol+opx+cpx±plag±sp* saturated liquids from 1 atm to 20 kbar projected from PLAG onto the plane OL-DI-SIL, based on the results of this study, of Green et al. (1979), and of Presnall et al. (1979). The results of Presnall et al. (1979) apply to the model system $CaO-MgO-Al_2O_3-SiO_2$, while those of this study and of Green et al. (1979) apply to the complex natural system. *Open squares* are from Presnall et al. (1979). *Asterisks* are from Green et al. (1979), transferred from Fig. 4. *Solid circles* are from this study, transferred from Fig. 2. *Numbers* next to the symbols indicate the pressure in kilobars. The 0 kbar *solid circle* is from Walker et al. (1979)

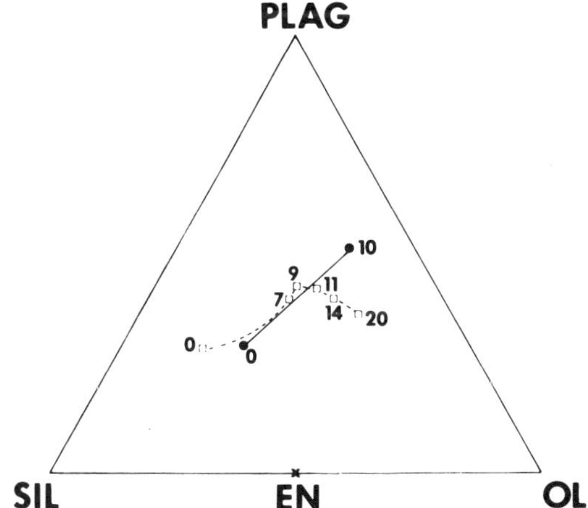

Fig. 6. Same comparison as in Fig. 5, showing the changes in compositions of *ol+opx+cpx±plag±sp* saturated liquids with pressure, projected from DI onto the plane OL-PLAG-SIL. No results are available on the *plag* and/or *sp* saturated liquids in the natural system above 10 kbar, so the *solid circles* are only given at 0 and 10 kbar

and those of Presnall et al. (1979) shown in Figs. 5 and 6 can be summarized as follows: At all pressures, the results on natural samples have higher OL/SIL ratios than those from the simplified system. In other words, all of my cotectics are further to the left in Fig. 5 than equivalent cotectics based on the results of Presnall et al. (1979). While at 20 kbar, Presnall et al. (1979) show the *ol+opx+cpx+sp*-saturated liquid to be hypersthene-normative, in the natural system it appears that by 20 kbar, liquids on this cotectic are nepheline-normative. This difference between the natural and simplified system was anticipated by Presnall et al. (1979). Other differences are not as striking. The normative diopside content of liquids on the *ol+opx+cpx* cotectic increases from 1 atm to 10 kbar in the natural system and then levels off, being essentially constant from 10 kbar to 20 kbar. In contrast, the 10 and 20 kbar points in the simplified system are slightly lower in diopside content than the 1 atm point. The absence of detail in the natural system relative to the simplified system may be due to the less detailed coverage of the natural system, or to the dampening of these effects in natural systems by the addition of other components.

The major similarity between the results presented in this work and of Presnall et al (1979) is in the amount of movement of the $ol + opx + cpx$ saturated "point" between 1 atm and 20 kbar. The vectors between the 1 atm and 10 kbar points in the natural and simplified systems in Figs. 5 and 6 are very similar. The same is also true, though less so, between 10 and 20 kbar in Fig. 5. The same slowing of the left-ward movement of this point in Fig. 5 between 10 and 20 kbar observed in the simplified system is also observed in the natural system. This is more apparent in the results of Green et al. (1979) than in mine due to the discrepancy in our pressure calibrations.

In Fig. 2, I show temperature dropping on the $ol + cpx$ cotectic in the direction of nepheline-normative compositions at 10 and 15 kbar; that is, from the $ol + opx + cpx$ cotectic toward and finally beyond the OL-DI join. From a petrogenetic point of view, this means that in this pressure range, tholeiitic liquids will fractionate to alkali olivine basalt liquids by fractionation of olivine + clinopyroxene at 10–15 kbar. Presnall et al. (1978) maintain that if spinel joined this crystallization sequence, the liquid line of descent would reverse itself and proceed from alkalic to tholeiitic compositions. There are, however, no data currently available in complex systems to test this suggestion.

The directions of the arrows in Figs. 2 and 3 imply that orthopyroxene is in reaction relation with the liquid to form olivine + clinopyroxene between 10 and 20 kbar. Three pieces of evidence suggest that this reaction relation and the directions of falling temperature in Figs. 2 and 3 are valid. First, the glasses in 519-11 and 519-51 are indeed nepheline normative. It is difficult to see how they could have arrived at such a condition if the direction of falling temperature were different from that shown in Figs. 2 and 3. Second, the $ol + cpx + liquid$ three-phase triangles shown in Fig. 7 generally corroborate this conclusion. Pyroxene-glass pairs in my work (Fig. 7a), from Kushiro and Thompson (1972) (Fig. 7b), and from Green et al. (1979) (Fig. 7c) all suggest falling temperature from tholeiitic to alkalic compositions. The results of Kushiro (1973) and Kushiro and Thompson (1972) are, however, somewhat ambiguous since although the 3-phase triangles in Fig. 7b indicate a tholeiitic →alkalic trend during fractionation at 8 kbar, the 8 kbar residual liquid composition given by Kushiro (1973) is more hypersthene-normative than the T-87

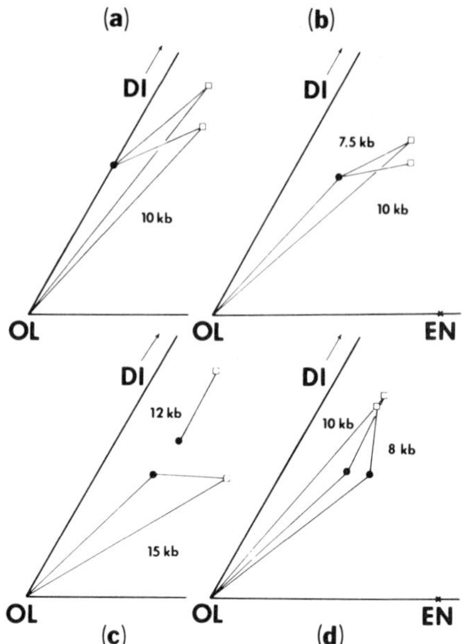

Fig. 7a–d. Coexisting $liquid + cpx \pm ol$ assemblages in natural MORB systems at pressures of 8–15 kbar, projected from PLAG onto the plane OL-DI-SIL. *Solid circles* are liquid compositions; *open squares* are coexisting clinopyroxenes. (a) this study (Run No. 519-11); (b) from Kushiro and Thompson (1972); (c) from Green et al. (1979); and (d) from Bender et al. (1978). Note that (a)–(c) all suggest that fractionation of $ol + cpx \pm plag$ from tholeiitic liquids would produce alkalic residual liquid compositions, but (d) suggests the opposite

starting composition. However, as Presnall et al (1979) point out, this type of effect could be due to analytical uncertainty since small changes in Na_2O or SiO_2 translate to large uncertainties in the locations of points on these diagrams. The results of Bender et al. (1978) suggest that at 10 kbar, temperature falls from alkalic to tholeiitic compositions. It is not clear why the results of Bender et al. (1978) on a similar liquid composition should differ from the other studies shown in Fig. 7. None of these pyroxene compositions are reversed, however, and all should be regarded with caution. Curiously, Bender et al. (1978) do not observe opx in their 10 kbar crystallization sequence, which they should if their high-Ca pyroxene composition is valid. The third bit of evidence that Figs. 2 and 3 are drawn appropriately is that in many basalt crystallization experiments at 10–20 kbar, orthopyroxene disappears from the crystallization sequence after cpx and olivine join the crystallization sequence (e.g., Green and Ringwood 1967) and that usually if orthopyroxene has not joined the crystallization sequence before clinopyroxene, it never does, suggesting that $ol + cpx$ crystallization drives residual liquids away from *opx* saturation, as shown in Figs. 2 and 3.

At some pressure below 10 kbar, orthopyroxene will no longer be in reaction relation with the liquid at the *ol + opx + cpx + liq* "point" in the natural system. This will happen when the composition of the *ol + opx + cpx*-saturated liquid is colinear in Fig. 2 (or 7) with the olivine-clinopyroxene tie line. At lower pressures, the multiply saturated liquid composition will be to the right of the line; under these conditions, liquids to the left of the tie line will fractionate to alkalic compositions while liquids to the right will fractionate to more hypersthene-normative compositions, ultimately precipitating low-Ca pyroxene along with olivine and clinopyroxene. When the *ol − cpx* tie line approximately coincides with the OL-DI join, a tholeiitic→alkalic fractionation trend on the *ol + cpx* cotectic will no longer be possible. This geometry holds at 1 atm, where temperature falls away from the *ol + cpx + plag* join toward tholeiitic compositions (see arrows on 1 atm cotectics in Figs. 2 and 3). It is not possible to tell from currently available data at what pressure between 1 atm and 10 kbar this geometry is achieved.

Discussion

Are Any MORB Samples Primary?

Armed with the phase boundaries presented in Figs. 2 and 3, it is possible to evaluate critically the various hypotheses that have been advanced to account for the petrological characteristic of mid-ocean ridge basalts. In Figs. 8 and 9, I have superimposed the phase boundaries from Figs. 2 and 3 on projections of the compositions of approximately 1,000 analyses of MORB glasses from a compilation by Melson et al. (1977). As pointed out first O'Hara (1968a), and recently re-emphasized by Walker et al. (1979), the majority of MORB samples are clustered about the olivine + clinopyroxene + plagioclase cotectic at low pressure (∼ 1 atm). This suggests that most of these samples have been modified by low pressure fractionation, during which residual liquids inevitably move to low-pressure cotectics. The increase in TiO_2 content from samples at the upper left hand end to the lower right hand end of the cluster in Fig. 8 also suggests that the cluster represents a liquid line of descent, from primitive samples at the silica-poor end to more fractionated samples at the silica-rich end. The samples studied by me, by Green et al. (1979), and by Bender et al. (1978) all plot near the silica-poor end of the cluster.

It is apparent from Fig. 8 that although the *ol +*

opx + cpx cotectic comes close to the cluster of MORB compositions it does not reach it. The multiply saturated liquid composition comes closest to the MORB cluster at slightly less than 10 kbar, at the middle of the cluster, where samples have intermediate TiO_2 contents and Fe/(Fe + Mg) ratios. The *ol + opx + cpx* cotectic diverges from the cluster at higher pressures and is significantly displaced from the primitive, SiO_2-poor end of the cluster of MORB compositions. This multiply saturated "point" does approach the cluster and overlap with its fringes at slightly less than 10 kbar, explaining why Kushiro and Thompson (1972) and Fujii and Kushiro (1977) found low-Ca pyroxene saturation in some MORB samples at 7.5–8 kbar.

Although the phase boundaries in Figs. 8 and 9 must be confirmed and more tightly constrained by future work, they are consistent with recent data from this work, Green et al. (1979), Bender et al. (1978), Fujii and Kushiro (1977), and Kushiro (1973) and are unlikely to be grossly in error. Consequently, they provide a strong endorsement of the analysis of O'Hara (1968a), now over a decade old, that suggested that the main cluster of MORB compositions is not low-Ca pyroxene-saturated at any pressure.

Presnall et al. (1979) have suggested that some of the elongation of the MORB cluster shown in Figs. 8 and 9 reflects analytical uncertainty in published MORB analyses, especially in Na_2O and SiO_2. If all of the points plotting at the primitive end of the cluster did so because of this effect, this would undermine the statements made above that the primitive end of the cluster is displaced from the 10 kbar *ol + opx + cpx* cotectic. It is, however, apparent from the projected locations of replicate analyses of the starting material used in this study, from a comparison of the projected locations of these analyses with the location of the analysis of the same sample reported by Langmuir et al. (1977), from the coherence at the primitive end of the cluster of the projected locations of the compositions studied by me, by Bender et al. (1978), and by Green et al. (1979), and from the experiments of Green et al. (1979), that some, and in fact probably many of the compositions plotting at the primitive end of the cluster are indeed properly located and, most importantly, are displaced from the 10 kbar *ol + opx + cpx* cotectic.

Figures 8 and 9 thus argue against the hypothesis of MORB genesis advanced by Presnall et al. (1979). Contrary to their suggestion, the *ol + plag + cpx + opx* cotectic at approximately 9 kbar does not coincide with the primitive end of the MORB cluster in the complex, natural composition space. To be sure, the closest approach of the multiply saturated point to the cluster does occur near 9 kbar, but it is toward

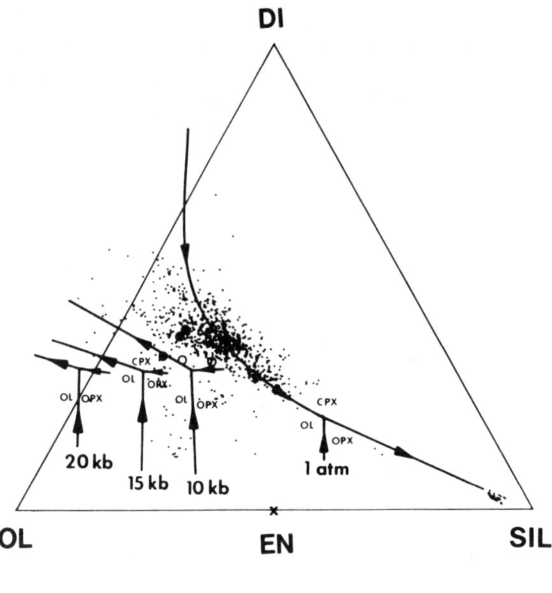

Fig. 8. Compositions of 960 MORB glass analyses compiled by Melson et al. (1977) compared with the deduced phase boundaries between 0–20 kbar from Fig. 2 projected from PLAG onto OL-DI-SIL. *Large closed circles* are primitive basalt compositions studied by Bender et al. (1978). Green et al. (1979). and in this work. none of which show *opx* crystallizing at their liquidi at any pressure. *Open circles* are compositions from Kushiro (1973) and Fujii and Kushiro (1977) that are low-Ca pyroxene-saturated at 7.5–8 kbar

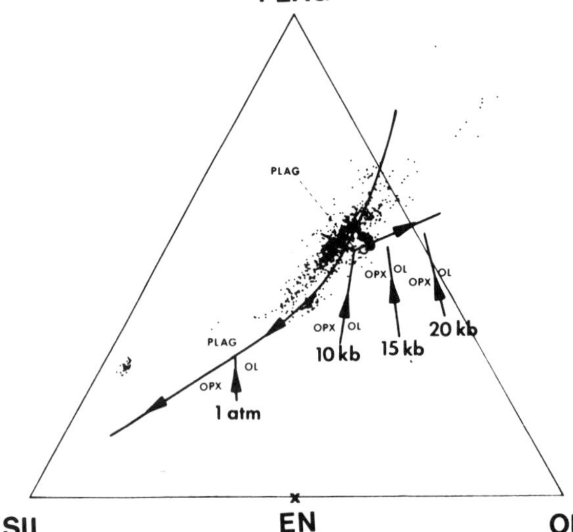

Fig. 9. Similar to Fig. 8. only projected from DI onto OL-PLAG-SIL

the middle, rather than the most primitive end of the cluster. Thus, despite the elegance of the concept of MORB genesis proposed by Presnall et al. (1979), the data presented here do not appear to support this model.

Based solely on the data presented in Figs. 8 and 9, modifications of the Presnall et al. (1979) hypothesis might be suggested. For example, note that the trend of multiply saturated liquids from about 5–15 kbar approximately parallels the MORB cluster.

One might propose a spectrum of primary magmas generated between these pressures, which then undergo differing amounts of olivine fractionation en route to the surface, yielding liquids arrayed in an elongated cluster such as that shown in Figs. 8 and 9. This would, however, provide no ready explanation for the coherence from the silica-poor to the silica-rich end of the cluster of increasing TiO_2 and Fe/(Fe + Mg) ratio, which points to a single fractionation trend from a primitive, low TiO_2, low Fe/(Fe + Mg)

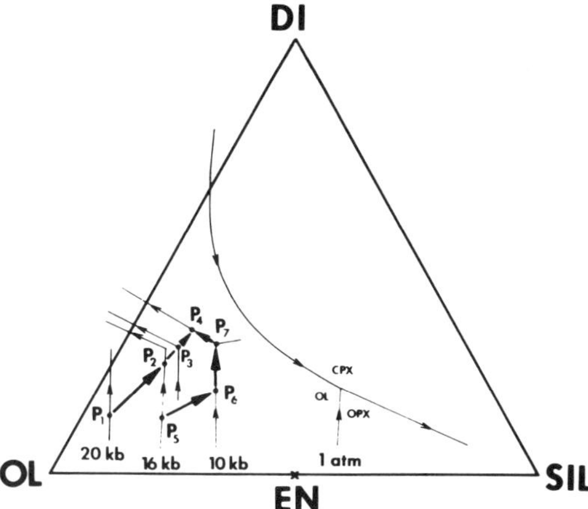

Fig. 10. Liquidus phase diagram with cotectics from Fig. 1 showing, schematically, a range of possible primary liquid compositions and fractionation paths for generating primitive MORB compositions. See text for discussion

liquid at the SiO_2-poor end of the cluster. Scenarios incorporating aspects of the Presnall et al. (1979) model can undoubtedly be constructed, but it appears that their concept of a single 9 kbar primary liquid for the suite corresponding to the most primitive end of the cluster, is inadequate.

Possible Scenarios for MORB Genesis

Based on the preceding discussion, the following ground rules apply to any permissible model of MORB petrogenesis. It is assumed that both olivine and orthopyroxene were left as residual phases in the source regions of MORB primary magmas and thus, any permissible primary magma must project on an olivine + orthopyroxene cotectic in both Figs. 8 and 9. By some fractionation path, residual liquids must arrive at the silica-poor end of the MORB cluster; subsequent low pressure fractionation can yield the rest of the MORB cluster. Although not narrowly restrictive, these ground rules do constrain the range of permissible models of MORB petrogenesis.

The usual and indeed simplest solution to the MORB petrogenesis problem has been to assume that the primitive end of the MORB cluster (Figs. 8 and 9) evolved from primary liquids only by fractionation of olivine. Thus, as shown in Fig. 10, Green et al. (1979) propose that a liquid corresponding to a primitive MORB sample (e.g., P_4: Fig. 10), plus 17% Fo90 olivine (e.g., P_2: Fig. 10) could be the primary liquid from which the primitive end of the MORB cluster evolved by fractionation of olivine only. This liquid

would indeed be saturated with olivine + orthopyroxene at about 16 kbar (20 kbar according to Green et al. 1979; in the remainder of this discussion, my pressure values will be used, although at this point there is no reason to prefer one pressure calibration over another) and could yield the primitive end of the MORB cluster by fractionation of olivine polybarically en route to the surface or entirely or in part in near surface magma chambers. This model is, however, by no means unique. At some lower pressure, probably about 14–15 kbar, a liquid with slightly less added olivine (e.g., P_3: Fig. 10) would be saturated with *ol + opx + cpx* and could equally well evolve by olivine fractionation alone to the primitive end of the MORB spectrum of compositions. Thus, a third type of pyrolite model, in which a slightly less olivine-rich magma is added to lherzolitic residue should have been considered by Green et al. (1979). The Green-type model also permits a *more* olivine-normative primary liquid leaving behind an olivine + orthopyroxene residue at pressures *greater* than 16 kbar (e.g. P_1: Fig. 10). Thus, if it is postulated that only olivine fractionated from primary liquids to generate the primitive MORB samples, no unique primary liquid composition or set of source region conditions can be specified, but limits on permissible scenarios can be set: The primary magma cannot segregate from the source region at pressures less than about 15 kbar, and at this minimum permissible pressure a residue consisting of olivine + orthopyroxene + clinopyroxene is implied. Such a liquid would have somewhat less than 27% olivine in its norm, but greater than 21% olivine in the norm (Green et al. 1979). More olivine-normative magmas are permitted; however: the greater the normative olivine content, the greater the depth of segregation from an olivine + orthopyroxene residue. It should be pointed out that O'Hara (1968ab) maintained that with increasing pressure beyond about 20–25 kbar the multiply saturated liquid composition in Figs. 8 and 9 moves back from the alkali basalt volume to the tholeiite volume (i.e., from the left to the right of the OL-DI join in Fig. 8 and from the right to the left of the OL-AN join in Fig. 9). If confirmed, and my data give no insight into this point, higher pressure genesis of this type of MORB primary liquids, leaving *diopside ± garnet* as residual phases in addition to *ol + opx*, would be permitted.

Despite the relative simplicity of the O'Hara (1968a) and Green et al. (1979) models involving olivine-only fractionation in the genesis of primitive MORB samples, other models are permissible. As an example, consider the composition P_5 in Fig. 10. This could be a primary liquid generated by partial melting of a mantle peridotite at 16 kbar, leaving only olivine and orthopyroxene in the residue. Suppose this liquid segregates from its source region and rises to a magma chamber at a pressure of 10 kbar and subsequently fractionates at this pressure. The fractionation path would be $P_5 \rightarrow P_6 \rightarrow P_7 \rightarrow P_4$, P_4 representing an example of a primitive MORB sample (in this case, the composition studied by Bender et al. 1978). From $P_5 \rightarrow P_6$, only olivine would fractionate; from $P_6 \rightarrow P_7$, olivine and orthopyroxene would crystallize; from $P_7 \rightarrow P_4$, olivine and clinopyroxene would crystallize, orthopyroxene being lost from the crystallization sequence when it goes into reaction relation at P_7. Such a path for derivation of primitive MORB sample is indeed complex and has nothing in particular to recommend it over the olivine-only fractionation models of O'Hara (1968a) or Green et al. (1979), except perhaps the suggestion that some of the clinopyroxene xenocrysts in MORB samples are manifestations of an earlier high pressure stage of fractionation of primitive MORB liquids (Bender et al. 1978).

Despite its complexity, the $P_5 \rightarrow P_4$ fractionation sequence in Fig. 10 demonstrates several important points. First, there is a lower limit to the pressure from which MORB primary magmas can develop; somewhere below 10 kbar pressure, the $ol + cpx$ cotectic will slope toward the $ol + opx + cpx$ cotectic with decreasing temperature, rather than away from it. When this occurs, it becomes impossible for $ol + opx$ saturated liquids to fractionate to the primitive end of the MORB cluster. Second, there are essentially an infinite number of possible fractionation paths by which primitive MORB samples could evolve from $ol + opx$ saturated liquids. Some of these are more complex than others, and it remains to be seen whether complex fractionation paths could be reproduced time after time the world over to produce similar primitive MORB compositions in all mid-ocean ridge environments.

Picritic Primary Magmas: Conditions of Formation

Several rocks have been identified whose MgO-rich compositions have been proposed to represent the picritic primary magmas that O'Hara (1968a) first suggested may be parental to the MORB suite. These rock compositions, along with two model compositions, one that Elthon (1979) suggested represents the average composition of the oceanic crust and the other that Irvine (1977) calculated assuming that olivine was the only phase that crystallized from primary magmas to produce primitive MORB compositions, are plotted in Figs. 11 and 12. Although the validity of individual compositions as representatives of prim-

itive liquids (as opposed to cumulate or phenocryst-enriched compositions) has been controversial (e.g., Hart and Davis 1978, 1979; Clarke and O'Hara 1979; Elthon and Ridley 1979), it is nevertheless valuable to consider the conditions under which such compositions could be generated by dry melting of mantle material.

The first point to be made from Figs. 11 and 12 is that there is a coherence among these actual and theoretical rock compositions, suggesting that although they are rare, picritic compositions of this sort may indeed be a reproducible rock type produced the world over. There is, of course, selection bias involved here. In response to the work of O'Hara (1968ab), many geologists searched for such compositions and if they were identified tended to suggest them as primary liquid compositions. The discriminating factor in such selection bias was, however, high MgO content, not the coherence illustrated in Figs. 11 and 12, and it may therefore be significant.

It is clear from Figs. 11 and 12 that the suggestion of Elthon (1979) that these compositions could equilibrate with olivine and orthopyroxene between 5 and 10 kbar is incorrect. On the other hand, these compositions could equilibrate with olivine and orthopyroxene and thus be consistent with primary magmas at pressures between 15–20 kbar. Indeed, they are similar to the preferred primary magma composition of Green et al. (1979) that was proposed to have formed under similar conditions. After generation by partial melting and segregation from mantle source regions at 15–20 kbar leaving $ol + opx$ ($+$ possibly cpx and/or sp for some of the compositions in Fig. 11) in the residue, they could, by fractionation of olivine during ascent to the ridge, evolve to the primitive end of the MORB cluster of compositions.

Why, if picritic magmas are the primary liquids of MORB suites, are they so rare (none have yet been found among MORB samples) and why do they always fractionate approximately the same amount prior to eruption? If the average oceanic crust is indeed approximately picritic in composition as suggested by Elthon (1979) this could indicate that picritic liquids are delivered essentially unmodified to magma chambers in the oceanic crust, and that the extensive olivine fractionation occurs there. Two factors, probably among many others, could result in rare eruption of these picritic compositions and in the constancy in composition of the liquids that do erupt. The existence of a magma chamber, possibly in approximately a steady state, can account for the rarity of these liquids and the constancy of erupted products' compositions (e.g., O'Hara 1977; Elthon 1979; Walker et al. 1979). Only early in a ridge's history, before a magma chamber is well developed and has

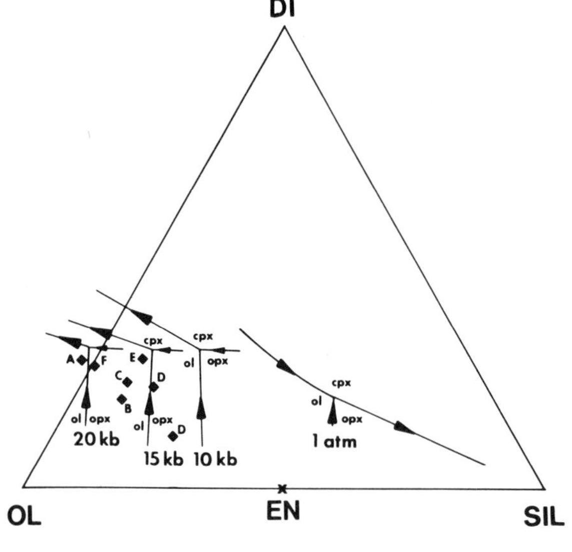

Fig. 11. Phase boundaries from Fig. 2 compared with actual and theoretical picritic compositions that may be parental to the MORB suite. *A* lavas from Svartenhuk (Clarke 1970); *B* Baffin Bay picrite (Clarke 1970); *C* "komatiite" from Gorgona Island (Gansser et al. 1979); *D* dikes from Tortuga, Chile (Elthon 1979); *E* "average" oceanic crust (Elthon 1979); *F* parental liquid of MORB suite (Irvine 1977)

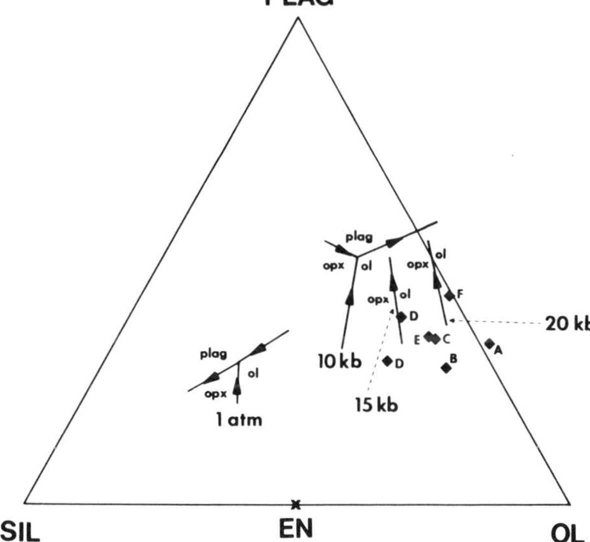

Fig. 12. Phase boundaries from Fig. 3 compared with actual and theoretical picritic compositions that may be parental to the MORB suite. Compositions as given in the caption of Fig. 11

reached a steady state, or in the waning stages of the magma chamber, can picritic magmas reach the surface; once a magma chamber has developed, it will trap the ascending picrite and assimilate it into an already fractionated magma. Another possible contributing factor, discussed by Stolper and Walker (1980), is that these picritic compositions are significantly denser than the average MORB composition when molten, and possibly even denser than the solidified oceanic crust. Thus, these picrites could be trapped beneath the crust or the magma chamber itself, with little tendency to erupt until they fractionated to approximately a typical MORB composition, at which point the residual liquid would have decreased in density sufficiently to permit eruption.

Summary

The phase diagrams presented in Figs. 2 and 3, showing the locations of olivine, orthopyroxene, and clinopyroxene-saturated liquids in MORB systems in two projections, provide a basis for assessing the various hypotheses for MORB petrogenesis. Although the orthopyroxene-volume comes closest to the cluster of MORB compositions near 9 kbar, it only skims it and is nearer the middle rather than the primitive end of the cluster. Since orthopyroxene + olivine saturation is presumed to be a characteristic of primary magmas, the hypothesis of Presnall et al. (1979) (at least in its simplest form) that primitive MORB compositions are essentially primary, generated by melting of $ol + opx + cpx + plag + sp$ at ~ 9 kbar, is called into question, as is the applicability of the concept of a cusp in generating these magmas.

Picritic compositions that have been proposed as primary magmas for MORB petrogenesis occupy a narrow region of composition space. Contrary to the suggestion of Elthon (1979), these picritic compositions could not represent liquids equilibrated with $ol + opx$ at 5–10 kbar. These picrites can, however, have been produced by melting in the mantle, leaving a residue of $ol + opx \pm cpx \pm sp$ (i.e., a harzburgite or a lherzolite) at pressures between 15–20 kbar, largely in agreement with the conclusions of Green et al. (1979). The data presented here cannot be used to assess the hypothesis of O'Hara (1968 ab) that these picrites could equilibrate with $ol + opx + cpx + garnet$ at 25–30 kbar. If so, other approaches (e.g., geochemical considerations) would be needed to distinguish between these two models.

Acknowledgements. The experiments reported in this paper were conducted at Harvard University in the laboratory of Professor J.F. Hays and were supported by NSF grants EAR 79-06321, EAR 79-23977; NASA grant NGL 22-007-247; and the Committee on Experimental Geology and Geophysics of Harvard University. I thank W.B. Bryan, A.J. Irving, and D. Walker for making samples available to me and D. Eggler and D.C. Presnall for their reviews of the manuscript. Special thanks to D. Walker for preparing the projections of the MORB compositions from the catalog of Melson et al. (1977) that were used in Figs. 8 and 9. Contribution Number 3370, Division of Geological and Planetary Sciences, California Institute of Technology, Pasadena, CA 91125.

References

Bond WL (1951) Making small spheres. Rev Sci Instrum 22:344–345

Bender JF, Hodges FN, Bence AE (1978) Petrogenesis of basalts from the Project FAMOUS area: experimental study from 0 to 15 kbars. Earth Planet Sci Lett 41:277–302

Boyd FR, England JL (1960) Apparatus for phase equilibrium experiments at pressures up to 50 kbars and temperatures up to 1.750° C. J Geophys Res 65:741–748

Cawthorn RG, Ford CE, Biggar GM, Bravo MS, Clarke DB (1973) Determination of the liquid composition in experimental samples: discrepancies between microprobe analysis and other methods. Earth Planet Sci Lett 21:1–5

Chodos AA, Albee AL, Gancarz AJ, Laird J (1973) Optimization of computer-controlled quantitative analysis of minerals. Proc 8th Nat Conf on Electron Probe Analysis, New Orleans, LA

Clarke DB (1970) Tertiary basalts of Baffin Bay: Possible primary magma from the mantle. Contrib Mineral Petrol 25:203–224

Clarke DB, O'Hara MJ (1979) Nickel, and the existence of high-MgO liquids in nature. Earth Planet Sci Lett 44:153–158

Elthon D (1979) High magnesia liquids as the parental magma for ocean floor basalts. Nature 278:514–518

Elthon D, Ridley WI (1979) Comments on: "The partitioning of nickel between olivine and silicate melt", by: SR Hart and KE Davis. Earth Planet Sci Lett 44:162–164

Fujii T, Kushiro I (1977) Melting relations and viscosity of an abyssal tholeiite. Carnegie Inst Washington Yearb 76:461–465

Gansser A, Dietrich VJ, Cameron WE (1979) Paleogene komatiites from Gorgona Island. Nature 278:545–546

Green DH, Ringwood AE (1967) The genesis of basaltic magmas. Contrib Mineral Petrol 15:103–190

Green DH, Hibberson WO, Jaques AL (1979) Petrogenesis of mid-ocean ridge basalts. In: MW McElhinney (ed) The Earth: Its Origin, Structure and Evolution. Academic Press, London

Hart SR, Davis KE (1978) Nickel partitioning between olivine and silicate melt. Earth Planet Sci Lett 40:203–219

Hart SR, Davis KE (1979) Reply to DB Clarke and MJ O'Hara, "Nickel, and the existence of high-MgO liquids in nature". Earth Planet Sci Lett 44:159–161

Irvine TN (1977) Definition of primitive liquid compositions for basic magmas. Carnegie Inst Washington Yearb 76:454–461

Johannes W, Bell PM, Mao HK, Boettcher AL, Chipman DW, Hays JF, Newton RS, Siefert F (1971) An interlaboratory comparison of piston-cylinder pressure calibration using the albite breakdown reaction. Contrib Mineral Petrol 32:24–38

Kesson S (1975) Mare basalts: Melting experiments and petrogenetic interpretations. Proc Lunar Sci Conf 6th. Geochim Cosmochim Acta, Suppl 6:921–944

Kushiro I, Thompson RN (1972) Origin of some abyssal tholeiites from the Mid-Atlantic Ridge. Carnegie Inst Washington Yearb 71:403–406

Kushiro I (1973) Origin of some magmas in oceanic and circumoceanic regions. Tectonophysics 17:211–222

Langmuir CH, Bender JF, Bence AE, Hanson GN, Taylor SR (1977) Petrogenesis of basalts from the FAMOUS area: Mid-Atlantic Ridge. Earth Planet Sci Lett 36:133–156

Longhi J, Walker D, Hays JF (1978) The distribution of Fe and Mg between olivine and lunar basaltic liquids. Geochim Cosmochim Acta 42:1545–1558

Melson WG, Byerly GR, Nelen JA, O'Hearn T, Wright TL, Vallier T (1977) A Catalog of the major element chemistry of abyssal volcanic glasses. Smithson Contrib Earth Sci 19:31–60

Mysen BO, Kushiro I (1977) Compositional variations of coexisting phases with degree of melting of peridotite in the upper mantle. Am Mineral 62:843–865

O'Hara MJ (1968a) Are any ocean floor basalts primary magma? Nature 220:683–686

O'Hara MJ (1968b) The bearing of phase equilibria studies in synthetic and natural systems on the origin and evolution of basic and ultrabasic rocks. Earth Sci Rev 4:69–133

O'Hara MJ (1977) Geochemical evolution during fractional crystallization of a periodically refilled magma chamber. Nature 266:503–507

Presnall DC, Dixon SA, Dixon JR, O'Donnell TH, Brenner NL, Schrock RL, Dycus DW (1978) Liquidus phase relations on the join diopside-forsterite-anorthite from 1 atm to 20 kbar: their bearing on the generation and crystallization of basaltic magma. Contrib Mineral Petrol 66:203–220

Presnall DC, Dixon JR, O'Donnell TH, Dixon SA (1979) Generation of Mid-ocean Ridge Tholeiite. J Petrol 20:3–35

Roeder PL (1974) Activity of iron and olivine solubility in basaltic liquids. Earth Planet Sci Lett 23:397–410

Stolper E (1980) A phase diagram for mid-ocean ridge basalts. EOS 61:405

Stolper E, Walker D (1980) Melt density and the average composition of basalt. Contrib Mineral Petrol 74:7–12

Takahashi E (1980) Olivine/liquid nickel partitioning at high pressures: experiments with an olivine capsule. EOS 61:397

Walker D, Longhi J, Lasaga AC, Stolper EM, Grove TL, Hays JF (1977) Slowly cooled microgabbro 15555 and 15065. Proc Lunar Sci Conf 8th. Geochim Cosmochim Acta, Suppl 8:1521–1547

Walker D, Shibata T, Delong SE (1979) Abyssal tholeiites from the Oceanographer Fracture Zone II. Phase equilibria and mixing. Contrib Mineral Petrol 70:111–125

Watson EB (1980a) Apatite saturation in basic to intermediate magmas. Geophys Res Lett 12:937–940

Watson EB (1980b) Apatite saturation in magmas at high pressures. EOS 61:396

Received January 24, 1980; Accepted June 24, 1980

Part III

BASALTS AND MANTLE EVOLUTION

Editors' Comments
on Papers 34 Through 37

34   JAMIESON and CLARKE
Excerpts from *Potassium and Associated Elements in Tholeiitic Basalts*

35   HART et al.
Abstract from *Ancient and Modern Volcanic Rocks: A Trace Element Model*

36   JAHN
Trace Element Geochemistry of Archean Volcanic Rocks and Its Implication for the Chemical Evolution of the Upper Mantle

37   SUN and HANSON
Evolution of the Mantle: Geochemical Evidence from Alkali Basalt

Because basalts are either wholly derived from the mantle or, at least, have a significant mantle component, they can be used to make inferences about the present conditions of the mantle and its past evolution. As discussed in Part IV, even in the apparently homogeneous environment of the ocean basins there is considerable evidence that compositional heterogeneity in the mantle is required to explain the variation in properties of MORB and related rocks. Major compositional differences have been proposed for magmas generated from plume sources and those generated in a depleted low-velocity layer (e.g., Schilling, 1975); smaller-scale variations, including transitions between the plume and nonplume environments, have been studied extensively. On a broader scale is the question of whether the suboceanic and subcontinental mantles differ in composition. Even more speculative is the question of whether, or how, the composition of the mantle has varied through time, a question that can be approached by investigating variation in the nature of basaltic volcanism through time.

A very large number of papers have been written about these problems, and with great difficulty we have chosen four to present here. The paper by Jamieson and Clarke (1970, Paper 34) is one of the earliest papers that investigates the significance of lithophilic elements (principally K) in basalt evolution; these authors conclude that oceanic and continental basalts have different compositions because of crustal contamination. A more recent discussion of possible differences between suboceanic and subcontinental mantle has been provided by Leeman (1977).

The remaining three papers deal with the question of mantle evolution through time. One of the earliest efforts to use basalt composition to investigate this problem is the paper by Hart et al. (1970, Paper 35); they concluded that Archean basalts were normally formed from a mantle more enriched in lithophilic elements than the present mantle. Jahn (Paper 36), however, concluded that Archean and modern basalts are not sufficiently different to prove such variation. Sun and Hanson (Paper 37) demonstrated long-term heterogeneities in the mantle based on evidence from alkali basalts.

REFERENCES

Hart, S. R., C. Brooks, T. E. Krogh, G. L. Davis, and D. Nava, 1970, Ancient and Modern Volcanic Rocks: A Trace Element Model, *Earth and Planetary Sci. Letter* **10:**17–28.

Jamieson, B. G., and D. B. Clarke, 1970, Potassium and Associated Elements in Tholeiitic Basalts, *Jour. Petrology* **11:**183–204.

Leeman, W. P., 1977, Comparison of Rb/Sr, U/Pb, and Rare Earth Characteristics of Sub-continental and Sub-oceanic Mantle Regions, in Dick, H. J. B., ed., *Magma Genesis, Oregon Dept. Geology and Mineral Industries Bull. 96,* pp. 149–168.

Schilling, J. -G., 1975, Rare-Earth Variations across "Normal Segments" of the Reykjanes Ridge, 60° to 53°N, Mid-Atlantic Ridge, 29°S, and East Pacific Rise, 2° to 19°S, and Evidence on the Composition of the Underlying Low-Velocity Layer, *Jour. Geophys. Research* **80:**1459–1473.

34

Reprinted from pages 193-200, 203, and 204 of *Jour. Petrology* **11**:183-204 (1970)

POTASSIUM AND ASSOCIATED ELEMENTS IN THOLEIITIC BASALTS

B. G. Jamieson and D. B. Clarke

[*Editors' Note:* In the original, material precedes and follows this excerpt.]

POTASSIUM ENRICHMENT PROCESSES

The following processes are assessed below as possible factors leading to the enrichment of K and associated elements in tholeiites; the first four factors being discussed from the standpoint of a chemically homogenous mantle.

(*a*) Degree of partial melting of mantle peridotite.

(*b*) Crystal–liquid fractionation.

(*c*) Mantle wall-rock reaction.

(*d*) Crustal wall-rock reaction.

(*e*) Lateral and/or vertical mantle inhomogeneity with respect to K and associated elements.

242

Degree of partial melting

Gast (1968), Griffin & Murthy (1968*b*), and O'Hara (1968*a*) have recently estimated the degree of partial melting which occurs during primary magma genesis to be 5–30 per cent. Green & Ringwood (1967), on the other hand, consider that between 20 and 40 per cent of the mantle peridotite is melted. As Table 1 indicates, it is probable that 90 per cent of the total K and associated elements is in the liquid phase when 10 per cent melting has occurred. This represents a 9–10 times enrichment and further melting can only effect a dilution of these elements in the primary magma. In contrast, the effect on the major oxides of increased melting, say from 10 to 20 per cent, will be much less marked. The results of Ito & Kennedy (1967) indicate that, with increased melting, the primary picritic magma which is produced at 30 kb pressure from a garnet lherzolite mantle will become enriched in *ol* and *hy*.

TABLE 6

*Average tholeiites of Kilauea and Mauna Loa**

	K_2O	P_2O_5	Ba	Sr	Zr
Kilauea	0·54 %	0·30 %	184	652	202
Mauna Loa	0·38 %	0·24 %	59	481	144

* From Macdonald & Katsura (1964, table 9, nos. 6 and 7). Trace elements in ppm from Prinz (1967, table IV).

Harris (1967, p. 312) has drawn attention to the importance of the degree of melting on concentration levels of K and associated elements.

We interpret the relative compositional differences between the average basalts of Kilauea and Mauna Loa as reflecting different degrees of partial melting (Table 6). No low-pressure fractionation scheme, which is plausible in terms of the recorded appearance of crystalline phases, can account for the major oxide composition *and* the relative concentrations of K and associated elements in these basalts.

Crystal–liquid fractionation

It has been demonstrated that low-pressure crystal–liquid fractionation cannot, alone, account for the range of K and associated elements in parental tholeiites. This is likely to be true of any crystal–liquid fractionation scheme which is indexed by the MgO content of the residual liquid; the change in major oxide composition is too rapid to allow a sufficient build-up of K and associated elements in residual liquids which are still basic in composition. This consideration led Green & Ringwood (1967) to conclude that no crystal–liquid fractionation scheme could account for the enrichment levels of K and associated elements observed in alkali basalts and some tholeiites.

However, one proposed crystal–liquid fractionation scheme—eclogite fractionation from primary picritic magmas at pressures in excess of *c.* 25 kb (O'Hara

243

& Yoder, 1967)—could enrich residual liquids in K and associated elements without markedly reducing the contents of MgO or *ol*. A detailed analysis of the likely geochemical features of high-pressure eclogite fractionation is given by O'Hara & Yoder (1967, pp. 107–14). The K/Na ratio is perhaps the best index of eclogite fractionation.

One result of such fractionation would be a reduction of *hy* in the residual liquids. However, because of the similarity between the bulk composition of the primary picritic magma and the bimineralic crystalline extract, the movement of residual liquids towards *ne*-normative compositions would be slow and it might be possible to double K enrichment factors before this condition was reached (O'Hara, 1968a, p. 118).

Fractional crystallization during the ascent of magmas which had been subjected to small, but varying degrees of high-pressure eclogite fractionation, would produce, at the surface, tholeiites with broadly similar major oxide compositions but with varied relative enrichment factors of K and associated elements which had been inherited from the high-pressure event.

The petrogenetic scheme of O'Hara (1968a, p. 118), which involves only closed-system partial melting and fractional crystallization, can account for enrichment factors for K and associated elements in tholeiites in the range 8–66.

Harris (1967) has discussed the role of crystal–liquid fractionation. However, he considered only the case when the equilibrium involving 4 solid phases and liquid is a eutectic. Fractional crystallization of a 'pseudo-eutectic' liquid at elevated pressure would certainly increase enrichment factors of K and associated elements but would also produce residual magmas with unusually high Fe/Mg and Na/Ca ratios when compared with other basic magmas.

Reaction between ascending magma and mantle wall-rock

The zone-refining process applied by Harris (1957) involves the continuous solution of wall-rock peridotite by a basic magma and precipitation of the major crystalline phases of the peridotite. Elements with solid/liquid distribution coefficients < 1 would be progressively concentrated in the liquid phase. Harris considered such a process might be relevant to the genesis of K-rich mafic lavas.

Green & Ringwood (1967) envisaged this process operating on a much broader scale and they considered all alkali basalts and some tholeiites to have had their contents of 'incompatible' elements enhanced by reaction between magma and mantle wall-rock.

We accept the principle of mantle wall-rock reaction but find it very difficult to assess the importance of the process. The following two pertinent observations have been made recently with regard to this process:

(*a*) '. . . although it might work reasonably well for the first batch of magma to ascend through a conduit, it is difficult to envisage it working with equal or increased efficiency for succeeding batches . . .' (O'Hara, 1968*a*, p. 117).

(*b*) '. . . the opportunity for enrichment will decrease rapidly as small conduits [10–100 cm] coalesce into larger channels' (Gast, 1968, p. 1081).

These observations appear very reasonable to the authors and are taken to imply that reaction between peridotite wall-rock and magma is, in general, of limited importance as a factor giving rise to enrichment in K and associated elements. It is important to realize that this process is not haphazard but is controlled by distribution coefficients and the rather obscure phase relations during the initial stages of partial melting of peridotite—stage 1 (O'Hara, 1968*a*). The magma composition produced after extensive wall-rock reaction would be similar to that produced by small amounts ($<$ 3 per cent) of partial melting of peridotite.

Crustal wall-rock reaction

Engel *et al.* (1965) considered that all basalts with contents of K and associated elements higher than those of deep oceanic tholeiites were derivatives of a primary magma with the composition of low-K, deep oceanic tholeiites. The differentiation processes they envisaged were crystal–liquid fractionation, alkali transfer, and crustal contamination—this last process being an important factor in the continental environment. The crustal contamination process was rather poorly defined. However, as Si was included amongst the elements (Si, K, Ba, Cs, Rb, Sr, Th, U) which are 'scavenged from sialic inclusions and from the walls of crustal conduits' (Engel *et al.*, p. 727), this process is rather different from the selective wall-rock reaction which was proposed by Green & Ringwood (1967, p. 177) as an unusual, but possible, contamination process.

A consideration of two features of Table 2 casts serious doubts on the role of crustal contamination as envisaged by Engel *et al.* (1965).

(*a*) In the Rhodesian olivine-rich tholeiites (Table 2, No. 9) the enrichment factors of K and associated elements relative to deep oceanic tholeiite are: K_2O—10, TiO_2—1·6, P_2O_5—2·5, Ba—50, Rb—30, Sr—6, and Zr—3. Even relative to the average oceanic tholeiite (Table 7), K_2O is enriched by three times. However, these enrichment factors are not reflected in SiO_2—deep oceanic tholeiite 49·9 per cent, oceanic tholeiite 49·3 per cent, Rhodesian olivine-rich lavas 49·8 per cent.

It is unlikely that wholesale assimilation of crustal material, or its partial melt, as proposed by Engel *et al.* (1965), could account for these chemical features.

(*b*) The olivine-rich basalts of West Greenland–Baffin Island (Table 1, nos. 3 and 2) were extruded in a continental environment, yet they have very low contents of K and associated elements. Because of differences in MgO content, the concentrations of K and associated elements are not directly comparable with those in deep oceanic tholeiites. However, the Na/K ratio of the Baffin Island lavas is indicative of a concentration level broadly similar to that found in deep oceanic tholeiites.

This paucity of K and associated elements strongly suggests that crustal contamination has made no significant contribution to the observed concentration levels of these elements in the West Greenland–Baffin Island province.

TABLE 7

Average analyses of continental and oceanic tholeiites

	282 oceanic tholeiites	946 continental tholeiites
SiO_2	49·3	51·5
TiO_2	2·4	1·2
Al_2O_3	14·6	16·3
Fe_2O_3	3·2	2·8
FeO	8·5	7·9
MnO	0·17	0·17
MgO	7·4	5·9
CaO	10·6	9·8
Na_2O	2·2	2·5
K_2O	0·53	0·86
P_2O_5	0·26	0·21

Data from Manson (1967, table IV).

Having demonstrated that the lavas which we have studied in detail are unlikely to have been modified by sialic contamination, we continue with a more general appraisal of the importance of crustal contamination in the development of continental tholeiites. In particular, the important evidence from Sr isotopic studies is considered.

In their discussion of crustal contamination, Engel *et al.* (1965) argued from the fact that continental tholeiites contain more SiO_2 and K_2O than deep oceanic tholeiites. Table 7 shows that this is also true when continental tholeiites are compared with all oceanic tholeiites—including both the deep oceanic tholeiites and tholeiites from oceanic islands. When the average continental and oceanic tholeiites (submarine and subaerial) are plotted in Fig. 3 it is clear that the continental average analysis is more evolved, both in terms of MgO and K_2O, than the oceanic average analysis. On the basis of the low-pressure differentiation trends in Fig. 3, it is concluded that near-surface crystal–liquid fractionation cannot produce the continental average as a residual liquid composition from a magma with the composition of the average oceanic tholeiite. The average continental tholeiite appears to be not only more evolved in terms of low-pressure differentiation but is also inherently more potassic, at comparable stages of evolution.

These are not unexpected findings. The passage of basic magma through the continental crust offers an extended period for loss of thermal energy and for fractionation. The existence of a continental crust may have an inhibiting effect on the ascent of magma through the underlying mantle and thus encourage mantle K-enrichment processes such as wall-rock reaction or crystal–liquid fractionation. The nature of oceanic and continental geothermal gradients implies that, in general, continental primary magmas may be generated at greater depth than oceanic magmas.

We prefer, therefore, to consider that the real compositional differences between average oceanic and continental tholeiites reflect the different physical conditions in the two environments and that crustal contamination is not of general significance.

Strontium isotopic ratios would seem to offer the best means of resolving a differentiation *versus* contamination controversy. The pronounced positive skew to the frequency distribution of Sr^{87}/Sr^{86} initial ratios in continental basalts—in sharp contrast to the near-Gaussian distribution of Sr^{87}/Sr^{86} in oceanic basalts—has been taken as evidence of contamination of basic magma by crustal material (Gast, 1967; Hedge, 1966). However, for the following two reasons we feel this conclusion should not be uncritically accepted:

(*a*) Because of the probable distribution of Rb/Sr ratios, and hence Sr^{87}/Sr^{86} ratios, in the continental crust, it is difficult to account for the anomalously high Sr^{87}/Sr^{86} ratios of some continental tholeiites (0·708–0·712) by contamination of basic magma with average crustal material. Faure & Hurley (1963) estimated that the upper, sialic, continental crust has an average Rb/Sr ratio of 0·25 and a Sr^{87}/Sr^{86} ratio of 0·725. Assuming equal Sr contents in a basic magma and the contaminant, some 25 parts of average upper crustal material, or its partial melt, must be added to 100 parts of basic magma to increase the Sr^{87}/Sr^{86} ratio from 0·704 to 0·708.

The alternative to this gross contamination is a smaller degree of contamination by granite, or its partial melt. Because of a high Rb/Sr ratio—0·73 (Hedge, 1966)—granitic material must be much richer in radiogenic Sr than the average upper continental crust. The limited availability of granitic material in the crust implies that this contamination process must be restricted to high crustal levels—an environment in which marginal reaction and wholesale assimilation would be relatively limited.

(*b*) One aspect of initial Sr^{87}/Sr^{86} ratios in basic rocks which is not fully understood is the apparent sympathetic relationship between radiogenic Sr enrichment and the SiO_2 content. This relationship is recorded in Antarctica (Compston *et al.*, 1968) and in the Tertiary volcanic areas of the Great Basin and Rocky Mountains (Hedge, 1966). More recently Pankhurst (1969) has drawn attention to this relationship in the Newer Gabbros of North-east Scotland. In these continental environments this evidence could be interpreted as being favourable to a crustal contamination theory. Unfortunately for that model, the same relationship is found in the following oceanic islands—St.

Helena, Ascension, and perhaps Samoa and the Hawaiian Islands (Gast, 1967; Gast *et al.*, 1964; Hedge, 1966; Powell *et al.*, 1965).

The interpretation of the isotopic results from the Insch basic mass of Northeast Scotland (Pankhurst, 1969, p. 136) is particularly relevant to this discussion. Pankhurst concludes that neither the major element geochemistry nor K and associated elements have been appreciably modified by assimilation of country rock. Nevertheless the initial Sr^{87}/Sr^{86} ratios of the Insch cumulates increase, with differentiation, from 0·703 to 0·712. To account for this Pankhurst has proposed a process of isotopic equilibration between the fractionating magma and the country rock. If of general occurrence, such a process undermines the basic assumptions in the conventional interpretation of Sr isotopic data.

In summary, the evidence from two continental tholeiitic provinces is not favourable to the crustal contamination hypothesis of Engel *et al.* (1965). Nevertheless, the average continental tholeiite is more evolved than its oceanic counterpart. To bring about this additional degree of evolution we prefer an essentially closed scheme involving crystal–liquid fractionation in the crust, while the more evolved K_2O content is attributed to mantle processes. On occasions— e.g., in the case of Tasmanian and Antarctic dolerites (Compston *et al.*, 1968) and the Tertiary basic and ultrabasic rocks of Skye (Moorbath & Welke, 1968) —this model might be supplemented by reaction between crustal wall-rock and magma. Sr^{87}/Sr^{86} isotopic studies may be of limited application to this problem. Even if one does not accept the conclusions of Pankhurst (1969), some explanation is required of the relationship between SiO_2 and Sr^{87}/Sr^{86} ratios in some oceanic lavas.

Mantle inhomogeneity

Possible processes of K-enrichment have been discussed above from the standpoint of a mantle which is chemically and mineralogically homogenous. This is an over-simplification since, for example, the mineral assemblage at the beginning of melting of a chemically homogenous mantle will vary with depth because of P- and T-dependent solid solution effects. Thus there will be small, but possibly significant, changes in distribution coefficients. The stabilities of amphibole and phlogopite in the uppermost mantle are likely to be critically dependent on P, T, and volatile content (Kushiro *et al.*, 1967; Lambert & Wyllie, 1968). Dickinson & Hatherton (1967) considered that the sympathetic relationship between the K_2O content of andesites and the depth of seismic activity in the Benioff zone is due to melt–solid distribution coefficients varying with pressure.

Gast (1968, p. 1077) has proposed that deep oceanic tholeiites may be generated from peridotite which has been subjected to an earlier partial melting episode. However, this is difficult to reconcile with a dynamic spreading seafloor hypothesis which requires that fresh mantle is being brought up under the oceanic ridges.

Reference has been made to the K_2O content of the average continental tholeiite and it was concluded that continental tholeiitic magma, before it is subjected to low-pressure differentiation in the upper crust, is inherently more potassic than its oceanic equivalent. This feature could be taken as evidence of the mantle under the continents being enriched in K_2O. This proposal is difficult to reconcile with continental drift models which involve crust–mantle decoupling and we prefer to conclude that the presence of a continental crust encourages K-enrichment processes in the mantle.

Barium in the average Rhodesian tholeiites (Table 2, nos. 8 and 9) shows an enrichment factor of 56 relative to deep oceanic tholeiites and of 650–200 relative to a mantle content of 1·2–3·9 ppm. This range of mantle Ba contents was obtained by taking the chondritic Ba content—4 ppm—and assuming that there is no Ba in the Earth's core and that between 33 and 80 per cent of the Earth's Ba is in the crust (see Taylor, 1964). Enrichment factors above 70–80 appear to be beyond the range which can be produced by a closed-system partial melting and fractional crystallization scheme (O'Hara, 1968a). The concentration of Ba in the Rhodesian tholeiites, therefore, implies that wall-rock reaction and/or mantle inhomogeneity with respect to K and associated elements were factors in the genesis of these basalts.

In conclusion, we feel it is premature to assume the upper mantle displays marked lateral chemical heterogeneity. Although it is tempting to consider that parts of the oceanic mantle are depleted in K and associated elements relative to the mantle under the continents, there are many combinations of the variable factors described above and an interplay of these processes might account for the observed range of contents of K and associated elements. Until there is a better understanding of these and other processes we feel it is more realistic to assume a mantle which is, on a broad scale, laterally homogenous.

[*Editors' Note:* Only the references cited in the preceding excerpt are reproduced here.]

REFERENCES

COMPSTON, W., McDOUGALL, I., & HEIER, K. S., 1968. Geochemical comparison of the Mesozoic basaltic rocks of Antarctica, South Africa, South America and Tasmania. *Geochim. cosmochim. Acta*, **32**, 129–49.

DICKINSON, W. R., & HATHERTON, T., 1967. Andesitic volcanism and seismicity around the Pacific. *Science, N.Y.* **157**, 801–3.

ENGEL, A. E. J., ENGEL, C. G., & HAVENS, R. G., 1965. Chemical characteristics of oceanic basalts and the upper mantle. *Bull. geol. Soc. Am.* **76**, 719–34.

FAURE, G., & HURLEY, P. M., 1963. The isotopic composition of strontium in oceanic and continental basalts: application to the origin of igneous rocks. *J. Petrology*, **4**, 31–50.

GAST, P. W., 1967. Isotope geochemistry of volcanic rocks, pp. 325–58 in *Basalts: the Poldervaart treatise on rocks of basaltic composition*, eds. Hess, H. H., & Poldervaart, A. New York: Interscience.

—— 1968. Trace element fractionation and the origin of tholeiitic and alkaline magma types. *Geochim. cosmochim. Acta*, **32**, 1057–86.

—— TILTON, G. R., & HEDGE, C. E., 1964. Isotopic composition of lead and strontium from Ascension and Gough Islands. *Science, N.Y.* **145**, 1181–5.

Green, D. H., & Ringwood, A. E., 1967. The genesis of basaltic magmas. *Contr. Miner. Petrol.* **15,** 103–90.

Griffin, W. L., & Murthy, V. R., 1968*b*.   Distribution of K, Rb, Sr and Ba in some minerals relevant to basalt genesis. *Preprint.*

Harris, P. G., 1957. Zone-refining and the origin of potassic basalts. *Geochim. cosmochim. Acta,* **12,** 195–208.

—— 1967. Segregation processes in the upper mantle, pp. 305–17 in *Mantles of the earth and terrestrial planets,* ed. Runcorn, S. K. New York: Interscience.

Hedge, C. E., 1966. Variations in radiogenic strontium found in volcanic rocks. *J. geophys. Res.* **71,** 6119–26.

Ito, K., & Kennedy, G. C., 1967. Melting and phase relations in a natural peridotite to 40 kilobars. *Am. J. Sci.* **265,** 519–38.

Kushiro, I., Syono, Y., & Akimoto, S., 1967. Stability of phlogopite at high pressures and possible presence of phlogopite in the Earth's upper mantle. *Earth planet. Sci. Lett.* **3,** 197–203.

Lambert, I. B., & Wyllie, P. J., 1968. Stability of hornblende and a model for the low velocity zone. *Nature, Lond.* **219,** 1240–1.

Manson, V., 1967. Geochemistry of basaltic rocks: major elements, pp. 215–69 in *Basalts: the Poldervaart treatise on rocks of basaltic composition,* eds. Hess, H. H. & Poldervaart, A. New York: Interscience.

Moorbath, S., & Welke, H., 1968. Lead studies on igneous rocks from the Isle of Skye, northwest Scotland. *Earth planet. Sci. Lett.* **5,** 217–30.

O'Hara, M. J., 1968*a*. The bearing of phase equilibria studies in synthetic and natural systems on the origin and evolution of basic and ultrabasic rocks. *Earth-Sci. Rev.* **4,** 69–133.

—— & Yoder, H. S., Jr., 1967. Formation and fractionation of basic magmas at high pressures. *Scott. J. Geol.* **3,** 67–117.

Pankhurst, R. J., 1969. Strontium isotope studies related to petrogenesis in the Caledonian basic igneous province of N.E. Scotland. *J. Petrology,* **10,** 115–43.

Powell, J. L., Faure, G., & Hurley, P. M., 1965. Strontium 87 abundance in a suite of Hawaiian volcanic rocks of varying silica content. *J. geophys. Res.* **70,** 1509–13.

Taylor, S. R., 1964. Trace element abundances and the chondritic Earth model. *Geochim. cosmochim. Acta,* **28,** 1989–98.

35

ANCIENT AND MODERN VOLCANIC ROCKS: A TRACE ELEMENT MODEL

S.R. HART

*Department of Terrestrial Magnetism,
Carnegie Institution of Washington, USA*

C. BROOKS

*Department of Geology, University of
Montreal, Canada*

T.E. KROGH and G.L. DAVIS

*Geophysical Laboratory, Carnegie Institution
of Washington, USA*

and

D. NAVA

*Goddard Space Flight Center, Greenbelt,
Maryland, USA*

Received 3 August 1970
Revised version received 19 October 1970

In terms of the relative abundance of K, Rb, Cs, Sr and Ba, modern oceanic and island-arc basalts may be viewed as a gradational family ranging from alkali basalt (enriched in these elements) to low-K-tholeiite, such as ocean-floor basalt (depleted in these elements). In island arcs, the contents of these trace-elements in basalts are proportional to the depth to the seismic zone. Our model for this relationship suggests derivation of alkali basalts from non-depleted mantle by low degrees of partial melting at relatively great depths; the low-K tholeiites are satisfactorily explained as products of high degrees of partial melting of residual or "depleted" mantle at relatively shallow depths. Some of these tholeiites are shown to be contaminated with oceanic sediment, and the usefulness of Cs and Ba concentrations to detect such contamination is established.

New trace-element analyses reveal that the basalts of the Archaean (2.7 by) greenstone belts of Canada average 2100 ppm K, 6 ppm Rb, 0.4 ppm Cs, 175 ppm Sr and 70 ppm Ba. These trace-element abundances show that the Archaean basalts are most closely related to the modern low-K tholeiites of island arcs. This analogy leads us to propose that Archaean basalts are derived by high degrees of melting at shallow depths. We find no evidence for the presence of residual or "depleted" mantle during the Archaean. Our model implies a thin lithosphere during the Archaean and we suggest that Archaean volcanics evolved in an environment dominated by multiple thin oceanic plates with arc-trench boundaries and shallow subduction zones.

251

36

Trace element geochemistry of Archean volcanic rocks and its implication for the chemical evolution of the upper mantle

by Bor-Ming JAHN *

Mots clés. — Analyse mineurs, Roche volcanique, Archéen, Protérozoïque inf., Actuel, Fusion partielle, Manteau sup. Régénération.

Monde.

Résumé. — Les roches volcaniques archéennes (âge $\geq$ 2,5 b.a.) apparaissent habituellement sous forme de ceintures de roches vertes (« greenstone belts ») et on les trouve dans tous les continents. Elles sont en général entourées par de vastes étendues de roches granitiques. Dans beaucoup de ces ceintures de roches vertes, les basaltes à composition tholéiitique dominent largement sur les volcanites à compositions andésitique et felsitique. A la partie inférieure des ensembles volcaniques, on trouve fréquemment des roches komatiitiques riches en Mg. Ces laves à composition de komatiite péridotitique impliquent l'existence, à l'intérieur du globe, d'un régime thermique beaucoup plus élevé à l'Archéen qu'à l'heure actuelle.

Les quantités d'éléments traces (K, Rb, Sr, Ba, Ni, Cr et Terres Rares) dans les roches volcaniques archéennes et dans leurs équivalents actuels sont passées en revue. De cet examen, il ressort que les spectres de répartition des éléments traces ne sauraient constituer de bons marqueurs du cadre géotectonique dans lequel se sont formées les volcanites archéennes. S'il est bien établi que les ceintures archéennes de roches vertes correspondent à un environnement océanique, certaines d'entre elles sont apparues dans un contexte de ride océanique, d'autres dans des contextes différents.

Bien que la région source des basaltes des rides médio-océaniques (MORB) et de certaines roches archéennes soit appauvrie en certains éléments traces, globalement la quantité en éléments traces ne semble pas avoir varié de l'Archéen à l'Actuel. Dans ces conditions, il est possible que dans les 200 km supérieurs du Manteau, là où se produisent en majorité la fusion partielle et l'individualisation des magmas, la composition globale du manteau soit, en gros, restée constante. Pour expliquer cette constance, quatre hypothèses sont proposées. Le modèle retenu fait intervenir à la fois un apport d'éléments traces en provenance du manteau profond ($>$ 200 km) et un recyclage du système croûte-manteau supérieur.

Abstract. — Archean volcanic rocks ($\geq$ 2.5 b.y. of age), commonly occur as greenstone belts, are distributed in all continents. They are usually surrounded by vast areas of granite rocks. In most greenstone belts, basalts of tholeiitic composition predominate, andesitic and felsic volcanic rocks constitute the remainder. Locally, high-Mg komatiitic rocks occur ; usually in the lower part of volcanic sequence. The occurrence of peridotitic komatiite lava suggests a much higher thermal regime in the Archean than the present time.

Trace element abundances (K, Rb, Sr, Ba, Ni, Cr. and rare earth elements) in Archean volcanic rocks and their modern analogues are reviewed. It is concluded that trace element distribution patterns are not a good indicator of tectonic environments in which Archean volcanic rocks were formed. It is certain that Archean greenstone belts represent some kind of oceanic environment and that some of them were formed in ocean-ridge type of environment and some in other types.

Although the source of modern mid-ocean ridge basalts (MORB) and some sources of Archean rocks are demonstrated to be depleted in some trace elements, the trace element abundances in Archean and modern volcanic rocks are similar in a general fashion. It is suggested that the top 200 km of the upper mantle, in which partial melting and magma segregation most commonly take place, has essentially maintained a gross compositional constancy throughout the geologic age. Four model interpretations are presented for this phenomenon. It is concluded that the gross constancy is resulted from the interplay of replenishment of trace element from below the 200 km depth and some degree of recycling of the crust-mantle system.

---

* Centre armoricain d'étude structurale des socles (C.N.R.S.). Univ. de Rennes, Inst. de géologie, 35042 Rennes Cedex, France.

Note présentée oralement le 6 décembre 1976 ; manuscrit définitif remis le 28 novembre 1977.

I. — INTRODUCTION.

Trace element abundances of Archean volcanic rocks have often been compared with those of their modern equivalents, and therefore, various tectonic schemes have been proposed for the Archean greenstone belts [e.g. Hart *et al.*, 1970 ; White *et al.*, 1971 ; Jahn *et al.*, 1974 ; Condie, 1976].

The distribution and nature of Archean crust have been summarized by Goodwin [1976]. Age determinations and isotopic characteristics of Archean crust are well reviewed by Moorbath [1977]. An overall review, in particular on models of continental evolution, was given in a book by Windley [1977].

In this paper I shall review some important geochemical characteristics of Archean volcanic rocks, and then examine the applicability of these chemical data to the assignments of ancient tectonic environments.

II. — GEOCHEMICAL CHARACTERISTICS.

There are at least four tectonic environments which have been proposed to represent the environments in which Archean greenstone belts were formed. These are : (*a*) oceanic ridges ; (*b*) island arcs ; (*c*) marginal basins ; (*d*) continental rift zones. Quite predictably, none of them alone can satisfy all the criteria of geochemical characteristics and field relationships in the Archean greenstone belts. In terms of major element content, the most significant difference between Archean volcanic rocks and their modern equivalents are the higher Fe/Mg ratios and lower Al_2O_3 contents of the Archean volcanics, a trend which persists across the spectrum of volcanic rocks [Glikson, 1971, 1976]. The Fe/Mg ratios of Archean tholeiites resemble those of olivine-depleted oceanic tholeiites [Miyashiro *et al.*, 1969 ; Shido *et al.*, 1971] and island arc tholeiites [Jakes and Gill, 1970]. Thus it appears that the relatively high Fe/Mg may be a characteristic of the primary undifferentiated magma, possibly signifying lesser degree of iron segregation to form the earth's core [Glikson, 1971]. The lower Al contents of Archean tholeiites ($\sim$ 14-15 % Al_2O_3) may suggest a shallow level of magma equilibrium, because of the breakdown of plagioclase under pressures greater than 5 kb [Green and Ringwood, 1967] could lead to Al enrichment in the magma. The trace element characteristics for Archean volcanic rocks is reviewed below :

A) *K, Rb, Sr and Ba abundances.*

In general, K, Rb, Sr and Ba are considered and have been demonstrated to be more mobile than the more refractory rare earth elements and some transition metals during the processes of sea-water alteration and low-grade metamorphism [Hart R. A., 1970, 1973 ; Hart, 1969, 1971 ; Hart *et al.*, 1974 ; Frey *et al.*, 1974]. In their study of basaltic glass and crystalline portions of DSDP cores, Frey *et al.* [1974] observed and summarized the following chemical effects on these rocks during the process of sea-water alteration : (1) The conversion of glass to palagonite results in a much greater loss of Na, Ca, Mg, Mn and Si and larger enrichment in K, H_2O, total Fe, and Fe^{3+} than the alteration of crystalline rocks. Larger change of trace element abundance were also found in glasses than in crystalline rocks. (2) Sea-water alteration causes enrichment in K, Pb, Cu, Ba, Sr, B, Li and Rb ; Cr, Ni and V tend to be slightly depleted ; but Co, Y, Zr, Hf and Sc remain unchanged. (3) HREE tend to be depleted by alteration, but there is no relative fractionation among HREE. LREE are fractionated by glass alteration in some cases, particularly in palagonites. Even in crystalline portions, LREE (especially La and Ce) are subjected to change up to more than a factor of two, thus alteration processes could have changed the REE distribution from LREE depleted pattern to LREE enriched pattern and thus tholeiitic basalts would appear to be more alkalic.

According to these observations, except for a few elements like Co, Zr, Hf and Sc, almost all trace elements would be subject to various degrees of change in sea-water alteration. This makes it difficult to asses precisely the original magma composition for all volcanic rocks deposited in oceanic environments, regardless of geologic ages. However, the situation may not be so severe, when depth and degree of alteration is not great. Regarding Archean volcanic rocks, erosional processes may have exposed fresher parts that were not initially subject to chemical alteration. The commonly observed flat and parallel chondrite-normalized REE patterns in Archean tholeiites may represent the magmatic REE distribution. The prevailing low-grade metamorphism poses another problem for trace element redistribution. How trace elements would behave in low-grade metamorphism after sea-water alteration is a long-standing problem. In terms of K, Rb, Sr and Ba, both enrichment and depletion phenomena may have occurred in Archean greenstones, as to be seen below. For individual samples, open system behaviour for these four elements has been demonstrated both in concentration and in isotopic composition [see figs. 3 and 4 ; Jahn and Murthy, 1975 ; Jahn and Condie, 1976]. However, if the coverage of sample analysis is large enough the positive and negative fluxes of elemental migration (i.e. enrichment and depletion) may have cancelled out each other, and the average may

Sample No	K (ppm)	Rb (ppm)	Sr (ppm)	Ba (ppm)
(A) Tholeiites				
(i) Rhodesia				
RH-12	1741	5.47	103.7	50.0
RH-38	809	2.21	107.9	40.4
RH-78	1222	2.05	81.2	47.8
RH-26	1910	5.14	357.9	123.8
RH-33	1935	5.25	251.9	54.3
RH-168	17.4	0.066	126.8	4.55
RH-170	561	1.65	373.7	82.2
RH-147	1639	5.57	113.8	32.4
RH-152	2023	3.95	172.1	58.5
RH-154	80.7	0.239	247.3	30.8
RH-51	3615	17.49	47.1	140.1
RH-53	1009	1.545	88.9	28.7
RH-56	5922	35.6	16.0	17.2
RH-59	2053	5.74	88.6	17.2
RH-60	1314	1.144	106.1	16.4
RH-64	1159	0.704	98.6	14.8
RH-67	1086	3.24	114.6	62.1
RH-68	1219	2.44	97.7	49.0
RH-70	795	0.683	72.0	186.4
RH-116	765	2.48	99.1	15.6
RH-117	92.7	0.263	14.3	17.0
RH-118	1931	7.17	87.8	95.4
RH-119	4698	22.6	119.2	150.2
RH-122	2265	6.33	137.9	163.1
RH-124	2850	8.51	77.2	135.1
RH-146	4050	12.3	183.0	158.0
(ii) NE Minnesota				
EG-16	1352	2.28	108.6	18.1
EG 26	2645	10.23	147.8	59.5
7201	245.4	0.58	172.4	44.3
7228	1841	7.34	86.1	48.9
7251	1969	1.21	102.4	80.5
7258	428	0.38	117.3	5.5
7360a	3229	11.88	137.8	82.0
BJ-9	2491	5.52	120.6	–
7152	1902	3.22	289.0	35.7
7194	1583	5.07	220.4	208
BJ-2	1790	4.29	107.5	51.0
MEAN	1790	5.73	135	67.4
(B) Andesites and felsic volcanic rocks				
(i) Rhodesia				
RH-23	4956	23.4	219.1	204.7
RH-29	5683	26.5	327.0	163.1
RH-169	11003	38.0	167.1	190.8
RH-140	9152	31.7	275.0	253.2
RH-143	7584	20.7	152.1	260.3
RH-151	17360	61.5	281.0	379.0
RH-153	18245	43.5	209	633
RH-52	6181	32.6	200	201.2
RH-63	6392	35.3	227.0	196.0
RH-120	8112	25.6	164	244.0
RH-126	14740	21.2	22.9	187.4
(ii) NE Minnesota				
7288	14146	31.7	136.5	231
BJ-5	14755	37.7	133.9	–
BJ-6	5913	19.2	165.4	–
BJ-13	8661	45.6	51.9	–
BJ-31	14181	59.3	252.8	507
BJ-1	6450	24.7	337	176
BJ-3	9208	34.5	186.3	–
BJ-12	11053	31.0	110.8	188
BJ-32	5063	18.9	128.3	310
MEAN	9942	33.1	187.3	270

TABLE I. — K, Rb, Sr and Ba abundances of Arcehan volcanic rocks *.

* All these values have been determined by the isotopic dilution method. Uncertainties in all cases are better than 3 %. Rhodesian data are new, NE Minnesota data are from Jahn *et al.* [1974].

approach the values of original magma composition. This approach has been taken by Hart *et al.* [1976] and Jahn *et al.* [1974] in their studies of Archean volcanic rocks. Although not strickly valid, the conclusions derived from the behaviour of these four elements are essentially consistent with those derived from the more refractory REE and transition metal distributions.

Data of K, Rb, Sr and Ba abundances in Rhodesian and NE Minnesota greenstone belts are presented in table I. The abundances of these four elements in some volcanic rocks of different tectonic environments are shown in table II for comparison. Figure 1 illustrates the K vs Rb relationship in volcanic rocks from these two greenstone belts. As expected, the data points are scattered. If the mean values were considered only, it would be quite surprising to notice that the mean values of Rhodesian greenstone belts are very close to those of NE Minnesota samples previously estimated by Jahn *et al.* [1974]. The combined means of Rhodesian and Minnesota rocks (table I) are also very similar to that determined on a 70 sample composites from the Canadian shield by Hart *et al.* [1970]. These values are also close to the grand average of more than 500 analyses of Archean volcanic rocks [Jahn *et al.*, 1974]. The proximity of this value to Rhodesian and Minnesota rocks, and other greenstone belts, probably indicates that these two belts are quite typical of Archean greenstone belts, and their close values are not fortuitous.

	K (ppm)	Rb (ppm)	Sr (ppm)	Ba (ppm)	K/Rb	Rb/Sr	Sr/Ba	K/Ba	Ni (ppm)	Cr (ppm)	Ni/Cr	Références
Archean basalts (A)	2006	5.84	161	69	343	0.036	2.33	29.1				1
Archean basalts (B)	2330	6.38	152	96	365	0.042	1.58	24.3	120	320	0.38	2
MORB	1160	1.10	136	10.5	1050	0.008	12.95	110	140	480	0.29	3
Oceanic alkali basalt	13800	33	815	498	518	0.040	1.64	27.7	15	20	0.75	4
Low-k tholeiites of Island arcs	2170	3.2	250	100	680	0.012	2.5	21.7				3
Basalt of Island Arc tholeiite series (IATS)	5000	5	200	75	1000	0.025	2.67	66.7	30	50	0.60	5
Marginal Basin basalts	2660	3.1	163	34	858	0.019	4.79	78.2	121	309	0.39	6
Andesite of calc-alkaline series	12900	30	385	270	430	0.078	1.43	47.8	18	25	0.72	5
Archean andesites + felsic rocks	9942	33.1	187.3	270	300	0.177	0.694	36.8				7

TABLE II. — K, Rb, Sr and Ba abundances and ratios of volcanic rocks from different tectonic environments.

References. 1 : Weighted average from « Grand basic » composite of 70 samples of Hart *et al.* [1970] and Table I of this paper. All values determined by the isotopic dilution method ; 2 : Average of 525 Archean basalts from Canadian Shield, W. Australian shield and Rhodesian craton ; Ni and Cr values estimated by Tarney *et al.* [1976] ; 3 : Hart *et al.* [1970] ; Ni and Cr values from Frey *et al.* [1974] ; 4 : Engel *et al.* [1965] ; Ni and Cr values are estimated from various sources ; 5 : Jakes and White [1972] ; 6 : Average of Mariana [Hart *et al.*, 1972] and Lau Basin [Hawkins, 1976]. Ba value taken from Hart *et al.* [1972] only ; 7 : From (B) of Table I.

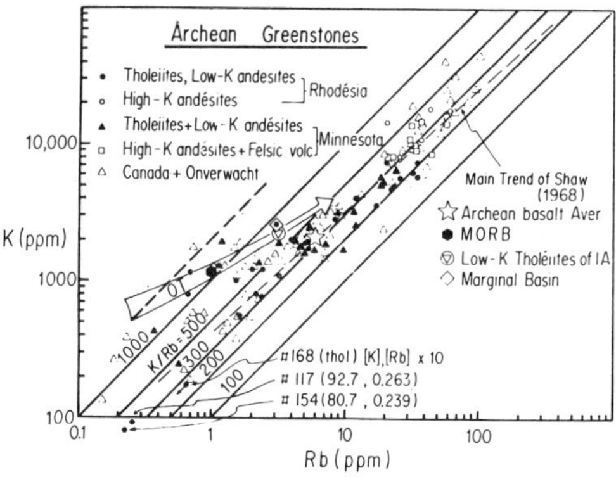

FIG. 1. — K vs Rb variation diagram for Archean volcanic rocks from Rhodesia, Minnesota, Canada and Onverwacht, S. Africa.

All values are determined by the isotopic dilution method (uncertainties < 3 %). The data clearly show severe depletion of K and Rb in some rocks (lower left end of the diagram). Enrichment is not easily detected, but must exist in some rocks. Data sources : Hart *et al.* [1970], Jahn *et al.* [1974], Jahn and Condie [1976], Jahn (unpublished). Data sources for average values, see Table I. Abbreviations : OT : oceanic trend ; LKT : low-K tholeiites of island arcs (I.A) ; MORB : mid-ocean ridge basalts.

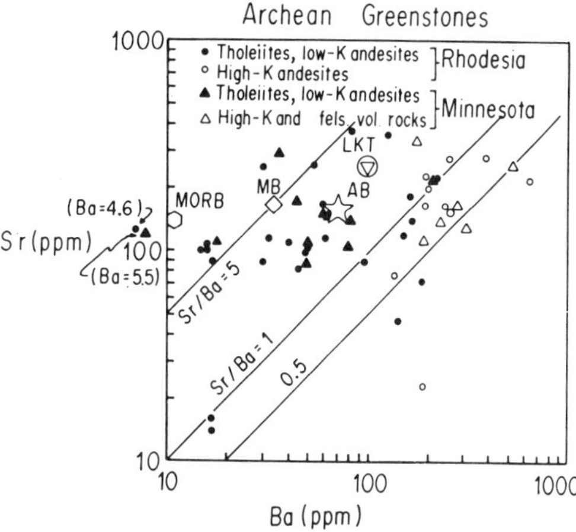

FIG. 2. — Sr vs Ba variation diagram for Archean volcanic rocks from Rhodesia and Minnesota.

Like in the case of K and Rb variation, depletion of Sr and Ba is clearly seen in some cases ; enrichment is again not easily detected but must exist. Abbreviations : MB : marginal basin basalts ; AB : Archean basalts ; others, see figure 1. Data sources : Jahn *et al.* [1974], Jahn and Condie [1976], Jahn (unpublished). For average values, see Table I.

As mentioned earlier, K, Rb, Sr and Ba tend to be enriched in seawater alteration processes and the higher values of K and Rb as compared to MORB (Mid-Ocean ridge basalt or ocean floor basalt) may be explained as due to this effect. However, a scrutiny of data in table I and figure 1 will reveal that many rocks have in fact been depleted in these elements. For example, K values in RH-168, RH-154, RH-117, 7201 are 17.4, 80.7, 92.7 and 245 ppm respectively ; Rb values in these rocks are 0.066, 0.239, 0.263, and 0.58 ppm, respectively. No modern basaltic rocks would possess such depleted abundances. Therefore, it is clear that both enrichment and depletion may have occurred in Archean volcanic rocks.

Ba is geochemically similar to K, and this is expected to be more mobile than Sr. Sr has been shown to enrich a little or not at all in seawater alteration [Hart *et al.*, 1974 ; Frey *et al.*, 1974]. Thus in low-grade Archean greenstones, Sr would be expected to scatter less than Ba if any post-crystallization alteration has been imposed upon the rocks. Indeed this is shown in figure 2. However, extremely low Sr (< 25 ppm) and Ba contents (< 6 ppm) observed in some rocks give a strong evidence of Sr and Ba depletion.

Other data points are scattered, but it is extremely difficult to assess if they reflect their original compositions or those that have been modified. Recent weathering probably does not affect the trace element contents significantly. Or isotopic ages would be severely disturbed and become unobtainable. The close ages obtained for greenstone belts and for surrounding granitic rocks attest to this point [Jahn and Murthy, 1975 ; Hawkesworth *et al.*, 1975 ; Jahn and Condie, 1976]. Therefore, if large amount of fresh samples from large geographic areas are collected and carefully analysed, a closed system value might be approached.

In both figures 1 and 2, the average values of volcanic rocks from different tectonic environments are also shown for comparison. It is seen that average Archean basalts generally distinguish themselves from MORB by having very different Rb and Ba contents. Average Archean basalts are chemically more similar to the low-K tholeiites of island arc [Hart *et al.*, 1974 ; Jahn *et al.*, 1974] or marginal basin basalts.

B) *Ni and Cr abundances.*

Because of the large partition coefficient of Ni, and perhaps Co as well, in olivine relative to silicate liquids [Hart *et al.*, 1976], the abundances of Ni, Co and Cr (controlled by pyroxenes ?) in volcanic rocks have been used to indicate possible olivine fractionation in the subduction zone of island arc environments. Modern volcanic rocks of island arc generally show lower Ni and Cr contents than MORB or marginal basin basalts (table II). Archean basalts closely resemble marginal basin basalts and, to lesser degree, MORB, but they are very different from basalts of the island arc series. Archean andesites, however, are quite similar to island arc andesites in this respect. The abundances of Ni and Cr and the Ni/Cr ratio in some Archean basalts probably suggest that olivine fractionation has not played a significant role in their generation. In other cases, particularly in rocks of komatiitic series, olivine fractionation is quite evident [e.g. Arndt *et al.*, 1977 ; Blais *et al.*, 1977].

It has been proposed that Archean greenstone belts may be formed in island arc type environments [Hart *et al.*, 1970 ; Jahn *et al.*, 1974]. Based on the lower contents of Cr, Ni and Co in recent island arc volcanics than in Archean volcanic rocks, Hawkesworth and O'Nions [1977] argued that Archean volcanic rocks cannot be formed in island arc environments. This argument may be correct but is not entirely sound, for (1) the upper mantle of Archean time and present day may be different in transition element contents, thus the partial melts produced in similar tectonic environments but different times may be different in composition ; (2) more importantly, the K_D values of Ni, and possibly Co and Cr, is strongly compositionally dependent and, to lesser degree, temperature dependent [Hart *et al.*, 1976]. Hart *et al.* [1976] have shown that D_{Ni} ($= K_D$ of Ni) of olivine changes logarithmically with MgO content of the silicate

liquid ; the higher the MgO content, the lower the D_{Ni} value. D_{Ni} also decreases slightly with increasing temperature of equilibrium. Thus, if a primary liquid is MgO rich, such as in picrites or picritic basalts, some olivine fractionation (D_{Ni} will be as low as 4) may still not drastically deplete Ni content of the liquid. That is to say, in a high Mg liquid of Archean time (also of higher geothermal gradient), olivine fractionation may be allowed to occur in an island arc type environment, but there will be no substantial Ni, and possibly Co and Cr, depletion.

C) *Rare Earth Elements.*

Rare earth elements (REE) are a group of geochemically coherent and refractory elements. Their partition coefficients between crystals and silicate liquids are better known than that of most other elements.

REE distribution in some Archean low-K tholeiitic basalts are shown in figure 3. All samples show relatively flat REE patterns, with both negative and positive Eu anomalies. Generally, the REE contents are 6 to 15 times chondritic abundances. Condie [1976] recently divided Archean tholeiites into (1) DAT (depleted Archean tholeiites), believed to be comparable to MORB and island arc tholeiites, and (2) EAT (enriched Archean tholeiites), believed to be comparable to ocean island or calc-alkaline tholeiites. He noted that DAT is the most common rock type in most greenstone belts, comprising 50

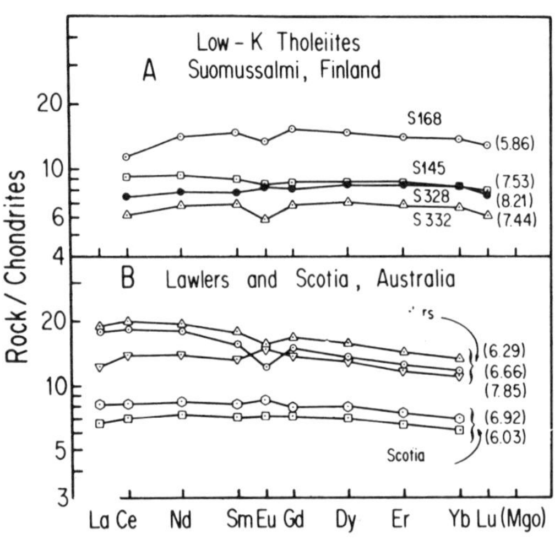

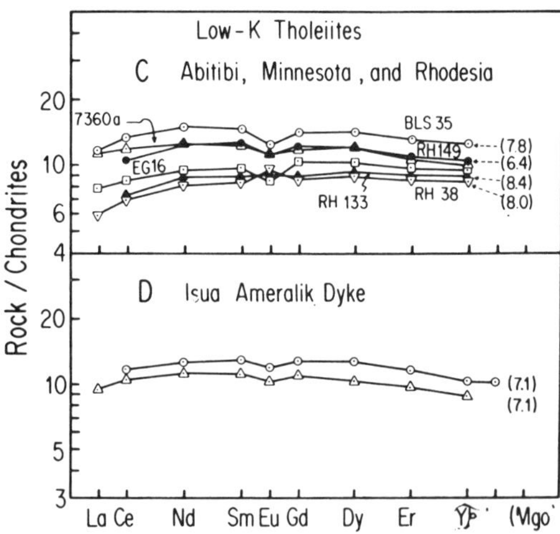

Fig. 3. — REE patterns for Archean low-K tholeiites.
MgO contents are also included. Data sources : Sun and Nesbitt [1978], Jahn *et al.* [1974], Jahn and Sun [1978] and Hawkesworth and O'Nions [1977]. For internal consistency, all values are normalized by the same chondrite values.

to 90 % of typical sections. Nevertheless, it is rather clear that all these REE characteristics are similar to those of modern MORB, island arc or marginal basin basalts. There may be a slight difference in Ce/Sm or La/Sm ratio between Archean tholeiites and MORB [Jahn *et al.*, 1974 ; Sun and Nesbitt, 1977 *a*, *b*], but this is expected because the source of MORB has been depleted in many LIL elements such as K, Rb, or Ba and LREE since about two billion years ago [Gast, 1968 *a*, *b* ; Jahn and Nyquist, 1976]. This depleted nature is also reflected in isotopic composition [Jahn and Nyquist, 1976].

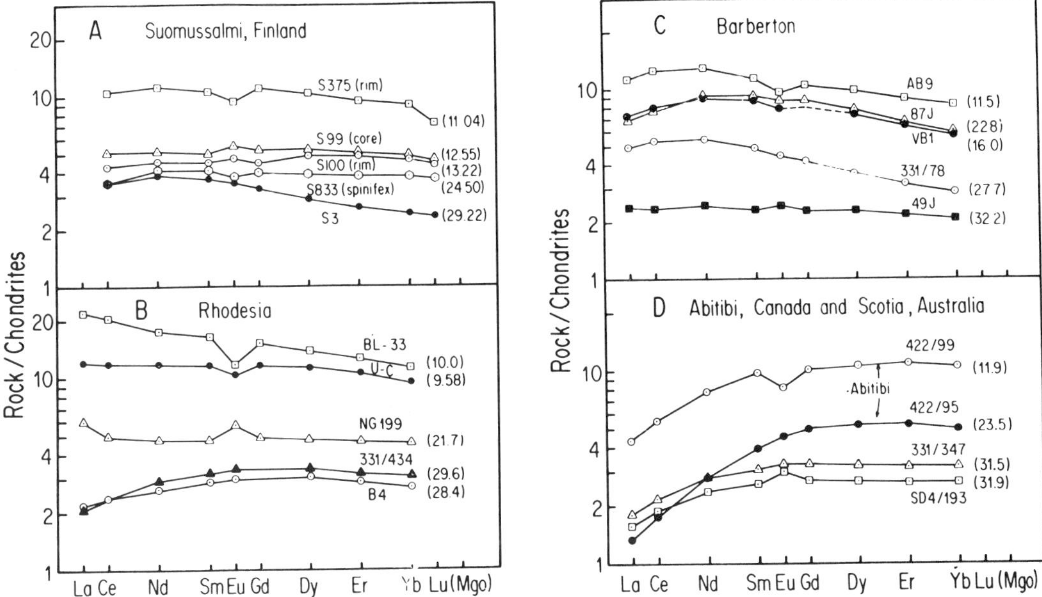

Fig. 4. — REE patterns for Archean high-Mg volcanic rocks : basaltic and peridotitic komatiites.
REE abundances generally increase with decreasing MgO contents. Data sources : Jhan and Sun [1978, for 4 a], Sun and Nesbitt [1978, for 4 b, 4 c, and 4 d].

Figure 4 illustrates the REE distribution in Archean komatiites. Komatiites of both basaltic and peridotitic varieties commonly exhibit quench textures (spinifex) [Viljoen and Viljoen, 1969 *a*, 1969 *b* ; Nesbitt, 1971 ; Pyke *et al.*, 1973]. They are Mg-rich, and commonly posses high CaO/Al_2O_3 ratios of > 1.0, but their TiO_2 contents are in general less than 1.0 % [Brooks and Hart, 1972, 1974]. Nesbitt and Sun [1976] recently argued that high CaO/Al_2O_3 ratios can be achieved by a combination of factors, such as garnet separation, Al loss or Ca addition during metamorphism, and therefore the ratio cannot be used strictly as a criterion for identifying komatiite. Nevertheless, all chemical characteristics (Ti-Zr-Nb-Y-REE) suggest that spinifex-texture peridotites (STP) represent liquids produced by very large degrees (70-80 %) of partial melting of an Archean mantle source [Nesbitt and Sun, 1976].

Generally, basaltic komatiites possess overall REE characteristics similar to Archean low-K tholeiites (fig. 3). Peridotitic komatiites are characterized by flat REE patterns, but with $(La/Sm)_N < 1.0$. Their total abundances are about 2 to 5 times chondrites.

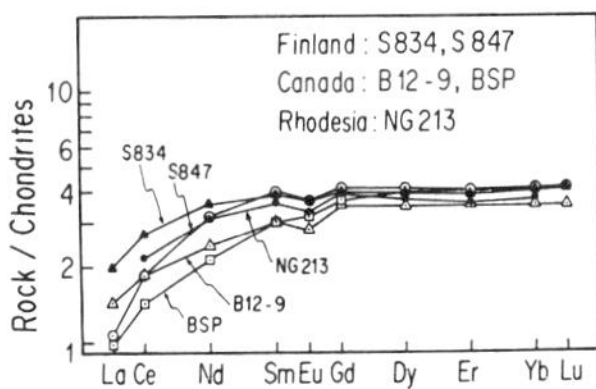

Fig. 5. — REE patterns for some « depleted » komatiites.
Strong LREE depletions are clearly demonstrated. Data sources : Jahn *et al.* (in preparation, for Finland and Canada), Hawkesworth and O'Nions [1977, for Rhodesia].

Note that in figure 4 D, two samples from Abitibi show patterns of flat HREE but strong depletion of LREE. In addition, similar data from Noranda, Canada, and Finland, are shown in figure 5. REE patterns for some komatiitic rocks from the Munro Township [Arth *et al.*, 1977] and from Rhodesia [NG 213; Hawkesworth and O'Nions, 1977] also exhibit the similar features. This observation leads to a strong implication that the depletion in the upper mantle is rather worldwide and the mantle heterogeneity existed at least 2.7 b.y. ago or earlier.

For Archean andesitic rocks, Condie [1976] has given a detailed review of their trace element geochemistry. He concluded that the association of DAT, EAT, andesite and siliceous volcanics in Archean terrains (Rhodesian craton, for example) can be best matched by the products of modern island arcs evolving at convergent plate boundaries on or near oceanic crust.

The siliceous volcanic rocks and sediments constitute only a minor part of Archean greenstone belts. Their REE characteristics and genetic implications have recently been reviewed by Jahn and Sun [1978]. They suggested that Archean siliceous volcanic rocks were not derived by fractional crystallization of basaltic magmas; rather, by a separate melting event in which garnet plays an important residual role in the source region.

III. — TECTONIC ENVIRONMENTS OF ARCHEAN GREENSTONE BELTS.

Based upon the above trace element characteristics, it is difficult to determine whether a greenstone belt was formed in any particular tectonic environment. Jahn and Sun [1978] attributed the failure of using trace element characteristics to identifying ancient tectonic setting to the following factors : (1) insufficient understanding of geochemical characteristics of trace elements or simply due to indiscriminate LIL and REE patterns in modern volcanic rocks from different tectonic settings; (2) insufficient understanding of geochemical behaviors of some transition element during partial melting and crystal fractionation under different P, T, X (composition) conditions; (3) possible evolutionary change of the upper mantle composition through time; (4) existence of mantle heterogeneities in both the Archean and the present times; and finally, (5) insufficient understanding in tectonic styles during the Archean.

Volcanic rocks in a single greenstone belt may also exhibit the chemical characteristics of more than one tectonic environment. For example, in the Onverwacht group, South Africa, the lower three formations (ultramafic unit) are comparable to the products formed on the ocean floor, but the upper three formations (mafic-felsic unit) are comparable to those formed in island arc environments. Different greenstone belts usually have different proportions of rock types. In some cases, andesites are important, but in other cases, they are almost totally lacking. It is conceivable that Archean volcanic rocks may have formed in various tectonic environments, just like their modern equivalents.

Hawkesworth and O'Nions [1977] recently prefered that the Rhodesian greenstone belts were formed in areas of continental rifting or incipient rift. Continental rift zones as tectonic environments may be partially consistent with the geologic data of greenstone belts resting upon a substantially older granitic basement, such as in Rhodesia, but they certainly cannot be the environments for many other greenstone belts with which massive contemporaneous tonalitic or granodioritic rocks are associated. The interfingering stratigraphy characterized by many Archean greenstone-granite terrains, such as in NE Minnesota [Sims, 1972] would not be easy to be explained by this rift zone model. Besides, the volcanic rocks related to modern rift zones are commonly highly alkalic in nature, such as in East Africa, and this is in quite a contrast with the composition of Archean volcanic rocks.

IV. — POSSIBLE SIGNIFICANCE ON CHEMICAL EVOLUTION OF THE UPPER MANTLE.

In the forgoing discussion I have implicitly assumed that all Archean and modern tholeiites are generated by partial melting of the upper mantle source(s). In recent years, work on the partition coefficients of REE in crystal-melt systems has shown that Sm/Nd ratios tend to decrease in liquid during partial melting events. That is, melts with flat REE patterns (e.g. some island arc tholeiites) may be derived from LREE depleted sources (e.g. some LREE-depleted peridotites), or melt with LREE-enriched pattern (e.g. alkali basalts) may be derived from sources of flat REE patterns (e.g. garnet peridotites with a few times chondritic abundances). Whatever the source is, partial melting events would tend to increase Sm/Nd ratio in the residual source. Considering the total volume extracted from the upper mantle in the last 3.5 b.y. via volcanic and plutonic activities, one would imagine that Sm/Nd ratio in the upper mantle would be increased gradually through time.

However, a recent Nd isotopic study on terrestrial samples of different geologic ages indicates that old rocks and recent basalts have evolved essentially in a mantle source of nearly constant chondritic (Sm/Nd) ratio [Depaolo and Wasserburg, 1976].

Data reported by Depaolo and Wasserburg [1976] and Richard *et al.* [1976] consistently show

that MORB have evolved in a higher Sm/Nd source
(i.e. LREE-depleted source) for some period of
time (1-2 b.y.). At present, available data suggest
that except for MORB sources, the upper mantle
has remained a nearly constant Sm/Nd ratio over
the last 3.5 b.y.

The fact that alkali basalt data lie nearly on or
slightly above the chondritic evolution line means
that alkali basalts have not evolved in a mantle
source with LREE-enriched pattern for a long
period of time, say > 1 b.y. In this respect, most
alkali basalts from either continental or oceanic
region are probably derived by very small partial
melting of mantle source with flat REE patterns
[about 2-5 times chondrites, Kay and Gast, 1973].
If the source has been enriched in LREE as advocated by Sun and Hanson [1975 a], the enrichment
eventhas to have occurred in a fairly recent time.

Partial melting not only tends to increase Nd/Sm
ratio but almost all trace elements in the melt.
After several billions of years of igneous activities,
the modern mantle source would be greatly deprived
of trace elements and thus appear more barren than
Archean mantle. However, when we compare the
trace element contents of Archean volcanic rocks
with modern analogues, except for a minor feature
exhibited in MORB, we find a gross compositional
constancy in the volcanic rocks of all ages. That is,
the modern upper mantle is generally not so depleted in trace elements when compared to the Archean
mantle. This apparent gross compositional constancy of the upper mantle must be explained.

In their study of Rb-Sr systematics, Jahn and
Nyquist [1976] have shown that many young
continental basalts and oceanic alkali basalts have
evolved essentially along the Main Path (with
Rb/Sr = 0.026 — 0.034) along with Archean basalts.
The Sr isotopic compositions of many continental
basalts, being somewhat higher than that of MORB,
have been argued as not being due to crustal contamination, but rather reflecting the true nature of
their source material [Brooks et al., 1976]. Thus,
except for some depleted sources (e.g. MORB,
Minnesota Archean), the upper mantle has conserved
essentially a rather constant Rb/Sr ratio (0.026 —
0.034).

All the above arguments point to a single question :
how can the average upper mantle (< 200 km, in
which partial melting takes place) maintain grossly
constant trace element abundances and ratios
throughout geologic time ? Some model interpretations are illustrated in figure 6. In the following,
the upper mantle of interest is strictly meant as
the portion (< 200 km) of upper mantle in which
partial melting commonly takes place. It is not the
whole upper mantle (< 650 km) of geophysical
definition.

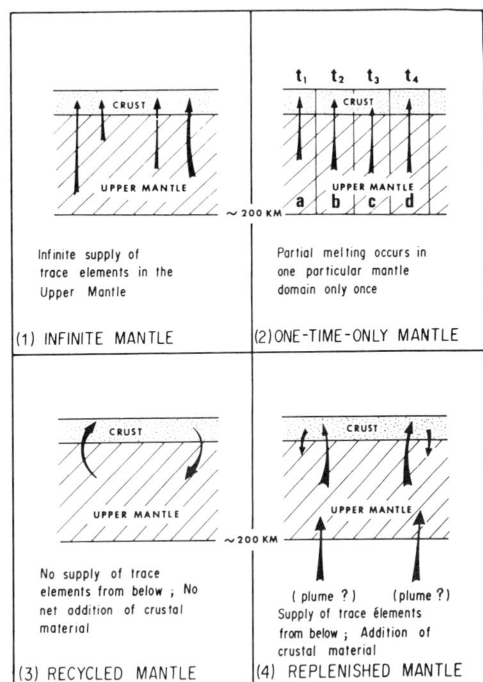

FIG. 6. — Model interpretations for the gross constancy of
the upper mantle trace element compositions.

1) *Infinite mantle* (fig. 6-1).

The upper mantle serves as an infinite reservoir
for trace elements. So no matter how much volcanic
liquid has been extracted, the upper mantle always
appears pristine. This model encounters a difficulty
because certain depleted mantle domains have been
recognized. For instances, the source regions of
MORB, the Finnish and the Superior Archean greenstone belts (figs. 4 and 5). The depletion of present
MORB source probably took place about 2 b.y.
ago based on both Sr and Pb isotopic evidence
[Jahn and Nyquist, 1976 ; Sun and Hanson, 1975 b].

2) *One-time-only mantle* (fig. 6-2).

In this model, the upper mantle is not an infinite
reservoir. Volcanic liquid is only tapped once from
each single domain of pristine mantle. That is,
domain a (fig. 6-2) from which volcanic liquid
was extracted at time t_1 will not be tapped again
at any later time. This model is dynamically unac-
ceptable due to the convective motion of the upper
mantle. Moreover, the modern MORB is a clear
example that liquid has been tapped from a source
where previous tapping has taken place long before.

3) *Recycled mantle* (fig. 6-3).

In this model, the crust-mantle system is considered closed and there is no supply of trace elements from below. The crust and the upper mantle have undergone recycling process throughout geologic time [Jahn *et al.*, 1974]. In this case, no net addition of crustal material is implied. This closed system recycling process can effectively keep the trace element abundances relatively constant in the upper mantle source. However, this by no means suggests that the recycling process is thorough and complete, because mantle heterogeneity has always existed during the past 3.5 b.y.

Recent isotopic studies of modern volcanic rocks from some island arcs indicate that recycling of continental materials in the generation of arc magmas is important [Tatsumoto, 1969; Armstrong, 1968; Armstrong and Hein, 1973]. In contrast, the studies of the Cascade Mountains and the Mariana island arc system indicate that sediment involvement in the generation of volcanic rocks is very limited [$< 2\%$, Church and Tilton, 1973; Meijer, 1976]. However, it was also argued that the source of arc volcanics may be some altered oceanic crust in the subducting zone. In a sense, the process of alteration followed by subduction to produce arc volcanics implies some degree of recycling. Of course, direct recycling of lighter continental mass back into the upper mantle would present severe physical difficulty. Shaw [1976] recently presented strong arguments favoring recycling processes for crustal evolution. At present, a proper and effective recycling mechanism is not yet understood, but the role of recycling seems to exist and it is consistent with isotopic characteristics of the crust-mantle system [Armstrong, 1968; Armstrong and Hein, 1973; Shaw, 1976].

4) *Replenished mantle* (fig. 6-4).

The crust-mantle system is no longer closed to the transport of trace elements and crustal materials. In this model, supply of trace elements from below continues throughout geologic time. This also implies a net addition of crustal materials to the system. The loss of trace elements by partial melting and subsequent removal (magma segregation) of the melts to the crustal regime is continuously replenished by the addition of these elements from lower parts of the mantle. This process may be equated to the ascending of mantle plumes. Limited recycling between the upper mantle and the crust is also allowed.

It is concluded that the replenished mantle model coupled with limited recycling process provides the best mechanism to maintain the gross compositional constancy of the upper mantle throughout geologic time.

The highly depleted nature of the modern MORB, as evidenced by their low K, Rb, Ba, LREE and very high model age, was explained as due to continuous or multistage extraction of trace elements during previous episodes of partial melting and volcanism [Gast, 1968 *b*; Tatsumoto *et al.*, 1965; Jahn and Nyquist, 1976]. This and the recognition of depleted mantle sources in the Archean [figs. 4 and 5; Jahn and Nyquist, 1976; Condie, 1976] emphasize that the convective process has not been very effective in homogenizing the upper mantle composition.

Hawkesworth and O'Nions [1977] have exercised a simple calculation and argued that the upper mantle remained essentially as an infinite reservoir for trace elements. The method of their calculation and the regime of the upper mantle of interest are somewhat misleading because in their calculation (1) possible recycled continental mass was not taken into account, thus the amount of trace elements « lost » from the upper mantle was under-estimated, and (2) the regime of the upper mantle in which partial melting and magma segregation commonly occur is more likely to be < 200 km, rather than the entire upper mantle of geophysical definition (650 km) or the entire mantle (2.900 km). Therefore, although their conclusion of compositional constancy is correct, the values reported in their table 5 are strongly under estimated.

Finally, we may consider the more general conditions regarding magma generation in Archean time. If the hypothesis that the hydrosphere and atmosphere were produced by continuous degassing of the mantle [Rubey, 1951] is correct, the Archean mantle would be considerably richer in water content. Assuming 20 per cent of the present ocean water was released from the upper mantle since ~ 3.5 b.y. ago, and if we add this water back to the top 200 km of the mantle, the upper mantle would be enriched by 0.2% water in Archean time. This additional water certainly would lower the solidus temperature of the mantle peridotites in the partial melting [Green, 1971]. Furthermore, the higher geotherm, as due to 2 to 4 times higher of heat production depending on the model chosen, in Archean time coupled with higher H_2O content would lead to magma generation at higher (shallower) levels and magmas of tholeiitic affinity produced by larger degree of partial melting. This may also explain the scarcity of alkali volcanic rocks in the Archean terrains.

At present, the available geochemical data cannot be positively used to identify the tectonic styles (plate or non-plate tectonics) prevailing in the Archean. The gross constancy of trace element abundances in the upper mantle should be taken into consideration in the future modeling of the mantle evolution.

Acknowledgments.

Discussions of Archean problems with my colleagues of this Institute (Geology/Rennes) have been very useful. I am grateful for the efficient typing of Mlle Cécile Dalibard and drafting of M. M. Lautram.

References

ARNDT N. T., NALDRETT A. J. and PYKE D. R. (1977). — Komatiitic and iron-rich tholeiitic lavas of Munro Township, Northeast Ontario. *J. Petrol.*, vol. 18, p. 319-369.

ARTH J. G., ARNDT N. T. and NALDRETT A. J. (1977). — Genesis of Archean komatiites from Munro Township, Ontario : Trace element evidence. *Geology*, vol. 5, p. 590-594.

ARMSTRONG R. L. (1968). — A model for the evolution of strontium and lead isotopes in a dynamic earth. *Rev. Geophysics.*, v. 6. 175-199.

ARMSTRONG R. L. and HEIN S. M. (1973). — Computer simulation of Pb and Sr isotope evolution of the Earth's crust and upper mantle. *Geochim. Cosmochim. Acta*, vol. 37, p. 1-18.

BLAIS S., AUVRAY B., CAPDEVILA R., JAHN B. M., BERTRAND J. M., and HAMEURT J. (1977). — The Archean greenstone belts of Karelia (eastern Finland) and their komatiitic and tholeiitic series. Archean Geochemistry, India, Nov. 1977 (in press).

BROOKS C. and HART S. (1972). — An extrusive basaltic komatiite from a Canadian Archean metavolcanic belt. *Can. J. Earth. S.*, vol. 9, p. 1250-1253.

BROOKS C. and HART S. R. (1974). — On the significance of komatiites. *Geology*, vol. 2, p. 107-110.

BROOKS C., JAMES D. E. and HART S. R. (1976). — Ancient lithosphere : its role in young continental volcanism. *Science*, vol. 193, p. 1086-1094.

CONDIE K. C. (1976). — Trace-element geochemistry of Archean greenstone belts. *Earth Sc. Rev.*, vol. 12, p. 393-417.

CONDIE K. C. and HARRISON N. M. (1976). — Geochemistry of the Archean Bulawayan Group. Midlands greenstone belt, Rhodesia. *Precam. Res.*, vol. 3, p. 253-271.

CHURCH S. E. and TILTON G. R. (1973). — Lead and strontium isotopic studies in the Cascade Mountains : bearing on andesite genesis. *Bull. Geol. Soc. Amer.*, vol. 84, p. 431-454.

DEPAOLO D. J. and WASSERBURG G. J. (1976). — Nd isotopic variations and petrogenetic models. *Geophys. Res. Lett.*, vol. 3, p. 249-252.

ENGEL A. E., ENGEL C. G. and HAVENS R. G. (1965). — Chemical characteristics of oceanic basalts and the upper mantle. *Bull. Geol. Soc. Amer.*, vol. 76, p. 719-734.

FREY F. A., BRYAN W. B. and THOMPSON F. (1974). — Atlantic ocean floor : geochemistry and petrology of basalts from Legs 2 and 3 of the deep-sea drilling project. *J. Geophys. Res.*, vol. 79, p. 5507-5527.

GAST P. W. (1968 a). — Trace element fractionation and the origin of tholeiitic and alkaline magma types. *Geochim. Cosmochim. Acta*, vol. 32, p. 1057-1086.

GAST P. W. (1968 b). — Upper Mantle chemistry and evolution of the earth's crust. *In :* The History of the Earth's Crust. R. A. Phinney ed. Princeton University Press, 15-27.

GLIKSON A. Y. (1971). — Primitive Archean element distribution patterns : chemical evidence and geotectonic significance. *Earth Planet. Sc. Lett.*, vol. 12, p. 309-320.

GLIKSON A. Y. (1976). — Stratigraphy and evolution of primary and secondary greenstones : significance of data from shields of the southern hemisphere. *In :* The early history of the earth. B. F. Windley ed. John Wiley and Sons. London, 257-277.

GOODWIN A. M. (1976). — Giant impacting and the development of continental crust. *In :* The early history of the earth. B. F. Windley ed. London, John Wiley and Sons, 77-95.

GREEN D. H. (1971). — Magmatic activity as the major process in the chemical evolution of the earth's crust and mantle. *In :* The Upper Mantle, A. R. Ritsema ed. *Tectonophysics*, vol. 13, p. 47-71.

GREEN D. H. and RINGWOOD A. E. (1967). — The genesis of basaltic magmas. *Contr. Min. Petrol.*, vol. 15, p. 103-190.

HART R. A. (1970). — Chemical exchange between seawater and deep ocean basalts. *Earth Planet. Sc. Lett.*, vol. 9, p. 269-279.

HART R. A. (1973). — A model for chemical exchange in the basalt-seawater system of oceanic layer 2. *Can. J. Earth Sc.*, vol. 10, p. 799-816.

HART S. R. (1969). — K, Rb, Cs contents and K/Rb, K/Cs ratios of fresh and altered submarine basalts. *Earth Planet. Sc. Lett.*, vol. 16, p. 295-303.

HART S. R. (1971). — K, Rb, Cs, Sr and Ba contents and Sr isotope ratios of ocean floor basalts. *Phil. Trans. Roy. Soc. London*, Ser. A, vol. 268, p. 573-587.

HART S. R., DAVIS K. E., KUSHIRO I. and BRUCE E. (1976). — Partitioning of nickel between olivine and silicate liquid. (abs.). *Ann. Meet. Geol. Soc. Amer.*, vol. 8, p. 906-907.

HART S. R., BROOKS C., KROGH T. E., DAVIS G. L. and NAVA D. (1970). — Ancient and modern volcanic rocks : a trace element model. *Earth Planet. Sc. Lett.*, vol. 10, p. 17-28.

HART S. R., GLASSLEY W. E. and KARIG D. E. (1972). — Basalts and sea floor spreading behind the Mariana island arc. *Earth Planet. Sc. Lett.*, vol. 15, p. 12-18.

HART S. R., ERLANK A. J. and KABLE E. J. D. (1974). — Sea floor alteration : some chemical and Sr isotopic effects. *Contr. Min. Petrol.*, vol. 44, p. 219-230.

HAWKESWORTH C. J., MOORBATH S., O'NIONS R. K. and WILSON J. F. (1975). — Age relationships between greenstones belts and « granites » in the Rhodesian Archean craton. *Earth Planet. Sc. Lett.*, vol. 25, p. 251-262.

HAWKESWORTH C. J. and O'NIONS R. K. (1977). — The petrogenesis of some Archean volcanic rocks from southern Africa. *J. Petrol.*, vol. 18, p. 487-520.

HAWKINS J. W. (1976). — Petrology and geochemistry of basaltic rocks of the Lau Basin. *Earth Planet. Sc. Lett.*, vol. 28, p. 283-297.

JAHN B.-M., SHIH C. Y. and MURTHY V. R. (1974). — Trace element geochemistry of Archean volcanic rocks. *Geochim. Cosmochim. Acta*, vol. 38, p. 611-627.

JAHN B.-M. and MURTHY V. R. (1975). — Rb-Sr ages of the Archean rocks from the Vermilion district, northeastern Minnesota. *Geochim. Cosmochim. Acta*, vol. 39, p. 1679-1689.

JAHN B.-M. and NYQUIST L. E. (1976). — Crustal evolution in the early earth-moon system : constraints from Rb-Sr studies. *In :* The Early History of the Earth. B. F. Windley ed. London, John Wiley and Sons, p. 55-76.

JAHN B.-M. and CONDIE K. C. (1976). — On the age of Rhodesian greenstone belts. *Contr. Min. Petrol.*, vol. 57, p. 317-330.

JAHN B. M. and SUN S. S. (1978). — Trace element distribution and isotopic composition of Archean greenstones. Second symp. on the origin of the elements. U.N.E.S.C.O., Paris (in press).

JAKES P. and GILL J. B. (1970). — Rare earth elements and the island arc tholeiitic series. *Earth Planet. Sc. Lett.*, vol. 9, p. 17-28.

JAKES P. and WHITE A. J. R. (1972). — Major and trace element abundances in volcanic rocks of orogenic areas. *Bull. Geol. Soc. Amer.*, vol. 83, p. 29-40.

KAY R. and GAST P. W. (1973). — The rare earth content and origin of alkali-rich basalts. *J. Geol.*, vol. 81, p. 653-682.

MEIJER A. (1976). — Pb and Sr isotopic data bearing on the origin of volcanic rocks from the Mariana island-arc system. *Bull. Geol. Soc. Amer.*, vol. 87, p. 1358-1369.

MIYASHIRO A., SHIDO F. and EWING M. (1969). — Diversity and origin of abysal tholeiites from the Mid-Atlantic ridge near 24° and 30° north latitude. *Contr. Min. Petrol.*, vol. 23, p. 38-52.

MOORBATH S. (1977). — Ages, Isotopes and evolution of Precambrian continental crust. *Chem. Geol.*, vol. 20, p. 151-187.

NESBITT R. W. (1971). — Skeletal crystal forms in the ultramafic rocks of the Yilgarn Block, western Australia : evidence for an Archean Ultramafic liquid. *Geol. Soc. Aus. Spec. Publ.*, vol. 3, p. 331-347.

NESBITT R. W. and SUN S. S. (1976). — Geochemistry of Archean spinifex-textured peridotites and magnesian and low-magnesian tholeiites. *Earth Planet. Sc. Lett.*, vol. 31, p. 433-453.

PYKE D. R., NALDRETT A. J. and ECKSTRAND O. R. (1973). — Archean ultramafic flows in Munro Township, Ontario. *Bull. Geol. Soc. Amer.*, vol. 84, p. 955-978.

RICHARD P., SHIMIZU N. and ALLÈGRE C. J. (1976). — ^{143}Nd/^{146}Nd, a natural tracer : an application to oceanic basalts. *Earth Planet. Sc. Lett.*, vol. 31, p. 269-278.

RUBEY W. W. (1951). — Geological history of sea water. *Bull. Geol. Soc. Amer.*, vol. 62, p. 1111-1148.

SHAW D. M. (1968). — A review of K-Rb fractionation trends by covariance analysis. *Geochim. Cosmochim. Acta.*, vol. 32, p. 573-602.

SHAW D. M. (1976). — Development of the early continental crust. Part. 2 : Prearchean, Protoarchean and later eras. *In :* The early history of the earth. London, John Wiley and Sons, p. 33-53.

SHIDO F., MIYASHIRO A. and EWING M. (1971). — Crystallization of abyssal tholeiites. *Contr. Min. Petrol.*, vol. 31, p. 251-266.

SIMS P. K. (1972). — Vermilion district and adjacent areas, *in :* Geology of Minnesota. A centennial volume, P. K. Sims and G. B. Morey ed. *Minnesota Geol. Survey.*, 49-62.

SUN S. S. and HANSON G. N. (1975 *a*). — Origin of Ross Island Basanitoids and limitations upon the heterogeneity of mantle sources for alkali basalts and nephelinites. *Contr. Min. Petrol.*, vol. 52, p. 77-106.

SUN S. S. and HANSON G. N. (1975 *b*). — Evolution of the mantle : geochemical evidence from alkali basalt. *Geology*, vol. 3, p. 297-302.

SUN S. S. and NESBITT R. W. (1977). — Chemical heterogeneity of the Archean mantle, composition of the bulk earth and mantle evolution. *Earth Planet. Sc. Lett.*, vol. 35, p. 429-448.

SUN S. S. and NESBITT R. W. (1977 *b*). — Petrogenesis of Archean ultrabasic and basic volcanics : evidence from rare earth elements. *Contr. Min. Petrol.*, vol. 65, p. 301-325.

TARNEY J. (1976). — Geochemistry of Archean high-grade gneisses, with implications as to the origin and evolution of the precambrian crust. *In :* The early history of the earth. B. F. Windley ed. London, John Wiley and Sons, p. 405-417.

TATSUMOTO M. (1965). — Potassium, rubidium, strontium, thorium, uranium and the ratio of strontium-87 to strontium-86 in oceanic tholeiitic basalts. *Science*, vol. 150, p. 886-889.

TATSUMOTO M. (1969). — Lead isotope study of volcanic rocks and possible ocean-floor thrusting beneath island arcs. *Earth Planet. Sc. Lett.*, vol. 6, p. 369-376.

VILJOEN M. J. and VILJOEN R. P. (1969 *a*). — The geology and geochemistry of the Lower Ultramafic Unit of the Onverwacht Group and a proposed new class of igneous rock. *Geol. Soc. South Africa. Spec. Publ.*, vol. 2, p. 55-85.

VILJOEN R. P. and VILJOEN M. J. (1969 *b*). — Evidence for the existence of a mobile extrusive peridotitic magma from the Komati Formation of the Onverwacht Group. *Geol. Soc. S. Africa. Spec. Publ.*, vol. 2, p. 87-112.

WHITE A. J. R., JAKES P. and CHRISTIE P. M. (1971). — Composition of greenstones and the hypothesis of sea-floor spreading in the Archean. *Spec. Publ. Geol. Soc. Aust.*, 3, p. 47-56.

WINDLEY B. F. (1977). — The evolving continent. London, John Wiley and Son, 385 p.

37

Evolution of the mantle:
Geochemical evidence from alkali basalt

Shine Soon Sun, Gilbert N. Hanson
Department of Earth and Space Sciences
State University of New York at Stony Brook
Stony Brook, New York 11794

ABSTRACT

A careful consideration of the heterogeneities in the trace-element concentrations and lead and strontium isotope ratios for primary alkali basalt and nephelinite shows that the isotopic heterogeneities reflect real geochemical heterogeneities of their mantle source that have existed variously from about 1,000 to 3,000 m.y. ago. It is suggested that at present there is not mantle-wide convection and that the mantle source for alkali basalt is deeper than the presently convecting mantle. During Archean time, mantle convection was much deeper than at the present time, and this deeper convection led to hot-line tectonics that may have determined the tectonic style found in Archean greenstone belts.

INTRODUCTION

A systematic comparison of the major-element compositions of alkali basalt by Schwarzer and Rogers (1974) shows that alkali basalt in ocean islands, on continents, and back of island arcs is a primary type of magma with a similar parentage regardless of its tectonic environment. This does not include alkali basalt and high-potassic basalt typically found in the east African rift valleys or the Rhine graben, as these are geochemically distinct (S. Kesson, 1974, oral commun.). Compared to the sources for ocean-ridge basalt, the sources for alkali basalt are more enriched in the incompatible elements, that is, K, Ba, Rb, and light rare-earth elements, and the alkali basalt appears to have been derived from a less depleted mantle than the ocean-ridge basalt (Gast, 1968). Kay and Gast (1973) studied the concentrations of the rare-earth elements, Ba, Rb, and Sr in alkali basalt and nephelinite and concluded that they result from several percent or less of partial melting of a garnet-bearing mantle with a chondritic rare-earth element (REE) pattern and REE concentrations of 2 to 5 times chondrite at depths greater than 60 km and that the chemical differences between nephelinite and alkali basalt can be explained by nephelinite having formed from the same mantle source but from lower extents of partial melting. We (S. Sun and G. Hanson, in prep.) have reevaluated the evidence for the REE composition of the mantle source for alkali basalt and nephelinite and concluded that alkali basalt results from 7 to 15 percent melting, and nephelinite results from 3 to 7 percent melting, of a garnet-bearing mantle with a light REE-enriched pattern relative to chondrite, in which La is about 14 times chondrite and Yb is about 3 times chondrite. Lead and strontium isotopes from alkali basalt and tholeiite on ocean islands are generally more variable and more radiogenic than those from ocean-ridge basalt (Gast and others, 1964; Gast, 1967; Tatsumoto, 1966a, 1966b; Cooper and Richards, 1966; Doe, 1968; Oversby and Gast, 1970; Hedge and Peterman, 1970). The isotopic variations in the alkali basalt on ocean islands have been explained to be a result of contamination by pelagic sediment, frequent addition of conti-

nental crust material to the mantle, mixing of alkali-basalt source with ocean-ridge basalt source, isotopic disequilibrium during partial melting, or real heterogeneities within the mantle source that have existed for at least many hundreds of millions of years (Gast and others, 1964; Cooper and Richards, 1966; Tatsumoto, 1966b; Gast, 1967; Armstrong, 1968; Oversby and Gast, 1970; Armstrong and Hein, 1973; O'Nions and Pankhurst, 1973). The sources of alkali basalt and nephelinite may be attributed to melting of mantle in the low-velocity zone, deep-mantle plumes (Morgan, 1972), or less depleted portions of present-day convection cells.

This paper is an outgrowth of our study of the alkali basalt at Ross Island, Antarctica, as part of the Dry Valley Drilling Project (S. Sun and G. Hanson, in prep.) and the study of lead isotopes from oceanic volcanic rocks (S. Sun, in prep.). New ways of looking at the geochemical data resulted when we asked the question, "How similar in trace elements and Sr and Pb isotopes is the source of the volcanic rocks that occur on a continental crust to similar volcanic rocks on ocean islands?" With recent high-quality data from the literature (S. Sun and G. Hanson, in prep.; S. Sun, in prep.), it is possible to evaluate the proposed models for the geochemical heterogeneities found in alkali basalt and nephelinite. We find the most viable model to be one in which the isotopic heterogeneities are in the source; this has implications for present-day convection in the mantle as well as for convection during Precambrian time. The depth of convection may have had an effect on controlling the style of Archean and later tectonics.

TRACE-ELEMENT AND ISOTOPE DATA

We will assume that during melting the major and trace elements in the melt are in at least local equilibrium with the residual minerals. This seems reasonable, as Kay and Gast (1973) have explained the enrichment in light REE and incompatible elements and the relative depletion of the heavy REE by using mineral-melt distribution coefficients for an experimentally determined garnet peridotite mineralogy necessary to produce alkali basalt and nephelinite (Green, 1973). Also, if there were large and varying extents of disequilibrium, it would be difficult to explain why the major and trace elements in alkali basalt and nephelinite are so similar.

Figure 1 compares REE for primary[1] alkali basalt from Ross Island, Antarctica, with highest precision REE data from other primary alkali basalt from ocean islands and continents. The total variation for an REE for primary alkali basalt is less than a factor of 2.5 and may reflect variations in (1) REE abundances of mantle source, (2) mineralogic composition of mantle source, or (3) extents of partial melting.

---

[1] To reduce the effect of fractional crystallization, primary alkali basalt and nephelinite are considered to be those with $Mg/(Mg+Fe^{+2})$ (atomic) >0.64 and nickel contents >100 ppm. Irving (1971) showed that primary basalt equilibrated with mantle olivine (Fo_{88-92}) should have $Mg/(Mg+Fe^{+2})$ ratios of 0.68 to 0.77. Since olivine has a mineral-melt distribution coefficient for Ni of about 10 (Gast, 1968), primary melts of a mantle with about 2,000 ppm Ni (a common value for peridotite nodules) should have about 300 ppm Ni. For 10-percent fractional crystallization of equal amount of olivine and clinopyroxene, the liquidus minerals for alkali basalt will reduce the $Mg/(Mg+Fe^{+2})$ ratio in the melt by about 5 percent of the original ratio and the Ni content by about 40 percent of the original Ni content. Thus the Ni concentration in particular is very sensitive for indicating small percentages of fractional crystallization of olivine and clinopyroxene.

Figure 2 presents the lead data for ocean-island volcanic rocks and the continental Ross Island volcanic rocks and ocean-ridge basalt. The ocean-ridge basalt occupies a restricted field on the diagram, whereas the ocean-island volcanic rocks vary linearly, and the lines for each island or island group are sub-parallel. The linear relation for the ocean-island volcanic rocks may be interpreted in two ways: (1) a result of mixing, such as an alkali-basalt component mixing with an ocean-ridge basalt or a sediment component either before or after melting, or (2) the alkali basalt sources having a history of isolation, with long-term mantle heterogeneity in the $^{238}U/^{204}Pb$ ratio, the heterogeneity having existed for approximately the last 2,000 m.y. Data for a precise $^{238}U/^{204}Pb$ versus $^{206}Pb/^{204}Pb$ isochron age are sparse; however, the available data plot with very large scatter about an isochron of approximately 1,000 m.y.

Figure 3 is a $^{87}Sr/^{86}Sr$ versus $^{87}Rb/^{86}Sr$ plot of primary alkali basalt and nephelinite. For comparison purposes, a 4,600-m.y.-old isochron with an initial ratio of 0.699 and a 2,000-m.y.-old isochron with an initial ratio of 0.7015 are plotted. The primary alkali basalt and nephelinite can be seen to plot about the 2,000-m.y. isochron. For both the lead and strontium data (Figs. 2, 3), individual volcanic groups have slopes suggesting a heterogeneity of ages from about 1,000 to 3,000 m.y. but with a mode of about 2,000 m.y.

The approximate 2,000-m.y. age for the Pb and Sr isotope data would suggest that inhomogeneities in the $^{238}U/^{204}Pb$ and Rb/Sr ratios were imposed about 2,000 m.y. ago if they are not a result of recent processes. In order to determine if mixing may be a viable model, $^{206}Pb/^{204}Pb$ is plotted against $^{87}Sr/^{86}Sr$ in Figure 4. If there is any correlation between $^{206}Pb/^{204}Pb$ and $^{87}Sr/^{86}Sr$, it is an inverse relationship, and the data do not lie on a mixing line between ocean-ridge basalt and a primitive alkali basalt. For Iceland, however, where there is mixing with ocean-ridge basalt magma (Schilling, 1973; O'Nions and Pankhurst, 1973; Sun and others, in prep.), the data for a suite of volcanic rocks plot along a line between the ocean-ridge basalt and alkali basalt. Gast and others (1964) have shown that sedimentary contamination sufficient to

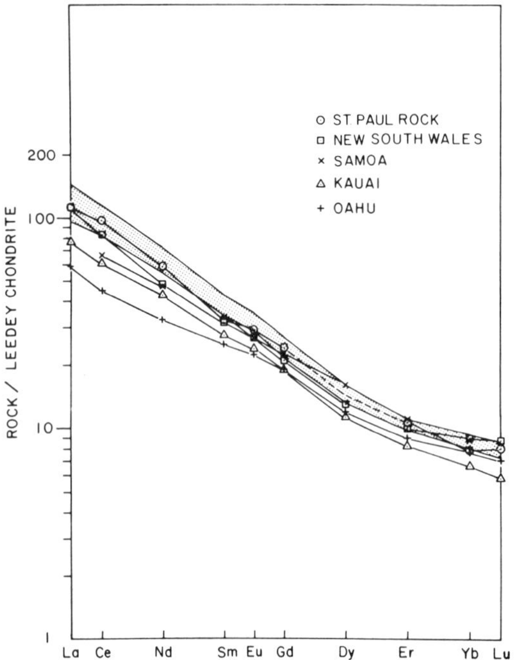

Figure 1. Comparison of REE for primary alkali basalt covering the total range in REE. The shaded field is for Ross Island, Antarctica. Data are normalized to Leedey chondrite (Masuda and others, 1973). REE data are from Frey (1970), Kay and Gast (1973), Hubbard (1971), and S. Sun and G. Hanson (in prep.).

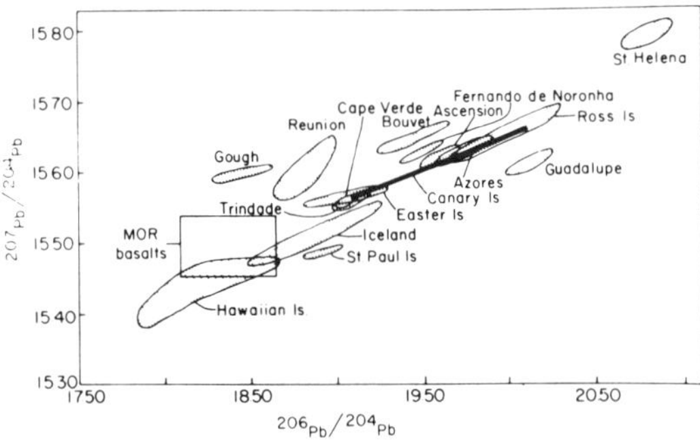

Figure 2. Plot of $^{207}Pb/^{204}Pb$ versus $^{206}Pb/^{204}Pb$ fields for ocean-island volcanic rocks not restricted to primary alkali basalt or nephelinite and for ocean-ridge basalt. Data from S. Sun (in prep.), S. Sun and G. Hanson (in prep.), Oversby (1972), and Tatsumoto (1966a).

change significantly the $^{87}Sr/^{86}Sr$ ratio for alkali basalt requires so much sediment that it would greatly change the major-element chemistry and mineralogy and would completely dominate the lead-isotope composition.

O'Nions and Pankhurst (1973) have suggested that the variations in the strontium isotope ratios might be explained by isotopic disequilibrium during partial melting of a homogeneous mantle. They suggest that during melting the minerals that melt at low temperature, presumably amphibole, phlogopite, and apatite, give their isotopic character to the melt; that is, the melt is not in isotopic equilibrium with the total residue. This leads to melts with a more radiogenic isotopic composition than the total parent. Assuming that amphibole, phlogopite, and apatite in the mantle can retain a radiogenic isotopic composition for hundreds of millions of years at mantle temperatures, the lead data in Figure 2 suggest that the mantle source has maintained at least mineral-isotopic heterogeneities for about 2,000 m.y. In the melting process presumably more radiogenic lead would first be removed from a mineral such as apatite, and as melting proceeded the less radiogenic lead from phlogopite or amphibole would be added, or vice versa. In either case, the lead-isotopic data would be along an isochron defining the time when the minerals were last isotopically homogenized. If disequilibrium were an important process, the strontium-isotope ratios in melts from one region should show an inverse relation to the extent of partial melting. In the Hawaiian Islands, however, tholeiite as well as alkali basalt, and nephelinite formed by very different extents of partial melting (Kay and Gast, 1973; Hubbard, 1969), give essentially the same $^{87}Sr/^{86}Sr$ ratio. Thus we cannot seriously consider that isotopic disequilibrium during partial melting is the main source of the isotopic variations found in alkali basalt and nephelinite. Rather, the variations in the $^{238}U/^{204}Pb$ and Rb/Sr ratios seem to be in the mantle source and have existed for 1,000 to 3,000 m.y.

Trace-element data have been used by Kay and Gast (1973) to confirm that alkali basalt and nephelinite are derived by low percentages of partial melting of garnet peridotite. It

should also be possible to test for the presence of the isotopically important minerals apatite and phlogopite in the residue at the time the melt left the residue. The cerium content can be used to determine the extent of partial melting if the mantle source for alkali basalt is relatively homogeneous, if there is no apatite in the residue, and if the numerical value for the fraction of melting is greater than the bulk distribution coefficient for the mantle.[2] If apatite is a minor phase in the mantle during melting, it will control the Ce content of the melt, as apatite has a mineral-melt distribution coefficient for Ce of about 5 (inferred from Frey and Green, 1974), whereas the other presumed mantle minerals have a bulk distribution coefficient of less than 0.01 (Kay and Gast, 1973). If there is apatite in the residue, the P_2O_5 concentrations of the melts would be independent of the extent of melting, because P is a stoichiometric element in apatite, whereas the Ce concentration would be inversely proportional to the extent of melting and percent of apatite present. However, if there is no phosphate mineral in the residue, P_2O_5 and Ce should be covariant because

---

[2]The concentration of trace elements in the magma relative to the original mantle (C_1/C_0) as a result of simple partial melting is given by Shaw (1970) as $C_1/C_0 = \dfrac{1}{D(1-F) + F}$ where D is the bulk distribution coefficient of the residue at the moment of removal of liquid from the residue. $D = \sum\limits_{a}^{n} X_a K_d^{a}$ where X_a is the weight fraction of each phase and K_d^{a} is the mineral-melt distribution coefficient. If D is very large for small percentages of melting, $C_1/C_0 = 1/D$. If D approaches zero, $C_1/C_0 = 1/F$. If zone refining is an important process, it will increase the concentration of the trace elements for alkali basalt to the limit of $C_1/C_0 = 1/D$ (Harris, 1974). In most cases the effect of zone refining cannot be distinguished from the effects of small extents of partial melting. Whether zone refining occurs or not has little effect on the conclusions derived regarding the mantle source for alkali basalt or nephelinite.

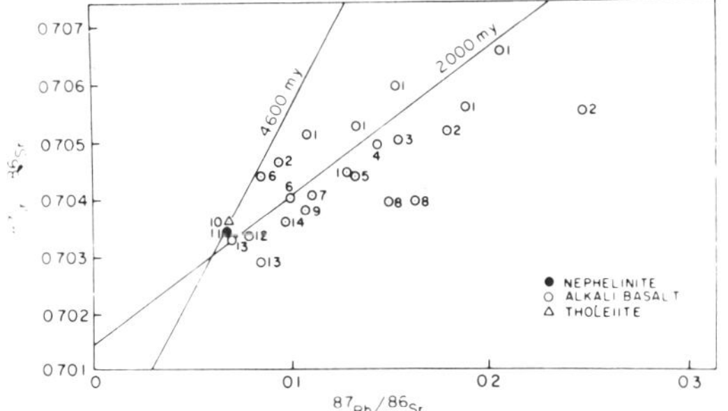

Figure 3. Plot of $^{87}Sr/^{86}Sr$ versus $^{87}Rb/^{86}Sr$ for primary alkali basalt and nephelinite for ocean islands. 1, Samoa; 2, Kerguelen; 3, Gough; 4, Tristan de Cunha; 5, Tahiti; 6, Reunion; 7, Amsterdam; 8, Crozet; 9, Eniwitok; 10, Kilauea, Hawaii (tholeiite); 11, Oahu, Hawaii; 12, St. Helena; 13, Galapagos; 14, Guadalupe. Data from Stuckless (1974, oral commun.), Hedge and Peterman (1970), Hedge and others (1973), Hedge (1974, written commun.), McBirney and Williams (1969), Peterman and Hedge (1971), Hedge and others (1972), Hart (1973), S. Sun and G. Hanson (in prep.), and O'Nions and Pankhurst (1974).

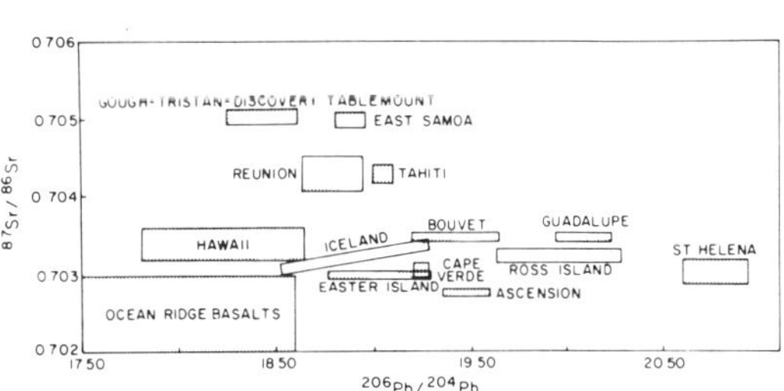

Figure 4. Plot of $^{87}Sr/^{86}Sr$ versus $^{206}Pb/^{204}Pb$ for volcanic rocks (not restricted to primary volcanics) from ocean islands and Ross Island. Data from references in Figure 2 and Figure 3 plus S. Sun (in prep.), Swainbank (1967), O'Nions and Pankhurst (1974), Sun and others (in prep.), McDougall and Compston (1965), Stuckless (1974, oral commun.), Grant and others (1973), and Klerkx and others (1974).

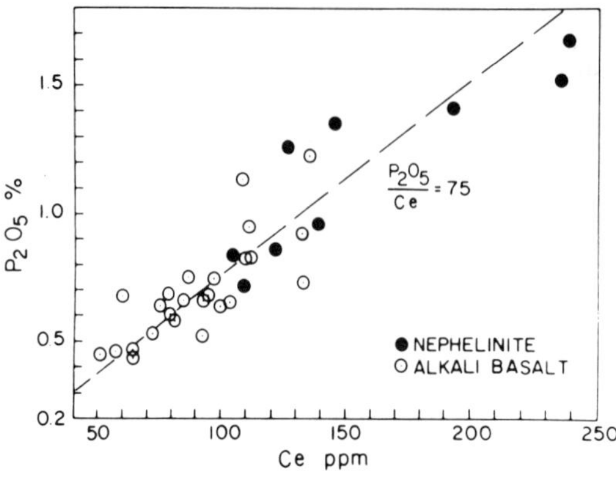

Figure 5. Plot of P_2O_5 versus cerium for primary alkali basalt and nephelinite. Data from Kay and Gast (1973), Stice (1968), Hubbard (1971), Price and Taylor (1973), Frey and Green (1974), Kesson (1973), and Goldich (1975, written commun.).

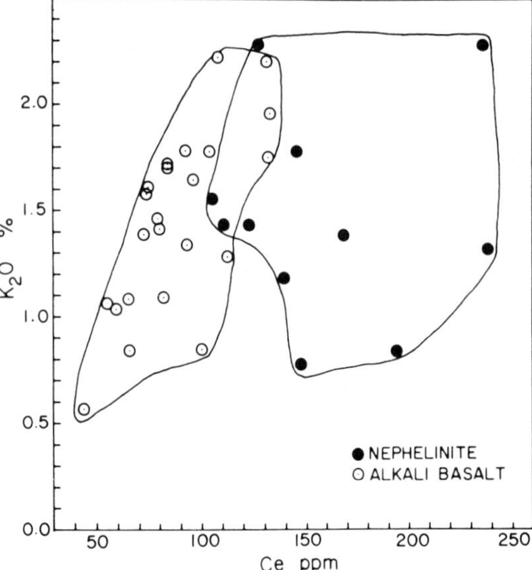

Figure 6. Plot of K_2O versus cerium for primary alkali basalt and nephelinite. Data from Philpotts and others (1972) and from references in Figure 5.

the concentration of both will be inversely proportional to the extent of partial melting as a result of dilution. Figure 5 is a plot of P_2O_5 versus Ce showing that both are roughly covariant for both alkali basalt and nephelinite. Much of the variation about the line in Figure 5 may be a result of relatively large analytical uncertainties, particularly for P_2O_5. However, from these data it can be suggested that the P_2O_5/Ce ratio in the mantle source is 75 ± 15. We conclude that apatite is probably not a mineral phase in the residue at the time of magma separation and that it should be possible to use Ce concentration as an indicator of the extent of partial melting, assuming alkali basalt and nephelinite were produced from relatively homogeneous mantle sources.

Figure 6 is a plot of K_2O versus Ce for primary alkali basalt and nephelinite. The mineral-melt distribution coefficients for K are low for all possible mantle minerals except phlogopite, for which K is stoichiometric, and perhaps amphibole. Thus if phlogopite is absent and amphibole is minor, there should be an inverse relationship between K_2O and the extent of melting, leading to a covariance of K_2O and Ce. This is the case for the alkali basalt. The nephelinite data, however, are more scattered, suggesting that K is either much more variable in the mantle source for the nephelinite or that a phase controlling K content, for example, phlogopite (Yoder and Kushiro, 1969), is a residual mineral in the mantle for nephelinite melts. The suggestion that nephelinite is a result of lower degrees of partial melting than is the alkali basalt (Green, 1973) is in agreement with the higher Ce concentration for nephelinite. For the nephelinite we interpret the K_2O-versus-Ce plot to indicate that a K-rich mineral such as phlogopite is in the residue and that the variation in the K content in nephelinite may be a function of mantle and melt composition (particularly volatiles such as F, Cl, and H_2O) as well as temperature and total pressure, which would control the stability of phlogopite and thus the concentration of potassium in the melt.

MANTLE EVOLUTION

We suggest that magma mixing, sediment contamination, or isotopic disequilibrium during partial melting are not controlling factors in determining the isotopic composition of Pb and Sr for alkali basalt and nephelinite. Rather they are controlled by heterogeneities in the mineralogy and chemistry of the mantle and history of the mantle. It is also suggested that alkali basalt and nephelinite have a mantle source which is similar to and independent of the environments in which they are found. The ocean-ridge basalt, derived presumably from the convecting mantle, has Rb/Sr ratios too low to support its ^{87}Sr/^{86}Sr ratios, lower Ba/Sr ratios relative to alkali basalt, depleted light REE concentrations relative to heavy REE, and high K/Rb ratios relative to alkali basalt (Gast, 1968). These geochemical properties suggest that the mantle sources for ocean-ridge basalt have undergone more extensive melting than have the sources for alkali basalt and nephelinite. The sources of alkali basalt and ocean-ridge basalt must differ, and mantle convection must have a geometry that isolates these sources. Figure 7 is a graphical representation showing possible mantle activity and its surficial volcanic expression.

One possible source for the alkali basalt is near the middle of a convection cell shown by plume 1, which may not have as complicated a history as the outer parts of the cell. This source would restrict alkali basalt to regions away from plate margins and therefore is not a good source for all alkali basalt and nephelinite. Another possible source is the low-velocity zone. Leeds and others (1974) suggested that the lid of the low-velocity zone comes essentially to the base of the oceanic crust near the ridge and becomes deeper away from the ridge. Many ocean islands with alkali basalt occur near ridges (source 2 in Fig. 7), where the low-velocity zone would be too shallow to have garnet present as a stable phase (Ringwood, 1969) and where a depleted mantle would be involved. The low-velocity

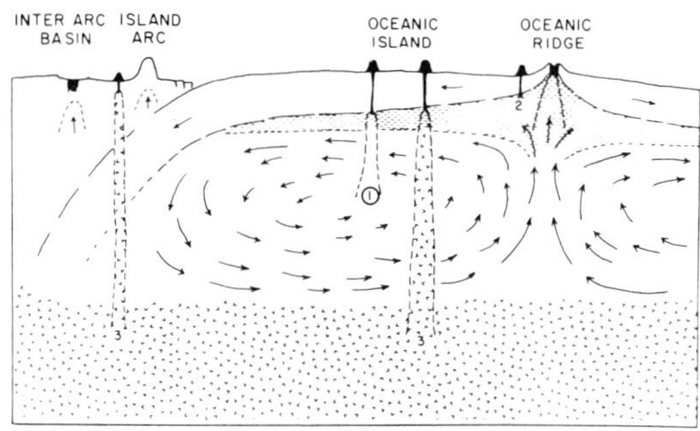

Figure 7. Diagrammatic representation of present-day mantle.
1 is possible source of alkali basalt and nephelinite in center of convection cell, 2 is possible source in low-velocity zone, and 3 is most probable source in mantle below present-day convection.

zone, therefore, is probably not the source. Unless there is some way to maintain an isolated closed system for long periods of time within a large part of the convecting mantle, we suggest that a mantle plume (Morgan, 1972; Wilson, 1973; 3 in Fig. 7) originating below the presently convecting mantle is the only viable alternative for a relatively similar source available under all tectonic environments.

If the alkali basalt is a result of mantle plumes, then this deeper mantle was involved in magma generation prior to some 1,000 to 3,000 m.y. ago, suggesting deeper convection and steeper geothermal gradients. Presumably, then, in early Pre-cambrian time the convection was much deeper, and high percentages of melted materials from this less depleted mantle were the source of basalt in the Archean greenstone belts. This suggestion is supported by the trace-element geochemistry of tholeiite in Archean greenstone belts, whose mantle sources have strong affinities to the mantle sources of present-day alkali basalt (Hart and others, 1970). Richter (1973) suggested that with steeper geothermal gradients a second-order (or may-be even a higher order) convection may occur, forming hori-zontal rolls normal to the major convection. This results in hot lines, which may be capable of breaking up a continent, rather than hot spots. Tectonically this might be similar in some ways to present-day rift valleys or interarc basins on continents

The hot lines would have associated with them massive amounts of tholeiitic magma resulting from high percentages of melting of the mantle. Basalt would be forming along hot lines on the oceanic crust as well as on the continental crust. If at the same time there was an oceanic crust acting somewhat like that of today, with ridges, island arcs, and subduction zones, the volcanic rocks associated with the hot lines on the oceanic crust would be returned to the mantle, whereas those volcanic rocks formed adjacent to or within the pre-existing continental crust would be preserved.

This model might explain why the Archean terrane is made up of low-grade greenstone belts and higher grade and sometimes older gneiss belts (Hepworth, 1971; Windley, 1973). The gneiss belts would represent the pre-existing continental crust derived by processes probably unrelated to greenstone belts (Bridgewater and others, 1974). The greenstone belts

would represent the uplifted and rifted parts that are filled with volcanics and mainly volcanogenic sediment. The later granitic rocks intrusive into the greenstones are derived by partial melt-ing of the deeper parts of the volcanic and sedimentary pile or are derivatives of partial melting of material at mantle depths (Arth and Hanson, 1975). The lowered continental crust would be the basement for sedimentary basins in which the pre-existing continental crust and sediment are metamorphosed to granulite grade by the high heat flux. Due to the high heat flux, there may be melting of the lower continental crust, resulting in synkinematic granitic intrusions at higher levels. In the later stages of the cycle, with reduction of the heat flux, lower ex-tents of melting would be represented in the greenstone belts by alkali basalt (Cooke and Moorhouse, 1969; Ridler, 1970) and by late alkalic stocks (Anhauesser and others, 1969; Arth and Hanson, 1975). At the end of the tectonic cycle, with isostatic readjustment, the rifted zones are lowered and only slightly eroded, leaving the relatively low-grade metavolcanic and metasedimentary rocks in the greenstone belts. The conti-nental crust would rebound and be more deeply eroded, ex-posing the high-grade metamorphic rocks of the gneiss belts.

This model explains many of the major features found in an Archean terrane (Anhauesser and others, 1969) and also explains the lack of some features necessary for island-arc analogies, for example, the lack of blueschist-facies rocks, Alpine-type peridotite, and major compressional tectonic features in greenstone belts.

Perhaps because of loss of heat due to extensive vol-canism during Archean time, the mantle cooled, convection became shallower, and there was separation of the present-day mantle source of ocean-ridge basalt and the mantle source of alkali basalt. The Rb-Sr and Pb-Pb ages would suggest that some parts of the mantle were isolated only for the last 1,000 m.y. Perhaps the depth of convection grew shallower until at least that time. The 1,000- to 3,000-m.y. ages for the present-day mantle source for alkali basalt and nephelinite may thus repre-sent the last time of convection or beginning of isolation of different parts of the Earth's mantle, which may or may not be related to a time of special tectonic activity. In any case, it would appear that after Archean time, the style of convec-tion changed and became more similar to that of the present.

REFERENCES CITED

Anhaeusser, C. R., Mason, R., Viljoen, M. J., and Viljoen, R. P., 1969, Reappraisal of some aspects of Precambrian shield geology: Geol. Soc. America Bull., v. 80, p. 2175–2200.

Armstrong, R. L., 1968, A model for the evolution of strontium and lead isotopes in a dynamic earth: Rev. Geophysics, v. 6, p. 175–199.

Armstrong, R. L., and Hein, S. M., 1973, Computer simulation of Pb and Sr isotope evolution of the Earth's crust and upper mantle: Geochim. et Cosmochim. Acta, v. 37, p. 1–18.

Arth, J. G., and Hanson, G. N., 1975, Geochemistry and origin of the early Precambrian crust of northeastern Minnesota: Geochim. et Cosmochim. Acta, v. 39, p. 325–362.

Bridgewater, D., McGregor, V. R., and Myers, J. S., 1974, A horizontal tectonic regime in the Archean of Greenland and its implications for early crustal thickening: Precambrian Research, v. 1, p. 179–197.

Cooke, D. L., and Moorhouse, W. W., 1969, Timiskaming volcanism in the Kirkland Lake area, Ontario, Canada: Canadian Jour. Earth Sci., v. 6, p. 117–132.

Cooper, J. A., and Richards, J. R., 1966, Lead isotopes and volcanic magmas: Earth and Planetary Sci. Letters, v. 1, p. 259–269.

Doe, B. R., 1968, Lead and strontium isotopic studies of Cenozoic volcanic rocks in the Rocky Mountain region: A summary: Colorado School Mines Quart., p. 149–174.

Frey, F. A., 1970, Rare earth and potassium abundances in St. Paul's rocks: Earth and Planetary Sci. Letters, v. 7, p. 351–360.

Frey, F. A., and Green, D. H., 1974, The mineralogy, geochemistry, and origin of lherzolite inclusions in Victorian basanites: Geochim. et Cosmochim. Acta, v. 38, p. 1023–1059.

Gast, P. W., 1967, Isotope geochemistry of volcanic rocks, *in* Hess, H. H., and Poldervaart, A., eds., Basalts: New York, Interscience Pubs., p. 325–358.

———— 1968, Trace element fractionation and the origin of tholeiitic and alkaline magma types: Geochim. et Cosmochim. Acta, v. 32, p. 1057–1086.

Gast, P. W., Tilton, G. R., and Hedge, C. E., 1964, Isotopic composition of lead and strontium from Ascension and Gough Islands: Science, v. 145, p. 1181–1185.

Grant, N. K., Powell, J. L., Walther, J. V., and Burkholder, F. R., 1973, Isotopic composition of strontium in lavas from the island of St. Helena, South Atlantic: EOS (Am. Geophys. Union Trans.), v. 54, p. 501–502.

Green, D. H., 1973, Experimental melting studies on a model upper mantle composition at high pressure under water-saturated and water-undersaturated conditions: Earth and Planetary Sci. Letters, v. 19, p. 37–53.

Harris, P. G., 1974, Origin of alkaline magmas as a result of anatexis, *in* Sorensen, H., ed., The alkaline rocks: New York, John Wiley & Sons, Inc., p. 427–436.

Hart, S. R., 1973, Submarine basalts from Kilauea rift, Hawaii: Non-dependence of trace element composition on extrusion depth: Earth and Planetary Sci. Letters, v. 20, p. 201–203.

Hart, S. R., Brooks, C., Krogh, T. E., Davis, G. L., and Nava, D., 1970, Ancient and modern volcanic rocks: A trace element model: Earth and Planetary Sci. Letters, v. 10, p. 17–28.

Hedge, C. E., and Peterman, Z. E., 1970, The strontium isotopic composition of basalt from the Gordo and Juan de Fuca Rises, northeastern Pacific Ocean: Contr. Mineralogy and Petrology, v. 27, p. 114–120.

Hedge, C. E., Peterman, Z. E., and Dickinson, W. R., 1972, Petrogenesis of lavas from western Samoa: Geol. Soc. America Bull., v. 83, p. 2709–2714.

Hedge, C. E., Watkins, N. D., Hildreth, R. A., and Doering, W. P., 1973, $^{87}Sr/^{86}Sr$ ratios in basalts from islands in the Indian Ocean: Earth and Planetary Sci. Letters, v. 21, p. 29–34.

Hepworth, J. V., 1971, Reappraisal of some aspects of Precambrian shield geology: Discussion: Geol. Soc. America Bull., v. 82, p. 803–804.

Hubbard, N. J., 1969, A chemical comparison of oceanic ridge, Hawaiian tholeiitic and Hawaiian alkalic basalts: Earth and Planetary Sci. Letters, v. 5, p. 346–352.

———— 1971, Some chemical features of lavas from the Manu'a Islands, Samoa: Pacific Sci., v. 25, p. 178–187.

Irving, A. J., 1971, Geochemical and high pressure experimental studies of xenoliths, megacrysts, and basalts from southeastern Australia [Ph.D. thesis]: Canberra, Australian National Univ.

Kay, R. W., and Gast, P. W., 1973, The rare earth content and origin of alkali-rich basalts: Jour. Geology, v. 81, p. 653–682.

Kesson, S. E., 1973, The primary geochemistry of the Monaro alkaline volcanics, southeastern Australia—Evidence for upper mantle heterogeneity: Contr. Mineralogy and Petrology, v. 42, p. 93–108.

Klerkx, J., Deutsch, S., and DePaepe, P., 1974, Rubidium, strontium content and strontium isotopic composition of strongly alkalic basaltic rocks from the Cape Verde Islands: Contr. Mineralogy and Petrology, v. 45, p. 107–118.

Leeds, A. R., Knopoff, L., and Kausel, E. G., 1974, Variations of upper mantle structure under the Pacific Ocean: Science, v. 186, p. 141–143.

Masuda, A., Nakamura, N., and Tanaka, T., 1973, Fine structures of mutually normalized rare-earth patterns of chondrites: Geochim. et Cosmochim. Acta, v. 37, p. 239–248.

McBirney, A. R., and Williams, H., 1969, Geology and petrology of the Galapagos Islands: Geol. Soc. America Mem. 118, 197 p.

McDougall, I., and Compston, W., 1965, Strontium isotope composition and potassium-rubidium ratios in some rocks from Reunion and Rodriguez, Indian Ocean: Nature, v. 207, p. 252–253.

Morgan, W. J., 1972, Deep mantle convection plumes and plate motions: Am. Assoc. Petroleum Geologists Bull., v. 56, p. 203–213.

O'Nions, R. K., and Pankhurst, R. J., 1973, Secular variation in the Sr-isotope composition of Icelandic volcanic rocks: Earth and Planetary Sci. Letters, v. 21, p. 13–21.

———— 1974, Petrogenetic significance of strontium isotope variations in volcanic rocks from the mid-Atlantic [abs.]: Paris, International meeting for geochronology, cosmochronology, and isotope geology.

Oversby, V. M., 1972, Genetic relations among the volcanic rocks of Reunion: Chemical and lead isotopic evidence: Geochim. et Cosmochim. Acta, v. 36, p. 1167–1179.

Oversby, V. M., and Gast, P. W., 1970, Isotopic composition of lead from oceanic islands: Jour. Geophys. Research, v. 75, p. 2097–2114.

Peterman, Z. E., and Hedge, C. E., 1971, Related strontium isotopic and chemical variations in oceanic basalts: Geol. Soc. America Bull., v. 82, p. 493–499.

Philpotts, J. A., Schnetzler, C. C., and Thomas, H. H., 1972, Petrogenetic implications of some new geochemical data on eclogitic and ultrabasic inclusions: Geochim. et Cosmochim. Acta, v. 36, p. 1131–1166.

Price, R. C., and Taylor, S. R., 1973, The geochemistry of the Dunedin volcano, East Otago, New Zealand: Rare earth elements: Contr. Mineralogy and Petrology, v. 40, p. 195–205.

Richter, F. M., 1973, Dynamical models for sea floor spreading: Rev. Geophysics and Space Physics, v. 11, p. 223–287.

Ridler, R. H., 1970, Relationships of mineralization to volcanic stratigraphy in Kirkland-Larder Lakes area, Ontario: Geol. Assoc. Canada Proc., v. 21, p. 33–42.

Ringwood, A. E., 1969, Composition and evolution of the upper mantle, *in* Hart, P. J., ed., The earth's crust and upper mantle: Am. Geophys. Union Mon. 13, p. 1–17.

Schilling, J-G., 1973, Iceland mantle plume: Geochemical study of Reykjanes Ridge: Nature, v. 242, p. 565–571.

Schwarzer, R. R., and Rogers, J.J.W., 1974, A world-wide comparison of alkali olivine basalts and their differentiation trends: Earth and Planetary Sci. Letters, v. 23, p. 286–296.

Shaw, D. M., 1970, Trace element fractionation during anatexis: Geochim. et Cosmochim. Acta, v. 34, p. 237–243.

Stice, G. D., 1968, Petrography of the Manu'a Islands, Samoa: Contr. Mineralogy and Petrology, v. 19, p. 343–357.

Swainbank, J. G., 1967, The isotopic composition of lead and strontium from the volcanic rocks of the islands of the South Pacific [Ph.D. thesis]: New York, Columbia Univ., 114 p.

Tatsumoto, M., 1966a, Isotopic composition of lead in volcanic rocks from Hawaii, Iwo Jima, and Japan: Jour. Geophys. Research, v. 71, p. 1721–1733.

———— 1966b, Genetic relations of oceanic basalts as indicated by lead isotopes: Science, v. 153, p. 1094–1101.

Wilson, J. T., 1973, Mantle plumes and plate motions: Tectonophysics, v. 19, p. 149–164.

Windley, B. F., 1973, Crustal development in the Precambrian: Royal Soc. London Philos. Trans., v. A273, p. 321–341.

Yoder, H. S., and Kushiro, I., 1969, Melting of a hydrous phase: phlogopite: Am. Jour. Sci., v. 267-A, p. 558–582.

ACKNOWLEDGMENTS

Reviewed by R. L. Armstrong, F. Barker, N. Carter, S. S. Goldich, C. E. Hedge, R. Kay, M. Lanphere, and Z. Peterman.

Supported mainly by the Office of Polar Programs, National Science Foundation, Grant VO–40762, and to a lesser extent by the Geology Section, National Science Foundation, Grant AO–30747.

D. Weidner, R. Kay, S. Kesson, D. H. Lindsley, and F. Hodges provided fruitful discussions, and S. S. Goldich supplied P_2O_5 analyses for some of the samples.

MANUSCRIPT RECEIVED JANUARY 29, 1975

MANUSCRIPT ACCEPTED APRIL 2, 1975

Part IV

OCEANIC BASALTS

Editors' Comments
on Papers 38 Through 45

38 KAY, HUBBARD, and GAST
Excerpts from *Chemical Characteristics and Origin of Oceanic Ridge Volcanic Rocks*

39 WHITE and BRYAN
Sr-isotope, K, Rb, Cs, Sr, Ba, and Rare-Earth Geochemistry of Basalts from the FAMOUS area

40 LANGMUIR et al.
Abstract from *Petrogenesis of Basalts from the FAMOUS Area: Mid-Atlantic Ridge*

41 SHIBATA, THOMPSON, and FREY
Excerpts from *Tholeiitic and Alkali Basalts from the Mid-Atlantic Ridge at 43°N*

42 SUN, NESBITT, and SHARASKIN
Abstract from *Geochemical Characteristics of Mid-Ocean Ridge Basalts*

43 THOMPSON et al.
Petrology and Geochemistry of Basalts and Related Rocks from Sites 214, 215, 216 DSDP Leg 22, Indian Ocean

44 SCHILLING
Abstract from *Azores Mantle Blob: Rare-Earth Evidence*

45 CLAGUE and BEESON
Excerpts from *Trace Element Geochemistry of the East Molokai Volcanic Series, Hawaii*

About a hundred years ago, there was a rock called basalt. During the later 1800s, different varieties of basalts were discovered, and the simplest subdivision was into alkalic basalts (oceanic) and subalkalic basalts (continental). Later, geologists obtained adequate descriptions of different sequences, both alkalic and tholeiitic, within oceanic islands, the only oceanic material available for sampling at that time. Deep sea exploration then produced dredged and drilled samples of the oceanic crust. By that time it was clear that there were differences between the basalts of oceanic islands and those of mid-ocean spreading centers. More sampling then demonstrated that the MORB were not even uniform in lithology but showed subtle variations in composition, mineralogy, and other properties from ridge to ridge and even within local sampling sites.

Now, after approximately one-quarter century of modern study of basalts from oceanic islands and ridges, the diversity of rock types is matched only by the diversity of explanations for their existence. The number of published papers is in the hundreds (thousands?), and it is impossible for us even to list the names of all of those workers who have made significant contributions to the field. We have chosen, therefore, excerpts from eight papers that will provide an overview of major concepts of the evolution of oceanic basaltic rocks. The papers discuss various points of view concerning the processes or source materials that control the nature of the basalt formed at some time and place within the ocean basins. Experimental petrology relevant to MORB genesis was discussed in Part II.

Six of the papers presented here are concerned primarily with typical oceanic ridge volcanism, and only one specifically discusses oceanic islands. The relationship between typical ridge segments and island platforms (such as Iceland) developed on them is a matter of particular controversy, and we call attention to a paper by Schilling et al. (1983) that attributes platform volcanism to the activity of mantle plumes.

The paper by Kay, Hubbard, and Gast (1970, Paper 38) is one of the earliest comprehensive studies of MORB. It documents the effects of low-pressure fractionation on the development of the MORB suite. Three of the following papers are from the home of that felicitous acronym, the FAMOUS area of the mid-Atlantic ridge. This area is in the transition zone between topographically normal (i.e., low-elevation) portions of the ridge and the elevated, partly emergent, Azores plateau. It thus provides an opportunity to investigate different kinds of basalts and to test the possibility that the Azores and similar features might consist of volcanic products derived from different source volumes, like plumes, than from the source for typical MORB. White and Bryan (Paper 39) indicated that the various rock types could be derived from a homogeneous source that has undergone different degrees of partial melting and later fractionation of the derived melts. Langmuir et al. (1977, the abstract from which appears as Paper 40) also proposed *dynamic* partial melting of a homogeneous source, followed by fractionation, to account for the basalt variability. Shibata, Thompson, and Frey (1979, Paper 41) described the differences between basalts of the axial valley of the ridge and those of a fracture zone and propose that elemental relationships are too complex to explain by partial melting and closed-system fractionation of a homogeneous source; they indicated the possibility of mantle heterogeneity, open-system crystallization, or magma mixing.

In addition to these papers on the mid-Atlantic ridge, we present the abstract (Paper 42) of a general summary of the chemical properties of MORB by Sun, Nesbitt, and Sharaskin (1979) and a paper on the Ninetyeast ridge by Thompson et al. (Paper 43). The Ninetyeast ridge, which is not a simple spreading center, contains basalts that are transitional between MORB and a more alkalic variety.

We should mention a major complication in the study of MORB. Considerable work has been done on interactions between erupted basalts and sea water, which is part of the question of hydrothermal activity at ridge crests. Two recent papers that review the subject are by Mottl and Seyfried (1980) and Hajash and Archer (1980). We also refer readers to the papers in Rona and Lowell (1980).

We mentioned previously that the FAMOUS area was adjacent to the Azores plateau and cited a paper by Schilling et al. (1983) concerning the importance of plumes in the development of oceanic island plateaus. An initial proposal of this concept was in an earlier paper on the Azores by Schilling (1975), and we present the abstract here (Paper 44).

Our final paper of this section deals with Hawaii, where the basalt lavas are presumably derived from source material involved in an upwelling plume. Clague and Beeson (Paper 45) discussed the relationships between the tholeiitic sequences that constitute most of the islands, the transitional and alkalic basalts that may also have been formed as separate melts from the mantle, and the differentiation products of these original magmas.

REFERENCES

Hajash, A., and P. Archer, 1980, Experimental Seawater/Basalt Interactions— Effects of Cooling, *Contrib. Mineralogy and Petrology* **75:**1–13.

Kay, R., N. J. Hubbard, and P. W. Gast, 1970, Chemical Characteristics and Origin of Oceanic Ridge Volcanic Rocks, *Jour. Geophys. Research* **75:**1585–1613.

Langmuir, C. H., J. F. Bender, A. E. Bence, and G. N. Hanson, 1977, Petrogenesis of Basalts from the FAMOUS Area: Mid-Atlantic Ridge, *Earth and Planetary Sci. Letters* **36:**133–156.

Mottl, M. J., and W. E. Seyfried, 1980, Sub-Seafloor Hydrothermal Systems: Rock- and Seawater Dominated, in *Seafloor Spreading Centers: Hydro-*

thermal Systems, Peter A. Rona and Robert P. Lowell, eds., Benchmark Papers in Geology, vol. 56, Dowden, Hutchinson & Ross, Stroudsburg, Pa., pp. 66–82 (Paper 4).

Rona, P. A., and R. P. Lowell, eds., 1980, *Seafloor Spreading Centers: Hydrothermal Systems,* Benchmark Papers in Geology, vol. 56, Dowden, Hutchinson & Ross, Stroudsburg, Pa., 424p.

Schilling, J. G., M. Zajac, R. Evans, T. Johnston, W. White, J. D. Devine, and R. Kingsley, 1983, Petrologic and Geochemical Variations along the Mid-Atlantic Ridge from 29°N to 73°N, *Am. Jour. Sci.* **283:**510–586.

Shibata, T., G. Thompson, and F. A. Frey, 1979, Tholeiitic and Alkali Basalts from the Mid-Atlantic Ridge at 43°N, *Contr. Mineralogy and Petrology* **70:**127–141.

Sun, S. -S., R. W. Nesbitt, and A. Ya. Sharaskin, 1979, Geochemical Characteristics of Mid-Ocean Ridge Basalts, *Earth and Planetary Sci. Letters* **44:**119–138.

38

Reprinted from pages 1585 and 1606-1611 of *Jour. Geophys. Research*
75:1585-1613 (1970)

Chemical Characteristics and Origin of Oceanic Ridge Volcanic Rocks

R. Kay, N. J. Hubbard, and P. W. Gast

Lamont-Doherty Geological Observatory of Columbia University
Palisades, New York 10964

The major element, alkali metal, alkaline earth, rare earth, and nickel content of mid-ocean-ridge rocks from 30° North on the mid-Atlantic ridge, the Gorda rise, the Juan de Fuca ridge, and the equatorial East Pacific rise are presented. The mid-ocean ridge volcanic rocks are characterized by olivine tholeiite normative compositions with variable Al_2O_3 and total iron contents. Chondritic rare earth element (REE) relative-abundance patterns, with some depletion in La and Ba, are found in almost all samples. Negative europium anomalies are found to correlate with high total rare earth (RE) contents. The observed variations in chemical composition are ascribed to shallow differentiation dominated by plagioclase and olivine. The major characteristics described above are ascribed to extensive partial melting ($\sim$30%) at shallow depths (15–25 km) in mantle material being intruded under the axial valley of the mid-ocean ridges. The extent of partial melting may be controlled by the total water content of the mantle.

[*Editors' Note:* Material has been omitted at this point.]

Origin of Observed Chemical Characteristics

The characteristic occurrence and chemistry of the ocean-ridge basalts suggest that these liquids may be the result of processes quite distinct from those that produce other volcanic rocks. In this section we shall discuss two aspects of the origin of these rocks: (1) low-pressure or shallow-fractional crystallization processes, and (2) partial-melting processes that produce the parent liquid of the oceanic-ridge basalts.

Green and Ringwood [1967] noted the high Al in some ocean-ridge basalts and proposed that crystallization of orthopyroxene at intermediate pressures (9 kb) or varying degrees of partial melting at this pressure could create a continuum of basalts from olivine tholeiite to high Al_2O_3 basalt. The fractionation trends calculated by Green and Ringwood at various pressures are shown on Figures 3 and 4. The relatively constant CaO content and the iron enrichment of some ocean-ridge basalts that have low alumina are not explained very well by these fractionation trends. Furthermore, neither the crystallization of orthopyroxene nor of olivine can explain the variations in total REE content and the europium anomalies in these basalts.

Nicholls [1965] and *Yoder and Tilley* [1962]

point out that resorption of plagioclase could create the high-alumina basalts from olivine tholeiites. However, in most of our examples, basalts higher in Al_2O_3 are generally higher in MgO also. Thus resorption of olivine would also be necessary. In addition, ocean-ridge basalts high in Fe and large cations would not be explained by resorption. Furthermore, if the primary magmas have no Eu anomalies (note that negative Eu anomalies might be caused by residual plagioclase in the mantle-source region), the lack of positive Eu anomalies in the basalts with higher Al_2O_3 puts a limit on the amount of plagioclase resorption.

We suggest that the sequence: high alumina basalt, olivine–tholeiite, iron-rich quartz tholeiite can be explained by crystallization of olivine and plagioclase, the common phenocryst minerals in ocean-ridge basalts, as suggested by *Melson et al.* [1968]. Experimental work by *Green and Ringwood* [1967], *Kushiro and Yoder* [1966], and *Cohen et al.* [1967] puts limits on the pressure at which olivine and plagioclase crystallize from a basalt similar to sample 5A (despite some phenocrysts, we believe 5A is a reasonable primary magma), although we note that experimental work on natural basalts has been done on samples with lower CaO and Al_2O_3 contents than 5A, and therefore that 5A can crystallize a larger amount of plagioclase before other phases appear on the liquidus.

The effect of plagioclase and olivine crystallization is shown in Figures 3 and 4, on the plots of CaO versus Al_2O_3 and FeO versus MgO. A qualitative comparison of the $CaO–Al_2O_3$ and FeO–MgO diagrams with the observed range of compositions immediately suggests that the variations, i.e., high FeO/MgO ratios, low Al_2O_3, high Ti, and high REE contents of samples such as 1A, 2A, and V2023 can be explained in terms of olivine and plagioclase crystallization. The low Ni and Eu negative anomalies are also consistent with this hypothesis. In order to evaluate the magnitude of this crystallization and in order to ascertain whether the observed effects are quantitatively consistent with each other, we shall assume that samples like 2A and 1A are derived from a liquid similar to 5A, i.e., a liquid with high Al_2O_3 content and low Fe/Mg ratio.

Using the scheme of *Philpotts and Schnetzler* [1968], we can estimate the extent of plagioclase crystallization required to generate a 10% negative europium anomaly as about 35%. Although the plagioclase phenocrysts in ocean-ridge basalts are commonly more calcic, we find empirically that 35% An_{70}, when separated from 5A together with 8% Fo_{90} produces a liquid resembling 2A in several important features: SiO_2 remains almost constant, FeO increases, MgO decreases slightly, CaO remains almost constant, Na_2O increases slightly, and TiO_2, K_2O, and the REE increase by almost a factor of two since they are excluded from crystallizing phases. Ni is lower, as is expected due to olivine crystallization: $(K_{Ni}{}^{1}/\text{olivine})$ determined by *Häkli and Wright* [1967] predicts that Ni should be a factor of three lower in the derived liquid. Sr is slightly lower, as is expected due to plagioclase crystallization: $(K_{Sr}{}^{1}/\text{Plag})$ is about 1.5 [*Philpotts and Schnetzler*, 1969]. Figures 3 and 4 show the similarity of ocean-ridge basalt and Skaergaard fractionation trends.

We conclude from the chemical data summarized here that shallow fractional crystallization, yielding Fe-enriched differentiates, has occurred in some of the ocean-ridge basalts (especially those from the Juan de Fuca ridge).

The shallow differentiation trend noted here is most evident in sample V2140, a glassy residue. The large Eu anomaly clearly indicates that extensive plagioclase has crystallized. Similarly, the extremely low Ni content indicates extensive olivine crystallization. We note that in spite of the evidence for extensive fractional crystallization, there is no change in the slope of the REE abundance pattern. It appears that shallow fractionation is an unlikely mechanism for fractionating the REE from each other. This is in accord with the work of *Haskin and Haskin* [1968] on the Skaergaard intrusive.

O'Hara [1968b] has postulated that all ocean-ridge basalts have in general lost >10% of olivine (and possibly only olivine) in order to accommodate the experimental work of *Ito and Kennedy* [1967], *Green and Ringwood* [1967] and *O'Hara and Yoder* [1967], with the assumed depth of origin of these liquids. The rather sharp cutoff in Ni and MgO contents observed here at ~200 ppm Ni and 9% MgO is in conflict with the assumption that these liquids are in general derived from more Mg- and Ni-rich liquids. We suggest that the liquids with MgO contents and Ni contents of 8–9% MgO and 180–250 ppm Ni are relatively undifferentiated. An explanation giving an origin consistent with the experimental data will be given below.

MID-OCEAN RIDGES AND ORIGIN OF BASALTIC LIQUIDS

If fractional crystallization is indeed the explanation for the high FeO, low Al_2O_3 ocean-ridge basalts, then we conclude that the parent liquids are critically saturated (hypersthene and olivine normative) and have relatively high Al_2O_3 contents (16–17.5%). The parent liquids also have extremely low contents of elements with large cations and unfractionated (chondritic) REE patterns. It is unlikely that other liquids, e.g., alkali olivine basalts (as proposed by *Engel et al.* [1965a]), or quartz tholeiites with high large ion contents, can be derived from such a liquid. *Green and Ringwood* [1967] and *Aumento* [1967] have suggested that the varying bulk-element chemistry of rocks in the vicinity of the ocean ridges may result from different degrees of partial melting (or fractional crystallization of orthopyroxene) and different depths of melting. *Gast* [1968] has suggested that the degree of partial melting is, in general, responsible for the different dispersed-element contents of undersaturated and oceanic ridge liquids, and, in particular, that the ocean-ridge basalts are produced by extensive partial melting during an adiabatic decompression associated with a convection cell under the mid-ocean ridges, whereas alkali basalts result from partial melting and segregation at lesser degrees (3–5%) of partial melting. With the present summary of chemical characteristics, it will be possible to examine these hypotheses in more detail.

In doing so, it will be important to note the effect of water on the degree of partial melting. *Harris* [1967] discussed certain topological relationships that result from introducing water into a peridotite during melting. *Kushiro et al.* [1968a, b] reported experimental results which validate the earlier discussion of Harris. The experimental work of Kushiro et al. showed, first, that when $PH_2O = P$ total, the solidus is depressed by more than 200° at pressures in excess of 20 kb and, second, that water-saturated basaltic liquids in equilibrium with olivine and orthopyroxene probably contain normative quartz. Thus for systems that contain some water, we can infer: (1) that the initial liquid will be silica saturated, i.e., quartz normative, at pressures lower than 25–30 kb, and (2) that the extent of partial melting will be largely controlled by the amount of water in the system, if hydrous phases are unstable at the basalt solidus. The experimental work of *Green and Ringwood* [1967] indicates that liquids produced by dry melting of peridotite at $P < 15$ kb will contain olivine and hypersthene in the norm. In addition, we recall that the MgO (i.e, normative olivine content) and Al_2O_3 content of liquids produced on the dry solidus will vary with the pressure where liquid and solid are equilibrated. At constant partial melting, low pressure (<9 kb), liquids have high Al_2O_3 contents and low MgO contents relative to high-pressure liquids. The absence of normative quartz in primary ocean-ridge basalts thus leads to the inference that melting has proceeded to the point where the available water was diluted by the silicate liquid, or where it occurred in a dry system.

Let us consider now the melting behavior of a volume element during an adiabatic decompression from 100 km to 15 km depth. The melting-point gradient for basalt has been determined by several workers [*Cohen et al.*,

1967; *Green and Ringwood*, 1967; *Ito and Kennedy*, 1968]. The slope found by these workers ranges from 10° to 13° per kb. *Ito and Kennedy* [1967] and *Kushiro et al.* [1968b] have also determined the basalt solidus by measuring the onset of melting in lherzolitic and garnet peridotite compositions. The gradient for the 'basalt solidus' in these two is identical and similar to that determined for basaltic compositions. *Kushiro et al.* [1968b] have also determined the temperatures for the onset of melting for a water saturated lherzolite composition. As in the case of pure phases, olivine and enstatite [*Kushiro and Yoder*, 1969], the onset of melting is depressed by several hundred degrees at pressures in excess of 20 kb. This result shows that if melting takes place in the presence of vapor, i.e., an H_2O-bearing gas phase, (1) very water-rich liquids will be formed, and (2) the amount of such a liquid phase will be a very sensitive function of the water content, of the gas phase. If, for example, water solubility ranges from 10–20% in the basaltic liquid, the amount of liquid formed for a total water content of 0.2% ranges from 1–2%. If we assume that the depression of the solidus is similarly proportional to water content at all pressures, it is readily seen that even one hundred or more degrees above the wet solidus, the extent of melting is still only 1.5–3%. That is, the extent of partial melting is insensitive to temperature over wide temperature ranges at constant pressure. This is illustrated in Figure 13, for one example. Also, if a volume element, say at point B in Figure 13, is subject to an adiabatic decompression, we see that it will undergo very little further melting initially because the solubility of H_2O in a basaltic liquid becomes more or less independent of total water pressure at high pressures. *Yoder and Kushiro* [1969] have demonstrated that water-undersaturated liquids may be formed where a hydrous phase is stable at temperatures in excess of ~1100°. Their example again involves a eutectic composition that is much richer in water than the total system. The water content of basaltic liquids cannot be inferred from their work. However, if we arbitrarily choose 5% as the water content of eutectic formed by the breakdown of the appropriate hydrous phases, we infer that ~4% of such liquid can be formed. If the water-depressed eutectic and dry melting temperatures differ by 200–300°, we can again infer that further melting will be buffered as above. Thus an adiabatic decompression of a volume element containing this amount of liquid will again undergo very little further melting until PT conditions near the solidus, i.e , 4% water-content curve, are reached. Thus both of these cases have the property that the proportion of liquid is maintained constant at fairly low values <5% over wide ranges of temperatures and pressures if the initial bulk water content of the mantle is significant but low, i.e., less than ~0.5%. Clearly, the extent of melting of the rising solid is further limited by requiring that the heat of fusion be obtained from the enthalpy of the solid and liquid. For heat capacities of 0.3 cal/g deg and from the heat of fusion of basalt, 100 cal/g, it is readily seen that in the region of 1–5% partial melting, the extent of partial melting is limited by water content since each 10° difference between adia-

bat and melting-point gradient can melt 3% of the volume element. In the vicinity of the dry solidus, the extent of melting will be controlled by the availability of heat, i.e., it will lag behind that inferred from melting-point temperature. Furthermore, the extent of melting will clearly be determined by the bulk composition, i.e., by the proportion of basalt that can be obtained from the peridotite. For the adiabats chosen in Figure 13, the heat available at 15 km is sufficient to melt approximately 30% of the initial solid. The illustrated case shows that melting may take place at pressures from 3–5 kb lower than the pressures inferred from the dry solidus because of the limitations that result from the thermal balance. The liquid should be critically saturated with respect to silica and contain virtually all the large ions in the original mantle.

McKenzie [1967], *Langseth et al.* [1966] and *Oxburgh and Turcotte* [1968] have presented

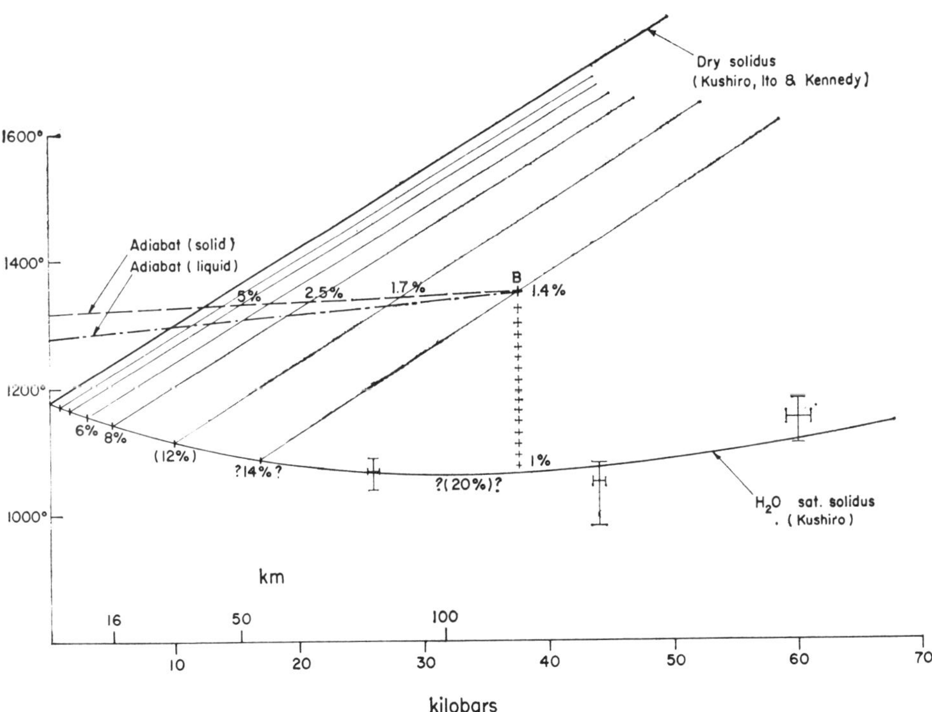

Fig. 13. Melting-point gradients for peridotite systems under dry and water-saturated conditions. Crosses on lower curved line are the experimental data of *Kushiro et al.* [1968b]. The points labeled 1%, 1.4%, 2.5%, etc. are the extent of partial melting, for a bulk H_2O content of 0.2%, if the solubility of H_2O in basaltic liquid is 20%, 14%, 8%, etc., and if the depression of the solidus is proportional to water content in the water-undersaturated region. It is also assumed that hydrous phases break down at temperatures in excess of 1,050°C.

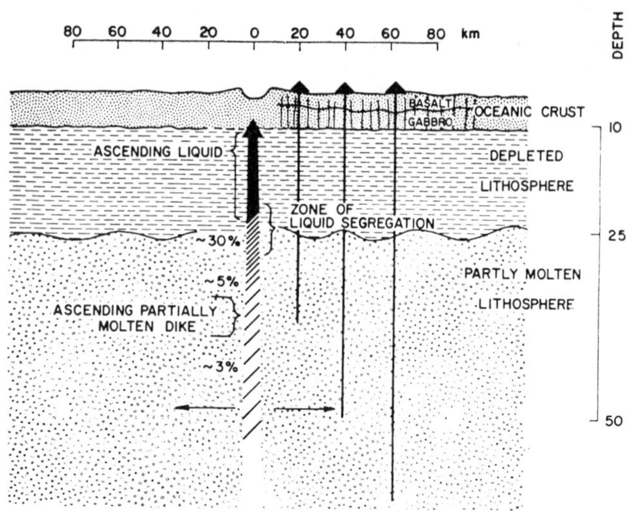

Fig. 14. Schematic of intrusion model. It is difficult to picture a dynamic process in a single picture. The width of the ascending material is clearly somewhat exaggerated in this picture. For a half spreading rate of 1.5 cm/yr, one can picture the growth of the lithosphere in terms of one 3-km dike intruded per 10^5 years. Lines of constant percentage melting are inferred from isotherms as given by *Sleep* [1969]. They are shown to emphasize the possible transitional character between asthenosphere and lithosphere in the axial zone underneath the ridges.

two somewhat different interpretations of the geophysical observations made on the mid-ocean ridges. The model of McKenzie involves the injection of sheets of material from the low-velocity zone or asthenosphere into the zone beneath the axial valley or zone of the mid-ocean ridges. In this model (intrusive) the lithosphere literally grows from a plane. *Sleep* [1969] in a later study shows that the initial temperatures chosen by McKenzie may be taken considerably higher and still satisfy the observed heat flow. The alternative model, based on more or less laminar flow of a column of ascending mantle material, involves growth of the lithosphere over a wider zone of mantle (upwelling model). The region of extensive melting due to adiabatic decompression will also be much broader in this model.

The present observation that rocks with the characteristic ocean-ridge composition may be limited to the narrow axial zone of the mid-ocean ridges and to the sharp boundaries between magnetic anomaly bands suggests that the source of volcanic liquids that are responsible for formation of most of layers 2 and 3 in the oceans is restricted to a rather narrow zone (<10 km wide) in the upper mantle. This

situation is somewhat easier to reconcile with a model similar to that of McKenzie, i.e., intrusion of dikes, than with a region of upwelling. Figure 14 shows in rather schematic form an explanation of volcanic processes around the mid-ocean ridges. The model illustrated is clearly oversimplified. The direction of flow lines below 25-km depth cannot at present be completely specified from either volcanological or geophysical restrictions. The essential features are (1) extensive melting at 25- to 15-km depth; (2) separation of liquid and solid at this depth, followed by more rapid ascent of liquid than surrounding solid, i.e., vertical segregation of liquid; and (3) retention of liquid in the partially molten region below 25 km. The designation of this partially molten region as lithosphere or asthenosphere is clearly a semantic question.

The model illustrated here, in common with most models of the mid-ocean ridges, has a broad region at depth that is partially molten below the central portion of the ridge. Several chemical features of this model should be noted. First, most of the liquid is produced at unusually low pressures, <9 kb. This is consistent with the high Al_2O_3 contents, [cf. *Green and*

Ringwood, 1967], and low MgO contents [cf. *Cohen et al.*, 1967; *Ito and Kennedy*, 1967; *Green and Ringwood*, 1967; *O'Hara*, 1968a]. The range of Al_2O_3 contents of 15–18% that can be inferred for the primary liquids implies Al_2O_3 contents ranging from 3.6–0.7% in residual orthopyroxene in a harzburgitic (60% olivine, 40% opx) residue, in a system consisting of 30% liquid and 70% residue having an initial Al_2O_3 content of 5.6%. Second, the extensive melting at depths of 15–25 km that is inferred in our model will prevent the ascent of liquid from greater depths (40–90 km underneath the axial zone of the ridge). Since the volumes of such liquids will be substantially less than those produced at shallower depths, they would generally be mixed with the shallower liquids if they actually came up. Third, if the 1–10% partial melting liquids remained with their source rocks as spreading from the plane center proceeded, a source for the liquids observed in the flanks of the ridges would be provided. The variation in chemistry as a function of distance from the ridge, reported by *McBirney and Gass* [1967] and *Aumento* [1968], could result from tapping successively greater depths, i.e., smaller degrees of partial melting, or from the partial crystallization of the lithospheric liquid at depth as it moved away from the axis of the ridge. Both the more silica-undersaturated characteristics of the seamounts and the dispersed-element characteristics can be inferred from the higher pressure [*Green and Ringwood*, 1967; *Bultitude and Green*, 1967] and smaller degree of partial melting [*Gast*, 1968] respectively.

Finally, we should note that none of the processes postulated here explain the high K/Rb or the low light REE + Ba content consistently observed in ocean-ridge basalts. *Gast* [1968] postulated that these, like the Rb/Sr and U/Pb fractionations inferred to explain variations in Sr^{87}/Sr^{86} and Pb^{206}/Pb^{204} ratios, are the result of previous partial melting episodes. The wide range of Ba/Sr ratios and Rb/Sr or K/Rb ratios observed is an indication of the extent of this effect. Among the elements investigated here, the alkali metals, particularly Rb and Cs, and Ba, appear to be most depleted in this earlier mantle differentiation. We shall not discuss details of this early mantle depletion here. It should be stated, however, that the mantle systems required to produce the inferred resi-

dues must contain phases that fractionate K and Rb, and Ba and Sr, rather effectively. The common minerals that may be involved that accomplish this are hornblende, phologopite, and plagioclase. In the context of the present model, this requires that the mantle material at depths in excess of 60–80 km have previously undergone chemical fractionation. This hypothesis also implies that such depletion of the mantle is very widespread since depletion in Ba and Rb is very common in the ocean-ridge basalts.

The model proposed here also implies a rather heterogeneous upper mantle. The upper part, Moho to $\sim$25 km, is depleted in basalt and H_2O and thus should be predominantly olivine and orthopyroxene, i.e., harzburgite. Perhaps some gabbroic dikes and sills remain in this region. The lower portions may retain much of the original basaltic fraction. However, much of this fraction, $\sim$5–10%, may be segregated into intrusive bodies. The mineral assemblage will depend on the extent to which basaltic fraction has been segregated from the peridotite; that is, it may consist of eclogite and harzburgitic peridotite rather than lherzolite or spinel peridotite. Mineralogical and chemical models of the mantle based on the occurrence of accidental xenoliths, or occasional intrusions, may be strongly biased toward the depleted layer, if the present model is correct. It seems likely that the upper 150 km of the mantle is significantly richer in Ca and Al than indicated by the average lherzolite composition inferred from nodules occurring in undersaturated basalts.

[*Editors' Note:* Only the references cited in the preceding excerpt are reproduced here.]

REFERENCES

Aumento, F., Magmatic evolution on the Mid-Atlantic ridge, *Earth Planet. Sci. Lett., 2,* 225, 1967.

Aumento, F., The mid-Atlantic ridge near 45°N, 3, Basalts from the Confederation peak, *Can. J. Earth Sci., 5,* 1, 1968.

Bultitude, R. J., and D. H. Green, Experimental study at high pressures on the origin of olivine nephelinite and olivine melilite nephelinite magmas, *Earth Planet. Sci. Lett., 3,* 325, 1967.

Cohen, L. H., K. Ito, and G. C. Kennedy, Melting and phase relations of an anhydrous natural basalt to 40 kilobars, *Amer. J. Sci., 265,* 475, 1967.

Engel, A. E. J., C. G. Engel, and R. G. Havens, Chemical characteristics of oceanic basalts and the upper mantle, *Bull. Geol. Soc. Amer.*, *76*, 719, 1965a.

Gast, P. W., Trace-element fractionation and the origin of tholeiitic and alkaline magma types, *Geochim. Cosmochim. Acta*, *32*, 1057, 1968.

Green, D. H., and A. E. Ringwood, The genesis of basalt magmas, *Contrib. Mineral. Petrol.*, *15*, 103, 1967.

Häkli, T. A., and T. L. Wright, The fractionation of nickel between olivine and augite as a geo-thermometer, *Geochim. Cosmochim. Acta*, *31*, 877, 1967.

Harris, P. G., Segregation processes in the upper mantle in *Mantles of the Earth and Terrestrial Planets*, edited by K. Runcorn, Interscience, London, 1967.

Haskin, L. A., and M. A. Haskin, Rare-earth elements in the Skaergaard Intrusion, *Geochim, Cosmochim. Acta*, *32*, 433, 1968.

Ito, K., and G. C. Kennedy, Melting and phase relations in the plane tholeiite–lherzolite–nepheline basanite to 40 kilobars with geological implications, *Contrib. Mineral. Petrol.*, *19*, 177, 1968.

Kushiro, I., and H. S. Yoder, Jr., Anorthite–forsterite and anorthite–enstatite reactions and their bearing on the basalt–eclogite transformation, *J. Petrol.*, *7*, 337, 1966.

Kushiro, I., and H. S. Yoder, Jr., Melting of forsterite and enstatite at high pressures under hydrous conditions, *Carnegie Inst. Wash. Yr. Book 67*, p. 153, 1969.

Kushiro, I., H. S. Yoder, Jr., and M. Nishikawa, Effect of water on the melting of enstatite, *Bull. Geol. Soc. Amer.*, *79*, 1685, 1968a.

Kushiro, I., Y. Syono, and S. Akimoto, Melting of a peridotite nodule at high pressures and high water pressures, *J. Geophys. Res.*, *73*, 6023, 1968b.

Langseth, M. G., X. Le Pichon, and M. Ewing, Crustal structure of mid-ocean ridges, 5, Heat flow through the Atlantic Ocean floor and convection currents, *J. Geophys. Res.*, *71*, 5321, 1966.

McBirney, A. R., and I. G. Gass, Relations of oceanic volcanic rocks to mid-oceanic rises and heat flow, *Earth Planet. Sci. Lett.*, *2*, 265, 1967.

McKenzie, D. P., Some remarks on heat-flow and gravity anomalies, *J. Geophys. Res.*, *72*, 6261, 1967.

Melson, W. G., G. Thompson, and T. H. van Andel, Volcanism and metamorphism in the Mid-Atlantic ridge, 22°N latitude, *J. Geophys. Res.*, *73*, 5925, 1968.

Nicholls, G. D., Basalts from the deep-ocean floor, *Mineral. Mag.*, *34*, 373, 1965.

O'Hara, M. J., The bearing of phase-equilibrium studies in synthetic and natural systems on the origin and evolution of basic and ultrabasic rocks, *Earth Sci. Rev.*, *4*, 69, 1968a.

O'Hara, M. J., Are ocean-floor basalts primary magma? *Nature*, *220*, 683, 1968b.

O'Hara, M. J., and H. S. Yoder, Jr., Formation and fractionation of basic magmas at high pressures, *Scot. J. Geol.*, *3*, 67, 1967.

Oxburgh, E. R., and D. L. Turcotte, Mid-ocean ridges and geotherm distribution during mantle convection, *J. Geophys. Res.*, *73*, 2643, 1968.

Philpotts, J. A., and C. C. Schnetzler, Europium anomalies and the genesis of basalt, *Chem. Geol.*, *3*, 5, 1968.

Sleep, N. H., Sensitivity of heat flow and gravity to the mechanism of sea-floor spreading, *J. Geophys. Res.*, *74*, 542, 1969.

Yoder, H. S., Jr., and I. Kushiro, Melting of a hydrous phase: Phylogopite, *Carnegie Inst. Wash. Yr. Book 67*, p. 161, 1969.

Yoder, H. S., Jr., and C. E. Tilley, Origin of basaltic magmas: An experimental study of natural and synthetic rock systems, *J. Petrol.*, *3*, 342, 1962.

39

Sr-isotope, K, Rb, Cs, Sr, Ba, and rare-earth geochemistry of basalts from the FAMOUS area

WILLIAM M. WHITE *Graduate School of Oceanography, University of Rhode Island, Kingston, Rhode Island 02881*
W. B. BRYAN *Woods Hole Oceanographic Institution, Woods Hole, Massachusetts 02543*

This article is one of a series appearing in the April and May issues of the Geological Society of America *Bulletin* on the scientific results of Project FAMOUS. These studies were undertaken in the axial area of the Mid-Atlantic Ridge between approximately 36°30′ and 37°N latitudes.

ABSTRACT

Ten basalt samples recovered from the FAMOUS area were selected so as to obtain representatives of a wide geographical and compositional range. The samples were analyzed for $^{87}Sr/^{86}Sr$, K, Rb, Cs, Sr, Ba, and rare-earths. Sr-isotope ratios fall in the narrow range of 0.70288 to 0.70307, which implies that these samples were derived from an isotopically homogeneous source. The FAMOUS area lies in a geochemical transition zone between the Azores Plateau and "normal" ridge areas south of lat 33°N. The LIL (large-ion-lithophile) and Sr-isotope geochemistry of FAMOUS basalts is thus influenced by the Azores mantle plume; this results in higher Sr-isotope and LIL concentrations in these basalts than is typical of Mid-Atlantic Ridge basalts. Trace-element distributions in FAMOUS area basalts cannot be entirely accounted for by fractional crystallization models that are based on major-element chemistry. The LIL distribution in FAMOUS basalts could be due to variable extents of partial melting. Zonation within the magma chamber may result from incomplete mixing of successive batches of magma entering the chamber and could be further enhanced by fractional crystallization. The variation in partial melting would require significant increases in mantle temperature over a relatively short period of time. According to this model, the Mount Pluto magma represents the highest degree of partial melting and may mark the initiation of a new cycle of eruptive activity in the median valley.

INTRODUCTION

Rare-earth patterns in basalts erupted along the Mid-Atlantic Ridge progressively change from light-rare-earth enriched on the Azores Plateau (lat 39°N) to light-rare-earth depleted south of lat 33°N (Schilling, 1975). Although a relatively wide range of major-element chemistry is observed at any given latitude, magmatic processes such as fractional crystallization or variable partial melting can account for only small variations in rare-earth patterns; these processes cannot account for the general trend observed. Schilling (1975) attributed the observed large-scale geochemical gradients to the presence beneath the Azores of an upwelling mantle plume that mixes southward with the asthenospheric source of normal ridge basalts. In a subsequent test of this model, White and others (1975, 1976) found progressive increases in Sr-isotope ratios, in alkali-metal and alkaline-earth trace elements, and in Rb/Sr, Rb/K, and Ba/K ratios along the Mid-Atlantic Ridge toward the Azores Plateau.

The FAMOUS area lies within the geochemical transition zone between the LIL- (large-ion-lithophile) depleted and low-$^{87}Sr/^{86}Sr$ basalts of the normal ridge area south of lat 33°N and the LIL-enriched and high-$^{87}Sr/^{86}Sr$ basalts erupted along the Mid-Atlantic Ridge transect of the Azores Plateau to the north (Fig. 1). In view of

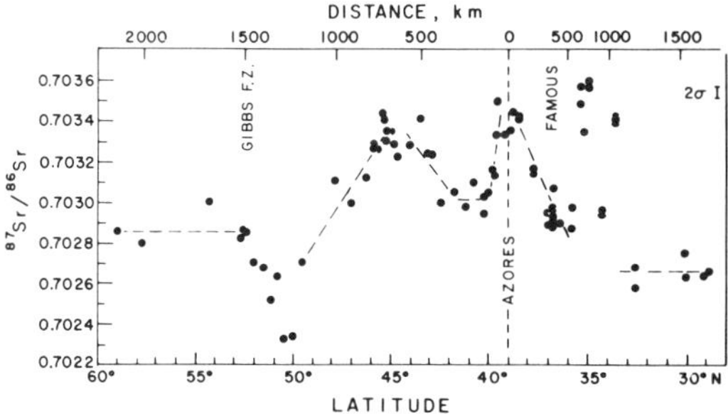

Figure 1. $^{87}Sr/^{86}Sr$ in basalts from the Mid-Atlantic Ridge versus their radial distance from the center of the Azores Plateau (lat 39°N). The FAMOUS area lies in the geochemical transition zone between the Azores Plateau and "normal" ridge areas to the south. An error bar is shown in the upper right. (After White and others, 1975.)

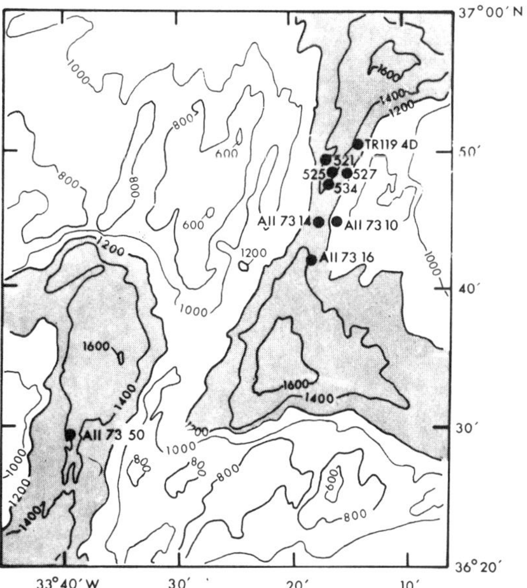

Figure 2. Bathymetric map of the FAMOUS area showing sampling sites. Contours are in fathoms; areas deeper than 1,200 fm are shaded.

281

these regional geochemical variations along the Mid-Atlantic Ridge, it is of interest to examine the geochemical variability on a local scale. The extensive and well-documented suite of basalts recovered by dredge and submersible in the FAMOUS area provides an excellent opportunity for such a study.

We report $^{87}Sr/^{86}Sr$ and concentrations of K, Rb, Cs, Sr, Ba, and rare earths for ten samples selected from dredge and submersible collections in and near the dive area. Locations are shown in Figure 2. The purpose of this paper is to present preliminary data for Sr isotopes and selected trace elements; these data are critical to relating the rock samples from the FAMOUS area to these regional patterns. Also, we will consider the implications of these data for the petrogenetic model outlined by Bryan and Moore (1977).

SAMPLING AND ANALYTICAL PROCEDURE

Samples recovered by the submersible *Alvin* were selected so as to obtain representatives from both the central volcanic areas (Mount Pluto) and the flank areas and to obtain the greatest petrologic variation. Details of sample sites and sampling methods for the *Alvin* collection are discussed by Ballard and van Andel (1977) and Bryan and Moore (1977). Dredged samples were obtained by R/V *Atlantis II* of Woods Hole Oceanographic Institution and R/V *Trident* of the University of Rhode Island.

$^{87}Sr/^{86}Sr$ analyses were performed on the 6-in. solid-source mass spectrometer at the Department of Terrestrial Magnetism, Carnegie Institution of Washington. All values have been normalized to $^{86}Sr/^{88}Sr = 0.11940$ and are reported relative to a value of 0.70800 for the Eimer and Amend standard. Actual measured values for the Eimer and Amend and the NBS-SRM 987 standards were 0.70792 ± 0.00002 and 0.71017 ± 0.00005 (±2 standard errors of the mean). K, Rb, Cs, and Ba were analyzed by isotope dilution, using the DTM 9-in. mass spectrometer. Sr was analyzed by x-ray fluorescence at the Geophysical Laboratory, Carnegie Institution of Washington. Precision for these elements is estimated at better than ±2%, except Cs, for which the precision is better than ±10%. Details of the analytical techniques have been reported by Hart and Brooks (1974). Rare earths were analyzed by instrumental neutron activation using the facilities of Rhode Island Nuclear Science Center. Precision for these analyses varies from 5% to 10%, depending on the element. The analytical procedure has been discussed by Schilling and others (1977). Major-element chemistry for all samples is reported elsewhere (Bryan and Moore, 1977; J.-G. Schilling and W. M. White, in prep.).

RESULTS AND DISCUSSION

Table 1 lists the trace-element and $^{87}Sr/^{86}Sr$ analyses along with the Thornton and Tuttle differentiation index (D.I.—sum of normative quartz, alkali feldspar, and feldspathoids) and the $\Sigma FeO/(\Sigma FeO + MgO)$ ratio. These latter are based on whole-rock analyses of the dredged samples and analyses of glass in the *Alvin* samples. Trace-element data in all cases represent whole-rock analyses. Because these basalts are nearly aphyric (generally less than 5% to 10% phenocrysts), the effect of this difference should be small. Major-element chemistry for samples 527-6-1 and 534-3-1 is unavailable, and we have used differentiation indices based on samples from the same stations (527-6-3 and 534-3-2) instead. Available data indicate that variations between samples at specific sites are generally minor (Bryan and Moore, 1977).

The Sr-isotope ratios of all samples except AII 73 10-19 fall in the narrow range of 0.70288 to 0.70297, which is barely significant at the 5% level (based on the Student's t test). There is no significant difference between central and flank lavas. AII 73 10-19 was recovered from the east wall of the rift valley, and judging from its inferred spreading age and appearance, it is significantly older than the other samples. The higher $^{87}Sr/^{86}Sr$ ratio may represent either small secular variations in the chemistry of the mantle-source that feeds the volcanism in the rift valley or a small degree of posteruptional alteration. The K/Cs ratio, a good indicator of alteration (Hart, 1970), suggests that the latter is not the case. The $^{87}Sr/^{86}Sr$ ratios of a single sample from south of fracture zone B (AII 73 50-6) is not significantly different from those from north of the fracture zone. It thus appears that the FAMOUS basalts have been derived from an isotopically homogeneous source. The uniformity of Sr-isotope ratios reported here implies that chemical variability in FAMOUS basalts reported by Bryan and Moore (1977) is due to some differentiation process rather than to source heterogeneity.

TABLE 1. Sr-ISOTOPE AND LIL ANALYSES OF FAMOUS BASALTS

	TR119* 4D-1B	TR119* 4D-4	AII 73* 10-19	AII 73* 14-17	AII 73* 16-2	AII 73* 50-6	521-4-3	525-5-3	527-6-1	534-3-1
$^{87}Sr/^{86}Sr'$	0.70295	0.70289	0.70307	0.70290	0.70293	0.70290	0.70297	0.70296	0.70294	0.70288
K (ppm)	2,280	2,605	1,750	1,130	1,590	2,090	2,330	1,010	2,170	1,590
Rb (ppm)	5.3	4.6	4.5	2.9	4.1	5.1	5.4	2.5	6.7	3.7
Cs (ppm)	0.055	0.089	0.063	0.033	0.052	0.057	0.083	0.027	0.069	0.039
Sr (ppm)	112	114	111	94	111	101	121	74	107	108
Ba (ppm)	63	91	51	36	49	57	56	30	68	46
La (ppm)	7.3	8.5	5.7	4.9	6.0	5.8	7.1	3.8	7.9	5.3
Ce (ppm)	..	..	..	..	..	..	15.0	7.4	18.9	12.5
Sm (ppm)	3.6	3.3	2.6	2.4	3.0	2.5	3.0	2.1	3.9	2.9
Eu (ppm)	1.1	1.1	0.87	0.97	0.95	0.88	0.97	0.71	1.2	0.95
Tb (ppm)	0.59	0.79	0.51	0.53	0.73	0.62	0.72	0.51	0.81	0.72
Yb (ppm)	3.6	2.9	1.9	2.5	2.1	2.4	2.5	2.7	2.8	1.9
Lu (ppm)	0.63	0.44	0.30	0.41	0.43	0.40	0.43	0.38	0.47	0.29
$[La/Sm]_{E.F.}$ §	1.42	1.82	1.53	1.40	1.41	1.62	1.65	1.27	1.43	1.38
$\dfrac{\Sigma FeO}{\Sigma FeO + MgO}$	0.579	0.533	0.536	0.529	0.562	0.552	0.544	0.465	0.561	0.534
D.I.*	20.6	20.6	19.1	18.8	19.0	19.8	22.3	17.4	22.1	20.7

Note: TR designates R/V *Trident* dredge stations, AII designates R/V *Atlantis II* dredge stations, 500 series is *Alvin* dive, station, and sample number.

* Rare-earth data for these samples are from Schilling (1975).

Isotopic ratios are ±0.00006, except for AII 73 10-19 and 527-6-1, which are ±0.00005 and ±0.00007, respectively (±2 standard errors of the mean).

§ $[La/Sm]_{E.F.}$ = chondrite-normalized ratio of La to Sm; E.F. = "enrichment factor."

* D.I. = differentiation index, based on whole-rock analyses for dredged samples and on glass analyses for *Alvin* samples.

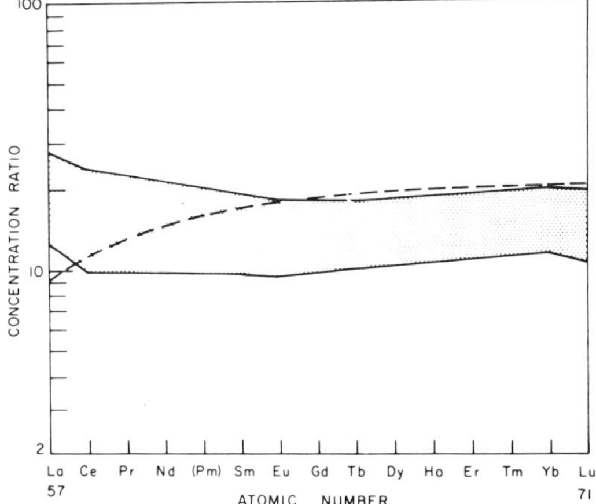

Figure 3. Field of chondrite-normalized rare-earth patterns of FAMOUS basalts (shaded) compared with an average mid-oceanic ridge pattern reported by Schilling (1971). FAMOUS basalts are slightly light-rare-earth enriched, whereas typical mid-oceanic ridge basalts are light-rare-earth depleted.

Table 2 lists (1) the range of trace-element concentrations, trace-element ratios, and Sr-isotope ratios found in FAMOUS samples, (2) the range of these values observed by White and others (1975) and Schilling (1975) for 39 basalts dredged from the Mid-Atlantic Ridge rift valley from lat 29° to 40°N, and (3) averages of these values for normal ridge areas of the Mid-Atlantic Ridge—that is, areas away from hot spots. Although alkali-metal, alkaline-earth, and rare-earth trace-element abundances in FAMOUS basalts are quite low in comparison to typical continental and oceanic island basalts, inspection of Table 2 clearly shows that they are not as depleted as "normal" Mid-Atlantic Ridge basalts. More importantly, ratios of large to small ions and Sr-isotope ratios are distinctly higher in FAMOUS basalts than in "normal" Mid-Atlantic Ridge basalts. This difference is further emphasized by Figure 3, which compares the field of rare-earth patterns for FAMOUS basalts with the pattern for an averaged mid-oceanic ridge basalt. Mid-oceanic ridge basalts are typically light-rare-earth depleted (Frey and others, 1968; Schilling, 1971), whereas FAMOUS basalts are all light-rare-earth en-

riched. We interpret these differences as being due to the effect of the nearby Azores mantle plume. Both the $[La/Sm]_{E.F.}$ and the Sr-isotope ratios reported here fit the regional pattern observed by Schilling (1975) and White and others (1975).

The geochemical transition zone and the LIL geochemistry of FAMOUS basalts can best be accounted for by mixing of two distinct mantle sources: the Azores mantle plume and the LIL-depleted asthenosphere (Schilling, 1975; White and others, 1976). This model not only explains the high LIL concentrations in FAMOUS basalts but also the fact that ratios such as Rb/K and Ba/K are nearly as high for the FAMOUS basalts as for the Mid-Atlantic Ridge basalts from the Azores Plateau, while FAMOUS basalt ratios such as $[La/Sm]_{E.F.}$ and $^{87}Sr/^{86}Sr$ are more intermediate (see Table 2). This peculiar geochemistry results from the fact that although trace-element concentrations in a mixture are linearly dependent on the proportions of the end members in the mixture, ratios are not since they depend on the relative concentrations of two elements in the end members. A more quantitative development of this problem will be presented elsewhere (White, in prep.). [The mixing equation for trace elements is similar to mixing equations for isotope ratios developed by Lancelot and Allegre (1974) and Sun and others (1975).] This mantle-source mixing postulated to explain the overall trace-element patterns and abundances and the Sr-isotope ratios is distinct from, and not to be confused with, possible mixing between magma batches derived by partial melting from the hybrid, well-mixed mantle source.

There is some evidence suggesting that plume flux may have varied with time. Schilling and others (1977) and Bryan and Thompson (1977) have shown that basalts from Leg 37, site 332 (lat 36°53'N, long 33°39'W, 34 km west of the median valley) have lower overall LIL concentrations than the FAMOUS basalts, although the site 332 basalts presumably originated in the median valley only a few million years ago. Older basalt at site 335 (lat 37°18'N, long 35°12'W, 182 km west of the median valley) is peculiar in that it shows distinct depletion in light rare earths, although concentrations of other LIL elements are similar to the more-fractionated FAMOUS samples. The relatively young age of overlying sediment (Melson, Aumento, and others, 1974) indicates that site 335 basalts may be a flank eruption and therefore could have tapped a separate mantle source.

Concentrations of selected trace elements, trace-element ratios, and $^{87}Sr/^{86}Sr$ are plotted versus $\Sigma FeO/(\Sigma FeO + MgO)$ in Figure 4. The trace-element and $^{87}Sr/^{86}Sr$ ratios do not vary systematically with differentiation index. Trace-element concentrations do appear to increase with differentiation index, but the scatter is considerable. Although this scatter may in part be due to the inadequacy of the

TABLE 2. COMPARISON OF FAMOUS BASALT GEOCHEMISTRY WITH REGIONAL
AND AVERAGE MID-ATLANTIC RIDGE GEOCHEMISTRY

	FAMOUS area		Lat 29° to 40°N		"Normal" Mid-Atlantic Ridge*
	max	min	max	min	
K (ppm)	2,605	1,010	9,050	430	855
Rb (ppm)	6.7	2.5	22.5	0.22	0.87
Cs (ppm)	0.089	0.027	0.283	0.0014	0.012
Sr (ppm)	121	74	298	55	103
Ba (ppm)	91	30	268	1.08	8.3
La (ppm)	8.5	3.8	30.7	0.53	2.7[§]
$[La/Sm]_{E.F.}$[†]	1.82	1.27	3.6	0.24	0.62[§]
Rb/K	0.0031	0.0018	0.0031	0.0005	0.0009
Ba/K	0.035	0.024	0.036	0.003	0.007
Rb/Sr	0.063	0.034	0.080	0.002	0.008
$^{87}Sr/^{86}Sr$	0.70307	0.70288	0.70349	0.70258	0.70266

Note: See text for explanation of sources.

* Average, based on 24 analyses, including data from Hart (1976).

[†] $[La/Sm]_{E.F.}$ = chondrite-normalized ratio of La to Sm; E. F. = "enrichment factor."

[§] From Schilling (1971).

iron-magnesium index (for example, plagioclase fractionation will not be reflected in this index), it does not appear that the variation in trace-element concentrations can be accounted for solely by fractional crystallization. As these samples were collected over a wide geographical area, this is not particularly surprising; however, even the variation in those samples from the Mount Pluto region cannot be accounted for by fractional crystallization alone.

Bryan and Moore (1977) have shown that in the Mount Pluto region, basalts become increasingly fractionated away from the central volcanic axis. Glass compositions all fall near the olivine-plagioclase-pyroxene cotectic, and this and other petrographic evidence suggest that fractional crystallization is an important factor in controlling composition. Bryan and Moore's computer modeling indicates that flank glasses can be derived by approximately 30% crystallization of the central lavas. However, they found that increases in K_2O, TiO_2, and H_2O are much greater than can be accounted for by this extent of crystal fractionation. Using computer programs to model fractional crystallization (Bryan and others, 1969; Wright and Doherty, 1970) and the mineral compositions reported by Bryan and Moore (1977), we find that about 23% crystallization (13% plagioclase, 4% pyroxene, 6% olivine) is required to derive our most differentiated lava (527-6-1) from our most primitive central lava (524-5-3). This much crystallization can account for a maximum trace-element enrichment of 30%. (The Raleigh fractionation law is $C = C_o f^{D-1}$, where C_0 and C represent the initial concentration of the element and its concentration in the residual liquid, respectively, f is the fraction of liquid remaining, and D is the solid/liquid partition coefficient. The enrichment, denoted by C/C_0, is maximized for a given value of f when $D = 0$.) The observed enrichment in K, Rb, Cs, Ba, and the light rare earths is 2 or greater, which requires more than 50% crystallization. Enrichments in Sr and the heavy rare earths are lower, but the higher partition coefficients for these elements allow a much lower percentage of enrichment for the same amount of crystallization. The nearly simul-

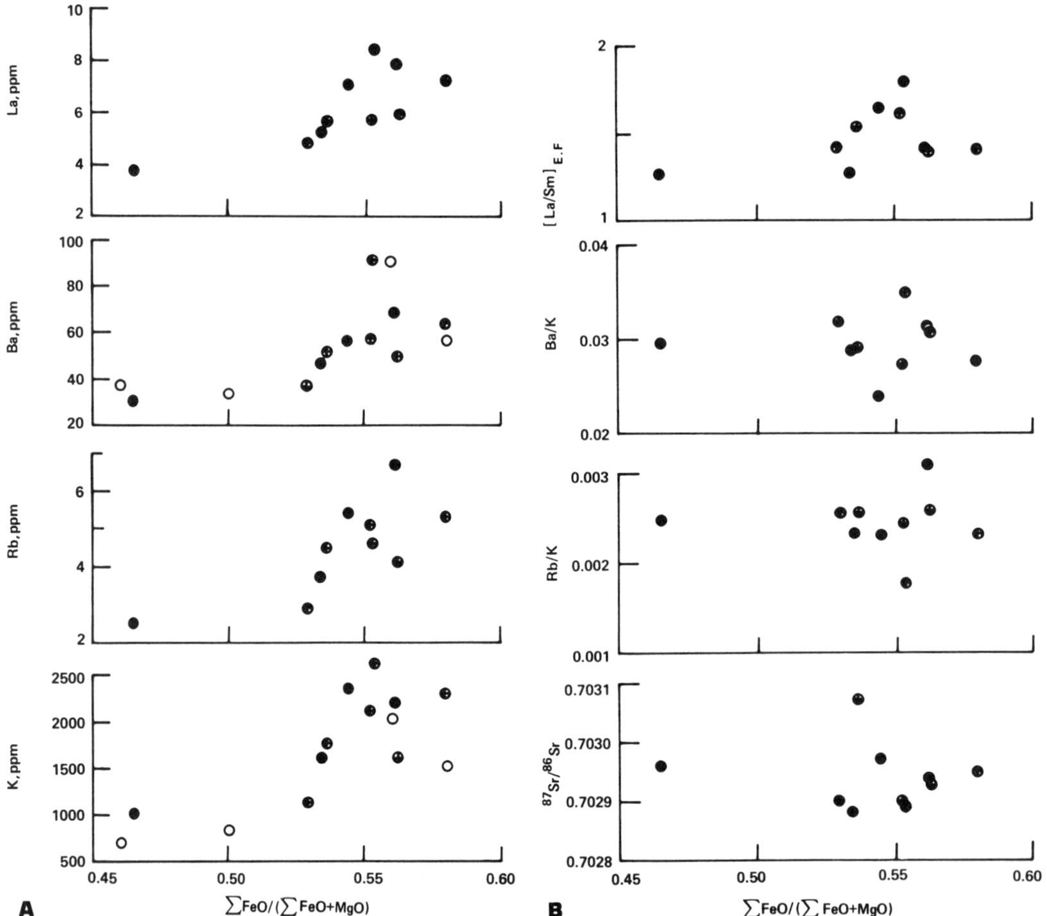

Figure 4. (A) Concentrations of K, Rb, Ba, and La plotted versus $\Sigma FeO/(\Sigma FeO + MgO)$. ΣFeO refers to total iron expressed as FeO. (B) Rb/K, Ba/K, $[La/Sm]_{E.F.}$, and $^{87}Sr/^{86}Sr$ plotted versus $\Sigma FeO/(\Sigma FeO + MgO)$. ● = *Alvin* samples, ⊕ = dredged samples, ○ = additional K and Ba x-ray fluorescence data provided by J. G. Moore, U.S. Geological Survey (L. Espos and B. Fabbi, analysts).

taneous eruption of flank and central lavas seems to preclude any secular geochemical variations (Bryan and Moore, 1977). The excess enrichment of alkali metals could perhaps be explained by the volatile transfer mechanism suggested by Bryan and Moore (1977). However, Ti, Ba, and the rare earths do not partition significantly into a hydrous fluid phase in equilibrium with a silicate melt (Holloway, 1971; Cullers and others, 1973).

The trends suggested by our data are similar to those that have been observed in data sets on other closely related sea-floor basalts. Bryan and others (1976) have shown that (1) excess enrichment in K_2O, TiO_2, La, and Ba is characteristic of specific data sets which otherwise appear compatible with simple fractional crystallization models and (2) LIL enrichments are in general very imperfectly correlated with major-element concentrations on a broad regional scale. This suggests that major elements and trace elements may be responsive to different processes.

Bryan and Moore (1977) have proposed a petrogenetic model in which volatile components and LIL elements are slowly accumulated along the relatively cool margins of a continuously crystallizing magma chamber maintained between the slowly separating lithospheric plates on either side of the median valley. This accumulation could result from rejection of these components from the crystalline phases accreting to the walls of the chamber, as well as from diffusion from the hot central zone toward the cooler walls. The principal uncertainties associated with this model involve the nature of volatile complexes, their diffusion rates, and their exact geochemical behavior, as well as the efficiency of convective mixing. We have already taken note of experimental work suggesting that Ba and the rare earths are not concentrated in hydrous fluid phases. The effects of complexing by other components such as the halogens or sulfur, however, are not known, and it may be premature to rule out volatile transfer. If the chamber can be maintained as an open system for several tens of millions of years, even very slow and inefficient concentration processes might be sufficient to produce the two or three times excess H_2O and LIL enrichment observed.

Variations in percentage of partial melting could also produce magma batches that are uniform in major-element chemistry and isotopic ratios, yet variable in LIL concentrations. Recent experimental work by Mysen and Kushiro (1976) shows that in at least some cases there may be little change in major-element melt composition over a relatively large range of melting of peridotite. This variable partial melting should be particularly efficient in producing variations in concentrations of LIL elements and also titanium, which is incompatible with most mantle phases (Gast, 1968). However, the systematic spatial relations of the median-valley lavas and their small range in age would require a carefully modulated increase in temperature with concurrent progressive increase in extent of melting over a relatively short time span. This model also would require successive injections of magma batches into the center of the chamber, each batch representing a slightly higher amount of partial melting than the previous one. These batches are similar in major-element chemistry but distinct in terms of trace-element concentrations. The zonation within the chamber could be established if each magma batch is incompletely mixed with the batch preceding it. The Mount Pluto magma may represent the greatest extent of partial melting and could represent an influx of new magma initiating a new cycle of volcanism in the median valley. A somewhat similar model has been proposed by Wright and Fiske (1971) to explain compositional variation in lavas of Kilauea. This magma-chamber zonation could, of course, be enhanced by fractional crystallization as outlined by Bryan and Moore (1977).

Variations in partial melting in the mantle imply fluctuations in temperature on the order of 50 to 100 °C over a time span of about 10,000 yr, according to the thermal gradients and lava ages cited by Bryan and Moore (1977) and Hekinian and others (1976). Unless the example encountered in the FAMOUS area is an isolated case, such episodes would have to occur repeatedly hundreds of times during the life span of a spreading ocean basin. It is not clear what might produce these temperature fluctuations, or whether the thermal inertia of mantle materials would allow fluctuations of this magnitude over such a limited time interval. There is, of course, no reason why either variable partial melting or prolonged crystallization in a steady-state chamber should be the *exclusive* process by which LIL enrichments could be produced. Because excess LIL enrichment appears to be a common feature of all sea-floor basalt suites, we believe the FAMOUS basalts are not a unique case and that the processes suggested to explain their trace-element variations may be generally applicable to all spreading ridges.

CONCLUSIONS

The uniformity of $^{87}Sr/^{86}Sr$ ratios in FAMOUS basalts implies that they are derived from a chemically homogeneous, uniformly mixed source.

The FAMOUS region lies within the geochemical transition zone between the LIL-enriched Azores Plateau to the north and the more typical LIL-depleted ridge area south of lat 33°N. LIL concentrations, ratios of large to small ions, and $^{87}Sr/^{86}Sr$ ratios are higher in FAMOUS basalts than in "normal" ridge basalts. This feature of FAMOUS basalt geochemistry can be explained by mixing of two mantle sources—the Azores mantle plume and the LIL-depleted asthenosphere.

Although FAMOUS basalts show a range in major-element chemistry nearly as large as the total range observed in mid-oceanic ridge basalts (Bryan and others, 1976), LIL concentrations and $^{87}Sr/^{86}Sr$ ratios in FAMOUS basalts fall well within the range observed in Mid-Atlantic Ridge basalts from lat 29° to 40°N.

Variations in LIL concentrations are too large to be satisfactorily explained by closed-system fractional crystallization alone. These variations may be due to variable partial melting of the mantle source region; if so, they would require systematic increases in the extent of partial melting with time. Magma batches arriving at the magma chamber must not completely mix with magma already present; thus zonation is produced within the chamber which could be enhanced by fractional crystallization of olivine, plagioclase, and pyroxene, as suggested by Bryan and Moore (1977). This model implies a transitory magma chamber that is periodically renewed by episodes of increased melting in the mantle source area.

ACKNOWLEDGMENTS

One of us (White) received a predoctoral fellowship from Carnegie Institution of Washington, which is gratefully acknowledged. We thank J-G. Schilling for critically reading this manuscript, S. R. Hart for enlightening discussions, F. DeMeglio and his staff for use of facilities at the Rhode Island Nuclear Science Center, and J. G. Moore for obtaining additional trace-element data. This work has been supported by the National Science Foundation.

REFERENCES CITED

Ballard, R. D., and van Andel, Tj. H., 1977, Project FAMOUS: Operational techniques and American submersible operations: Geol. Soc. America Bull., v. 88, p. 495–506.

Bryan, W. B., and Moore, J. G., 1977, Compositional variations of young basalts in the Mid-Atlantic Ridge rift valley near lat 36°49′N: Geol. Soc. America Bull., v. 88, p. 556–570.

Bryan, W. B., and Thompson, G., 1977, Basalts from DSDP leg 37 and the FAMOUS area: Compositional and petrogenetic comparisons: Canadian Jour. Earth Sci. (in press).

Bryan, W. B., Finger, L. W., and Chayes, F., 1969, Estimating proportions in petrographic mixing equations by least squares approximation: Science, v. 163, p. 926–927.

Bryan, W. B., Thompson, G., Frey, F. A., and Dickey, J. S., 1976, Inferred settings and differentiation in basalts from the Deep Sea Drilling Project: Jour. Geophys. Research, v. 81, p. 4285–4304.

Cullers, R. L., Medaris, L. G., and Haskin, L. A., 1973, Experimental studies of the distribution of rare earths as trace elements among silicate minerals and liquids and water: Geochim. et Cosmochim. Acta, v. 37, p. 1499–1512.

Frey, F. A., Haskin, M. A., Poetz, J., and Haskin, L. A., 1968, Rare earth abundances in some basic rocks: Jour. Geophys. Research, v. 73, p. 6085–6098.

Gast, P. W., 1968, Trace element fractionation and the origin of tholeiitic and alkaline magma types: Geochim. et Cosmochim. Acta, v. 32, p. 1057–1086.

Hart, S. R., 1970, Sea floor basalts: Carnegie Inst. Washington Year Book, v. 68, p. 403–408.

——1976, LIL element geochemistry, leg 34 basalts, *in* Yeats, R. S., Hart, S. R., and others, Initial reports of the Deep Sea Drilling Project, Vol. 34: Washington, D.C., U.S. Govt. Printing Office, p. 283–288.

Hart, S. R., and Brooks, C., 1974, Clinopyroxene matrix partitioning of K, Rb, Cs, Sr, and Ba: Geochim. et Cosmochim. Acta, v. 38, p. 1799–1806.

Hekinian, R., Moore, J. G., and Bryan, W. B., 1976, Volcanic rocks and processes of the Mid-Atlantic Ridge rift valley near 36° 49' N: Contr. Mineralogy and Petrology, v. 58, p. 83–110.

Holloway, J. R., 1971, Composition of fluid phase solutes in a basalt-H_2O-CO_2 system: Geol. Soc. America Bull., v. 82, p. 233–238.

Lancelot, J. R., and Allegre, C. J., 1974, Origin of carbonatitic magma in the light of the Pb-U-Th isotope system: Earth and Planetary Sci. Letters, v. 22, p. 233–238.

Melson, W. G., Aumento, F., and others, 1974, Leg 37 — The volcanic layer: Geotimes, v. 19, p. 16–18.

Mysen, B. O., and Kushiro, I., 1976, Melting behavior of peridotite at high pressure: EOS (Am. Geophys. Union Trans.), v. 57, p. 354.

Schilling, J-G., 1971, Sea-floor evolution: Rare earth evidence: Royal Soc. London Philos. Trans., ser. A, v. 268, p. 663–706.

—— 1975, Azores mantle blob: Rare earth evidence: Earth and Planetary Sci. Letters, v. 25, p. 103–115.

Schilling, J-G., Kingsley, R., and Bergeron, M., 1977, Rare earth abundances in DSDP sites 332A, 332B, 334, and 335, and inferences on the Azores mantle blob activity with time, *in* Melson, W. G., Aumento, F., and others, eds., Initial reports of the Deep Sea Drilling Project, Vol. 37: Washington, D.C., U.S. Govt. Printing Office (in press).

Sun, S. S., Tatsumoto, M., and Schilling, J-G., 1975, Mantle plume mixing along the Reykjanes Ridge axis: Lead isotope evidence: Science, v. 190, p.143–147.

White, W. M., Hart, S. R., and Schilling, J-G., 1975, Geochemistry of the Azores and the Mid-Atlantic Ridge: 29°N to 60°N: Carnegie Inst. Washington Year Book, v. 74, p. 224–234.

White, W. M., Schilling, J-G., and Hart, S. R., 1976, Evidence for the Azores mantle plume from strontium isotope geochemistry of the central North Atlantic: Nature, v. 263, p. 659–663.

Wright, T. L., and Doherty, P. C., 1970, A linear programming and least squares computer method for solving petrologic mixing problems: Geol. Soc. America Bull., v. 81, p. 1995–2008.

Wright, T. L., and Fiske, R. S., 1971, Origin of the differentiated and hybrid lavas of Kilauea Volcano, Hawaii: Jour. Petrology, v. 12, p. 1–65.

Manuscript Received by the Society November 26, 1975
Revised Manuscript Received August 30, 1976
Manuscript Accepted September 13, 1976
Contribution No. 3660 of the Woods Hole Oceanographic Institution

40

Reprinted from page 133 of *Earth and Planetary Sci. Letters* **36**:133–156 (1977)

PETROGENESIS OF BASALTS FROM THE FAMOUS AREA: MID-ATLANTIC RIDGE

C.H. LANGMUIR, J.F. BENDER, A.E. BENCE, G.N. HANSON

Department of Earth and Space Sciences, State University of New York, Stony Brook, N.Y. 11794 (USA)

and

S.R. TAYLOR

Research School of Earth Sciences, Australian National University, Canberra, A.C.T. 2600 (Australia)

Fresh basalt glasses most of which have $Mg/(Mg + Fe^{2+})$ of 0.66–0.72 from outcrops within 3 km of one other in the rift valley at the Project FAMOUS locality have been analyzed for major, minor and trace elements in order to determine their petrogenesis.

Transition metal abundances of the FAMOUS samples are similar to a wide variety of continental and oceanic basalts with high MgO and Ni, all of which show remarkably little variation, with the exception of Cu, Zn and Ti, on a chondrite-normalized plot. Modelling of these data suggests that the mantle beneath both continents and oceans is systematically fractionated relative to chondrites. This fractionation provides a constraint for models of earth formation and subsequent evolution.

The abundances of the rare earth and the incompatible elements, Ba, La, Th, U, and Nb, vary by more than a factor of three and the La/Yb and La/Sm ratios vary by factors of 3.1 and 1.6, respectively, in samples with similar, high $Mg/(Mg + Fe^{2+})$. There is no correlation between the degree of light-REE enrichment and the heavy-REE abundance. Furthermore, the trace element variations do not appear correlated with respect to location in the rift valley or to time of eruption. These trace element features demonstrate that successive eruptions in one small area of the rift valley can show wide variations in trace element chemistry over a short span of time; they preclude the derivation of these basalt glasses from a single magma chamber.

Despite the heterogeneities in REE and the variable trace element abundances, a homogeneous mantle source is suggested by the similarities among the samples in the incompatible element ratios of La/Ce, Ba/Th, Zr/Nb and K/Ba and the small range in $^{87}Sr/^{86}Sr$ isotope ratios observed in other samples from the FAMOUS region (White and Bryan, 1977). Thus, trace element heterogeneities appear to be generated by processes in the mantle during melting. However, processes such as batch partial melting, fractional fusion, fractional crystallization, zone refining, or mixing of magmas or sources acting alone are incapable of explaining the lack of correlation between the light and heavy REE.

It is suggested that the observed variations are a consequence of dynamic partial melting of a homogeneous mantle source region. This process includes varying degrees of partial melting of an uprising mantle source with continuous but incomplete removal of melt as melting proceeds, varying extents of batch partial melting, and zone refining. Dynamic melting can produce different melts from a homogeneous source which have different degrees of light-REE enrichment and crossing REE patterns. The variable trace element abundances which may be produced through dynamic melting may be the cause of the apparent decoupling of major and trace elements (Bryan et al., 1976) which previously has been suggested for the FAMOUS region (Bryan and Moore, 1977).

References

W.B. Bryan, G. Thompson, F.A. Frey and J.S. Dickey, Inferred geologic settings and differentiation in basalts from the Deep Sea Drilling Project, J. Geophys. Res. 81 (1976) 4285–4304.

W.B. Bryan and J.G. Moore, Compositional variation of young basalts in the Mid-Atlantic Ridge rift valley near 36°49′N, Geol. Soc. Am. Bull. (1977) in press.

W.R. White and W.B. Bryan, Strontium isotope, K, Rb, Cs, Sr, Ba and rare earth geochemistry of basalts from the FAMOUS area, Geol. Soc. Am. Bull (1977) in press.

41

Reprinted from pages 127 and 136–141 of *Contr. Mineralogy and Petrology*
70:127–141 (1979)

Tholeiitic and Alkali Basalts From the Mid-Atlantic Ridge at 43° N

Tsugio Shibata[1], Geoffrey Thompson[2], and Frederick A. Frey[3]

[1] Department of Earth Sciences, Okayama University, Tsushima, Okayama 700, Japan
[2] Woods Hole Oceanographic Institution Woods Hole, Massachusetts 02543, USA
[3] Department of Earth and Planetary Sciences, Massachusetts Institute of Technology,
Cambridge, Massachusetts 02139, USA

Abstract. Tholeiitic basalts dredged from the Mid-Atlantic Ridge (MAR) axis at 43° N are enriched in incompatible trace elements compared to the 'normal' incompatible element depleted tholeiites found from 49° N to 59° N and south of 33° N on the MAR. The most primitive 43° N glasses have $MgO/FeO^* =$ 1.2 and coexist with olivine (Fo_{90-91}) and chrome-rich spinel. The tholeiitic basalts from the MAR 43° N are distinct from the strongly incompatible trace element depleted tholeiites found elsewhere in the Atlantic, and have trace element features typical of island tholeiites and MAR axis tholeiites from 45° N. Petrographic, major, and compatible trace element trends of the axial valley tholeiites at 43° N are consistent with shallow-level fractionation; in particular, evolution from primitive liquids with forsteritic olivine plus chrome spinel as liquidus phases to fractionated liquids with plagioclase plus clinopyroxene as major crystallizing phases. However, each dredge haul has distinctive incompatible trace element abundances. These trace element characteristics require a hetrogeneous mantle or complex processes such as open system fractional crystallization and magma mixing.

Alkali basalts ($\sim 5\%$ normative nepheline) were dredged from a prominent fracture zone at 43° N. Typical of alkali basalts they are strongly enriched (compared to tholeiites) in incompatible elements. Their highly fractionated rare-earth element (REE) abundances require residual garnet during partial melting. The 43° N tholeiites and alkali basalts could be derived from a garnet peridotite source with REE contents equal to $2\times$ chondrites by $\sim 5\%$ and 1% melting, respectively. Alternatively, they could be derived from a moderately light REE enriched source by $\sim 25\%$ and 9.5% melting, respectively.

[*Editors' Note:* Material has been omitted at this point.]

V. Discussion

A. Genetic Relations Between the Two Basalt Types From 43° N

1. Shallow Level Fractionation. The petrographic and major element data for the MAR 43° N tholeiitic basalts are broadly consistent with the following shallow level fractionation model. The parent magma composition was such that olivine was the first phase to crystallize. Olivine crystallization thus forced the residual liquid to a cotectic where plagioclase and clinopyroxene started to precipitate almost simultaneously. Apparently, crystallization of magnesiochromite accompanied early olivine formation, but ceased as a result of liquid-spinel clinopyroxene reaction. The variation diagrams of Fig. 6, however, indicate that at given values of FeO^*/MgO, the three alkali basalts deviate from the trend defined by the tholeiitic basalts. Thus, it is not likely that simple shallow level fractionation involving olivine, plagioclase, and clinopyroxene could account for major and trace element variations in the three alkali basalts or could generate the MAR 43° N tholeiitic basalts

and alkali olivine basalts from a common parent magma. This inference is substantiated by the REE distributions (Fig. 7b) because Sample 3–10 is enriched in light REE but depleted in heavy REE with respect to Sample 12–7, the most mafic non-accumulative basalt. The bulk solid-liquid partition coefficients for light and heavy REE are much less than unity for mineral assemblage of olivine, plagioclase and clinopyroxene coexisting with a basaltic liquid; therefore, separation of olivine, plagioclase and clinopyroxene results in an essentially parallel upward shift of the REE pattern in increasingly fractionated residual liquids. Thus, the REE patterns of basalts 12–7 and 3–10 are not consistent with a shallow level fractional crystallization model.

Alternative mechanisms for relating these two basalt types are fractional crystallization at high pressures and partial melting processes of upper mantle peridotites. These models are examined below. Solid-liquid partition coefficients of REE and other trace elements used in the following model calculations are taken from Frey et al. (1978). Our model calculations consider the compositions of 12–7 (tholeiite) and 3–10 (alkali olivine basalt) as the most primitive (parental)

288

composition among each group of analysed basalts, and our models evaluate if these two compositions can be genetically related.

2. Fractional Crystallization at High Pressures. Possible mineral phases that crystallize from basaltic magma at pressures of 20 to 35 kb include garnet, clinopyroxene, and/or orthopyroxene (e.g., Yoder and Tilley, 1962; Green and Ringwood, 1967; O'Hara, 1968). High pressure fractionation of garnet and clinopyroxene ('eclogite fractionation') proposed by O'Hara and Yoder (1967) and O'Hara (1973) is an effective process to change the light/heavy REE ratio in a residual liquid without greatly changing the major element composition. In order to evaluate an eclogite fractionation model, we note that the partition coefficient of Ce between clinopyroxene and liquid is ~ 0.04, and that between garnet and liquid is ~ 0.021 (P.C. set 3 of Frey et al., 1978); hence, the bulk partition coefficient for a mixture of these two minerals in any proportion should be within the range given by these two values. In order to cause the observed enrichment factor for Ce (2.36), we may set a range with the above two Ce partition values for the fraction of eclogite to be removed by using the Rayleigh fractionation model $C_1/C_0 = F^{D-1}$ (C_1 is liquid concentration, C_0 is initial concentration, F is liquid weight fraction, and D is the partition coefficient). Solving this equation for the observed Ce concentration ratio in 3–10 and 12–7 basalts (i.e., 2.36) with $D = 0.04$–0.021, we obtain $F = 0.41$. This amount of eclogite fractionation constrains the range of Yb bulk partition coefficient values to bulk solid/liquid $D_{Yb} = \sim 1.3$; hence, with the Yb partition coefficients of garnet and clinopyroxene (P.C. set 3 of Frey et al., 1978), garnet must comprise approximately 30% of the cumulate assemblage in order to keep the Yb bulk solid-liquid partition coefficient ~ 1.3.

However, elements such as Sc, Cr, and Co are strongly partitioned into garnet and clinopyroxene, so garnet clinopyroxene fractionation rapidly depletes a residual liquid in these elements (Gast, 1968; Frey et al., 1974). Sc and Co abundances in 12–7 and 3–10 are similar, and there is no large depletion of Cr in 3–10 relative to 12–7. The $\sim 60\%$ removal of garnet and clinopyroxene from basalt 12–7, required by the REE calculation, would almost completely deplete a residual liquid in Sc, Cr, and Co. Consequently, we conclude that a REE distribution pattern similar to that of basalt 3–10 could be generated from basalt 12–7 by eclogite fractionation, but compatible trace element (sc, Cr, Co) abundances are not consistent with this model. This conclusion is valid for all reasonable choices of REE, Sc, Cr, and Co partition coefficients.

3. Partial Melting of Upper Mantle Peridotite. Upper mantle processes used to explain the genesis of basaltic magmas include batch (equilibrium) partial melting, fractional partial melting, zone refining (Harris, 1974), dynamic melting (Langmuir et al., 1977), and mixing (Schilling, 1973). These processes have been employed with various success to elucidate the genetic relation between alkali olivine basalts and LIL element-depleted abyssal tholeiites (e.g., Gast, 1968; Green, 1971) and that between LIL element-enriched and LIL element-depleted abyssal tholeiites (Schilling, 1975; White et al., 1976; Langmuir et al., 1977) and also to characterize to composition of the source regions from which these basaltic rocks are derived. However, all these model calculations must assume the type and relative proportions of phases in the initial mantle source, and also the phases in equilibrium with a liquid at a given degree of partial melting. Consequently, model calculations for upper mantle processes cannot give a unique solution. For this reason, we evaluate only if it is possible to generate the trace element compositions in the tholeiitic and alkali olivine basalts of the MAR 43° N region from a common mantle source by changing the degree of partial melting and/or the mineralogy of the source material. The equations developed by Gast (1968) with the modifications of Shaw (1970) were used in the following model calculations.

First, we note that the chondrite-normalized REE patterns of 12–7 and 3–10 intersect in the vicinity of Er (Fig. 7b). It is possible to generate intersecting REE patterns and complex relationships between the degree of light REE enrichment and heavy REE abundance by batch partial melting of a peridotite containing a phase that preferentially enriches heavy REE, for example, garnet peridotite. Moreover, if the mantle source for basalts 12–7 and 3–10 had nearly chondritic REE pattern, then it is likely that garnet was a residual mineral during generation of the strongly light REE-enriched pattern of basalt 3–10 (Kay and Gast, 1973). An approximately 1% equilibrium partial melt of a garnet lherzolite source (with REE abundances equal to $2 \times$ chondrites) has La/Yb equal to that of alkali olivine basalt 3–10, whereas a $\sim 5\%$ equilibrium melt has La/Yb similar to that of Sample 12–7 (Fig. 9). Therefore, the increasing degree of REE fractionation from the 43° N medium valley tholeiites to the fracture zone alkali basalt can be successfully modelled by varying degrees of equilibrium melting (~ 5 to 1%, respectively) of a homogeneous garnet lherzolite.

Partial melting of garnet lherzolite has also been used to successfully model other tholeiitic basalts with relative light REE enrichment compared to chondrites (Hawaii: Leeman et al., 1978; Iceland: Zindler et al.,

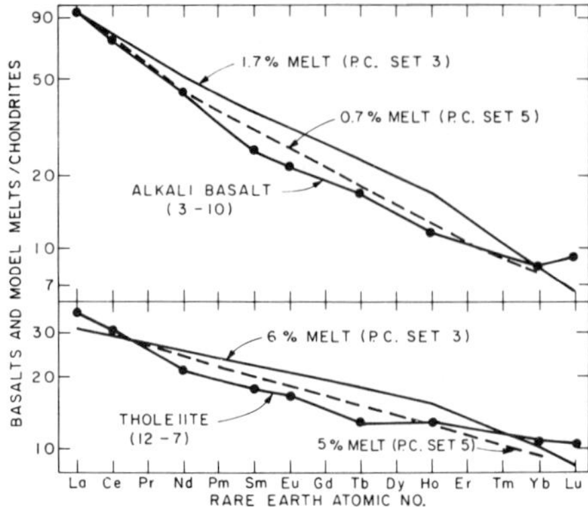

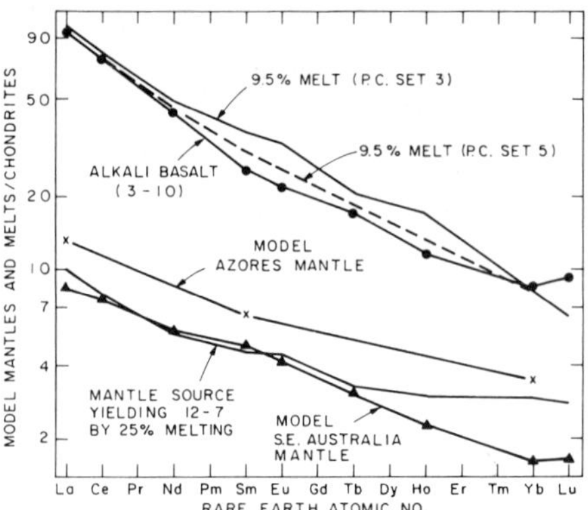

Fig. 9. REE abundances in model liquids compared to the 43° N basalts 3–10 and 12–7. *Upper:* alkali basalt 3–10 is approximated by equilibrium partial melts of garnet peridotite (initial wt.% proportions are 50% olivine, 25% orthopyroxene, 20% clinopyroxene and 5% garnet, and melt is composed of 5% olivine, 5% orthopyroxene, 45% clinopyroxene, and 45% garnet) with a REE content equal to 2× chondrites. Calculated 1.7% melt derived by using partition coefficient (*P.C.*) set 3 of Frey et al. (1978) and calculated 0.7% melt derived by using *P.C.* set 5 of Frey et al. (1978). *Lower:* tholeiite 12–7 compared to model liquids (same initial parameters as for 3–10 models) derived by 6% equilibrium melting (*P.C.* set 3 of Frey et al., 1978) and 5% melting (*P.C.* set 5 of Frey et al., 1978). Note these model tholeiitic liquids coexist with a garnet-bearing residue (~3% garnet)

Fig. 10. *Solid line* without data points in lower portion indicates REE abundances in a peridotite mantle source (olivine 50%, opx 25%, cpx 15%, garnet 10%) that can yield tholeiite 12–7 by 25% equilibrium melting with P.C. set of Frey et al. (1978). Up to 22.2% melting, the melt is 5% olivine, 5% opx, 45% cpx and 45% gt; after 22% melting, the melt is 5% olivine, 5% opx and 90% cpx. Shown for comparison are light REE enriched model mantles inferred for the Azores plume (×: White and Schilling, 1978) and Australian continental tholeiites (▲: Frey et al., 1978). *Upper part* of diagram shows REE abundances in alkali basalt 3–10 compared to 9.5% melts of the model peridotite which yields the REE abundances of 12–7 by 25% melting. Note these model alkalic liquids are in equilibrium with residual garnet (54.7% ol, 27.1% opx, 11.9% cpx and 6.3% gt) whereas the model tholeiite is in equilibrium with 64.5% ol, 31.5% opx, and only 4.0% cpx

1978). However, some experimental studies (Green, 1971) have concluded that tholeiitic basalts do not form in equilibrium with residual garnet. In this case, the fractionated REE patterns of the 43° N tholeiites could not be achieved by equilibrium melting of a peridotite source with relative REE abundances as in chondrites. Recent models for the mantle source of continental tholeiites (Fig. 10 of Frey et al., 1978) and mantle plumes (Azores: Fig. 9 of White and Schilling, 1978) have used sources with relative light REE enrichment (Fig. 10, this paper). For example, 25% melting (residue of 64.5% olivine, 31.5% orthopyroxene, and 4% clinopyroxene) of the light REE enriched source shown in Fig. 10 will yield a liquid with REE abundances equal to tholeiite Sample 12–7, and 9.5% melting (residue of 54.7% olivine, 27.1% orthopyroxene, 11.9% clinopyroxene, and 6.3% garnet) yields a model liquid with La/Yb ratios similar to alkali basalt Sample 3–10 (Fig. 10).

Thus, these calculations indicate that it is possible to derive the tholeiitic and alkali olivine basalts of the MAR 43° N region from a common mantle source; however, trace element abundances cannot uniquely define which model (Figs. 9 and 10) is appropriate. Moreover, isotopic data (Sr, Nd, and Pb) are required to further test the suitability of homogeneous source model for the 43° N sample suite.

B. Occurrence of LIL Element-Enriched Basalts Along the MAR Axis

The origin and processes generating LIL element-enriched tholeiites along the MAR are key factors in understanding the oceanic upper mantle. In areas such as Iceland and the Azores, such basalts are associated with anomalous topography, and the characteristic geological, geochemical and geophysical features of these settings have led to the model of mantle plumes (e.g., Schilling, 1973; White and Schilling, 1978). However, basalts with similar LIL element-enriched characteristics occur at 43° and 45° N on the MAR where bathymetric and geophysical features are similar to 'normal' MAR segments. In comparing tholeiites from 43° and 45° N, we note that basalts from 45° N have Sr$^{87/86}$ from 0.7033 to 0.7034, whereas Sample 12–8 from 43° N has a lower ratio of 0.7032 (White and Schilling, 1978); however, the degree of relative light REE enrichment is considerably greater in the 43° N tholeiites (Fig. 9). These results are not consistent with the covariation of REE fractionation and $Sr^{87/86}$ ratio found in the Reykjanes Ridge transect (Hart et al., 1973; Sun et al., 1975). Possibly, different mantle source compositions are required for the tholeiites from 43° and 45° N on the MAR.

VI. Conclusions

Tholeiitic basalts dredged from the MAR axis at 43° N are enriched in incompatible trace elements compared to the 'normal' abyssal tholeiites occurring from 49° N to 59° N and south of 33° N on the MAR. These results further establish the compositional hetrogeneity of basalts formed at the North Atlantic MAR axis and the systematic variations in composition as a function of latitude (White and Schilling, 1978). Basalts from each dredge haul have distinctive trace element abundances which preclude genetic relationships by simple fractional crystallization. Such hetrogeneity may be explained by complex melting and/or open-system crystallization models (e.g., Langmuir et al., 1977; O'Hara, 1977; Rhodes et al., 1979; Walker et al., 1979), but the systematic compositional variations with latitude must reflect horizontal and probably vertical heterogeneities in mantle composition.

Alkali olivine basalts at 43° N on the MAR form topographic highs similar to other oceanic occurrences of alkalic basalts, such as some seamounts (Batiza, 1977; Stebbins and Thompson, 1978) and in the Siqueiros Fracture Zone on the East Pacific Rise (Batiza et al., 1977). Their limited occurrence and spatial association with tholeiites suggests a close genetic relationship between alkali olivine basalts and tholeiites (Engel et al., 1965). However, we conclude that the alkali olivine basalts at 43° N are not related to nearby tholeiites by low or high pressure crystal fractionation. Based on trace element abundances the alkali olivine basalts may be derived from the same source as tholeiites but by smaller degrees of melting. However, in the Siqueiros Fracture Zone an alkali olivine basalt is isotopically ($Nd^{143/144}$) distinct from a tholeiite recovered in the same dredge haul (Carlson et al., 1978). Therefore, the two basalts could only be derived from the same source if disequilibrium melting occured (Carlson et al., 1978). Alternatively, this alkali olivine basalt originates from a source compositionally distinct from the source of spatially associated tholeiites.

The occurrence of alkali basalts in fracture zones is common, and we concur with Thompson and Melson (1972) that some of the oceanic crust in fracture zones is petrologically and compositionally distinct from the crust forming at the nearby spreading ridge crest. We suggest that this distinctive fracture zone crust results because magmas injected into transform faults (leaky transform fault, Menard and Atwater, 1969) segregate at greater mantle depths than tholeiites and possibly from a compositionally distinct mantle source. Presumably, it is the tectonics of a fracture zone which enable such alkalic magmas to reach the surface.

References

Batiza, R.: Age, volume, compositional and spatial relations of small isolated oceanic central volcanoes. Mar. Geol. **24**, 169–183 (1977)

Batiza, R., Rosendahl, B.R., Fisher, L.: Evolution of oceanic crust 3. Petrology and chemistry of basalts from the East Pacific Rise and the Siqueiros transform fault. J. Geophys. Res. **82**, 265–276 (1977)

Carlson, R.W., Macdougall, J.D., Lugmuir, G.W.: Differential Sm/Nd evolution in oceanic basalts. Geophys. Res. Lett. **5**, 229–232 (1978)

Engel, A.E.J., Engel, C.G., Havens, R.G.: Chemical characteristics of oceanic basalts and the upper mantle. Geol. Soc. Am. Bull. **76**, 719–734 (1965)

Frey, F.A., Bryan, W.B., Thompson, G.: Atlantic ocean floor: geochemistry and petrology of basalts from Legs 2 and 3 of the Deep-Sea Drilling Project. J. Geophys. Res. **79**, 5507–5527 (1974)

Frey, F.A., Green, D.H., Roy, S.D.: Integrated models of basalt petrogenesis: A study of quart tholeiites to olivine melilitites from southeastern Australia utilizing geochemical and experimental petrological data. J. Petrol. **19**, 463–513 (1978)

Gast, P.W.: Trace element fractionation and the origin of tholeiitic and alkaline magma types. Geochim. Cosmochim Acta **32**, 1057–1086 (1968)

Green, D.H.: Composition of basaltic magmas as indicators of conditions of origin: application to oceanic volcanism. Philos. Trans. R. Soc. Lond. [A] **268**, 707–725 (1971)

Green, D.H., Ringwood, A.E.: The genesis of basaltic magmas. Contrib. Mineral. Petrol. **15**, 103–190 (1967)

Harris, P.G.: Origin of alkaline magmas as a result of anatexis. In: The alkaline rocks (H. Sørensen ed.), pp. 427–436. New York: John Wiley & Sons 1974

Hart, S.R., Schilling, J.G., Powell, J.L.: Basalts from Iceland and along the Reykjanes Ridge: Sr isotope geochemistry. Nature Phys. Sci. **246**, 104 (1973)

Kay, R.W., Gast, P.W.: The rare earth content and origin of alkali-rich basalts. J. Geol. **81**, 653–682 (1973)

Langmuir, C.H., Bender, J.F., Bence, A.E., Hanson, G.N., Taylor, S.R.: Petrogenesis of basalts from the FAMOUS area: Mid-Atlantic Ridge. Earth Planet. Sci. Lett. **36**, 133–156 (1977)

Leeman, W.P., Murali, A.V., Ma, M.-S., Schmitt, R.A.: Mineral constitution of mantle source regions for Hawaiian basalts – rare earth evidence for mantle heterogeneity. Oregon Dept. Geol. Mineral. Indust. Bull. **96**, 169–183 (1978)

Menard, H.W., Atwater, T.: Origin of fracture zone topography. Nature **222**, 1037–1040 (1969)

O'Hara, M.J.: The bearing of phase equilibria studies in synthetic nad natural systems on the origin and evulution of basic and ultrabasic rocks. Earth-Sci. Rev. **4**, 69–133 (1968)

O'Hara, M.J.: Non-primary magmas and dubious mantle plume beneath Iceland. Nature **243**, 507–508 (1973)

O'Hara, M.J.: Geochemical evolution during fractional crystallization of a periodically refilled magma chamber. Nature **266**, 503–507 (1977)

O'Hara, M.J., Yoder, H.S. Jr.: Formation and fractionation of basic magmas at high pressure. Scott. J. Geol. **3**, 67–117 (1967)

Rhodes, J.M., Dugan, M.A., Blanchard, D.P., Long, P.E.: Magma mixing at mid-oceanic ridges: evidence from basalts drilled near 22° N on the Mid-Atlantic Ridge. Tectonophysics (in press, 1979)

Schilling, J.-G.: Iceland mantle plume. Nature **246**, 141–143 (1973)

Shaw, D.M.: Trace element fractionation during anatexis. Geochim. Cosmochim. Acta **34**, 237–243 (1970)

Stebbins, J., Thompson, G.: The nature and petrogenesis of intra-ocean plate alkaline eruptive and plutonic rocks: King's Trough north-east Atlantic. J. Volcanol. Geotherm. Res. **4**, 333–361 (1978)

Sun, S.S., Tatsumoto, M., Schilling, J.-G.: Mantle plume mixing along the Reykjanes Ridge axis: lead isotopic evidence. Science **190**, 143–147 (1975)

Thompson, G., Melson, W.G.: The petrology of oceanic crust across fracture zones in the Atlantic ocean: evidence of a new kind of sea-floor spreading. J. Geol. **80**, 526–538 (1972)

Walker, D., Shibata, T., DeLong, S.E.: Abyssal tholeiites from the Oceanographer Fracture Zone: II Phase equilibria and mixing. Contrib. Mineral. Petrol. **70**, 111–125 (1979)

White, W.M., Schilling, J.-G.: The nature and origin of geochemical variation in Mid-Atlantic Ridge basalts from central north Atlantic. Geochim. Cosmochim. Acta **42**, 1501–1516 (1978)

White, W.M., Schilling, J.-G., Hart, S.R.: Evidence for the Azores mantle plume from strontium isotope geochemistry of the central north Atlantic. Nature **263**, 659–663 (1976)

Yoder, H.S., Jr., Tilley, C.E.: Origin of basalt magmas: an experimental study of natural and synthetic rock systems. J. Petrol. **3**, 342–352 (1962)

Zindler, A., Hart, S.R., Frey, F.A., Jakobbson, S.P.: Nd and Sr isotope compositions and REE abundances in Reykjanes Peninsula basalts: evidence for mantle heterogeneity. EOS Trans. Am. Geophys. Union **59**, 410 (1978)

ERRATA

Page 137, line 13 right hand column should read: "also to characterize the composition. . . ."

Page 137, line 9 from the bottom, right hand column should read: "degree of REE fractionation from the 43°N media . . ."

GEOCHEMICAL CHARACTERISTICS OF MID-OCEAN RIDGE BASALTS

SHEN-SU SUN [1], ROBERT W. NESBITT

Department of Geology and Mineralogy, University of Adelaide, Adelaide, S.A. 5000 (Australia)

and

ANATOLY Ya. SHARASKIN

Vernadsky Institute of Geochemistry, U.S.S.R. Academy of Sciences, Moscow, 117334 (U.S.S.R)

Received July 26, 1978
Revised version received March 12, 1979

Major and trace element (Rb, Sr, Y, Zr, Nb, Ba, Sc, Ni, Co, Cr, V) data are presented on ten mid-ocean ridge basalts (MORB), together with a basalt sample from the offshore slope of the Yap Trench and a basalt from the Tertiary ophiolite complex of eastern Taiwan. Rare earth element (REE) and Sr isotope data are presented on eight of the samples, together with REE data on U.S.G.S. standard rock, BHVO-1. A strong correlation exists between percent TiO_2 (proportional to amount of melting) and Al_2O_3/TiO_2, CaO/TiO_2 ratios of these close to primary MORB. These ratios increase up to a maximum (about 20 and 17 respectively) as TiO_2 decreases, indicating a progressive release of Al and Ca from the mantle source. The limiting values of the ratios are close to chondritic and are reached at about 0.8% TiO_2. At this high degree of melting, the major Al- and Ca-bearing mineral phases are eliminated from the mantle residue. Model calculations on the major element data indicate that MORB with 0.7% and 1.5% TiO_2 could represent about 25 and 15% melting (respectively). This modelling is consistent with the abundances of the first transition series metals (Sc to Ni), which are also interpretable in terms of degree of partial melting and residual mineralogy.

Based on the REE data, the samples can be divided into depleted types (typical of "normal" MORB), and enriched types (the so-called "plume" and "transitional" types). Despite the range of $(La/Sm)_N$ in these samples (0.38–1.97) the Ti/Zr ratios remain close to chondritic (about 110) indicating that these two elements have a similar degree of incompatibility at this level of melting. Chondritic normalized patterns for typical "normal" (N)- and "plume" (P)-type MORB are presented for the REE and K, Nb, U, Th, Ba, Rb and Cs and it is suggested that this represents an increasing order of element incompatibility in MORB. The behaviour of Y, Ti, Zr, P and Sr relative to the REE is also discussed and it is shown that for most primitive MORB, Y = Ho, Ti = Eu, Zr = Sm, P = Nd and Sr is between Ce and Nd. In addition the Zr/Nb ratio correlates with the La/Sm ratio. This element correlation can be used to predict the general shape of REE patterns.

The Sr isotope data are discussed in terms of already published data from the three oceans. It is suggested that P-type MORB originate from depleted mantle sources which only recently (say ≤300 m.y.) were added to by an incompatible-element-rich phase. Various models proposed to explain both the depletion in N-type, enrichment in the P-type MORB sources and the proposed mantle heterogeneity are discussed. It is concluded that mantle depletion is due to a continuing process rather than an episodic event in the early Archaean or 1.6 b.y. ago.

---

[1] Present address: C.S.I.R.O. Division of Mineralogy, North
Ryde, N.S.W. 2113, Australia.

43

Reprinted from *Initial Reports of the Deep Sea Drilling Project* **22**:459–468 (1974)

19. PETROLOGY AND GEOCHEMISTRY OF BASALTS AND RELATED ROCKS FROM SITES 214, 215, 216, DSDP LEG 22, INDIAN OCEAN[1]

Geoffrey Thompson and W. B. Bryan, Woods Hole Oceanographic Institution, Woods Hole, Massachusetts
and
F. A. Frey and C. M. Sung, Department of Earth and Planetary Science, Massachusetts Institute
of Technology, Cambridge, Massachusetts

INTRODUCTION

Aspects of the chemistry, petrography, and genesis of eruptive rocks from Sites 214 and 216 on the Ninetyeast Ridge and from Site 215 from the central Indian Ocean basin are discussed. In conjunction with the major element data (Hekinian, Chapter 17) we present additional data for the trace elements B, Ba, Co, Cr, Cu, Ga, Li, Mo, Ni, Sc, Sr, U, V, Y, Zr and the lanthanide elements (REE). Petrographic study of the samples confirms the details outlined by Hekinian. Features which are believed to be especially relevant to the interpretation of the geochemical data and to the understanding the origin of the Ninetyeast Ridge are emphasized in this chapter.

Observations and interpretations of the chemical data lead to the suggestion that the volcanic rocks of Sites 214 and 216 on the Ninetyeast Ridge differ in composition, texture, and mineral paragenesis from typical mid-ocean ridge basalts. The petrography and chemistry are consistent with the hypothesis that the Ninetyeast Ridge samples are fractionated lavas which have cooled and crystallized in shallow magma chambers before extrusion. The basalts from Site 215 in the Central Indian Ocean Basin show textures and mineral paragenesis characteristic of deep-sea pillow basalts. However, they are generally more alkaline in character than either the Ninetyeast Ridge basalts or those from the mid-Indian Ocean Ridge.

SITE 215

Location

This drill hole is located in the Central Indian Ocean Basin at 8°07.30'S, 84°47.50'E. Water depth is 5319 meters; approximately 25 meters of basalt were penetrated beneath 151 meters of sediment.

Petrography

The recovered rocks are weathered to moderately fresh basalts and represent a succession of pillow lavas. Chilled margins and glass are common. Plagioclase appears to be the stable liquidus phase as indicated by well formed microphenocrysts. Olivine crystallized for a brief period as skeletal microphenocrysts. Groundmass plagioclase forms a network of randomly oriented skeletal crystals, many showing the typical hollow cores centered on the *a*-axis.

---

[1]Woods Hole contribution No. 3103.

Section zoning is evident in the more calcic cores of the microphenocrysts and in the hollow square shown by sections perpendicular to the *a*-axis of the groundmass grains. Pyroxenes form skeletal, barred intergrowths interstitial to the plagioclase. Magnetite appears only as tiny granules intergrown with pyroxene. Predominance of quench textures suggests that the magma contains a small amount of superheat as might be expected if it had risen rapidly from considerable depths. Thus, crystallization is delayed until the moment of outpouring and quenching.

Chemistry

Major and trace element analyses for five samples are given in Table 1. The concentrations of SiO_2 (47%-50%), Al_2O_3 (15%-17%), TiO_2 (1.6%-1.7%), and total iron as FeO (8%-9%) are similar to those in abyssal oceanic tholeiitic basalts. The relatively high ferric to ferrous iron ratios and high weight loss on ignition indicate weathering—that is, low temperature alteration with seawater. Typically this alteration also results in increased concentrations of K, B, and Li (Hart, 1971; Thompson and Melson, 1970). The sample with the highest water content (ignition loss), J51, has the highest K and Li concentrations. We have no major element data for the glass from Core 18, Section 3 in Table 1. It differs from the other basalts in the relatively high concentrations of Ba and Sr; we have typically found high Ba and Sr concentrations in those oceanic basalt glasses that have been palagonitized.

Comparison with Other Oceanic Rocks

Analyses of the Site 215 basalts with those from other oceanic regions are compared in Table 2. The Site 215 basalts differ from mid-ocean ridge basalts in their much higher concentrations of K_2O, P_2O_5, Ba, Sr, and Zr and their lower concentrations of Mg, Ca, Cr, Ni, and V. Although some potassium enrichment is a result of alteration phenomena, we believe that the high potassium abundance is a primary feature of these basalts. This conclusion results from the significantly different abundance data for trace elements which are less susceptible than potassium to alteration effects, e.g., Ba, Sr, Zr (Cann, 1970) and the observation that despite being more intensely altered the Site 214 basalts (4%-7% ignition loss, Table 3) have much lower potassium contents. We conclude that the basaltic magma at Site 215 had a chemical composition distinctly more "alkaline" than basalts typical of active mid-ocean ridges. Although more alkaline than mid-ocean ridge basalts, they are also clearly different in composition from the alkali basalts of oceanic regions (see Table 2). In

294

TABLE 1
Analyses of Basalts from Central Indian Ocean Basin, Site 215

Core-Section	18-2	18-3	19-1	20-2	18-3
Depth in Core (cm)	106-110	110-112	23-32	20-23	26-35
Lab No.	J51	J52	J53	J56	Glass
SiO_2	46.40	49.68	50.23	48.29	
Al_2O_3	15.76	16.51	16.58	16.21	
Fe_2O_3	8.50[a]	2.59	2.72	2.54	
FeO		5.63	5.69	5.77	
MgO	6.76	6.82	6.47	6.48	
CaO	10.64	10.52	10.39	10.24	
Na_2O		2.87	3.00	2.95	
K_2O	0.92	0.83	0.84	0.88	
TiO_2	1.65	1.63	1.66	1.67	
P_2O_5		0.28	0.29	0.30	
Ignition loss	3.04	2.59	2.74	2.63	
B	5	<2	4	5	<2
Ba	495	430	410	450	800
Co	43	53	43	50	44
Cr	260	265	265	270	315
Cu	65	70	65	55	65
Ga	18	18	18	15	18
Li	28	8	8	9	8
Mo	5	6	4	4	<2
Ni	100	105	95	100	100
Sr	515	380	320	360	750
U[b]	0.5	0.3	0.4	0.4	0.4
V	245	245	255	260	260
Y	35	38	35	36	42
Zr	170	160	160	155	135

Note: Major elements in percent weight; analyses by X-ray fluorescence taken from Hekinian (Chapter 17). Trace elements in ppm; analyses by emission spectrometry. See Thompson and Bankston (1969) for details on precision and accuracy.

[a]Total Fe as FeO.

[b]Analyzed by fission track technique (Fisher, 1970).

particular the Site 215 basalts have higher concentrations ot Mg, Ca, Co, Cr, and Ni and lower concentrations of K, P, Ba, Sr, Y, and Zr than the alkali basalts.

Although Site 215 is close to the Ninetyeast Ridge, the basalts are clearly different in composition from those at Site 214 (see Tables 2 and 3). Trace element data for the basalt from the East Indian Ocean Basin (Site 213) are not available, but the major element data for this basalt compare well with those for Site 215 (Chapter 17).

SITES 214 AND 216

Location

Sites 214 and 216 are located on the Ninetyeast Ridge at 11°20.21'S, 88°43.08'E and 1°27.73'N, 90°12.48'E, respectively. Water depth at Site 214 is 1665 meters, and approximately 56 meters of basement were penetrated beneath 440 meters of sediment. At Site 216 water depth is 2247 meters, and 27.5 meters of basement were penetrated beneath 457 meters of sediment.

Petrography

Basalts recovered at both sites were often vesicular and amygdaloidal. The upper portion of the igneous rocks at Site 214 (Cores 48 to 51) was composed of pilotaxitic trachytic-textured rocks differing in composition from the lower basalts. Only basalts were recovered at Site 216. However, all of the Site 214 and 216 lavas are similar in their textures and mineral paragenesis in spite of the contrasts in composition. The principal minerals are plagioclase, pyroxene, and magnetite. Apparently bulk compositional differences must be due to subtle variations in composition and proportions of these mineral phases. Difference in degree of alteration is a complicating factor; in general, the upper, more siliceous lavas of Site 214 are less altered than the lower basalts here and at Site 216.

Plagioclase and magnetite are the prominent phenocryst or microphenocryst phases in these lavas. Pyroxene does occur in minor amounts as coarser granules associated with the plagioclase and magnetite as noted by Hekinian (Chapter 17). Pyroxene and especially magnetite are very rare as phenocryst (liquidus) phases in ocean ridge basalts. Much of the magnetite has been marginally oxidized, but overall the relatively large subhedral crystals appear to be of primary magmatic origin and closely follow the precipitation of plagioclase and pyroxene. This suggests that the high total iron concentration in these lavas is a primary magmatic feature, although the relatively high ferric to ferrous iron ratio may have been enhanced by later oxidation. There is no evidence that olivine has ever crystallized in the Site 214 and 216 rocks, as its presence is not even indicated by crystal pseudomorphs.

The trachytic texture of the groundmass in these lavas is also typical. It implies that a considerable amount of crystalline material was in suspension in the liquid at the time of extrusion. This in turn implies that magmatic temperatures were at or below the liquidus. In contrast, typical ocean ridge pillow basalts show a predominance of randomly oriented groundmass minerals displaying delicate quench textures as described by Bryan (1972) and seen in the basalts of Site 215. Such delicate features could not survive shearing stresses produced by flow. Thus, the textures suggest that the Ninetyeast Ridge lavas reached thermal and chemical equilibrium in relatively shallow magma chambers.

Chemistry

Major and trace element analyses of the basalts from the lower part of Site 214 and the basalts from Site 216 are given in Table 3. Hekinian has divided the basalts of Site 214 into two groups—nonvesicular (Core 53 and Core 54, Section 2) and vesicular (Core 54, Section 3). No significant compositional differences between these two groups was found. However, within the basalts of Sites 214 and 216 a range of compositions that was believed to reflect both alteration and differentiation processes was noted.

TABLE 2
Comparative Analyses of Basalts from Site 215 and Those from Other Oceanic Regions

		Mid-Ocean Ridges			Indian Ocean			
	Site[a] 215	Indian[b]	Indian[c]	Atlantic[d]	Amirante[e] Ridge	Ninety-east[f] Ridge	Eastern[g] Basin	Oceanic[h] Alkali Basalts
SiO_2	49.40	49.14	50.47	49.21	49.17	45.29	48.37	47.41
Al_2O_3	16.43	16.22	16.96	15.81	16.62	14.00	16.18	18.02
Fe_2O_3	2.62	2.70	0.91	2.21	3.94	6.08	6.07	4.17
FeO	5.70	7.57	7.14	7.19	6.23	8.25	2.98	5.80
MgO	6.59	7.62	8.29	8.53	7.55	6.07	6.20	4.79
CaO	10.38	10.83	11.56	11.14	10.83	8.51	10.08	8.65
Na_2O	2.94	3.02	2.72	2.71	2.49	2.59	3.04	3.99
K_2O	0.85	0.18	0.16	0.26	0.20	0.35	0.85	1.66
TiO_2	1.65	1.65	1.04	1.39	1.00	2.21	1.63	2.87
P_2O_5	0.29		0.12	0.15	0.11	0.18	0.30	0.92
Ignition loss	2.65		0.22		1.48	5.83	4.00	2.45
B	3			7		46		
Ba	430			12	8	45		498
Co	49	73	26	41	33	65		25
Cr	267	347	325	296	225	38		67
Cu	63	90	62	87	165-	140		36
Ga	17	20	13	18	15	20		22
Li	8			8		11		11
Mo	5					3		
Ni	100	242	130	123	95	50		51
Sr	353	131	185	123	42	265		815
U	0.4					0.1		
V	253	340	150	289	305	525		252
Y	36	45	36	43	25	26		54
Zr	158	105	73	100	40	120		333

[a]Average J52, 53, 56 for major elements, J51, 52, 53, 56 for trace elements (see Table 1).

[b]Average 6 basalts from Carlsberg Ridge, from Cann (1969).

[c]Average 2 basalts from Mid-Indian Ocean ridge, from Engel and Fisher (1969).

[d]Average 33 basalts from Mid-Atlantic Ridge, from Melson and Thompson (1971).

[e]Average 3 basalts from Amirante Ridge, from Fisher et al. (1968).

[f]Average 5 basalts from Ninetyeast Ridge, Site 214 (see Table 3).

[g]Basalt from East Indian Ocean Basin, from Engel et al. (1965b).

[h]Average 10 oceanic alkali basalts, from Engel et al. (1965a).

Alteration Effects

The most obvious indicators of alteration are high weight loss on ignition and the extensive oxidation of iron. Those samples with the largest water content (ignition loss) also contain the highest concentrations of B and Li. Within the basalts the correlation of alteration with Li is fairly good and better than that with B. This may be due to some hydrothermal alteration as well as halmyrolysis; the former would tend to leach out boron (Thompson and Melson, 1970). In these particular samples there is no increase in U content with increasing degree of alteration, contrary to the observations made by Aumento (1971). In fact there may well be a negative correlation of U concentration with degree of alteration. Alteration also increases the alkali contents of basalts (Hart, 1969). However, these particular samples contain secondary smectites and calcite in the amygdales and vesicle walls (see Hekinian's descriptions, Chapter 17). These minerals would increase the K, Mg, and Sr concentrations (Thompson, 1972; Melson and Thompson, 1973).

Differentiation Effects

Differences attributable to fractional crystallization can be distinguished from alteration effects if one utilizes trace

TABLE 3
Analyses of Basalts from Sites 214 and 216, Ninetyeast Ridge

	Site 214						Site 216						
Core-Section	53-1	53-1	54-2	54-2	54-3	54-3	36-4	37-1	37-2	37-3	38-1	38	38-4
Depth in Core (cm)	26-30	97-100	28-30	117-127	0-4	42-50	33-40	98-104	80-87	85-90	117-122	cc	80-86
Lab No.	J42	J44	J45	J46	J47	J48	J22	J23	J25	J26	J27	J30	J32
SiO_2	45.50	43.96	47.71	45.32	43.94			45.76	48.14	48.07		48.67	
Al_2O_3	14.70	13.51	13.47	14.70	13.65			13.07	13.00	13.22		12.63	
Fe_2O_3	6.82	5.99	5.21	5.31	7.09			6.49	6.46	5.41		6.04	
FeO	7.69	8.96	8.00	8.00	8.61			8.00	8.00	8.00		7.15	
MgO	6.22	6.59	5.87	5.59	6.07			6.22	5.67	6.85		6.55	
CaO	8.88	8.37	9.98	8.50	6.84			8.16	9.18	7.42		9.10	
Na_2O	2.40	2.28	2.60	2.73	2.95			2.53	2.41	2.63		2.30	
K_2O	0.21	0.30	0.18	0.81	0.26			1.29	0.73	0.57		0.90	
TiO_2	2.12	2.03	2.25	2.14	2.50			2.71	2.65	2.76		2.50	
P_2O_5	0.17	0.16	0.18	0.18	0.21			0.22	0.21	0.21		0.19	
Ignition loss	5.80	7.16	4.31	5.28	6.62			4.26	2.29	4.90		3.71	
B	40	200	<2	6	20	10	5	8	3	<2	2	4	2
Ba	22	28	70	55	60	35	230	155	105	95	115	100	180
Co	57	78	61	49	80	66	62	50	50	53	56	47	52
Cr	40	50	30	40	35	30	165	25	20	20	22	25	35
Cu	50	140	185	175	160	125	110	55	80	210	60	70	85
Ga	19	18	20	21	20	20	22	21	17	20	17	19	20
Li	13	25	6	8	5	6	28	66	6	12	6	11	13
Mo	<2	3	7	4	2	4	3	5	7	6	6	8	5
Ni	55	60	45	45	50	45	65	40	35	40	35	45	45
Sr	165	185	150	355	365	370	610	220	145	115	150	170	240
V	450	580	430	450	680	555	415	540	415	455	470	395	415
Y	27	13	40	35	15	24	26	31	27	28	32	37	35
Zr	110	95	160	145	105	110	135	155	150	175	150	180	170
U[a]	0.1	0.04	0.1	0.1	0.07	0.2	0.1	0.3	1.0	0.2	0.3	0.2	0.1
Sc[b]	45			45									
Hf[b]	4			4.2									
Ta[b]	1.4			2.8									
La[b]	6.6			8.9									
Cd[b]	20.3			22.5									
Nd[b]	14.3			15.1									
Sm[b]	5.1			4.9									
Eu[b]	1.7			1.8									
Tb[b]	0.8			0.9									
Yb[b]	2.4			2.6									
Lu[b]	0.4			0.4									

Note: See note, Table 1.

[a]Analyzed by fission track technique.

[b]Analyzed by neutron activation.

element correlations which are believed to be unaffected by weathering. Within the basalts of Site 214 the increasing concentrations of Ba, Y, Zr, Ti, and P are correlated with decreasing concentrations of Cr, Co, Ni, V, and Mg. Samples J44 and 47 are the least fractionated and Samples J45 and 46 are the most fractionated (Figure 1). The basalts of Site 216 are more closely grouped in composition, but the same trend can be recognized with J30 as the most fractionated (high concentrations of Y and Zr). The relatively high concentrations of K, Ba, Y, and Zr and low concentrations of Co, Ni, and V in the Site 216 basalts compared with those at Site 214 indicate them to be slightly more differentiated. This is shown in Figure 1 where the average Site 216 basalt falls at the highly differentiated end of the trends.

Comparison with Other Rock Types

In Table 4 chemical analyses of the Site 214 and 216 basalts with those from other oceanic regions are compared. Although the Ninetyeast Ridge basalts are altered, much of their present chemical characteristics (particularly if recalculated on a water and CO_2-free basis) reflect the original magma composition. The Ninetyeast Ridge basalts are lower in Ca, Mg, Cr, and Ni concentrations and higher in Fe, Ti, K, Ba, Cu, Sr, V, and Zr concentrations that mid-ocean ridge basalts. Compared to basalts from Reunion Island in the Indian Ocean, the Ninetyeast Ridge basalts are higher in Fe and V concentrations but lower in Ca, K, Ba, Cr, Ni, and Sr concentrations. Compared to basalt from the Tonga Islands (which are also on a long linear ridge), the Ninetyeast Ridge basalts have higher concentrations of Fe, Na, K, Ti, P, Ba, Co, Cu, Ni, Sr, V, Y, and Zr and lower concentrations of Al, Ca, and Cr.

The basalts from Sites 214 and 216 are chemically similar to Icelandic tholeiitic basalts, particularly those from the Veidivötn region. In fact the Ninetyeast Ridge basalts differ from mid-Indian Ocean Ridge basalts in the same ways chemically that Jakobsson (1972) concluded Icelandic tholeiitic basalts differed from mid-Atlantic Ridge basalts in that, "Icelandic basalts are distinctly higher in Fe, Ti, K with lower Al, Mg and Na/K ratios." Also in Figure 2 the REE distributions are shown, relative to chondrites, for Site 214, Icelandic and Oceanic basalts. Normal mid-ocean ridge basalts are typically depleted in the light REE, while basalts from Site 214 are slightly enriched in these elements and have a very similar distribution and total REE content to tholeiitic basalts from Iceland.

Oceanic Andesites—Site 214

The upper lavas of Site 214 (Cores 48 to 51) differ compositionally from the lower basalts. Major and trace element analyses of these upper lavas are given in Table 5. They are characterized by high SiO_2 (54%-58%), Al_2O_3 (15%-16%), total iron as FeO (<11%), Na_2O (>3%-5%), and K_2O (>1%) and low MgO (<3%), CaO (<6%), and TiO_2 (<1%-5%). Hekinian has called these rocks intermediate differentiated rocks or oceanic andesites—the latter term is used here.

These andesites are much less altered than the basalts from the same site. Ignition loss, iron oxidation, and B and Li concentrations are generally lower. The sample with the

greatest ignition loss, J41, has the greatest Li concentration. Secondary minerals are also present in the vesicles, particularly calcite. Sample J41 has the highest Ca and Sr concentration attributable to the presence of a calcite vein.

Compositionally there is little variation within the group, although J41 appears to be the most differentiated (high Ba, Y, Zr, and P, with low Co and Mg abundances). The variations of Ba, Y, Zr, P, and Co in these oceanic andesites follow the same trend as in the basalts and the data are consistent with a process in which the basalts are parent magmas for the oceanic andesites. The close association in space and time (little sedimentation between flows) also supports a genetic relationship.

In Table 6 analyses of the Site 214 andesites with andesites or similar rocks from other oceanic regions are compared. Compared to the Tonga Island arc andesites, the Ninetyeast Ridge andesites have lower concentrations of Ca, Fe, Mg, Cr, Cu, Ni, and V and have higher concentrations of Na, K, Ti, P, Ba, Sr, and Zr. The Ninetyeast Ridge andesites have lower Fe concentrations than their parent basalts which is contrary to the Tonga andesites. A similar comparison can be made with the average circum-Pacific andesite computed by Taylor et al. (1969). Compared to the mugearites of Reunion Island in the Indian Ocean, the Ninetyeast Ridge andesites have lower concentrations of Ti, Na, K, Ba, Zr, and V.

Because the Ninetyeast Ridge basalts are similar in composition to Icelandic tholeiitic basalts and because a similar association of basalts and andesites occurs in Iceland (Thorarinsson, 1967; Jakobsson, 1972), analyses of andesites from Hekla, a volcano in the neovolcanic zone of Iceland, are shown. The similarity of these andesites to those of the Ninetyeast Ridge, particularly in major element composition, is evident in Table 6. The slopes of the REE distribution patterns (Figure 2) are also very similar. Neither the Ninetyeast Ridge andesites nor the Hakla andesites have REE abundances like calc-alkali andesites (Ewart et al., in press).

Some compositional differences between the Hekla and Ninetyeast Ridge andesites, e.g., the higher REE and Zr concentrations in Hekla samples were noted. Jakobsson (1972) has indicated that Hekla andesites may be derived from transitional Icelandic tholeiites which are slightly more alkaline than normal Icelandic tholeiites—those most closely comparable to the Ninetyeast Ridge basalts (see Table 4). Thus, the compositional differences between the Hekla and Ninetyeast Ridge andesites are consistent with slight compositional differences in their respective parent basalts. Nevertheless, the chemical similarity of these two andesitic suites is suggestive of a similar origin.

Origin of the Ninetyeast Ridge

The basalts and andesites from the Ninetyeast Ridge differ in texture, mineralogy, and chemistry from rocks extruded at spreading centers along mid-ocean ridges and from those found on the floors of the Indian Ocean basins. They also differ markedly in composition from lavas extruded in island arcs. Thus the Ninetyeast Ridge probably does not represent an old, previously active mid-ocean ridge and site of sea-floor spreading nor does it appear to be an island arc-type of structure.

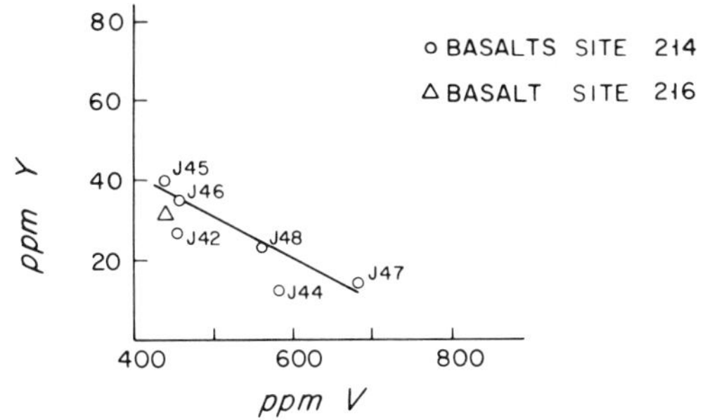

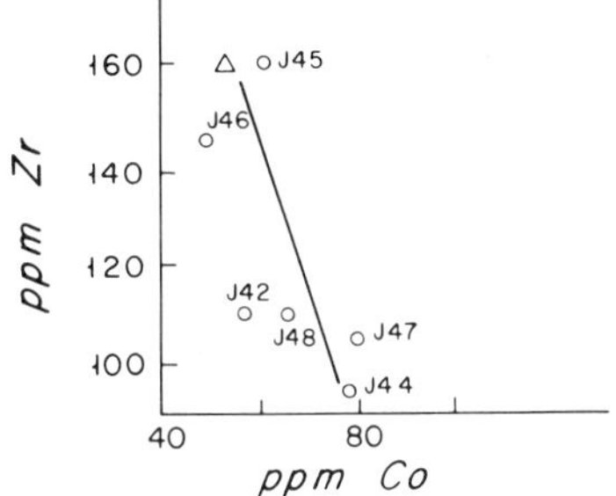

Figure 1. *Differentiation trends in the Site 214 basalts as indicated by the negative correlation of Y with V and Zr with Co. Circles represent Site 214 basalts, individual sample numbers are shown. The triangle represents the average composition of Site 216 basalts.*

Because of the similarity in mineralogy and chemical composition between basalts and andesites of the Iceland Plateau and Ninetyeast Ridge, a similar derivation is suggested. The Iceland Platform is presumed to represent the surface expression of a "hot-spot" or plume of molten material derived from deep in the mantle (Schilling, in press). Such a hypothesis for the Ninetyeast Ridge would lead to difficulties in explaining the linearity and present inactivity of the ridge and would require a change in the spreading rate of the Indian plate to be consistent with the hypotheses presented in the Regional Synthesis (Chapter 41).

The shipboard scientists conclude that the northern part of the Ninetyeast Ridge (e.g., Site 216) could be formed by passage of the Indian plate over a magma chamber. However, they argue that the southern part of the ridge (e.g., Site 214) may be better explained as a leaky transform fault (Chapter 41). Studies of such transform faults in the Atlantic (Thompson and Melson, 1972) indicate the eruptive rocks to be different from normal mid-ocean ridge basalts and more alkaline than the basalts of the Ninetyeast Ridge. Our data further indicate that the basalts at Site 214 are very similar in texture, mineralogy, and composition to those at Site 216, implying similar origins. It is suggested that the fractionated lavas of the Ninetyeast Ridge may represent the surface expression of a former mantle plume with consequent uplift of the lithosphere as it moved over the hot spot.

Hekinian has suggested that the Ninetyeast Ridge lavas are similar in composition to those found on St. Paul and New Amsterdam islands in the Indian Ocean. There are not enough data, particularly on trace element compositions, to make a thorough comparison. It would imply that these islands may be an extension of the Ninetyeast Ridge to the south, although they do appear to be offset from the main strike of the ridge.

ACKNOWLEDGMENTS

We thank Roger Hekinian for supplying the samples and giving us access to his major element data and thin sections. This work was supported by NSF Grant GA31244.

REFERENCES

Aumento, F., 1971. Uranium content of mid-oceanic basalts: Earth Planet. Sci. Lett., v. 11, p. 90-94.

Bryan, W. B., 1972. Morphology of quench crystals in submarine basalts: J. Geophys. Res., v. 77, p. 5812-5819.

Bryan, W. B., Stice, G. D., and Ewart, A., 1972. Geology, petrography, and geochemistry of the volcanic islands of Tonga: J. Geophys. Res., v. 77, p. 1566-1585.

Cann, J. R., 1969. Spilites from the Carlsberg Ridge, Indian Ocean. J. Petrology, v. 10, p. 1-19.

————, 1970. Rb, Sr, Y, Zr, Nb in some ocean floor basaltic rocks: Earth Planet. Sci. Lett., v. 10, p. 7-11.

Engel, A. E. J., Engel, C. G., and Havens, R. G., 1965a. Chemical characteristics of oceanic basalts and the upper mantle: Geol. Soc. Am. Bull., v. 76, p. 719-734.

Engel, C. G., Fisher, R. L., and Engel, A. E. J., 1965b. Igneous rocks of the Indian Ocean floor: Science, v. 150, p. 605-609.

Engel, C. G. and Fisher, R. L., 1969. Lherzolite, anorthosite, gabbro and basalt dredged from the mid-Indian Ocean Ridge: Science, v. 166, p. 1136-1141.

Ewart, A., Bryan, W. B., and Gill, J., in press. Mineralogy and geochemistry of the younger volcanic islands of Tonga, Southwest Pacific: J. Petrology.

Fisher, D. E., 1970. Homogenized fission track determination of uranium in whole rock geologic samples: Anal. Chem., v. 42, p. 414-416.

Fisher, R. L., Engel, C. G., and Hilde, T. W. C., 1968. Basalts dredged from the Amirante Ridge, western Indian Ocean: Deep-Sea Res., v. 15, p. 521-534.

Hart, S. R., 1969. K, Rb, Cs contents and K/Rb, K/Cs ratios of fresh and altered submarine basalts: Earth Planet. Sci. Lett., v. 6, p. 295-303.

————, 1971. K, Rb, Cs, Sr and Ba contents and Sr isotope ratios of ocean floor basalts: Phil. Trans. Roy. Soc. Lond. A, v. 268, p. 573-587.

Haskin, L. A., Helmke, P. A., Paster, T. P., and Allen, R. O., 1971. Rare earths in meteoritic, terrestrial and lunar materials: *In* Activation analysis in geochemistry and cosmochemistry, Brunfelt, A. O. (Ed.), E. Steinnes. Proc. NATO Advanced Studies Institute, Norway (Scand. Univ. Books), p. 201-208.

Jakobsson, S. P., 1972. Chemistry and distribution pattern of Recent basaltic rocks in Iceland: Lithos, v. 5, p. 365-386.

Melson, W. G., and Thompson, G., 1971. Petrology of a transform fault zone and adjacent ridge segments: Phil. Trans. Roy. Soc. Lond. A, v. 268, p. 423-441.

TABLE 4
Comparative Analyses of Basalts from Site 214 and 216, Ninetyeast Ridge and Those from Other Oceanic Regions

	Site[a] 214	Site[b] 216	Mid-Indian[c] Ocean Ridge	Mid-Atlantic[d] Ridge	Reunion[e] Island	Tonga[f] Island Arc	Icelandic Tholeiites Rekjanes[g] Peninsular	Icelandic Tholeiites Veidivötn[h] Region	Icelandic Transitional Torfajokull[i] Region
SiO_2	45.29	47.66	49.14	49.21	45.99	49.19	48.01	49.51	47.00
Al_2O_3	14.00	12.98	16.22	15.81	15.65	20.61	14.09	13.97	13.84
Fe_2O_3	6.08	6.10	2.70	2.21	5.95	2.93	2.69	2.15	2.74
FeO	8.25	7.79	7.57	7.19	6.50	5.63	10.06	11.37	12.68
MgO	6.07	6.32	7.62	8.53	7.66	5.93	8.29	5.97	5.96
CaO	8.51	8.46	10.83	11.14	12.60	11.78	11.77	10.76	9.74
Na_2O	2.59	2.47	3.02	2.71	2.39	1.21	2.17	2.71	2.98
K_2O	0.35	0.87	0.18	0.26	0.71	0.24	0.29	0.42	0.65
TiO_2	2.21	2.65	1.65	1.39	2.18	0.42	1.87	2.39	3.71
P_2O_5	0.18	0.21		0.15	0.32	0.04	0.19	0.23	0.33
Ignition loss	5.83	3.79				2.06	0.28	0.42	0.38
B	46	4		7					
Ba	45	140		12	310	14			
Co	65	53	73	41		30			
Cr	38	45	347	296	110	75			
Cu	140	96	90	87		51			
Ga	20	19	20	18		13			
Li	11	12		8					
Mo	3	6							
Ni	50	44	242	123	120	25			
Sr	265	235	131	123	390	115			
U	0.1	0.3							
V	525	445	340	289	340	230			
Y	26	31	45	43	22	12			
Zr	120	159	105	100	125	21			
La	7.7						8.6		
Sm	4.9						4.6		
Yb	2.6						2.7		

[a]Average 6 basalts from Table 3.

[b]Average 7 basalts from Table 3.

[c]Average 6 basalts from Carlsberg Ridge, from Cann (1969). See also other tabulations in Table 2.

[d]Average 33 basalts from Mid-Atlantic Ridge, from Melson and Thompson (1971).

[e]Average basalt, Reunion Island, Re 512G groundmass, from Upton and Wadsworth (1972).

[f]Basalt, Eua Island (111549-7), from Bryan et al. (1972).

[g]Average 12 basalts, Rekjanes Peninsula, from Jakobsson (1972); lanthanide elements from Rekjanes peninsula, from Schilling (in press).

[h]Average 11 basalts, Veidivötn Region, from Jakobsson (1972).

[i] Average 9 basalts, Torfajökull Region, from Jakobsson (1972).

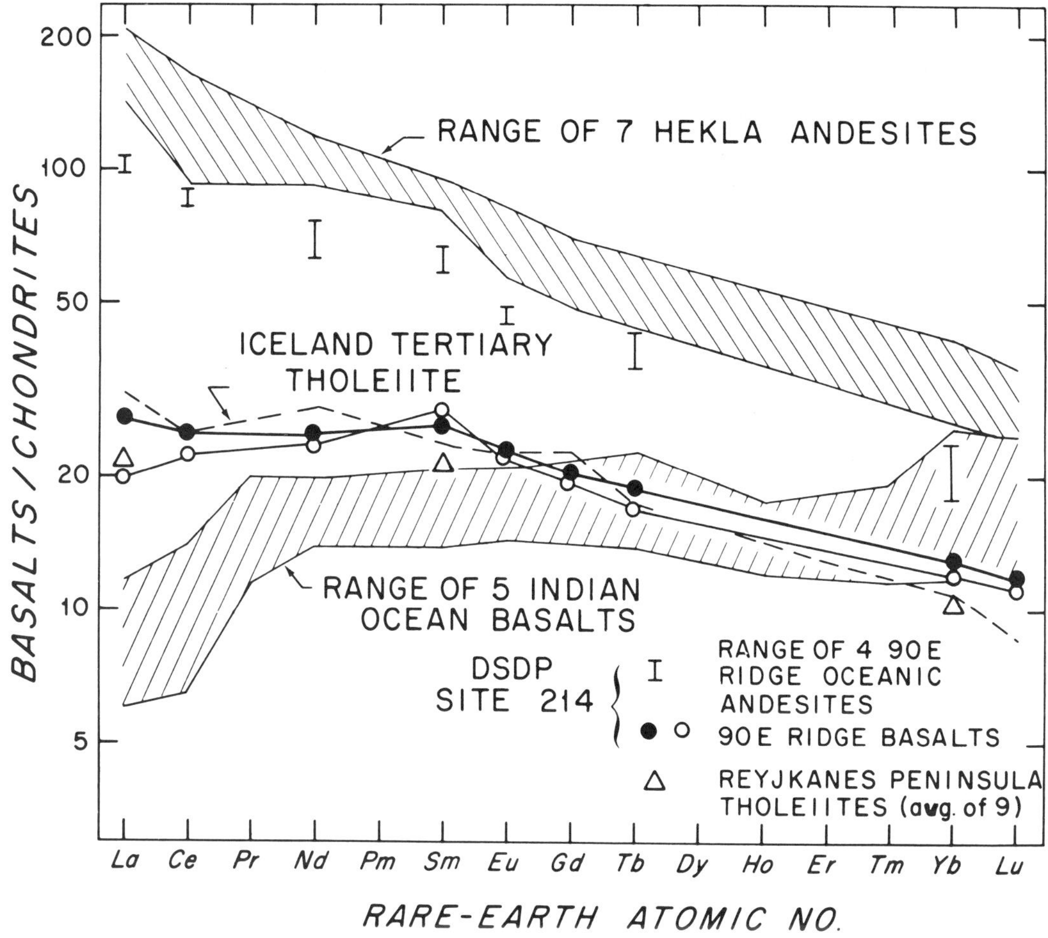

Figure 2. *Rare-earth element comparison plot showing data for two Site 214 Ninetyeast Ridge basalts (J46 filled circle and J42 open circle) and the range for four Site 214 Ninetyeast Ridge andesites (J35, J37, J38, J40). Also shown is the range for seven Hekla andesites (unpublished), the range for four mid-Indian Ocean ridge basalts (Schilling, 1971), the average of nine Reykjanes Peninsula tholeiites (Schilling, in press), and the average of 11 analyses of tertiary Icelandic tholeiitic basalts (Haskin et al., 1971).*

————, 1973. Glassy abyssal basalts, Atlantic sea floor near St. Paul's Rocks: petrography and composition of secondary clay minerals: Geol. Soc. Am. Bull., v. 84, p. 703-716.

Schilling, J. G., 1971. Sea-floor evolution; rare-earth evidence: Phil. Trans. Roy. Soc. Lond. A, v. 268, p. 663-706.

————, in press. Iceland mantel plume existence, and influence along the Reykjanes Ridge: geochemical evidence: Nature.

Taylor, S. R., Capp, A. C., Graham, A. L. and Blake, D. H., 1969. Trace element abundances in Andesites II. Saipan, Bougainville and Fiji: Contr. Min. Petrol., v. 12, p. 1-26.

Thompson, G., 1972. A geochemical study of some lithified carbonates from the deep-sea: Geochim. Cosmochim. Acta, v. 36, p. 1237-1253.

Thompson, G. and Bankston, D. C., 1969. A technique for trace element analysis of powdered materials using the d.c. arc and photoelectric spectrometry: Spectrochim. Acta, v. 24B, p. 335-350.

Thompson, G. and Melson, W. G., 1970. Boron contents of serpentinites and metabasalts in the oceanic crust: implications for the boron cycle in the oceans: Earth Planet. Sci. Lett., v. 8, p. 61-65.

————, 1972. The petrology of oceanic crust across fracture zones in the Atlantic Ocean: evidence of a new kind of sea-floor spreading: J. Geology, v. 80, p. 526-538.

Thorarinsson, S., 1967. Hakla and Kapla: *In* Iceland in Mid-ocean ridges, Bjornsson, S. (Ed.), Soc. Scient. Islandica, p. 190-199.

Upton, B. G. J. and Wadsworth, W. J., 1972. Aspects of magmatic evolution on Reunion Island: Phil. Trans. Roy. Soc. Lond. A, v. 271, p. 105-130.

TABLE 5
Analyses of Oceanic Andesites, Site 214 Ninetyeast Ridge

Core-Section	48-1	48-2	48-2	49-1	50-1	51-1
Depth in Core (cm)	76-83	9-13	117-123	137-144	145-150	108-114
Lab No.	J33	J35	J37	J38	J40	J41
SiO_2	56.69	55.70	54.59	57.41	55.70	51.34
Al_2O_3	16.10	15.40	15.67	15.81	15.30	18.73
Fe_2O_3	2.88	3.11	2.42	2.83	2.91	3.80
FeO	7.55	7.00	7.47	7.00	7.00	2.72
MgO	2.57	2.40	2.60	2.32	2.30	1.02
CaO	5.62	5.59	5.82	5.86	5.71	7.97
Na_2O	4.08	3.88	3.88	3.92	3.78	4.21
K_2O	1.56	1.55	1.27	1.58	1.49	1.83
TiO_2	1.40	1.40	1.43	1.46	1.44	1.56
P_2O_5	0.56	0.63	0.68	0.65	0.66	0.80
Ignition loss	1.22	1.15	1.99	1.18	1.42	5.06
B	<2	<2	<2	<2	<2	<2
Ba	640	600	530	590	530	780
Co	41	40	34	44	38	25
Cr	<5	<5	<5	<5	<5	<5
Cu	3	2	1	3	2	1
Ga	22	23	24	22	23	29
Li	6	7	8	4	4	11
Mo	<2	3	3	3	5	<2
Ni	<5	<5	<5	<5	<5	<5
Sr	800	625	615	615	580	1500
V	40	40	40	40	35	70
Y	63	67	73	62	62	85
Zr	235	255	290	245	235	380
U[a]	0.4	0.6	0.4	0.8	0.9	1.0
Sc[b]		17	15	17	17	
Hf[b]		8.7	7.8	8.1	8	
Ta[b]		10.4	7.4	10.3	11	
La[b]		30.8	34	33	33	
Ce[b]		85	73	78	73	
Nd[b]		461	38	41	44	
Sm[b]		10.9	11.8	11.3	11.1	
Eu[b]		3.35	3.02	3.21	3.1	
Tb[b]		1.98	1.57	1.76	1.9	
Yb[b]		4.85	3.66	4.46	4.6	
Lu[b]		0.97		0.88	1.0	
Th[b]					4.5	

Note: See note, Table 1.

[a]Analyzed by fission track technique.

[b]Analyzed by neutron activation.

TABLE 6
Comparative Analyses of Andesites from Site 214, Ninetyeast Ridge,
and Similar Rocks from Other Oceanic Regions

	Site[a] 214	Mugearite[b] Reunion Island	Andesite[c] Tonga Island Arc	Andesite[d] Circum- Pacific Type	Andesite[e] Hekla Iceland
SiO_3	56.02	52.1	57.57	59.5	55.57
Al_2O_3	15.66	17.1	14.14	17.2	15.12
Fe_2O_3	2.83	2.0	3.60		4.63
FeO	7.20	8.7	8.62	6.10*	7.14
MgO	2.44	2.6	3.29	3.42	2.32
CaO	5.72	6.9	8.42	7.03	6.61
Na_2O	3.91	4.9	2.42	3.68	4.07
K_2O	1.49	2.4	0.70	1.60	1.30
TiO_2	1.43	2.3	0.80	0.70	1.92
P_2O_5	0.64	0.7	0.12		
Ignition loss	1.39				
B	<2		18	<10	<10
Ba	578	800	165	270	395
Co	39		31	24	9
Cr	<5	8	6	56	<2
Cu	2	<10	215	54	10
Ga	23		19	16	25
Li	6			10	20
Mo	3				
Ni	<5	<10	9	18	7
Sr	647	650	225	385	485
U	0.6				
V	39	75	410	175	83
Y	65	47	25	21	39
Zr	252	390	40	110	450

[a]Average 5 andesites from Table 5, excluding J41.

[b]Average 6 mugearites from Reunion Island, from Upton and
Wadsworth (1972).

[c]Andesite, Late Island (111547-13), from Ewart et al. (in press).

[d]Average circum Pacific andesite, from Taylor et al. (1969).

[e]Average 3 andesites Hekla, Iceland (USNM 112046, 112030,
112050), unpublished.

303

44

AZORES MANTLE BLOB: RARE-EARTH EVIDENCE

J.-G. SCHILLING

Graduate School of Oceanography, University of Rhode Island, Kingston, R.I. (USA)

Received October 25, 1974
Revised version received January 2, 1975

Rare earths (RE) in basalts erupted within the rift of the Mid-Atlantic Ridge show a progressive change from light-RE enriched to depleted patterns from the Azores Platform (40°N) down to 33°30'N. South, the pattern remains light-RE depleted as along other "normal ridge" segments. A progressive increase in chemical variability of the basalts towards the Azores is also noted.

The latitudinal RE profile and corresponding $\Sigma FeO/\Sigma FeO + MgO$ variations, together, indicate that the origin of these basalts cannot be accounted for simply by considering variable extents of partial melting of a single mantle source and subsequent fractional crystallization during the ascent of the magmas. These two processes produce only second-order effects on the RE patterns. The data requires the presence of a distinct, light-RE richer, mantle source beneath the Azores Platform relative to that of south of 33°30'N and an intermediate zone where both mantle types mix. The relative contribution of the Azores mantle source to the mix appears to decrease fairly regularly southward along the ridge and becomes negligible at 33°30'N. Increasing chemical variability of the basalts towards the Azores is probably caused by correspondingly larger extent of fractional crystallization at shallow depth, and/or greater variability in the extent of partial melting, apparently subsequent to, and superimposed on the mixing of the two mantle sources.

The combined morphological, geophysical and RE evidence along the profile are consistent with a model suggesting upwelling of a major blob (plume) under the Azores Plateau; and reveal the present extent of the blob's overflow and mixing with the asthenosphere depleted in large ionic lithophile trace elements. The influence of the Azores blob is geochemically detectable up to 1000 km southwestward beneath the ridge axis.

304

45

TRACE ELEMENT GEOCHEMISTRY OF THE EAST MOLOKAI VOLCANIC SERIES, HAWAII

DAVID A. CLAGUE* and MELVIN H. BEESON

U.S. Geological Survey, 345 Middlefield Road,
Menlo Park, California 94025

ABSTRACT. Part of the East Molokai Volcanic Series crops out in a cliff above the Kalaupapa Peninsula and is accessible along the trail leading to the peninsula. Twenty-four of the more than 80 exposed flows were sampled and analyzed for major and trace elements. Microprobe analyses of the phenocryst and groundmass phases are presented elsewhere. The flows are mostly transitional to tholeiitic basalt near the base of the 490-m cliff but are nearly all alkalic basalt, hawaiite, and mugearite near the top. The transitional and alkalic flows are interbedded over at least 200 m of section. Ratios of the incompatible elements, particularly those involving either K or Ba, distinguish the alkalic basalts (K/Ba = 26) from the transitional basalts (K/Ba = 16). Pyrolite-based partial melting models indicate that the alkalic basalt is generated by about 18 percent partial melting and the transitional basalt by 23 to 25 percent partial melting. The mantle residuum of the alkalic basalt contains more clinopyroxene than that of the transitional basalt. The depth of magma segregation may be deeper for the alkalic basalt, as inferred by higher TiO_2/Al_2O_3 and Na_2O/CaO ratios. The distribution of K_2O and Ba suggests that the mantle residuum of the alkalic basalt contains residual phlogopite, but the residuum of the transitional basalt does not. Both the transitional and alkalic basalts have undergone slight to moderate shallow fractionation of olivine, clinopyroxene, and plagioclase. Models for the differentiation of alkalic basalt to hawaiite involve fractionation of titanomagnetite in addition to olivine, clinopyroxene, and plagioclase.

[*Editors' Note:* Material has been omitted at this point.]

* Department of Geology, Middlebury College, Middlebury, Vermont 05753

MAJOR-ELEMENT CONSTRAINTS ON BASALT GENERATION

The calculated parental magma compositions (table 3) can be used to estimate conditions of melting and mantle mineralogy. The transitional basalts have less Na_2O, K_2O, P_2O_5, and TiO_2 than the alkalic basalts; this is consistent with the transitional basalts being generated by a larger percentage of partial melting. In addition, the transitional basalts have lower ratios of TiO_2/Al_2O_3 (fig. 7) and higher ratios of CaO/Na_2O compared to the relatively undifferentiated alkalic basalts (Al_2O_3 < 15 percent). The higher CaO/Na_2O ratio in transitional basalts is consistent with a mantle residue containing a small amount of clinopyroxene with low jadeite component. The Na_2O content of mantle clinopyroxene tends to increase with increasing pressure (Green and Ringwood, 1967), suggesting the alkalic basalts are generated at slightly greater depths than the transitional basalts. Experimental evidence (MacGregor, 1969) shows that partial melting in the system MgO–SiO_2–TiO_2 yields liquids richer in TiO_2 with increasing pressure. Thus, the TiO_2/Al_2O_3 ratios support the interpretation that the alkalic basalts are generated at greater depths than the tholeiitic basalts. The calculated parental transitional and alkalic basalts both have Al_2O_3/CaO close to the chondritic ratio of 1.2, which suggests that the residual mantle is composed mostly of olivine plus orthopyroxene (Frey, Green, and Roy, 1978).

In general, the younger basalts in the Kalaupapa section appear to have been generated by smaller degrees of partial melting than the older ones, leaving a greater amount of clinopyroxene in the mantle residue. The melting interval also appears to have migrated to greater depth with time, so that the younger alkalic basalts are derived from deeper sources than the older transitional basalts. This trend is far from

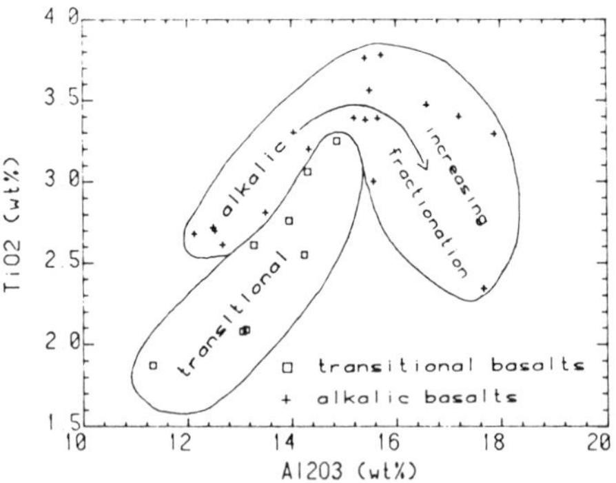

Fig. 7. TiO_2–Al_2O_3 variation diagram, showing lower TiO_2/Al_2O_3 ratio of the transitional compared to the unfractionated alkalic basalts. Note that TiO_2/Al_2O_3 increases and then decreases with increasing fractionation.

perfect, as the transitional and alkalic lavas are interbedded in the section.

TRACE-ELEMENT CONSTRAINTS ON BASALT GENERATION

It is possible that the three calculated parental magmas were generated by partial melting of a heterogeneous mantle. If this were the case, the percentages of partial melting could not be calculated because the model would have too many variables. If we assume that the sources for the alkalic and transitional basalts have the same trace-element composition, we can calculate the relative percentage of partial melting necessary to generate these magmas using an equilibrium melting model. The equation $C_l/C_o = [1/(D + F - PF)]$ where F is the fraction of melting, C_l is the trace-element concentration in the melt, C_o is the initial trace-element concentration in the solid, D is the bulk distribution coefficient of the starting solid, and P is the bulk distribution coefficient of the solid phases that actually melt simplifies to $C_l/C_o = 1/F$ as $D \rightarrow O$, as is the case for P_2O_5 (Shaw, 1970). Using a mantle (pyrolite) P_2O_5 concentration of 0.06 percent (Ringwood, 1966), we estimate that transitional basalt E (−12 percent olivine) results from 25 percent partial melting, transitional basalts G and H result from 23 percent partial melting, and alkalic basalt L (+6 percent olivine) result from 18 percent partial

TABLE 5

Key elemental ratios

	Transitional basalts		Alkali basalt	Chondritic ratios*	Mantle ratios**
	E − 12% ol	Avg. G + H	L + 6% ol		
Al_2O_3/CaO	1.2	1.3	1.2	1.2	—
CaO/Na_2O	6.5	4.7	4.4	—	—
TiO_2/Al_2O_3	0.16	0.16	0.22	—	—
TiO_2/Zr	160	145	140	175	—
TiO_2/Y	1250	710(?)	1030	517	—
TiO_2/V	94	95	90	—	—
TiO_2/Sc	~640	~560	~790	137	—
P_2O_5/Sr	6.4	6.3	7.1	163	9
P_2O_5/Zr	18	18	18	286	28
P_2O_5/Ba	27	21	16	1570	13
P_2O_5/K	1.8	1.3	0.6	3.7	0.5
K/Sr	4.0	4.9	12	44	15.5
K/Zr	10	14	30	77	50
K/Ba	15	16	26	420	23.3
Ba/Sr	0.23	0.30	0.46	0.10	0.66
Ba/Zr	0.67	0.87	1.2	0.18	2.1
Sr/Zr	2.9	2.8	2.5	1.7	3.2
Zr/Y	7.7	4.9(?)	7.3	3.0	—
K/Rb	—	—	430***	338	400

* Chondritic values from Wedepohl (1975) and Sun and Nesbitt (1976).

** Mantle ratios from Sun and Hanson (1975), for alkalic basalts from various locations, Zr ratios based on Frey, Green, and Roy (1978) data on Australian basalts. Sun and Hanson (1975) used $P_2O_5/Ce = 75$; ratios here calculated using $P_2O_5/Ce = 80$ based on 30 new analyses of the Honolulu series (Clague, Frey, and Natland, 1977).

*** Based on K and Rb values for hawaiite (Z), assuming K/Rb ratio has not changed during shallow fractionation (Shaw, 1968).

melting (table 6). These percentages are clearly dependent on the concentration of P_2O_5 chosen to represent the source region. We have assumed that the mantle contains approx 0.06 percent P_2O_5 (Ringwood, 1966), because this concentration of P_2O_5 yields percentages of partial melting consistent with experimental data for Hawaiian basalts (Green and Ringwood, 1967).

Calculations based on Zr indicate the same relative proportions of partial melting as those determined for P_2O_5. McCallum and Charette (1978) have shown that $D_{Zr}^{cpx} = 0.05$ to 0.22 for titaniferous lunar clinopyroxene; these values are probably larger than those for clinopyroxenes in these basalts since Zr probably substitutes for Ti in the M1 site. McCallum and Charette (1978) show that Zr is excluded from plagioclase ($D_{Zr}^{plag} < 0.01$), and we suggest that D_{Zr}^{ol} is very small also. For our calculations we have assumed that the bulk $D_{Zr}^{ol} = O$. With this assumption in mind, we calculate that the mantle concentration of Zr is about 32.5 ppm or 5.2 times that of chondrites. These calculations of other mantle concentrations clearly depend on the assumption that the source is homogeneous with respect to P_2O_5 and that it contains approx 0.06 percent P_2O_5.

Results using K or Ba are inconsistent with those obtained from P_2O_5 and Zr data. Frey, Green, and Roy (1978) noted the poor correlation between K and the incompatible oxide P_2O_5 or the light rare-earth elements for Australian basalts and suggested that the observed K_2O variation could be due to a heterogeneous mantle distribution of phlogopite or to the solubility of K in fluid phases and sensitivity to alteration. The samples analyzed here are all fresh; the observed K_2O variations are not due to alteration. The K_2O contents of samples L (0.69) and

TABLE 6

Partial melting calculations based on pyrolite P_2O_5 content

	P_2O_5	Zr	Ba	K_2O	Y	Sr	K/Ba
Mantle	0.06%	32.5 ppm*	38 ppm*	0.117%*	4.3 ppm*	94 ppm*	25.6
E − 12% OL	0.24	131	87	0.16	17	375	15.3
25% melt*	0.24	130	152	0.47	17	376**	25.6
Avg G+H	0.26	143	125	0.24	29***	410	16.0
23% melt*	0.26	141	165	0.51	19	409**	25.6
L+6% OL	0.33	178	211	0.65	24	460	25.6
18% melt*	0.33	181	211	0.65	24	522**	25.6

*Calculated using percentages of melting inferred from P_2O_5 abundances (see text). The mantle abundances are calculated based on all three parental magmas for Zr, on L+6% OL for Ba and K_2O, on E − 12% OL and L+6% OL for Y, and for E − 12% OL and Avg. G+H for Sr. If source is richer in P_2O_5 than assumed, all calculated mantle concentrations will increase accordingly.

** Assumes no residual clinopyroxene. If generation of the transitional basalts leaves behind clinopyroxene then the magma concentration of Sr is less than 522 ppm. The discrepancy between the 18% partial melt and alkalic basalt L+6% olivine suggests that the alkalic basalts leave more clinopyroxene in the residue.

*** Other transitional basalts (Sample B, for example) contain as little as 18 ppm. This could reflect either variable amounts of residual garnet or some irregularity in the Y data.

M (0.68) from near the bottom and top respectively of a 9 m thick flow
are not significantly different. In addition, there is no evidence from
these samples to suggest that K has been mobile due to fluids.

Because of the uncertainties involved in calculating mantle concentrations we have calculated elemental ratios for the transitional and
alkalic basalts and compared these ratios to those estimated for chondrites and the mantle (table 5). The transitional basalts are depleted in
K_2O relative to Ba (fig. 5), Sr (fig. 8A), Zr (fig. 8B), P_2O_5, and Na_2O
compared to the alkalic basalts. The transitional basalts are also depleted in Ba relative to Sr, Zr, and P_2O_5 compared to the alkalic basalts
but to a lesser degree than for K_2O. The ratios of P_2O_5/Sr, P_2O_5/Zr, and
Sr/Zr (table 5) are similar for both alkalic and transitional basalts, suggesting that the source regions for the basalts are homogeneous with
respect to these elements. However, this source is not identical to that
proposed by Sun and Hanson (1975). The variation of K_2O and Ba could
reflect vertical heterogeneity of mantle composition, or it could indicate
that the concentrations of these elements in the melts are controlled by
a K-bearing trace mineral in the mantle. If we model *equilibrium* partial melting, then only a *residual* K-bearing phase could control K_2O and
Ba (and possibly Rb). Both phlogopite and amphibole have been suggested as possible upper mantle phases (Wyllie, 1971). A magma in

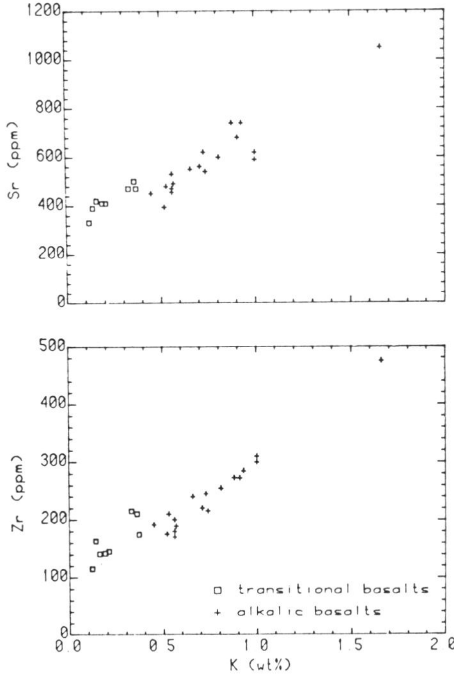

Fig. 8. Sr and Zr plotted against K for the transitional and alkalic basalts. Note
the depletion in K relative to both Sr and Zr of the transitional basalts.

equilibrium with residual phlogopite would have high K/Ba, K/Rb, and K/Sr (Griffin and Murthy, 1969), similar to the alkalic basalts from the Kalaupapa section. If the mantle contained residual amphibole, the equilibrium magma would have low K/Ba, K/Rb, and K/Sr, similar to the transitional basalts. This argument suggests that there may be residual phlogopite in the mantle even after as much as 18 percent partial melting, but that larger degrees of partial melting exhaust the phlogopite. An alternative interpretation is that the source is heterogeneous with respect to K_2O and Ba.

The model calculations in table 6 suggest that the mantle source for both the alkalic and transitional basalts contains 2 to 3 times the chondritic abundance of Y and about 5.2 times the chondritic abundance of Zr. This could indicate that the upper mantle does not have chondritic relative abundances, or that some of the Y is incorporated in residual garnet. The high concentrations of Sc in both the alkalic and transitional basalts argue against garnet in the mantle residuum (Frey, Green, and Roy, 1978).

V probably has a bulk solid/liquid partition coefficient between 0.2 and 0.3 for mantle mineral phases (Frey, Green, and Roy 1978). During partial melting, therefore, V varies inversely with the degree of partial melting. The V data for the calculated parental magmas in table 3 are consistent with the relative degrees of partial melting inferred from the incompatible trace elements.

CONCLUSIONS

1. Flows of transitional and alkalic basalts are interbedded along the trail leading to the Kalaupapa Peninsula on East Molokai.

2. Trace-element ratios such as K/Ba clearly distinguish the transitional and alkalic basalts, as do ratios of some major elements such as TiO_2/Al_2O_3.

3. Models based on incompatible trace elements indicate that the transitional basalts may have been generated by 23 to 25 percent partial melting and the alkalic basalts by about 18 percent partial melting of a pyrolitic mantle assumed to contain 0.06 percent P_2O_5.

4. The alkalic basalts were in equilibrium with mantle residuum containing more clinopyroxene than the mantle residuum after extraction of the transitional basalts. This conclusion supports Beeson's (1976) interpretation of the major-element data.

5. The alkalic basalts are probably generated at slightly greater depth than the transitional basalts, since they appear to be in equilibrium with clinopyroxene having a higher jadeite component.

6. The distribution of K_2O and Ba suggests that the alkalic basalts were in equilibrium with residual mantle phlogopite, whereas the transitional basalts were not, or that the source is heterogeneous, at least with respect to K_2O and Ba.

7A. If, as we have assumed, the magmas were generated from a homogeneous mantle (or one that changes composition continuously

310

downward), the melting does not occur at progressive levels, but rather melting episodes at shallower depths are interspersed with the generally deepening episodes.

7B. If, on the other hand, the mantle is heterogeneous in a manner that cannot be predicted, then we cannot estimate the depths at which the magmas are generated.

8. Shallow fractionation of the alkalic and transitional basalts modified the compositions of nearly all the analyzed samples. The transitional basalts display a sequence of differentiation, with the most differentiated basalt representing about 60 percent residual liquid with the remaining 40 percent removed as olivine, clinopyroxene, and plagioclase. The alkalic basalt differentiates include high FeO and TiO_2 basalts representing about 50 percent residual liquid generated by fractionation of clinopyroxene, olivine, and plagioclase. A hawaiite represents about 30 percent residual liquid; generation of the hawaiite requires late-stage removal of titanomagnetite in addition to the earlier fractionation of olivine, clinopyroxene, and plagioclase.

REFERENCES

[*Editors' Note:* Only the references cited in the preceding excerpts are reproduced here.]

Beeson, M. H., 1976, Petrology, mineralogy, and geochemistry of the East Molokai Volcanic Series, Hawaii: U.S. Geol. Survey Prof. Paper 961, 53 p.

Clague, D. A., Frey, F. A., and Natland, J. W., 1977, Petrogenesis of the Honolulu Volcanic Series: Constraints based on trace element data: Am. Geophys. Union Trans., v. 58, p. 533.

Frey, F. A., Green, D. H., and Roy, S. D., 1978, Integrated models of basalt petrogenesis: A study of olivine tholeiites to olivine melilitites from south eastern Australia utilizing geochemical and experimental petrologic data: Jour. Petrology, v. 19, p. 463-513.

Green, D. H., and Ringwood, A. E., 1967, The genesis of basaltic magmas: Contr. Mineralogy Petrology, v. 15, p. 103-190.

Griffin, W. L., and Murthy, V. R., 1969, Distribution of K, Rb, Sr, and Ba in some minerals relevant to basalt genesis: Geochim. et Cosmochim. Acta, v. 33, p. 1389-1414.

McCallum, I. S., and Charette, M. P., 1978, Zr and Nb partition coefficients: Implications for the genesis of mare basalts, KREEP and sea-floor basalts: Geochim. et Cosmochim. Acta, v. 42, p. 859-870.

MacGregor, I. D., 1969, The system MgO–SiO_2–TiO_2 and its bearing on the distribution of TiO_2 in basalts: Am. Jour. Sci., v. 267-A (Schairer V.), p. 342-363.

Ringwood, A. E., 1966, The chemical composition and origin of the Earth, *in* Hurley, P. M., ed., Advances in Earth Science: Cambridge, Mass., Mass. Inst. Technology Press, p. 287-356.

Shaw, D. M., 1968, A review of K–Rb fractionation trends by covariance analysis: Geochim. et Cosmochim. Acta, v. 32, p. 573-601.

———— 1970, Trace element fractionation during anatexis: Geochim. et Cosmochim. Acta, v. 34, p. 237-243.

Sun, S. S., and Hanson, G. N., 1975, Origin of Ross Island basanitoids and limitations upon the heterogeneity of mantle sources for alkali basalts and nephelinites: Contr. Mineralogy Petrology, v. 52, p. 77-106.

Wedepohl, K. H., 1975, The contribution of chemical data to assumptions about the origins of magmas from the mantle: Fortschr. Mineralogie, v. 52, p. 141-172.

Wyllie, P. J., 1971, Role of water in magma generation and initiation of diapiric uprise in the mantle: Jour. Geophys. Research, v. 76, no. 5, p. 1328-1338.

Part V

BASALTS OF SUBDUCTION ZONES

Editors' Comments
on Papers 46 Through 49

46   KUNO
Excerpts from *Lateral Variation of Basalt Magma Type across Continental Margins and Island Arcs*

47   DICKINSON
Potash-Depth (K-h) Relations in Continental Margin and Intra-Oceanic Magmatic Arcs

48   GREEN
Abstract from *Island Arc and Continent-Building Magmatism — A Review of Petrogenic Models Based on Experimental Petrology and Geochemistry*

49   CAWTHORN
Petrological Aspects of the Correlation between Potash Content of Orogenic Magmas and Earthquake Depth

A landmark paper by Benioff (1954) established the concept of inclined seismic zones under island arcs and continental margins. Because of much earlier investigation of deep seismicity under Japan (Wadati, 1935), some geologists refer to these inclined zones as Wadati-Benioff zones instead of simply Benioff zones. After the advent of the concept of sea-floor spreading (Hess, 1962), the seismic zones have almost universally been attributed to subduction.

Correlation of magmatism with the inclined seismic zones started even before the concept of subducting slabs was fully established (e.g., Kuno, 1959). We present the abstract of Kuno's (1966) paper and the summary diagram (Paper 46), which is a slight modification of Kuno's earlier work. Kuno (1959) originally proposed that different primary basalts (tholeiites, high-alumina basalts, and alkali olivine basalts) were derived by partial melting along the seismic zone and that the magmas rose directly to the surface. This simple system envisioned the formation of subalkaline basalts above about 200-km depth and of alkali basalts considerably below that depth. Early experimental studies on the basalt system (e.g., Yoder and Tilley, 1962), however, placed the transition between alkalic and subalkalic basalts at a much lower pressure, nearer to 100 km or shallower. Even to this day, there remains a discrepancy between the conclusions of field investigators and experimental petrologists concerning the depths of evolution of basaltic magmas.

Quantification of the relationship between depth to the subduction zone and composition of the overlying volcanic rocks has been attempted by a number of workers (e.g., Sugimura, 1968; Hutchison, 1975). An extremely important paper by Dickinson and Hatherton (1967) provided a method for estimating depth to subduction zones, including paleosubduction zones, from the K_2O contents of magmatic suites. Most of the quantification, however, is based on the compositions of andesites, which are more variable than the compositions of basaltic members of the suites. Several writers have concluded that the reltionships between rock chemistry and depth to subduction zone are not present or, at least, not as simple as is commonly believed (e.g., Nielson and Stoiber, 1973; Jaques, 1976).

Other factors than depth to underlying subduction zones are clearly related to basalt chemistry. Probably the most important is the nature and thickness of the crust overlying the subducted slab (e.g., see Condie, 1973). The paper by Dickinson (Paper 47) discusses various crustal configurations above subduction zones, thus elaborating on the paper by Dickinson and Hatherton (1967).

At present, there is a tendency to classify subduction-zone assemblages into three groups: (1) primitive island arc (island-arc tholeiite, low-K tholeiite), associated with the early stages of subduction of an oceanic slab under another oceanic crust; (2) later, calcalkaline, magmatism associated with more mature island arcs; and (3) continental-margin calcalkaline (high-K calcalkaline) activity. These various types are discussed in Part VII. Rogers and Novitsky-Evans (1977) and Rogers and Ragland (1980) discussed an assemblage proposed to have formed on a crust intermediate between that of a continental margin and an island arc.

A number of summary papers have been written explaining the origin of basaltic magmas along subduction zones, and we can call attention only to a few of them. Ringwood (1974) presented a summary of concepts developed to that date at the Australian National University. Gill (1974) explored the problems of determining source materials for calcalkaline magmas; he concluded that an ordinary oceanic crust of basaltic composition could not, by itself, partially melt to form a typical island-arc calcalkaline sequence. Both Gill (1974) and Ringwood (1974) discussed the effect of eclogite-gabbro transformations on partial melting processes. Green (1980, the abstract of which appears as Paper 48) presented a thorough discussion of island-arc and

315

continental-margin magmatism. Donnelly and Rogers (1980) explained primitive island-arc volcanic sequences as the result of partial melting of a hydrated mantle as subduction begins, whereas they attributed calcalkaline magmas to incorporation of early island-arc material in mantle sources. Cawthorn (Paper 49) discussed various models for the relationship between K_2O contents and depth to the subduction zone. The importance of amphibole in the generation of subduction-zone magmas was summarized by Cawthorn and O'Hara (1976), and Marsh (1979) discussed many of the physical problems associated with the generation and rise of magmas along subduction zones.

REFERENCES

Benioff, H., 1954, Orogenesis and Deep Crustal Structure—Additional Evidence from Seismology, *Geol. Soc. America Bull.* **65:**385–400.

Cawthorn, R. G., and M. J. O'Hara, 1976, Amphibole Fractionation in Calcalkaline Magma Genesis, *Am. Jour. Sci.* **276:**309–329.

Condie, K. C., 1973, Archean Magmatism and Crustal Thickening, *Geol. Soc. America Bull.* **84:**2981–2992.

Dickinson, W. R., and T. Hatherton, 1967, Andesitic Volcanism and Seismicity around the Pacific, *Science* **157:**801–803.

Donnelly, T. W., and J. J. W. Rogers, 1980, Igneous Series in Island Arcs—The Northeastern Caribbean Compared with Worldwide Island-Arc Assemblages, *Bull. Volcanol.* **43:**347–382.

Gill, J. B., 1974, Role of Underthrust Oceanic Crust in the Genesis of a Fijian Calc-alkaline Suite, *Contr. Mineralogy and Petrology* **43:**29–45.

Green, T. H., 1980, Island Arc and Continent-Building Magmatism—A Review of Petrogenic Models Based on Experimental Petrology and Geochemistry, *Tectonophysics* **63:**367–385.

Hess, H. H., 1962, History of Ocean Basins, in *Petrologic Studies: A Volume in Honor of A. F. Buddington,* A. E. J. Engel, H. L. James, and B. F. Leonard, eds., Geological Society of America, New York, pp. 599–620.

Hutchison, C. S., 1975, Correlation of Indonesian Active Volcanic Geochemistry with Benioff Zone Depth, *Geologie en Mijnbouw* **54:**157–168.

Jaques, A. L., 1976, High-K_2O Island-Arc Volcanic Rocks from the Finistere and Adelbert Ranges, Northern Papua New Guinea, *Geol. Soc. America Bull.* **87:**861–867.

Kuno, H., 1959, Origin of Cenozoic Petrographic Provinces of Japan and Surrounding Areas, *Bull. Volcanol.* **20:**37–76.

Kuno, H., 1966, Lateral Variation of Basalt Magma Type across Continental Margins and Island Arcs, *Bull. Volcanol.* **29:**196–221.

Marsh, B. D., 1979, Island Arc Development—Some Observations, Experiments, and Speculations, *Jour. Geology* **87:**687–713.

Nielson, D. R., and R. E. Stoiber, 1973, Relationship of Potassium Content in Andesitic Lavas and Depth to the Seismic zone: *Jour. Geophys. Research* **78:**6887–6892.

Ringwood, A. E., 1974, The Petrologic Evolution of Island-Arc Systems, *Geol. Soc. London Jour.* **130:**183–204.

Rogers, J. J. W., and J. M. Novitsky-Evans, 1977, The Clarno Formation of Central Oregon, U.S.A.—Volcanism on a Thin Continental Margin, *Earth and Planetary Sci. Letters* **34:**56–66.

Rogers, J. J. W., and P. C. Ragland, 1980, Trace Elements in Continental-Margin Magmatism: Part I. Trace Elements in the Clarno Formation of Central Oregon and the Nature of the Continental Margin on which Eruption occurred, *Geol. Soc. America Bull.* **91,** Part I:196–198, Part II, Card 3:1217–1292.

Sugimura, A., 1968, Spatial Relations of Basaltic Magmas in Island Arcs, in *Basalts: The Poldervaart Treatise on Rocks of Basaltic Composition,* H. H. Hess and A. Poldervaart, eds., Wiley-Interscience Publishers, New York, pp. 537–571.

Wadati, K., 1935, On the Activity of Deep-Focus Earthquakes in Japan Island and Neighbourhoods, *Geophys. Mag.* **8:**305–325.

Yoder, H. S., Jr., and C. E. Tilley, 1962, Origin of Basalt Magmas—An Experimental Study of Natural and Synthetic Rock Systems, *Jour. Petrology* **3:**342–532.

46

Lateral Variation of Basalt Magma Type Across Continental Margins and Island Arcs *

H. KUNO

Geological Institute, University of Tokyo, Tokyo, Japan

Abstract

Quaternary basalt magmas in the Circum-Pacific belt and island arcs and also in Indonesia change continuously from less alkalic and more siliceous type (tholeiite) on the oceanic side to more alkalic and less siliceous type (alkali olivine basalt) on the continental side. In the northeastern part of the Japanese Islands and in Kamchatka, zones of tholeiite, high-alumina basalt, and alkali olivine basalt are arranged parallel to the Pacific coast in the order just named, whereas in the southwestern part of the Japanese Islands, the Aleutian Islands, northwestern United States, New Zealand, and Indonesia, zones of high-alumina basalt and alkali olivine basalt are arranged parallel to the coast. In the Izu-Mariana, Kurile, South Sandwich and Tonga Islands, where deep oceans are present on both sides of the island arcs, only a zone of tholeiite is represented. Thus the lateral variation of magma type is characteristic of the transitional zone between the oceanic and continental structures. Because the variation is continuous, the physico-chemical process attending basalt magma production should also change continuously from the oceanic to continental mantle. Suggested explanations for the lateral variation assuming a homogeneous mantle are: 1) Close correspondence between the variations of depth of earthquake foci in the mantle and of basalt magma type in the Japanese Islands indicates that different magmas are produced at different depths where the earthquakes are generated by stress release: tholeiite at depths around 100 km, high-alumina basalt at depths around 200 km, and alkali olivine basalt at depths greater than 250 km. 2) Primary olivine tholeiite magma is produced at a uniform level of the mantle (100-150 km), and on the oceanic side of the continental margin, it leaves the source region immediately after its production and forms magma reservoirs at shallow depths, perhaps in the crust, where it undergoes fractionation to produce SiO_2-oversaturated tholeiite magma, whereas on the continental side, the primary magma forms reservoirs near the source region and stays there long enough to be fractionated to produce alkali olivine basalt magma, and in the intermediate zone, the primary magma forms reservoirs at intermediate depths where it is fractionated to produce high-alumina basalt magma.

---

* Paper read at the IAV International Symposium on Volcanology (New Zealand), scientific session of Nov. 25, 1965.

[*Editors' Note:* In the original, material precedes and follows Figure 18.]

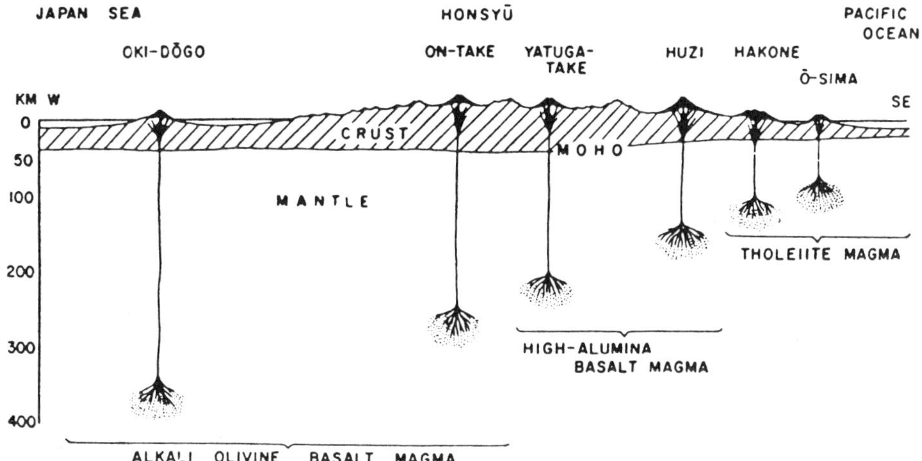

FIG. 18 - Schematic WNW-ESE cross-section through central Japan showing depths of magma production and of magma reservoirs based on KUNO and others' hypothesis.

47

Potash-Depth (K-h) Relations in Continental Margin and Intra-Oceanic Magmatic Arcs

William R. Dickinson
Geology Department
Stanford University
Stanford, California 94305

ABSTRACT

Plots of potash content against depth from volcanoes to inclined seismic zones for active magmatic arcs reveal slightly different relations for (a) continental margin arcs containing full or excessive thicknesses of continental crust overlying presumably old continental lithosphere, and (b) migratory intra-oceanic arcs that bound interarc basins of marginal seas and lack any known continental basement rocks. Detached intra-oceanic arcs containing continental basement rocks and stationary intra-oceanic arcs where back-arc spreading is absent display intermediate or equivocal relations. The significance of the difference in K-h correlation for continental margin and intra-oceanic magmatic arcs is uncertain but should be taken into account by theories for the petrogenesis of arc magmas.

INTRODUCTION

In previous papers, a general though imperfect correlation has been shown between the level of potash content in lava erupted from active volcanoes of modern magmatic arcs and the depth from the volcanoes to the inclined seismic zones that dip away from the associated trenches into the mantle beneath the arcs (Dickinson and Hatherton, 1967; Hatherton and Dickinson, 1969; Dickinson, 1970, 1973; Ninkovich and Hays, 1972; Lefevre, 1973). Recent evaluations of the correlation using rigorous hypocentral locations of earthquakes emphasize the crudeness of the correlation and thus highlight the ambiguities involved in estimating depths to seismic zones from petrologic characters observed at the surface, or vice versa (Nielson and Stoiber, 1973; James and others, 1973). The observation that levels of potash content in arc magmas also correlate roughly with local crustal thicknesses (Condie and Potts, 1969) encourages examination of the potash-depth relation separately for groups of arcs with grossly different crustal profiles. This paper is an attempt to clarify the potash-depth correlation by plotting two different curves for continental margin and intra-oceanic arcs using available published data.[1]

---

[1] Compilations of and references for data used to construct the figures in this article are available in tabular form by ordering supplementary material 75–9 from Documents Secretary, Geological Society of America, 3300 Penrose Place, Boulder, Colorado 80301, USA. There are 105 table lines and 105 references included in the data summary.

TYPES OF ARCS

The cumulative length of presently active magmatic arcs is roughly 40,000 km. On the basis of gross crustal relations, two main varieties can be recognized (Dickinson, 1974): (1) continental margin arcs (about 40 to 45 percent of the total), which stand on the edges of continental blocks; and (2) intra-oceanic arcs (about 55 to 60 percent of the total), which stand as elongate ridges between regions with oceanic water depths in excess of 1,500 m. Within these two broad categories of arc type, six main subtypes can be recognized on the basis of subsidiary geographic and structural criteria (percentages given are rough estimates of the relative abundance of each subtype as linear arc segments): (A) 25 percent, continental margin arcs with mainland volcanoes standing near the edges of contiguous continental landmasses (for example, the Andes). (B) 17.5 percent, continental margin arcs with fringing insular or peninsular volcanoes standing as island arcs with shallow shelf seas in the black-arc areas (for example, Sumatra-Java). (C) 7.5 percent, "migratory" intra-oceanic arcs with detached slivers of continental basement rocks within the arc structures, which are separated from the nearest continental blocks by interarc basins of marginal seas whose oceanic crust formed by back-arc spreading (for example, Japan). (D) 10 percent, "stationary" intra-oceanic arcs lacking known basement rocks and without clear-cut back-arc spreading but having marginal seas with oceanic crust in the back-arc areas (for example, the Aleutians). (E) 27.5 percent, "migratory" intra-oceanic arcs lacking known basement rocks but with adjacent interarc basins having oceanic crust formed by spreading in the back-arc areas (for example, the Marianas). (F) 12.5 percent, intra-oceanic arcs with so-called "reversed" polarity, in that the associated trench is located on the side toward the closest continental block (for example, the Solomons).

Figure 1 depicts reported crustal thicknesses for examples of the different types of arcs. Intra-oceanic arcs have crustal thicknesses intermediate between standard oceanic profiles and normal continental profiles, except where detached slivers of continental basement form part of the arc structures. In other cases, the crustal profiles of intra-oceanic arcs are presumably built entirely by arc magmatism. Continental margin arcs have crustal thicknesses comparable with or greater than those of ordinary continental blocks. In the Andean region, great crustal thicknesses may reflect major crustal telescoping related to the prominent foreland fold-thrust system of the Subandean belt along the rear side of the Andean

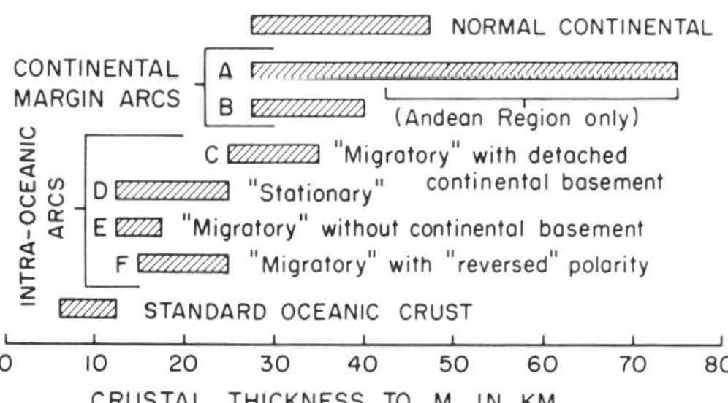

Figure 1. Crustal thicknesses in magmatic arcs and, for comparison, normal continental crust (after Condie, 1973, Fig. 7) and standard oceanic crust, including that of marginal sea basins (after Karig, 1971, Fig. 2; Packham and Falvey, 1971, Fig. 5). Types of magmatic arcs are keyed to text discussion by letter designation.

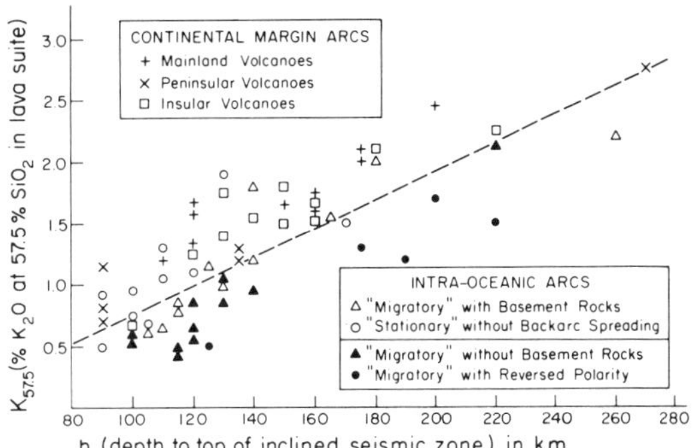

Figure 2. K-*h* plot for 64 selected volcanoes and volcano clusters in 14 different arc-trench systems (see text for derivation of $K_{57.5}$ and *h*). Dashed line separates points for undoubted continental margin arcs lacking island volcanoes (+'s and x's above line) and points for fully intra-oceanic arcs lacking known continental basement rocks and bordered by interarc basins of marginal seas (solid symbols below line). See text for discussion of arc types.

Ranges. The Subandean belt borders a pericratonic foreland basin that occupies a linear belt of subsiding continental basement lying between the arc and the continental interior.

POTASH-DEPTH RELATIONS

To develop or test the potash-depth correlation for magmatic arcs, indicators of level of potash content in lava and depth to seismic zone beneath volcanoes must be selected. Several procedures are possible in each case; the array of available data harbors inherent scatter, certainly from random errors and possibly from some systematic errors as well in both chemical analyses and hypocentral locations. In general, two main approaches are possible: (1) use of a maximum number of data points (Nielson and Stoiber, 1973), such as individual chemical analyses and hypocentral locations, thus avoiding masking implicit but unexpressed correlations in the result; and (2) use of preliminary plots of petrochemical trends in lava suites and of distributional trends for earthquakes related to a particular volcanic center or belt to derive group indices of potash content and hypocentral depth, thus simplifying the plot by reducing the random noise in the scatter of the data. The latter approach is adopted here. Figure 2 displays the potash-depth correlation in those arc segments for which data needed to determine the specific potash and depth indicators used in this paper are available.

For the depth *h* from volcano(es) to seismic zone, the indicative value plotted is the estimated distance from sea level to the upper part of the continuous or semicontinuous envelope of hypocenters defining the inclined zone at a point vertically beneath the volcano(es). The upper part of the seismic envelope is regarded as the best available index to the position of the upper surface of a slab of lithosphere descending into the mantle by plate consumption beneath the arc-trench system. Values of *h* were either scaled off transverse hypocentral profiles drawn to transect arcs at the appropriate points, or they were interpolated from contours drawn on the seismic zone. Although subjective judgment enters into derivation of *h* by either procedure, any systematic bias is unlikely to be significant in relation to inherent uncertainties in hypocentral locations.

The indicator chosen for potash content is termed $K_{57.5}$ and represents the K_2O content of a lava suite corresponding to an SiO_2 content of 57.5 percent on a simple Harker variation diagram of K_2O versus SiO_2. Points plotted on the variation diagram represent analyses recalculated to a volatile-free sum of weight percentages. To pick a value for $K_{57.5}$, a straight line was drawn by eye through the scatter of points, which show K_2O increasing with increasing SiO_2 in all arc-lava suites. The silica content of 57.5 percent was chosen because andesitic lava is the most widespread across the geographic and structural spectrum of arc types. The value 57.5 is roughly intermediate between more mafic andesite, common in intra-oceanic arcs, with silica contents near or below 55 percent and more felsic andesite, common in continental margin arcs, with silica contents near or above 60 percent. Use of $K_{57.5}$ as the potash index thus allows presentation of the potash-depth correlation for nearly all magmatic arcs on the same plot. Only those arc segments were used in Figure 2, however, for which $K_{57.5}$ could be read from a Harker variation diagram by interpolation between analytical data points without extrapolation beyond the range of silica contents for which analyses are available.

For any given value of $K_{57.5}$, the range in h is about 75 km, or nearly half the total spread in h for all arcs. Inspection of the plot reveals, however, that points representing segments of continental margin and intra-oceanic arcs tend to lie near the upper and lower fringes, respectively, of the scatter of points. The dashed line drawn in Figure 2 through the central region of the scatter achieves a full separation of two sets of points: (1) those representing continental margin arcs where volcanoes stand on contiguous parts of continental landmasses rather than on offshore islands, and (2) those representing migratory intra-oceanic arcs, of both normal and reversed polarity, where volcanic islands stand along barrier ridges lacking continental basement but separating broad ocean basins from interarc basins of marginal seas. From Figure 1, the former set comprises arcs with crustal profiles of undoubted continental dimensions and the latter set, arcs with profiles intermediate between standard oceanic and normal continental thickness. Points for stationary intra-oceanic arcs, migratory intra-oceanic arcs containing detached continental basement rocks, and insular arcs along the edges of partly drowned continental blocks all exhibit more equivocal distributions, perhaps reflecting variable crustal thicknesses, and lie to some extent on both sides of the dashed dividing line.

The dividing line thus implies different K-h correlations for "undoubted" continental margin arcs where volcanoes stand on integral parts of continental landmasses and for "fully" intra-oceanic arcs where back-arc spreading occurs, continental basement slivers are absent, and the arc structure is, in the broad sense, entirely volcanogenic. Overlapping relations remain for geographically and structurally hybrid arcs within the broad spectrum of arc types. Within each of the two discrete sets of arcs, however, the apparent range in h for a given value of $K_{57.5}$ is reduced to 25–50 km. Across the whole spread of h, the value of $K_{57.5}$ for a given value of h differs between fully intra-oceanic and undoubted continental margin arcs by an average of about 0.60 to 0.65 percent K_2O in andesitic rocks, whereas the total potash content in andesitic rocks of intra-oceanic arcs with low h may be of the order of only 0.5 percent K_2O. At high values of h, the difference is proportionally less dramatic but still significant. For example, a recently calculated average composition for volcanic rocks in continental margin andesitic arcs displays a potash content twice that of the comparable average andesite of island arcs (Gunn and others, 1974, Table 3).

In summary, the overall crude K-h correlation called attention to the way in which potash contents of arc lavas vary with depths to seismic zones in the mantle regardless of the nature of the crust beneath the arc volcanoes. The recognition of different trends of K-h correlation for continental margin and intra-oceanic arcs redirects attention to variations in the crustal structures of magmatic arcs. Moreover, differences in crustal profile may not be the only significant architectural variations among arcs, because differences in the nature of the immediately underlying lithosphere of the upper mantle above the low-velocity zone may be associated with any crustal variations present. All oceanic crust is part of relatively young lithosphere of minimal thickness, but old continental crust must be part of relatively old lithosphere that may perhaps attain greater thickness or evolve in other unknown ways.

IMPLICATIONS FOR PETROGENESIS

From plate tectonics, a variety of possible source materials from which magmas might be derived are presumed to occur at various levels in the vertical column of crust and mantle beneath magmatic arcs. Chief among these are (1) eclogitic rock formed at depth from initially basaltic-gabbroic rock in oceanic crustal layers near the upper surface of the slab of lithosphere descending along the inclined seismic zone where deformation coupled with juxtaposition against hotter mantle may achieve melting (Marsh and Carmichael, 1974); (2) peridotitic rock near the low-velocity zone in the mantle beneath the lithosphere in which the arc structure is embedded but above the inclined seismic zone, from which volatiles driven out of the descending slab may act as fluxing agents (Miyashiro, 1974); and (3) crustal rocks (Pichler and Zeil, 1972) of uncertain nature in the lower crust of the arc structure itself where heat flow may reach a critical level from the advective addition of magmas rising from below.

Primary intermediate magmas of broadly andesitic composition are generally the products postulated from melting of crustal sources, whether in arc roots or subducted slabs, but basaltic and rhyolitic as well as andesitic melts have been discussed as possibilities from melting of mantle (Yoder, 1973). Eichelberger (1974) suggested the likely importance of mingling diverse magmatic products from the same or multiple sources by mutual magma contamination within arc suites.

Using inferences from experimental petrology, Marsh and Carmichael (1974) have discussed possible phase equilibria that could explain the overall K-h correlation entirely in terms of partial melting of eclogitic materials in the descending slab. The slightly different trends of K-h correlation for con-

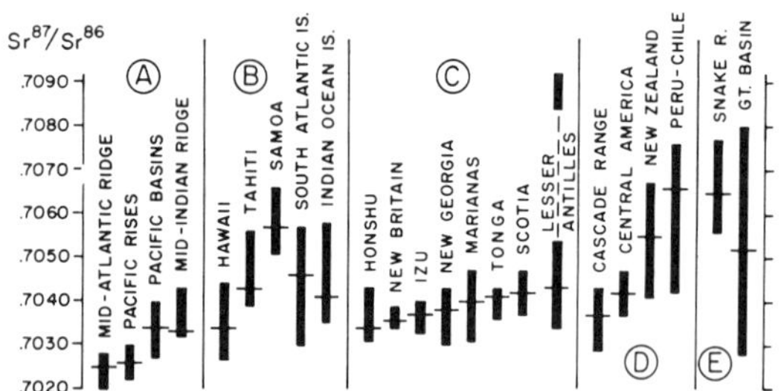

Figure 3. Ranges (bars) and medians (ticks) of initial Sr^{87}/Sr^{86} ratios in volcanic rocks of magmatic arcs and selected other areas: A, mid-oceanic rises and ocean floors; B, oceanic islands; C, intra-oceanic arcs; D, continental margin arcs plus New Zealand; E, intracontinental volcanic rocks in western United States.

tinental margin and intra-oceanic arcs suggest additional influences on the petrochemistry of arc magmas. Jakeš and White (1972) also noted that abundances of the trace elements Ba, Sr, Rb, Zr, Th, and U are greater in continental margin arcs than in island arcs, even for rocks of comparable K_2O contents.

The isotopic proportion of radiogenic strontium is often used as an indicator of magmatic sources, and particularly of the likelihood of crustal contributions, but it does not resolve uncertainties about the origin of arc magmas (see Fig. 3). Sr^{87}/Sr^{86} ratios in the lava suites of intra-oceanic arcs are slightly higher than those for many fresh basalts of mid-oceanic rises, but they are comparable with or less than those in lava suites of oceanic islands. Church (1973) concluded that any contributions from oceanic sedimentary deposits to these magmas must be low, although submarine weathering of ocean-floor basalt prior to plate consumption may be an important factor. He further showed that lead isotopes in arc magmas lie along or near the trend line for alkalic basalt of oceanic islands. The latter observation perhaps may be taken to suggest derivation of arc magmas either directly from similar mantle sources or by melting of seamount piles carried down the inclined seismic zone at the upper surface of the descending slab. Sr^{87}/Sr^{86} ratios in some continental margin arcs are similarly low but drift distinctly higher in the Andes (Pichler and Zeil, 1972; James and others, 1973). Even so, they are no higher than those for some types of intracontinental flood basalt, presumably derived from the mantle, such as those of the Snake River Plain. Hedge and Noble (1971) argued that high Sr^{87}/Sr^{86} ratios in intracontinental basalt of the Great Basin reflect derivation from unusually radiogenic strontium sources in mantle previously depleted by partial melting. The means by which the Andean lava flows acquired unusually high Sr^{87}/Sr^{86} ratios is thus unclear, because the materials contributing radiogenic strontium conceivably may lie within the subcontinental mantle rather than in the crust.

In summary, the petrogenetic problem posed by slightly different trends of K-h correlation in continental margin and intra-oceanic arcs remains open but apparently imposes a fresh constraint on theoretical reasoning. Some difference in origin seems implied—either in the nature of the source(s) and the parent magma(s), the evolution of the primary magma(s), or the mix of magmas derived from multiple sources or parents (also see Ringwood, 1974). Perhaps previously unsuspected factors are also involved. For example, continental margin arcs generally have more gently dipping seismic zones than do typical intra-oceanic arcs. As Lipman noted in reviewing this manuscript, the subsurface distance traveled by a descending slab of lithosphere to reach a given depth h thus may tend to be greater for continental margin arcs, and the characteristic subsurface residence time of a slab prior to passage beneath the arc also may be correspondingly greater. Post-review attempts to plot $K_{57.5}$ versus slant distance along inclined seismic zone from trench to point beneath volcanic chain showed no discernible correlation. However, it remains true that the mainland and peninsular volcanoes whose lava suites plot entirely above the dashed line of Figure 2 are associated with seismic zones whose average dip is perhaps 40°, whereas the fully intra-oceanic lava suites that plot entirely below the line are associated with seismic zones whose average dip is perhaps 60°. Improved understanding of facets of subcrustal as well as crustal geometry beneath magmatic arcs may thus be pertinent to the petrogenetic problem posed here.

REFERENCES CITED

Church, S. E., 1973, Limits of sediment involvement in the genesis of orogenic volcanic rocks: Contr. Mineralogy and Petrology, v. 39, p. 17–32.

Condie, K. C., 1973, Archean magmatism and crustal thickening: Geol. Soc. America Bull., v. 84, p. 2981–2992.

Condie, K. C., and Potts, M. J., 1969, Calc-alkaline volcanism and the thickness of early Precambrian crust in North America: Canadian Jour. Earth Sci., v. 6, p. 1179–1184.

Dickinson, W. R., 1970, Relations of andesites, granites, and derivative sandstones to arc-trench tectonics: Rev. Geophysics and Space Physics, v. 8, p. 813–859.

—— 1973, Reconstruction of past arc-trench systems from petrotectonic assemblages in the island arcs of the western Pacific, in Coleman, P. J., ed., The western Pacific island arcs, marginal seas, and geochemistry: New York, Crane-Russak, p. 569–601.

—— 1974, Sedimentation within and beside ancient and modern magmatic arcs, in Dott, R. H., Jr., and Shaver, R. H., eds., Modern and ancient geosynclinal sedimentation: Soc. Econ. Paleontologists and Mineralogists Spec. Pub. 19, p. 230–239.

Dickinson, W. R., and Hatherton, T., 1967, Andesitic volcanism and seismicity around the Pacific: Science, v. 157, p. 801–803.

Eichelberger, J. C., 1974, Magma contamination within the volcanic pile; origin of andesite and dacite: Geology, v. 2, p. 29–33.

Gunn, B. M., Roobol, M. J., and Smith, A. L., 1974, Petrochemistry of the Peléan-type volcanoes of Martinique: Geol. Soc. America Bull., v. 85, p. 1023–1030.

Hatherton, T., and Dickinson, W. R., 1969, The relationship between andesitic volcanism and seismicity in Indonesia, the Lesser Antilles, and other island arcs: Jour. Geophys. Research, v. 74, p. 5301–5310.

Hedge, C. E., and Noble, D. C., 1971, Upper Cenozoic basalts with high Sr^{87}/Sr^{86} and Sr/Rb ratios, southern Great Basin, western United States: Geol. Soc. America Bull., v. 82, p. 3503–3510.

Jakeš, P., and White, A.J.R., 1972, Major and trace element abundances in volcanic rocks of orogenic areas: Geol. Soc. America Bull., v. 83, p. 29–40.

James, D. E., Brooks, C., and Cuyubamba, A., 1973, Geochemical trends in Andean rocks: Carnegie Inst. Washington Year Book 72, p. 252–259.

Karig, D. E., 1971, Origin and development of marginal basins in the western Pacific: Jour. Geophys. Research, v. 76, p. 2542–2561.

Lefevre, C., 1973, Les caractères magmatiques du volcanisme plio-quaternaire des Andes dans le Sud du Pérou: Contr. Mineralogy and Petrology, v. 41, p. 259–272.

Marsh, B. D., and Carmichael, I.S.E., 1974, Benioff zone magmatism: Jour. Geophys. Research, v. 79, p. 1196–1206.

Miyashiro, A., 1974, Volcanic rock series in island arcs and active continental margins: Am. Jour. Sci., v. 274, p. 321–355.

Nielson, D. R.. and Stoiber, R. E., 1973, Relationship of potassium content in andesitic lavas and depth to the seismic zone: Jour. Geophys. Research, v. 78, p. 6887–6892.

Ninkovich, D., and Hays, J. D., 1972, Mediterranean island arcs and origin of high potash volcanoes: Earth and Planetary Sci. Letters, v. 16, p. 331–345.

Packham, G. H., and Falvey, D. A., 1971, An hypothesis for the formation of marginal seas in the western Pacific: Tectonophysics, v. 11, p. 79–109.

Pichler, H., and Zeil, W., 1972, The Cenozoic rhyolite-andesite association of the Chilean Andes: Bull. Volcanol., v. 35, p. 424–452.

Ringwood, A. E., 1974, The petrological evolution of island arc systems: Geol. Soc. London Jour., v. 130, p. 183–204.

Yoder, H. S., Jr., 1973, Contemporaneous basaltic and rhyolitic magmas: Am. Mineralogist, v. 58, p. 153–171.

ACKNOWLEDGMENTS

Reviewed by K. C. Condie, P. W. Lipman, J.W.H. Monger, and P. J. Wyllie.

Preliminary version presented at Symposium on Volcanic Geology and Mineral Deposits of the Canadian Cordillera held by Cordilleran Section of Geological Association of Canada in Vancouver.

MANUSCRIPT RECEIVED NOVEMBER 27, 1974
MANUSCRIPT ACCEPTED DECEMBER 5, 1974

48

ISLAND ARC AND CONTINENT-BUILDING MAGMATISM — A REVIEW OF PETROGENIC MODELS BASED ON EXPERIMENTAL PETROLOGY AND GEOCHEMISTRY

T.H. GREEN

School of Earth Sciences, Macquarie University, North Ryde, N.S.W. 2113 (Australia)

ABSTRACT

Detailed geochemical and petrological studies of igneous rock suites in island arc and continental marginal environments have revealed the complexity of magmatism in these regions. Rocks once collectively termed the orogenic calc-alkaline suite are now recognized as forming a continuum between tholeiitic, calc-alkaline and alkaline suites.

Although this diversity of rock types has inspired varied and overlapping hypotheses of origin, the application of experimental and geochemical data do limit the appropriate choice of a genetic hypothesis for individual magma suites in convergent plate regions. Also a number of hypotheses of origin may be linked to give an integrated model of the growth of island arcs and continental crust, and certain geochemical parameters are distinctive in tracing this progressive development. Minor phases such as titanochondrodite, titanoclinohumite and sphene have a critical bearing on key geochemical parameters.

Water plays a key role in the integrated model for generation of magmas in convergent plate regions. In the first stage, release of water through dehydration of phases in subducted material controls large-scale melting processes in the mantle, forming the early and most "primitive" tholeiitic magmas in the island arc suite (c.f. Ringwood, 1974), where olivine plays a dominating fractionating role. These magmas form a modified and thickened crust in plate collision regimes. Once this thickened crust has developed, subsequent calc-alkaline magmatism becomes more complex, involving a number of possible sources or processes, including (a) silicic melt from eclogite in the subduction zone, (b) melts from "contaminated" peridotite, and (c) relatively shallow level, hydrous, often amphibole-dominated fractionation in the thickened crust. Key trace element abundances can constrain the involvement of these sources or processes. Further magmatic activity in island arcs is related to deeper and low degrees melting, with mica breakdown making an important contribution, and jadeitic clinopyroxene remaining as a residual mantle phase. This produces high-K calc-alkaline rocks transitional into the alkaline suite. Finally, in both "mature" island arcs and continental margins where crustal thicknesses exceed 35 km, continued access of water from the subduction zone, and/or underplating of the crust by mantle- or subduction zone-derived magmas, causes partial melting of the crust. The resultant magmas are dominantly silicic and plutonic or ignimbritic.

An increasing amount of experimental, geochemical and mineralogical evidence indicates a prominent role for "mature" sedimentary material being involved in melting at depth and the generation of volcanic and plutonic magmas in evolved island arc or continental marginal environments. REE provide useful additional constraints on the genesis of these rocks, as they do for other rocks in the island arc suite.

REFERENCE

Ringwood, A. E., 1974. Petrological evolution of island arc systems. J. Geol. Soc. London, 130: 183-204.

49

Petrological aspects of the correlation between potash content of orogenic magmas and earthquake depth

R. GRANT CAWTHORN

Department of Geology, University of the Witwatersrand, Milner Park, Johannesburg 2001, South Africa

SUMMARY. The petrological interpretation of the variation of K_2O content with depth to the Benioff zone in orogenic magmas is complicated by several factors: (1) Magmas may not be true primary liquids but have undergone differentiation. K_2O contents, at a specific SiO_2 level, will depend upon the SiO_2 content of the parental magma and the nature of the crystallizing phases more than on original K_2O content. (2) The depth to the Benioff zone may not give a true reflection of the depth of melting. Magmas may not be derived from the uppermost layer of subducted oceanic crust but rather the overlying upper mantle. Earthquakes may not occur in the uppermost layer of the subducted plate, but in its colder, central parts. Hence, earthquake depth may not be related to the source of the magma. (3) It is difficult to envisage how the temporal increase in K_2O may occur in certain areas (*e.g.* Aleutian I., Fiji) if it is primarily controlled by the depth to the Benioff zone.

There may be some correlation between K_2O content of magmas and earthquake depth, but its cause is difficult to determine. None of the existing explanations is entirely satisfactory, and it is suggested that varying degrees of fractional crystallization may play an important role in controlling K_2O contents.

THE existence of systematic chemical variations in magma compositions across orogenic zones has been known for a long time. Sugimura (1968) provided an excellent review of the development of ideas and observations on this zonation. The work and interpretation of Kuno (1959) on the spatial distribution of pigeonitic, high-alumina basaltic, and alkaline volcanism and their relation to calc-alkaline magmas in Japan provided the fundamental model on which much subsequent work in orogenic belts has been based. One chemical parameter that shows a variation across orogenic belts is the K_2O content. Dickinson and Hatherton (1967) suggested a correlation between the K_2O content and the depth to the underlying Benioff zone ($= h$), which could be used to determine the position of volcanic rocks, both modern and ancient, within a tectonic region. The reliability of the correlation has been questioned (Nielson and Stoiber, 1973) and has required modification to a scheme whereby the different types of orogenic volcanic belts have distinct K–h correlation patterns (Dickinson, 1975).

This kind of systematic chemical variation is not unique to orogenic belts. McBirney (1967, fig. 5) suggested that there was a correlation between the horizontal distance from a mid-ocean ridge and the degree of alkalinity of volcanism.

Previous models

Several possible explanations have been presented to explain why the K_2O content apparently increases with the distance to the Benioff zone.

Increase in K_2O *content of the mantle with increasing depth.* This would appear to be the simplest model for explaining the K–h correlation. However, implicit in this model are the assumptions that phase relations remain identical over the depth range considered and that the degree of melting is constant. There is no other evidence for variations in K_2O content of the mantle and this has been rejected by most workers (Dickinson and Hatherton, 1967; Best, 1975).

K_D *between liquid and solid changes with depth.* Dickinson (1968) suggested that if the partition coefficient (K_2O of solid/K_2O of liquid) decreased with increasing pressure, higher K_2O

325

contents in magmas would be anticipated at greater depth (all other parameters remaining constant). Dickinson did not state to which minerals this hypothesis applied. Olivine and pyroxene are essentially free of K_2O. Hence, K_D must be extremely small (e.g. Delong, 1974) even at very low pressure. Thus a significant decrease in K_D for these phases with increasing pressure is not possible. The only plausible mantle phase that could have a significant K_D is phlogopite. If this phase were present in sufficient quantity at depth, its behaviour could produce the desired effect. Bravo and O'Hara (1975) demonstrated that the liquid in equilibrium with phlogopite and mantle minerals at 15 kb (= 1500 mPa) contains 1 % K_2O, while at 30 kb it is over 3 %. K_D, therefore, changes in the right direction to support Dickinson's (1968) hypothesis. However, while the synthetic liquid composition at 15 kb has similarities with andesites, at higher pressure it does not (Bravo and O'Hara, 1975). At pressures greater than 20 kb it is unlikely that any melting process in the mantle could produce a quartz-normative liquid (Kushiro, 1972; Nicholls and Ringwood, 1972; 1973). Some fractionation process would be required to produce andesite from these high-pressure primary magmas, which would cause changes in the original K_2O content of the magma.

Varying degrees of melting. It is likely that K_2O will be strongly partitioned into the liquid during melting; and K_2O-bearing phases, such as phlogopite and sanidine, will disappear very close to the solidus. Hence, with higher degrees of melting the K_2O content of the liquid will be diluted. If smaller degrees of partial melting were produced at higher pressure the K–*h* correlation might be explicable. However, dT/dP for hydrous peridotite is almost zero (Kushiro *et al.*, 1968; Green, 1972). This will probably be less than the geothermal gradient. Hence, provided melting under hydrous conditions is considered, as seems likely, to explain the properties of the low-velocity zone (Anderson, 1970) these gradients suggest that the degree of melting may increase with increasing depth. This is the reverse of that required to explain the variation in K_2O.

Green *et al.* (1967) envisaged the generation of magmas from diapirs initiated near the subduction zone. Depending on the distance up which they migrated, variable degrees of melting would be produced. However, diapirs from a deeper source would have the chance of greater upward movement with subsequent greater degree of melting and lower K_2O content; again producing the reverse of the desired effect.

Breakdown of hydrous phases. The extent to which hydration of oceanic crust occurs is unknown. However, assuming some water is subducted the formation of hornblende in the oceanic crust seems plausible. Conditions for the breakdown of hornblende in subduction zones have been located from experimental data (Lambert and Wyllie, 1968; Nishikawa *et al.*, 1971; Holloway and Ford, 1975) with a maximum at 30 to 35 kb. With its breakdown to pyroxene and garnet some K_2O as well as H_2O may escape into the overlying mantle. Phlogopite or sanidine may form during this process and may be carried to greater depth before breakdown or melting of these minerals to release K_2O from the subduction slab. These dehydration reactions may provide the driving force in orogenic magmatism as advocated by Fyfe and McBirney (1975). However, release of K_2O should occur at fairly discrete depths and hence produce distinct peaks rather than a systematic variation across tectonic belts.

Sanidine eclogite buffer. Marsh and Carmichael (1974) determined the stability of sanidine eclogite and liquid from the thermodynamic data and concluded that with increasing pressure the liquid should become enriched in K_2O. This may be perfectly plausible at the beginning of melting of oceanic crust. Green and Ringwood (1968) showed that 30 to 50 % melting of oceanic crust was required to produce andesitic liquids. It remains to be demonstrated whether, at such a high degree of melting, sanidine would still be a stable phase, but it seems unlikely from published data on melting of tholeiitic material (e.g. Green and Ringwood, 1968). The high-alumina quartz tholeiite used by Green and Ringwood (1968) contained 0·6 % K_2O, which is higher than that for oceanic tholeiite (Engel *et al.*, 1965), although it may be lower than for the altered material subducted, and yet they did not report sanidine from any of the anhydrous runs above or below the solidus. If sanidine is no longer in equilibrium with liquid, this mineral will not be able to buffer the K_2O content of the liquid.

Scavenging of incompatible elements. Best (1975) suggested that the incompatible elements may become enriched in the liquid by a zone-refining or wall-rock equilibration type of reaction. In view of the data presented by Jamieson and Clarke (1970) and Cox and Jamieson (1974) supporting a process of continued wall rock equilibration with magma, Best's hypothesis seems possible. The longer path-length of magmas from deeper sources would permit higher K_2O content to build up.

Crustal contamination. Somewhat similar to the previous model is the hypothesis of contamination of magma by crustal material. Dickinson (1975) has compiled isotope data ($^{87}Sr/^{86}Sr$) that suggest that orogenic magmas are not, in general, significantly richer in radiogenic Sr than ocean-floor tholeiites, and hence this process is unlikely to have played a major role in determining magma composition.

Fractionation effect

Implicit in the first five models discussed above is the assumption that the erupted magma is a true primary melt. The correlation between K_2O and depth has been attempted at various SiO_2 values. Dickinson and Hatherton (1967) used reference levels of 55 % and 60 % SiO_2, while Dickinson (1975) used a value of 57·5 % SiO_2. It would seem unlikely that magmas with these different SiO_2 values could all be true primary magmas.

One of the main factors that may affect the K_2O content of a magma at any particular SiO_2 level is fractionation. Nearly all orogenic volcanic suites show a linear chemical variation, which may satisfactorily be interpreted as the effect of fractional crystallization. The fact that basalt is subordinate to andesite in calc-alkaline suites has frequently been used to support the hypothesis that andesite is the primary magma. It may equally be argued that this relationship is excellent evidence for a fractionation hypothesis. That some basalt is erupted demonstrates that there was some material more basic than andesite present in the sub-volcanic region. The large volumes of andesite could indicate that fractionation has been extremely effective in allowing only a small volume of the parental composition (basalt) to erupt as the composition rapidly differentiated to the more evolved members of the series. Kuno (1968) came to a similar conclusion using his differentiation index (SI), stating (p. 674): 'This maximum frequency of eruption of the middle stage magmas (andesites) is interpreted as being caused by a balance between the decreasing amount of successive magmas produced by fractionation in magma reservoirs and the increasing tendency of the successive magmas to erupt at the surface because of gradual concentration of volatile substances. . . .'

In order to interpret the variation in K_2O across orogenic belts it may, therefore, be necessary first of all to remove the effect of fractionation from the K_2O values so that the K_2O values of the parental magmas, rather than of evolved magmas, may be compared with each other and with the depth to the subduction zone.

The complications that may arise due to fractionation may be conveniently discussed by reference to different chemical models. Suppose two parental magmas have different SiO_2 contents but comparable K_2O levels: What inferences could be made about the K–*h* correlation based on K_2O contents at a fixed SiO_2 value? This situation is indicated in fig. 1, where A and B represent the parental magmas. Let us suppose that the chemical variation in both suites is the result of crystallization of the same phases, having the bulk composition C. The two trends are indicated as A–A′ and B–B′. At the arbitrary reference SiO_2 value the K_2O contents of magmas from the two suites would be different because the degree of fractionation is not the same. Hence, applying the K–h correlation the interpretation would be that the depth to the subduction zone under suite A was greater than under suite B.

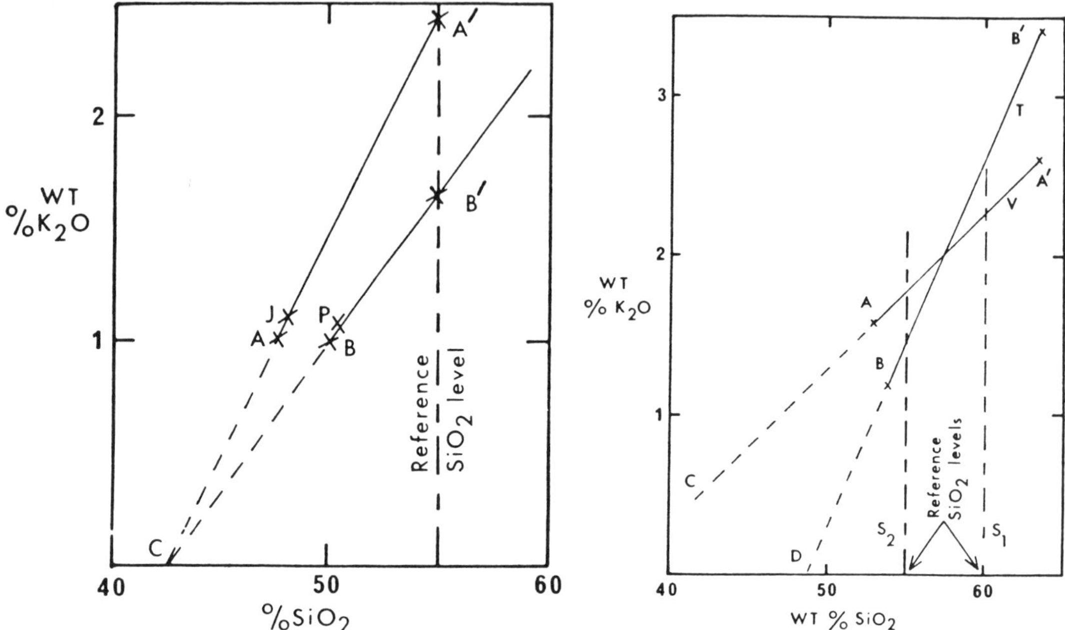

FIGS. 1 and 2: FIG. 1 (left). K_2O–SiO_2 plot to show the effect of differentiation on K_2O of two magmas with identical K_2O but different SiO_2 contents. J—parent to alkali olivine basalt series in Japan (Kuno, 1967). P—parent to high-K series in Papua (Gill, 1970). Solid lines are liquid evolution paths; dashed line indicates extract composition. FIG. 2 (right). Example of effect of differentiation of different crystalline extract compositions on magma series. T—Mt. Trafalgar; V—Mt. Victory differentiation trends from Jakes and Smith (1970). Solid lines are liquid evolution paths; dashed lines project back to hypothetical extract compositions.

To determine the parental magma of a suite is an extremely subjective matter. It is also a matter of personal prejudice whether or not one regards that parental magma as the primary melt produced during melting of some source rock. The terms parental magma and primary melt are used here as defined by Carmichael *et al.* (1974). While the least siliceous aphyric lavas may give a reasonable estimate of the parental magma composition, unless a very large number of samples have been analysed, one cannot be sure that the parental magma of a suite has been sampled.

One could cite, as an example of the effect mentioned in fig. 1, the high K_2O rocks of Eastern Papua (Gill, 1970; Jakes and Smith, 1970). The most basic lava analysed contains 50·5 % SiO_2 and 1·07 % K_2O (Gill, 1970; Fig. 8). For Japan, Kuno (1967) suggested the parental magma of the alkaline series contains 48·1 % SiO_2 and 1·13 % K_2O. If these were used as the two examples in fig. 1 that fractionated the same phases, the K_2O contents at 55 % SiO_2 (or whatever reference concentration one chose) would be significantly different. From the K–h correlation one might conclude that the Japanese samples were derived from greater depth, while from seismic data the Benioff zones are probably at comparable depths (Sugimura, 1968; Gill, 1970).

An example of the varying degree of differentiation undergone by the different suites may be taken from the data of Hatherton (1969). In his Fig. 12 he shows that the MgO values of various magma suites at 60 % SiO_2 decreases systematically with inferred increase in depth to the Benioff zone. MgO values decrease with increasing differentiation. Hence, it may be that the suites overlying progressively deeper parts of the Benioff zone have undergone a greater proportion of fractionation to reach the reference SiO_2 level. As K_2O increases with fractionation this extremely good negative correlation between K_2O and MgO suggests that fractionation may play a major role in determining K_2O contents of erupted magmas.

As an alternative geometrical complication of K–h correlation, consider the case where two parental magmas have the same SiO_2 contents, but different K_2O levels and suppose they crystallize different phases. The situation is shown in fig. 2 where parental magmas A and B crystallize phases with compositions C and D respectively. The resultant differentiation trends are A–A' and B–B' and they may cross each other. Depending on what SiO_2 content is chosen as the reference, it would be possible to infer that either suite A or B had the higher K_2O content. Hence if S_1 were the reference level, K_2O contents would imply that the distance to the Benioff zone were greater under suite B than A; while at S_2 exactly the reverse situation would be indicated.

An excellent example of this may be taken from the data of Jakes and Smith (1970). They analysed lavas from Mount Trafalgar and Mount Victory in Eastern Papua and presented a K_2O–SiO_2 variation diagram from which fig. 2 is taken. Mount Trafalgar is a recently extinct volcano while Mount Victory is still active. If one took 55 % SiO_2 as the reference composition it would be inferred from fig. 2 that Mount Victory had a higher K_2O content and hence that the depth to the subduction zone, h, might be increasing with time. However, if 60 % SiO_2 were chosen as the reference exactly the reverse inference would be obtained. Geographically these two volcanoes are so close together that one or both of these inferences must be invalid and one must therefore question the reliability of choosing arbitrary SiO_2 values for the K–h correlation.

Kuno (1968) cited an example in which he believed that two divergent suites were produced from the same parental magma by fractionation of different phases. In fig. 3a are plotted the average composition of the pigeonitic and hypersthenic series determined by Kuno (1968; Tables 3 and 4). From this graph it would seem that no single parental magma could give rise to the two series, although two very similar parental magmas could be envisaged (P' and P''). However, the second half of Kuno's hypothesis seems justifiable, that these magmas fractionated different phases to produce the observed trends. The exact mineralogy of these extract compositions, somewhere in the region of E' and E'', is not important here. That they are different is all that matters in this discussion. The nature of plausible crystallizing minerals has been reviewed by Cawthorn and O'Hara (1976).

The variation in K_2O values at different SiO_2 levels is only very minor in the above example. However, Kuno (1968) referred to another example where differences are much more significant. Holmes (1916) analysed samples from an alkaline and a hypersthenic suite of rocks that appear to be intimately associated. Kuno (1968) suggested that fractionation of different phases from a common parental composition may be responsible for the diversity of the trends plotted in fig. 3b. Inspection of this diagram suggests that there may not be a common parent to both series but that fractionation of the same phases from both series might have occurred. Perhaps instead of Kuno's (1968) interpretation, a model analogous to that discussed for fig. 1 might apply. However, the conclusion that may be drawn from the data is that distinctly different K_2O contents at specific SiO_2 levels may be deduced for suites that (according to Holmes) are spatially and temporally inseparable. The use of the K–h correlation model for these suites would demand the simultaneous existence of two different depths to a Benioff zone (or zones) at the same place. The author feels this is a rather unlikely situation.

The complications that could arise during fractionation due to the existence of different parental magmas having different K_2O and SiO_2 content and crystallizing different phases would make interpretation of the K–h correlation model extremely difficult.

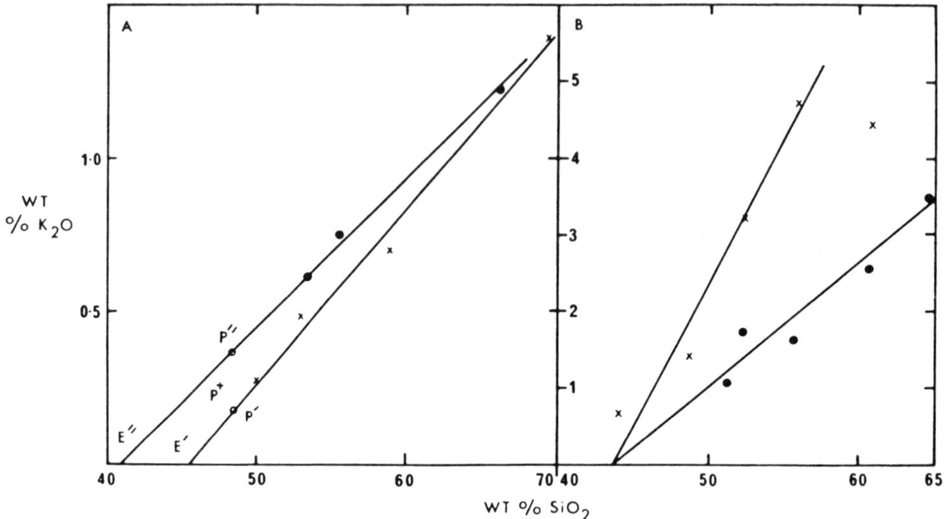

FIG. 3A (left). Differentiation trends of the pigeonitic series (crosses) and hypersthenic series (solid dots) from Japan (Kuno, 1968). P is parental magma to both series from Kuno (1950); P′ and P″ indicate more probable parental magma compositions based on a linear differentiation model with different phases crystallizing. 3B (right). Differentiation trends for alkalic (crosses) and hypersthenic series (solid dots) from Holmes (1916) indicating a common extract composition.

Variation in K₂O with time

The potassium variation recognized by Dickinson and Hatherton (1967) was essentially a spatial relationship. Subsequently, it has been demonstrated that K_2O also increases with time within a given area (Jakes and White, 1969). In terms of the K–h correlation hypothesis, this would imply that the distance to the Benioff zone from a specific point on the surface increased with time. This could be created by three different processes as shown schematically in fig. 4.

Fig. 4a depicts a situation where the dip of the subduction zone gradually increases with time, the position of the trench–arc relation remaining stationary. If this were the likely mechanism the oldest arcs would be expected to have the steepest slopes of the subduction zone. Such a trend would be difficult to demonstrate as there is no independent mechanism for determining the orientation of subduction zones in the past. Some tholeiitic to calc-alkalic regions have a very steep dip to the Benioff zone; for example, Tonga (Sykes, 1966) and Bougainville (Jakes and White, 1969). Hence, it would seem unlikely that more alkalic compositions could be erupted here by increasing the dip.

A second model involves the shifting of the trench relative to the arc, but not changing the angle of the subduction zone. One way this could occur is shown in fig. 4b and involves the locking of a subducted plate, breaking of the oceanic crust and the generation of a new subduction zone. The new subduction zone would be at a greater depth than the older one by an amount depending on the thickness of the old plate and the angle of subduction. Assuming the oceanic plates are approximately 70 km thick and taking a dip of 30°, the increased distance from Benioff zone to surface would be 100 km (fig. 4b). Hence, if a tholeiitic magma were initially erupted, with the new orientation the depth to subduction zone would be increased from 100 km to 200 km and shoshonitic or alkaline magmatism would be anticipated. This change is far more drastic than documented for the Aleutian Islands and Indonesia (Jakes and White, 1970; Gill, 1970). Hence, unless the new subducted material intrudes into the old plate rather than underneath it, as shown in fig. 4b, this mechanism would not produce the observed variation.

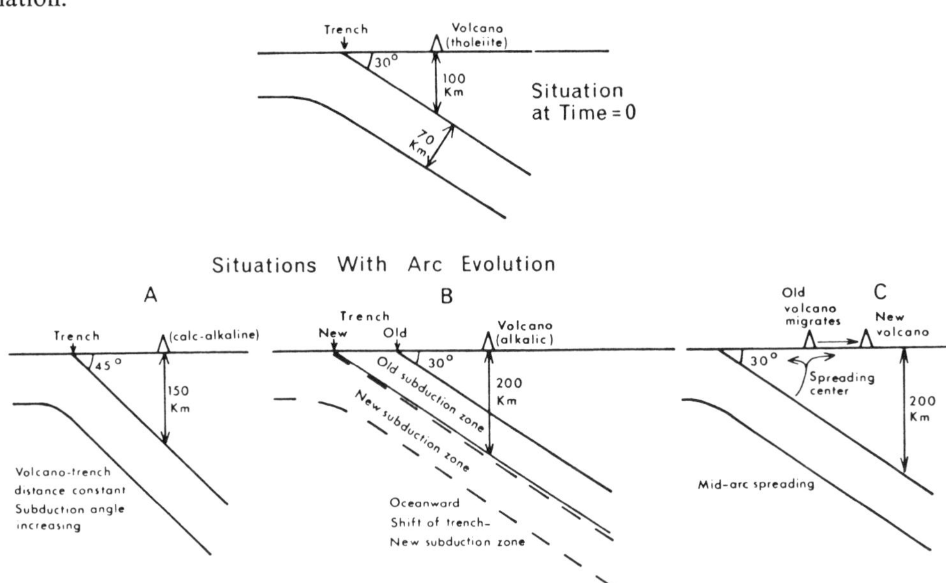

FIG. 4. Models to explain increasing depth to subduction zone during evolution of volcanic sequence causing increase in K_2O content of magmas with time. See text for explanation.

A third possibility could be termed 'mid-arc spreading' and is indicated on fig. 4c. In view of the immense volumes of batholiths exposed in the western Americas, it is tempting to suggest that some lateral expansion may occur to permit intrusion of these vast complexes. The main objection to this process is that the necessary volume of new crust is not always apparent. For example, in Tonga, the three stages of magmatism documented by Gill (1970) would require an increase in depth to the Benioff zone from 75 to 175 km. Assuming an angle of 45° for the subduction zone in this region (Sykes, 1966), this would imply an extension of 100 km in the continental crust to the south-east of Tonga for which there is no evidence.

Best's (1975) zone-refining model fitted the K–h correlation applied to the spatial variation in volcanic magmas. Can it be applied to the temporal relation? If we assume that the position of the Benioff zone does not shift appreciably with time (as discussed above) then suggesting that younger magmas may tap K_2O from a greater length of material cannot be considered. In fact, if the same material is being repeatedly processed one might even expect a decrease in K_2O contents with time, not an increase.

In conclusion, it is rather difficult to see how the K–h correlation model of Dickinson (1975) can be applied to the temporal variation observed in some orogenic areas (Jakes and White, 1969).

Orogenic magmas may be derived from mantle overlying the Benioff zone. If this is the case, then it might seem surprising that magmas become enriched in K_2O with time rather than depleted. The difference in minor element chemistry between tholeiitic and alkalic magmas has generally been attributed to varying degrees of melting (Gast, 1968) or varying degrees of fractionation (O'Hara, 1968) or a combination of the two effects. In an island arc environment there seems to be no reason why the degree of fractionation should increase with time to explain increasing K_2O contents, unless a thickening crust has greater resistance to penetration by magmas and hence causes a greater period of fractionation prior to final eruption. However, the degree of melting might well decrease with time if cold oceanic crust is constantly being introduced into the mantle. Small changes in temperature close to the beginning of melting may have a significant effect on the content of minor elements in magmas, which would increase in amount as the degree of melting decreased due to falling temperatures. This is different from the varying degree of melting mentioned earlier as this situation is envisaged as taking place at constant depth.

One consequence of decreasing degrees of melting for younger volcanics is that the actual volume of magma erupted ought to decrease as the magma becomes more potassic (if the degree of fractionation is the same for each suite). This has been demonstrated for the spatial relations between tholeiitic to alkalic lavas in Japan by Sugimura (1968), but there are no volumetric data available for an area that shows the same transition with time.

Significance of h values

The depth of the Benioff zone gives the depth to the plane where earthquakes are commonest. The data of Billington and Isacks (1975) showed that this plane can be located only as a zone at least 35 km thick (due either to experimental error or a real effect). As volcanoes usually occur in regions where the distance to the Benioff zone is between 80 and 180 km (Dickinson, 1975) there is a very significant relative error on all these measurements. However, far more serious than this is the assumption that the magma is derived from material within the earthquake zone. One of the preferred models for the origin of calc-alkaline magmatism is that it is derived from the mantle overlying the subduction zone (Kushiro, 1972; Boettcher, 1973; Fyfe and McBirney, 1975). Hence, the site of magma generation and the position of earthquake foci may be different. If there were some constant relation between the earthquake zone and the site of melting in the mantle the $K–h$ correlation might still be valid, although h values would be somewhat lower.

Even if magma were generated along the upper surface of the Benioff zone its relation to h may be questioned. Earthquakes may not, in general, occur along the contact plane between subducted plate and overlying mantle. Engdahl (1971) and Mitronovas *et al.* (1969) showed that intermediate and deep-focus earthquakes tend to occur in the cold centre of the subducted slab. Hence, if h is determined from seismic evidence, its relation to the upper surface of the plate, where melting may occur, is unknown. For example, if a plate is 70 km thick and is descending at $45°$, an earthquake 30 km down in the slab would be 45 km vertically below the upper surface of the slab (fig. 5). Thus it seems that the seismically determined value of h may bear very little resemblance to the depth at which melting actually occurs.

Conclusions

K_2O contents of orogenic volcanic suites do increase with distance to the Benioff zone, although the original linear correlation of Dickinson and Hatherton (1967) has been modified by Dickinson (1975). Both these authors in all their publications suggest that this correlation may be due either to the chemical characteristics (notably K_2O) of a suite reflecting the partial melting process at depth, or to variations in the nature of the eruption path (differentiation, contamination, etc.) between source and volcano. Nearly all the models that have been proposed to explain the correlation have assumed the former situation—that the K_2O content reflects that of the primary magma. The K_2O content for this correlation has been interpolated at various SiO_2 levels—55, 57·5, and 60 % SiO_2. It is unlikely that these can all be primary magmas and it seems that some degree of fractionation must be invoked to produce these SiO_2 values.

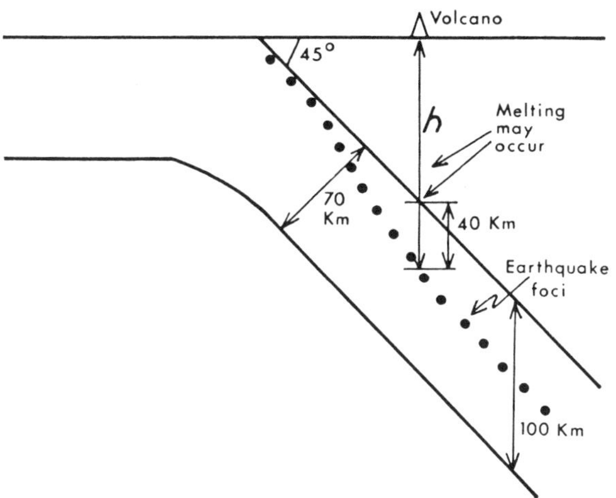

FIG. 5. Location of earthquake foci within the subduction zone in relation to possible different magma sources in the subducted plate or overlying mantle.

The alternative hypothesis—that of fractional crystallization—has been investigated by means of various theoretical K_2O–SiO_2 variation diagrams. A natural example analogous to each theoretical model has been documented. It is suggested that the K_2O content of a magma at a particular SiO_2 level may be more greatly influenced by the SiO_2 content of the parental magma and the nature of the crystallizing phases than the original K_2O content.

The value of h—the depth to the earthquake zone—may not indicate the depth of melting. Magmas may originate in the overlying mantle rather than the upper surface of the subducted slab and hence the depth of melting may be considerably less than that predicted from the value of h. The exact location of the earthquake foci may not coincide with the upper surface of the subducted slab along its entire length, but may occur *within* the slab. Hence, even if magmas were generated from the upper portion of the subducted oceanic crust the correlation with h is questionable.

The K_2O content of magmas also appears to increase with time in certain regions (e.g. Aleutians, Fiji). It is difficult to interpret this change in terms of a plate tectonic model whereby the depth to the earthquake zone is continually increasing. An alternative solution may be that the mantle is progressively chilled by the addition of cold subducted crust and that the degree of melting of the mantle consequently decreases, producing more potassic liquids.

Acknowledgements. The author thanks Drs. S. W. Richardson and A. McGarr for many useful discussions and comments.

REFERENCES

Anderson (D. L.), 1970. *Mineral. Soc. Am. Spec. Pap.* **3**, 85.
Best (M. G.), 1975. *Geology*, **3**, 429.
Billington (S.) and Isacks (B. L.), 1975. *Geophys. Res. Lett.* **2**, 63.

Boettcher (A. L.), 1973. *Tectonophysics*, **17**, 223.
Bravo (M. S.) and O'Hara (M. J.), 1975. *Physics and Chemistry of the Earth*, **9**, 845.
Carmichael (I. S. E.), Turner (F. J.), and Verhoogen (J.), 1974. *Igneous Petrology*. McGraw-Hill, New York, 739 pp.
Cawthorn (R. G.) and O'Hara (M. J.), 1976. *Am. J. Sci.* **276**, 309.
Cox (K. G.) and Jamieson (B. G.), 1974. *J. Petrol.* **15**, 269.
Delong (S. E.), 1974. *Geochim. Cosmochim. Acta*, **38**, 245.
Dickinson (W. R.), 1968. *J. Geophys. Res.* **73**, 2261.
—— 1975. *Geology*, **3**, 53–56.
—— and Hatherton (T.), 1967. *Science*, **157**, 801.
Engdahl (E. R.), 1971. *Science*, **173**, 1232.
Engel (A. E. G.), Engel (C. G.), and Havens (R. G.), 1965. *Geol. Soc. Am. Bull.* **76**, 719.
Fyfe (W. S.) and McBirney (A. R.), 1975. *Am. J. Sci.* **275-A**, 285.
Gast (P. W.), 1968. *Geochim. Cosmochim. Acta*, **32**, 1057.
Gill (J. B.), 1970. *Contrib. Mineral. Petrol.* **27**, 179.
Green (D. H.), 1972. *Tectonophysics*, **13**, 47.
Green (T. H.) and Ringwood (A. E.), 1968. *Contrib. Mineral. Petrol.* **18**, 105.
—— Green (D. H.), and Ringwood (A. E.), 1967. *Earth Planet. Sci. Lett.* **2**, 41.
Hatherton (T.), 1969. *N.Z. J. Geol. Geophys.* **12**, 436.
Holloway (J. R.) and Ford (C. E.), 1975. *Earth Planet. Sci. Lett.* **25**, 44.
Holmes (A.), 1916. *Q. J. Geol. Soc. Lond.* **72**, 222.
Jamieson (B. G.) and Clarke (D. B.), 1970. *J. Petrol.* **11**, 183.
Jakes (P.) and Smith (I. E.), 1970. *Contrib. Mineral. Petrol.* **28**, 259.
—— and White (A. J. R.), 1969. *Tectonophysics*, **8**, 223.
Kuno (H.), 1950. *Geol. Soc. Am., Bull.* **61**, 957.
—— 1959. *Bull. volcanol.* **20**, 37.
—— 1967. *in*: T. F. Gaskell, (ed.), *The Earth's Mantle: London, Academic Press*, 89–109.
—— 1968. *in*: H. Hess and A. Poldervaart, Eds., *Basalts*. J. Wiley, New York, **2**, 623.
Kushiro (I.), 1972. *J. Petrol.* **13**, 311.
—— Syono (Y.) and Akimoto (S.), 1968. *J. Geophys. Res.* **73**, 6023.
Lambert (I. B.) and Wyllie (P. J.), 1968. *Nature*, **219**, 1240.
McBirney (A. R.), 1967. *Geol. Rundschau.* **57**, 21.
Marsh (B. D.) and Carmichael (I. S. E.), 1974. *J. Geophys. Res.* **79**, 1196.
Mitronovas (W.), Isacks (B. L.), and Seeber (L.), 1969. *Bull. Seism. Soc. of Am.* **59**, 1115.
Nicholls (I. A.) and Ringwood (A. E.), 1972. *Earth Planet. Sci. Lett.* **17**, 243.
—— —— 1973. *J. Geol.* **81**, 285.
Nielson (D. R.) and Stoiber (R. E.), 1973. *J. Geophys. Res.* **78**, 6887.
Nishikawa (M.), Kushiro (I.), and Uyeda (S.), 1971. *Japanese J. Geol. Geogr.* **41**, 41.
O'Hara (M. J.), 1968. *Earth-Sci. Rev.* **4**, 69.
Sugimura (A.), 1968. *in*: H. H. Hess and A. Poldervaart, Eds., *Basalts*. J. Wiley, New York, 537–72.
Sykes (L. R.), 1966. *J. Geophys. Res.* **71**, 2981.

[*Manuscript received 5 July 1976*]

Part VI

CONTINENTAL BASALTS

Editors' Comments
on Papers 50 Through 53

50 WATERS
*Basalt Magma Types and Their Tectonic Associations: Pacific
Northwest of the United States*

51 HOOPER
The Columbia River Basalts

52 LIPMAN and MEHNERT
Excerpts from *Late Cenozoic Basaltic Volcanism and Develop-
ment of the Rio Grande Depression in the Southern Rocky
Mountains*

53 IRVING and GREEN
Excerpts from *Geochemistry and Petrogenesis of the Newer
Basalts of Victoria and South Australia*

Continental regions contain an enormous diversity of environments and volcanic products. Perhaps the most satisfactory classification would be threefold—rifts, plateaus, and others; this classification at least has the virtue of being comprehensive, if not informative. Before presenting appropriate papers, or their excerpts, we should illustrate some of the problems of classification.

Plateau basalts (flood basalts) are commonly regarded as tholeiitic, and most are. Nevertheless, alkali olivine basalts (mostly sodic) occur in some plateau provinces, and highly potassic basalts are present much less abundantly in a few places. As an example of the complexity, Mohr (1976) showed that mildly alkaline and mildly subalkaline basalts occur intermixed in the stratified sequences in both the Ethiopian rift valley (including Afar) and in the surrounding Ethiopian plateau; highly feldspathoidal differentiates, however, are restricted to the plateau. An interesting summary of plateau volcanism has been prepared by Cox (1980).

Complexities, in fact, began very early in the modern history of studies of basalts. In the famous Mull memoir (Bailey et al., 1924), the initial distinction of independent magma types was between plateau basalts and nonporphyritic central basalts (discussed and redefined by Kennedy, 1933; see Paper 1). Both basalt types, however, occur in the plateau areas. The plateau basalts were presumed to be alkali olivine basalts because they differentiated to trachyandesites, phonolites, and other alkalic rocks; the nonporphyritic central basalts were presumed to be tholeiites that differentiated to andesite and rhyolite. Unlike typical alkali olivine basalt of other areas, however, the Mull olivine basalts contain only 0.5% K_2O on the average, less than that of the associated nonporphyritic central basalts. Thus, there is a tendency for basalt terminology to obscure diversity in rock types.

We can add to the confusion by mentioning two other observations that appear to conflict with each other. First, as with the evidence mentioned earlier for the occurrence of highly alkalic rocks in the Ethiopian plateau, it is possible to propose a similar variation in basalt types around the Rio Grande rift of the southwestern United States. Lipman and Mehnert (1975) have suggested that the dominantly tholeiitic rocks of the rift grade away from the rift into more alkaline rock types. In fact, the Late Tertiary basalts of much of the Cordillera are typical alkali olivine basalts (e.g., Leeman and Rogers, 1970) or, in some areas, highly alkalic (including potassic) varieties (e.g., Moore and Dodge, 1980).

Second, tholeiitic volcanism in rifts, as contrasted to their surroundings, is not found in all places. Many geologists, in fact, tend to equate alkaline and peralkaline continental volcanism with extensional environments. The East African rift system, for example, is the home of widespread, highly alkalic, volcanic rocks even though it is tectonically an extension of the Ethiopian rift with its dominantly tholeiitic basalts (e.g., Lippard and Truckle, 1978; Williams, 1978). The best that can be said at the moment is that the relationships between plateau and rift volcanism are not clear.

Within our triad classification, the category of others is obviously enigmatic. Many continental areas contain basaltic volcanic rocks not clearly related to rift zones or large plateaus. To what are they related? Plumes? Incipient extension? Deep subduction from some distant continental margin? Controls and processes are uncertain. An excellent paper on the Eifel area of Germany, by Duda and Schmincke (1978), discusses some of these problems.

Also uncertain is the relationship of the basalts to other volcanic rocks types. Many continental volcanic suites are regarded as bimodal basalt-rhyodacite assemblages (e.g., part of the volcanic assemblage of the San Juan Mountains of Colorado; Lipman et al., 1978). A cogenetic relationship between basalts and accompanying rocks, however, has not been established for most suites, and in many areas, the basaltic volcanism seems to be unrelated to the formation of other magmatic rocks.

We present here four papers that give an indication of the varieties and complexities of continental basalts. The first is a classic investigation of the relationship between tectonic environment and varieties of basaltic volcanism by A. C. Waters (Paper 50). One of the rock types discussed by Waters is the Columbia River plateau basalt, and the paper by Hooper (Paper 51) summarizes a great deal of recent work on that important and well-studied sequence.

The paper by Lipman and Mehnert (Paper 52), mentioned previously, compares basaltic and other volcanism in the Rio Grande rift with volcanism outside the rift. We also refer readers to a related paper by Baldridge (1979) on the basalts of the rift. Because of intense interest of geologists in the subject of rifting, we should call attention to the many symposia on the subject, including Illies and Fuchs (1974), Pilger and Rosler (1975, 1976), Logatchev and Mohr (1978), Neumann and Ramberg (1978*a*, 1978*b*), Riecker (1979), Illies (1981), and Popoff and Tiercelin (1983).

Finally, Irving and Green (Paper 53) discussed the alkali olivine basalts (Newer Basalts) of southeastern Australia. This suite shows the interesting differentiation sequence to hawaiites, mugearites, and trachytes that has been found in other areas of alkali basalt volcanism.

REFERENCES

Bailey, E. G., H. H. Thomas, C. T. Clough, W. B. Wright, J. E. Richey, and G. V. Wilson, 1924, *Tertiary and Post-Tertiary Geology of Mull, Loch Aline, and Oban,* Geological Survey of Scotland Memoir, Edinburgh, 445p.

Baldridge, W. S., 1979, Petrology and Petrogenesis of Plio-Pleistocene Basaltic Rocks from the Central Rio Grande Rift, New Mexico, and Their Relation to Rift Structure, in *Rio Grande Rift: Tectonics and Magmatism,* R. E. Riecker, ed., American Geophysical Union, Washington, D.C., pp. 323-353.

Cox, K. G., 1980, A Model for Flood Basalt Volcanism, *Jour. Petrology* **21:**629-650.

Duda, A., and H. -U. Schmincke, 1978, Quaternary Basanites, Melilite, Nephelinites and Tephrites from the Laacher See Area (Germany), *Neues Jahrb. Mineralogie Abh.* **132:**1-33.

Illies, J. H., ed., 1981, Mechanism of Graben Formation, *Tectonophysics* **73** (special issue), 266p.

Illies, J. H., and K. Fuchs, eds., 1974, *Approaches to Taphrogenesis — Proceedings of an International Rift Symposium held in Karlsruhe, April 13-15, 1972,* Inter-Union Commission on Geodynamics Scientific Report No. 8, E. Schweizerbart'sche Verlagsbuchhandlung, Stuttgart, 460p.

Kennedy, W. Q., 1933, Trends of Differentiation in Basaltic Magmas, *Am. Jour. Sci.* **25,** 5th series, pp. 239-256.

Leeman, W. P., and J. J. W. Rogers, 1970, Late Cenozoic Alkali-Olivine Basalts of the Basin-Range Province, USA, *Contr. Mineralogy and Petrology* **25:**1-24.

Lipman, P. W., and H. H. Mehnert, 1975, Late Cenozoic Basaltic Volcanism and the Development of the Rio Grande Depression in the Southern Rocky Mountains, *Geol. Soc. America Mem.* **144:**119-154.

Lipman, P. W., B. R. Doe, C. E. Hedge, and T. A. Steven, 1978, Petrologic Evolution of the San Juan Volcanic Field, Southwestern Colorado—Pb and Sr Isotope Evidence, *Geol. Soc. America Bull.* **89:**59-82.

Lippard, S. J., and P. H. Truckle, 1978, Spatial and temporal variations in basalt geochemistry in the N. Kenya rift, in *Petrology and Geochemistry of Continental Rifts (Proceedings of NATO Advanced Study Institute in Oslo, Norway, 1977)*, E. R. Neumann and I. B. Ramberg, eds., D. Riedel Publishing Co., Dordrecht, pp. 123-131.

Logatchev, N. A., and P. Mohr, eds., 1978, Geodynamics of the Baikal Rift Zone, *Tectonophysics* **45** (special issue), 105p.

Mohr, P. A., 1976, A New Terminology for the Ethiopian Volcanics, with Especial Reference to Transitional Basaltic and Intermediate Lavas and Dikes, Harvard-Smithsonian Center for Astrophysics, Cambridge, Preprint Series 368, 39p.

Moore, J. G., and F. C. W. Dodge, 1980, Late Cenozoic Volcanic Rocks of the Southern Sierra Nevadas, California: I. Geology and Petrology, *Geol. Soc. America Bull.* **91,** Part I:515-518; Part II:1995-2038.

Neumann, E. -R., and I. B. Ramberg, eds., 1978a, *Petrology and Geochemistry of Continental Rifts—Volume One of the Proceedings of the NATO Advanced Study Institute, Paleorift Systems with Emphasis on the Permian Oslo Rift, held in Oslo, Norway, July 27–August 5, 1977,* D. Riedel Publishing Co., Dordrecht, 296p.

Neumann, E. -R., and I. B. Ramberg, eds., 1978b, *Tectonics and Geophysics of Continental Rifts—Volume Two of the Proceedings of the NATO Advanced Study Institute, Paleorift Systems with Emphasis on the Permian Oslo Rift, held in Oslo, Norway, July 27–August 5, 1977,* D. Riedel Publishing Co., Dordrecht, 444p.

Pilger, A., and A. Rosler, eds., 1975, *Afar Depression of Ethiopia—Proceedings of an International Symposium on the Afar Region and Related Rift Problems held in Bad Bergzabern, F. R. Germany, April 1-6, 1974, Volume 1,* Inter-Union Subcommission on Geodynamics Scientific Report no. 14, E. Schweizerbart'sche Verlagsbuchhandlung, Stuttgart, 416p.

Pilger, A., and A. Rosler, eds., 1976, *Afar Depression of Ethiopia—Proceedings of an International Symposium on the Afar Region and Related Rift Problems held in Bad Bergzabern, F. R. Germany, April 1-6, 1974, Volume 2,* Inter-Union Subcommission on Geodynamics Scientific Report no. 16, E. Schweizerbart'sche Verlagsbuchhandlung, Stuttgart, 216p.

Popoff, M., and J. -J. Tiercelin, eds., 1983, Rifts et fosses anciens, *Centre Recherches Pau Bull.* **7:**127-448.

Riecker, R. E., ed., 1979, *Rio Grande Rift—Tectonics and Magmatism,* American Geophysical Union, Washington, D.C., 438p.

Williams, L. A. J., 1978, The Volcanological Development of the Kenya Rift, in E. -R. Neumann and I. B. Ramberg, eds., *Petrology and Geochemistry of Continental Rifts; Proceedings of NATO Advanced Study Institute in Oslo, Norway, 1977,* E. -R. Neumann and I. B. Ramberg, eds., D. Riedel Publishing Co., Dordrecht, pp. 101-121.

50

Reprinted from *Geophysical Monograph No. 6, 1962, pp. 158–170*

Basalt Magma Types and Their Tectonic Associations: Pacific Northwest of the United States

Aaron C. Waters

Abstract—In the Pacific Northwest, during Miocene-Quaternary time, three varieties of tholeiitic basalt and one variety of high-alumina basalt erupted in enormous volume. Each is thought to represent a primary magma developed in the mantle, and not a product of the differentiation of a single parent magma. Differentiation did occur where some of these magmas rose to shallow levels and were stored temporarily in underground reservoirs. The ability to form near-surface reservoirs appears to depend on the tectonic conditions accompanying eruptions. The Yakima basalt, which is the most voluminous of the three tholeiitic varieties, poured from large fissures into a slowly subsiding basin. It shows no tendency to develop into central volcanoes, and consists almost wholly of thick nonporphyritic flows that do not vary in composition. The high-alumina basalt of the Oregon plateaus, on the other hand, was erupted during active block faulting; many of its feeder channels followed the faults. Although much of the early high-alumina magma escaped rapidly in fissure eruptions, simple cinder cones, large shield volcanoes. and even more complex volcanic centers have been built up above some of the fracture zones. These reveal a variety of differentiation products, but undifferentiated high-alumina magma was always available during their production for it continued to break through to the surface in nearby areas. Thus differentiation appears to have occurred only in the shallow chambers directly below volcanoes, not at great depth where the primary magmas were generated.

Introduction—Since the days of Harker's generalizations about the Atlantic and Pacific suites of igneous rocks there has been continuous interest and speculation on the connection between tectonic patterns and contemporaneous igneous activity. Nevertheless we are far from understanding the role that tectonic processes play in the formation of magmas and in their subsequent differentiation, if, indeed, they actually play any significant role at all. This paper does not pretend to offer answers to these difficult problems; it presents a description of the late Cenozoic igneous sequences in three adjacent parts of the Pacific Northwest, and summarizes some possible interrelations between magmatic evolution and contemporaneous folding, faulting, and other orogenic changes.

The Late Cenozoic volcanic provinces—Attention is confined mainly to the Miocene and post-Miocene volcanic rocks (Fig. 1) and the associated tectonic features in three provinces:

(1) The Columbia Plateau of eastern Washington, northern Oregon, and western Idaho; essentially a downsagging basin filled by three varieties of tholeiite that make up the Columbia River basalts. East-west folds characterize parts of this province, other parts are undeformed.

(2) The northern part of the Basin and Range Province in central and southeastern Oregon, where numerous fault block ranges reveal thick accumulations of high-alumina basalt, with minor amounts of rhyolite and andesite, and associated volcanic sedimentary rocks.

(3) The High Cascades, an area characterized by great numbers of hypersthene andesite stratovolcanoes, but with abundant lava fields, low shields, and cinder cones of high-alumina basalt as well. Many of the large volcanoes in this province have differentiated to dacite and rhyodacite. Somewhat older than these dominantly Pliocene-Quaternary lava fields and stratovolcanoes, are stocks, sill swarms, and small plutons of diorite and granodiorite which appear at the surface in parts of the central and northern Cascades.

Junctions between the three provinces—The

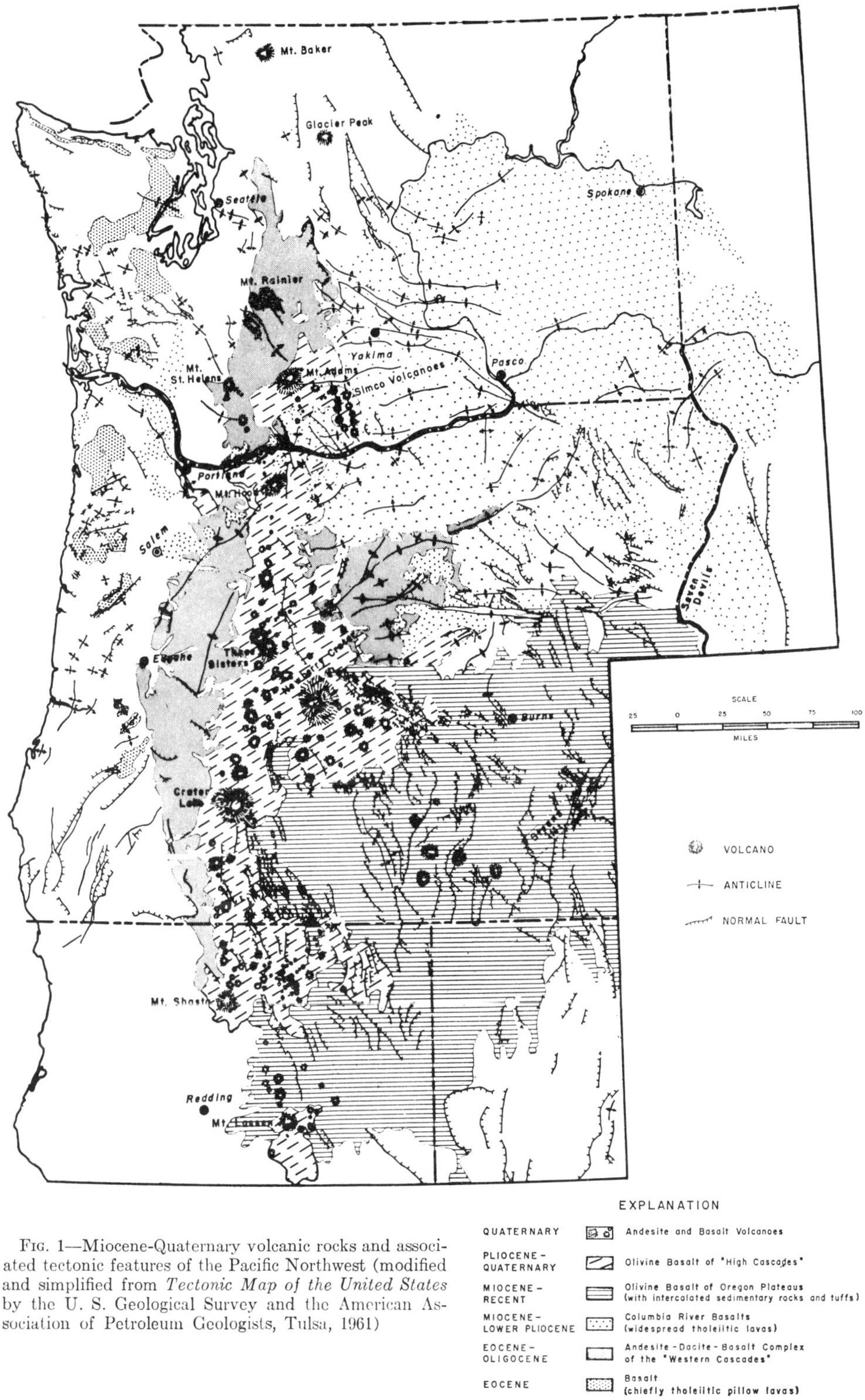

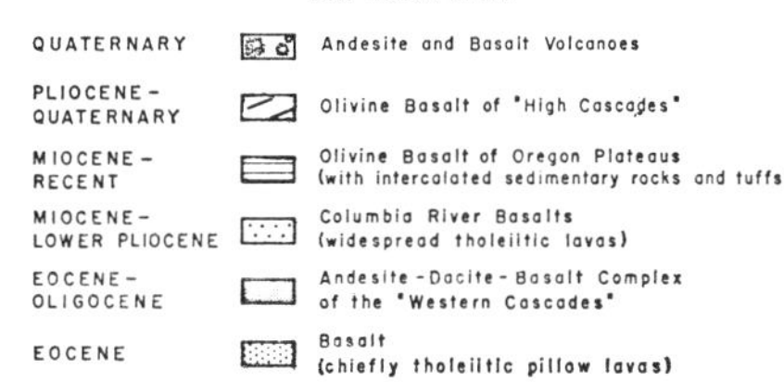

FIG. 1—Miocene-Quaternary volcanic rocks and associated tectonic features of the Pacific Northwest (modified and simplified from *Tectonic Map of the United States* by the U. S. Geological Survey and the American Association of Petroleum Geologists, Tulsa, 1961)

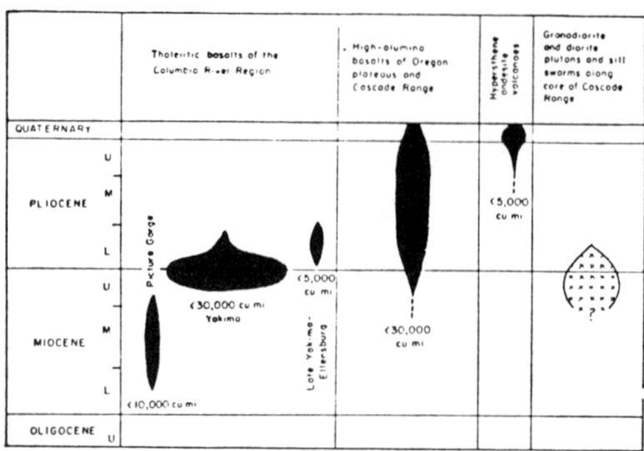

Fig. 2—Approximate volume and time relations, Miocene-Quaternary igneous rocks

Cascade Range, with its numerous Quaternary volcanoes, borders the Columbia Plateau and the Basin and Range Province on the west, but structurally it grades into them: folds, faults, and other structural features, as well as the volcanic sequences belonging to the two eastern provinces, extend westward without interruption beneath the volcanic cones of the High Cascades (Fig. 1).

The separation between the plateaus underlain by Columbia River basalt on the north, and the Basin and Range Province to the south, is also indefinite in position. The flood basalts of the Columbia River region lap southward against an old east-west tectonic and topographic high (generally called the Ochoco Mountains–Blue Mountains axis) which connects the Rocky Mountains in the Seven Devils region of Idaho with the Cascade Mountains east of the Three Sisters in central Oregon. This complex east-west tectonic zone consists of faulted and folded rocks of early Tertiary, Mesozoic, and Paleozoic ages, and also includes Mesozoic granitic intrusions, serpentines, and metamorphic rocks.

After the Columbia River tholeiitic basalts were extruded, renewed folding accentuated the older structures along this axis and produced new ones. Still later the faults of the Basin and Range Province encroached upon, and finally crossed these east-west folds. Simultaneously new centers of volcanic activity arose, but they produced high-alumina basalts and rhyolites characteristic of the Basin and Range Province, instead of tholeiitic basalts of the Columbia River varieties that had dominated the slightly earlier volcanic activity. Thus over a belt

about 50 miles wide along the Ochoco–Blue Mountains axis the geologic features of the two provinces overlap. Because of extensive uplift and erosion pre-Miocene rocks are also extensively exposed along this belt (Fig. 1).

The Columbia River basalts—The Columbia River basalts, of Miocene and early Pliocene age, comprise one of the largest areas of flood basalts in the world [*Russell*, 1893, 1901; *Merriam*, 1901; *Smith* 1901; *Lindgren*, 1901]. Recent work, as yet incompletely reported [*Waters*, 1961b], has shown that the Columbia River basalts are divisible into two main formations: an older Middle Miocene series of olivine-bearing basalts recently designated the Picture Gorge basalt, and an unconformably overlying and more siliceous unit, the Yakima basalt (Fig. 2). A third and slightly different variety of basalt magma, characterized by an unusually high content of iron and titanium, also made its appearance during the last stages of extrusion of the Yakima basalt formation, and it continued to pour forth, though in decreasing volume, during accumulation of the lower part of the overlying Ellensburg formation. Thus the Columbia River basalts can be separated into two stratigraphic divisions—the Picture Gorge basalt and the Yakima—but into three distinct magma subtypes with recognizable differences in mineralogy and chemical composition. These three magma subtypes have not been given formal names; they are referred to informally as the Picture Gorge magma, Yakima magma, and Late Yakima-Ellensburg magma after the stratigraphic divisions in which they are found.

The term tholeiitic basalt was popularized by *Kennedy* [1931, 1933] who adapted this name to volcanic rocks from the French term, *tholeiite*, which Steininger had proposed in 1841 for "doleritischen Trappgesteine" that occur at Schaumberg, near Tholey, in the Saar. These quartz-bearing dolerite dikes near Tholey have recently been investigated by Jung [1958].

Kennedy defined tholeiitic basalt as composed essentially of lime-poor pyroxene and labradorite, with subordinate iron ore and glass. He noted that olivine is either completely absent, or else present in very subordinate amount, and that there is a characteristic acidic residuum which is generally held in the glass, but crystallizes as quartz and alkali feldspar in

TABLE 1—*Minerals of the Columbia River basalts*

Mineral	Picture Gorge Basalt	Yakima Basalt	Late Yakima and Ellensburg Flows
Olivine	Zoned, phenocryst cores Fo_{80}, groundmass olivine and phenocryst rims generally in range Fo_{65}-Fo_{40}	Less than 2%, or absent; small resorbed grains near Fo_{80}; may be rimmed by pigeonite	Iron rich, generally Fo_{40}-Fo_{30}; some flows contain zoned phenocrysts with cores of Fo_{75}-Fo_{60}
Clinopyroxene	2V generally in range 52° to 44°	2V generally in range 47° to 40°; many flows contain pigeonite, 2V = 0° to 15° in addition	2V generally in range 50° to 37°; some flows also contain pigeonite, 2V = 0° to 10°
Plagioclase	Bytownite to labradorite	Labradorite to andesine	Labradorite to andesine
Opaques	Average 6%; well crystallized	Average 8%; chiefly in tachylyte	Average 7%; titaniferous skeletal crystals common
Chlorophaeite and its decomposition products	Average 5%	Average 2%	Average 3%
Glass	Average 5%, general range 1% to 20%; clear brown	Average 25%; general range 10% to 50%; black tachylyte	Average 15%; general range 2% to 30%
Crystallization	About half of flows contain inconspicuous phenocrysts of olivine and plagioclase; a few also have rare phenocrysts of clinopyroxene	Nonporphyritic, although small olivine grains were probably present in some flows before extrusion; abundant tachylyte	Textures variable; about half of flows contain phenocrysts of olivine and plagioclase

'pegmatitoides' (late segregations), and also within the final mesostasis of intrusive diabases. *Tilley* [1950] has extended and broadened Kennedy's definition to include a *tholeiitic series* of rocks, some of whose members are much richer in modal olivine: for example, the magmas that erupted to form the bulk of the shield-building lavas during all but the final stages of growth of the Hawaiian volcanoes. These rocks normally contain ten per cent or more of olivine, but nevertheless show normative quartz. Tilley also pointed out that tholeiitic basalts show large amounts of normative (not modal) hypersthene. *Kuno* [1960] has made further modifications in the definition of tholeiitic basalt, and has listed in tabular form a valuable summary of the distinctive features of the three main basalt magma types: tholeiites, high-alumina basalts, and alkali-olivine basalts. As thus enlarged by Tilley and Kuno, the term tholeiitic basalt will embrace all three varieties of the Columbia River basalts, although there are significant chemical and mineralogic differences between the three, and only the Yakima magma corresponds moderately well with Kennedy's definition.

The characteristics of the three kinds of Columbia River tholeiitic basalts can be listed in tabular form: Table 1 summarizes the nature of the chief minerals, Table 2 lists average chemical analyses, and Table 3 gives the average modal composition of the three varieties [see also *Waters*, 1961b]. Each of these three mineralogic variants of the Columbia River basalts apparently came from separate sites of magma generation, and the composition of each did not change appreciably throughout its period of volcanism. Flows of transitional composition between the three kinds appear to be absent, and individual flows from each group depart little in composition from the average of their particular type [*Waters*, 1961b]. For the purposes of this paper we are concerned mainly with the Yakima magma, which erupted in vastly greater volume than the other two, and which is also more uniform in composition.

Tectonic Movements Accompanying Extrusion of the Columbia River Basalts—The Picture Gorge basalt of middle Miocene age was erupted in volume from dike swarms within the Ochoco-Blue Mountain axis, and the lava

TABLE 2—*Average chemical composition of Columbia River*
tholeiitic basalts, and other tholeiitic basalts

	Columbia River Tholeiites (Waters, 1961b, p. 595)			Parental tholeiite magma of Hawaii [*Kuno* and others, 1957, p. 213]	Average Deccan basalt [*Sukheswala* and *Poldervaart*, 1958, p. 1487]	Average nonporphyritic central magma [*Walker* and *Poldervaart*, 1949, p. 649]
Compound	Picture Gorge type	Yakima type	Late Yakima-Ellensburg type			
	pct	pct	pct	pct	pct	pct
$Si\,O_2$	49.3	53.8	50.0	50.89	51.50	51.6
$Al_2\,O_3$	15.6	13.9	13.5	13.22	13.47	14.3
$Fe_2\,O_3$	3.5	2.6	1.9	2.03	3.10	3.5
$Fe\,O$	7.8	9.3	12.5	9.12	10.51	9.2
$Mg\,O$	6.5	4.1	4.4	8.02	5.28	5.3
$Ca\,O$	10.3	7.9	8.3	10.56	9.79	10.1
$Na_2\,O$	2.7	3.0	2.9	2.17	2.68	2.8
$K_2\,O$	0.5	1.5	1.4	0.43	0.81	1.1
$H_2\,O^+$⎱ $H_2\,O^-$⎰	1.8	1.2	0.9	0.27 0.11		
$Ti\,O_2$	1.6	2.0	3.2	2.79	2.37	1.6
$P_2\,O_5$	0.3	0.4	0.7	0.27	0.31	0.3
$Mn\,O$	0.2	0.2	0.25	0.14	0.18	0.3
$C\,O_2$	0.04		0.09			
No. of Analyses	16	8	4	32	18	11

filled a gradually deepening basin along this axis and to the north of it. Mountainous topography, especially on the south and east, walled in the lava floods, ponding the early extrusions in huge flows, many 200 feet or more thick. As the lava field grew in size, the individual flows became thinner and more widespread.

Eruption of the Picture Gorge basalt appears to have ceased as abuptly as it began. In the John Day Basin the upper surface of the Picture Gorge basalt was then buried beneath more than 1500 ft of sedimentary rocks—the Mascall formation of middle and late Miocene age. Moderate folding and erosion of the Mascall and underlying basalt then ensued before the great floods of Yakima basalt spread across this area.

The Yakima basalt, probably totaling far more than 30,000 cu mi in volume, was erupted in the late Miocene and early Pliocene. It spread from huge dike swarms located along the southern and western margins of the lava field [*Waters*, 1961b, Fig. 1]. The slowly subsiding basin in which the Yakima basalt accumulated on the east side of the Cascade Range is much larger than the one filled by the Picture Gorge basalt—it encompasses the entire basalt-cov-

ered area north of the Ochoco–Blue Mountains Axis and east of the Cascade Range (Fig. 1). On the west side of the Cascades, moreover, Miocene-Pliocene basalts of similar magma type broke forth from local fissures and inundated a smaller basin—the broadly synclinal trough of the lower Columbia River [*Wilkinson, Lowry,* and *Baldwin,* 1946; *Lowry* and *Baldwin,* 1952] and the interconnected parts of the Puget Sound downwarp between Salem, Oregon [*Thayer,* 1937] and Centralia, Washington [*Snavely* and others, 1958].

The most striking feature of the Yakima tholeiitic magma was its surprising uniformity in composition when we consider the great volume of lava that was extruded. Lava flows poured into the broad central basin east of the Cascade Range as rapidly as the basin lowered. In many measured sections, 2000 or more feet thick, there is no appreciable difference in the flows from top to bottom, and the same uniformity holds with respect to samples collected from widely spaced areas in the lava field.

Interbeds of sedimentary material are infrequent in the lower part of the Yakima basalt Toward the top of the sequence, however, thin layers of diatomite, shale, tuff, and coarse sand

TABLE 3—*Modes: Columbia River tholeiitic basalts*

Mineral	Picture Gorge type	Yakima type	Late Yakima-Ellensburg type
	pct	pct	pct
Plagioclase	41.0	34.7	41.0
Clinopyroxene	35.5	28.7	24.5
Olivine (including saponite)	5.2	0.2	8.3
Opaques	6.5	8.2	7.2
Chlorophaeite (including decomposition products)	4.7	2.4	2.8
Zeolites	0.6	0.0	0.5
Glass	6.5	25.8	15.7
Number of Specimens	51	79	14

occur between the flows. Saprolitic zones of weathered basalt, capped by ancient soils containing an abundance of fossil wood, are also common. Shallow lakes and marshes, later overwhelmed by new advances of lava, have left their record in extensive complexes of pillow basalt and palagonite breccia [*Fuller*, 1931a].

It was near the end of this period of relatively slow accumulation of lava and contemporaneous deposition of widespread sediments that flows of the iron-and-titania-rich Late Yakima-Ellensburg tholeiitic magma first made their appearance. They apparently are not differentiates of the Yakima variety of tholeiite, for they are of more basic composition, show no transitions to it, and are in places interbedded with the latest of the normal Yakima flows. Moreover, these late iron-and-titania-rich basalts were extruded under a different set of tectonic conditions than prevailed during the slow basining that accompanied the upbuilding of the Yakima basalt. Rapidly accelerating tectonic movements were beginning to break up this downsagging basin. Broad synclinal troughs had started to grow within it even before the latest flows of the normal Yakima magma were extruded. The anticlinal ridges, which are such conspicuous features of the Columbia Plateau today [*Smith*, 1903], had also begun to take form. These huge east-west folds, many are more than 50 miles long and 2 to 10 miles wide,

are not confined to the area of basining, they extend westward across the Cascade Range (Fig. 1).

The iron-and-titania-rich basalts, erupted during the early stages of this widespread deformation, are very minor in volume as compared with the Yakima basalt (Fig. 2) and they tend to cluster in the deeper synclinal troughs such as the Pasco Basin. They were erupted in part from dike swarms in the eastern part of the Cascade Range in central Washington. Therefore these lavas also are abundant along the eastern foothills of the Cascades, as in the Yakima district, but they occur at other points along the periphery of the Yakima basalt plateau as well; for example in northeastern Oregon and western Idaho where folding and faulting were also taking place at the same time as lava extrusion.

High-alumina basalt of the Oregon Plateaus and Cascade Range—Basalts that are rich in olivine and plagioclase, but poor in pyroxene, cover south-central Oregon and northeastern California, and extend westward beneath the large andesitic stratovolcanoes of Quaternary age in the Cascade Range. East of the Cascades these basalts are confined almost entirely to the area south of the Ochoco–Blue Mountains axis, but a few small lava shields and flows are found in block-faulted areas immediately north of this axis. Along the Cascade Mountains, however, these basalts have spread much farther to the north; the crest of the Cascades in Washington is drowned beneath them as far north as White pass (southeast of Mount Rainier), and they have also spilled eastward, overrunning the western edge of the Columbia River plateau in southern Washington. Large and complicated centers of eruption, such as the Simcoe Mountains volcanic area, lie entirely east of the Cascades upon the surface of the Yakima basalt.

In field habit, mode of eruption, and mineralogy these basalts are strikingly different from the Columbia River tholeiitic basalts. They form thin noncolumnar flows averaging less than 20 ft thick, although where ponded in fault troughs or other basins they attain greater thicknesses and show well-formed columns. Fissure eruptions account for many of the flows, but the dikes are not concentrated into the huge swarms typical of the tholeiitic feeders. Furthermore many dikes in south-central Oregon followed major

TABLE 4—*Minerals of the high-alumina basalts, Oregon plateaus and Cascade Range*

Mineral	Undifferentiated high-alumina basalt	Platy olivine basalt
Olivine	Sparse phenocrysts Fo_{82} to Fo_{70}; groundmass olivine mostly Fo_{65} to Fo_{40}	Mostly resorbed crystals near Fo_{75} to Fo_{65}
Clinopyroxene	2V ranges from 56° to 45°, mostly 54° to 48°	Variable; 2V range is 60° to 45°, mostly 53° to 50°
Hypersthene	. . .	Rare, about 10% of flows contain traces to 5%; 2V near 57°
Plagioclase	Labradorite to andesine	Labradorite to andesine
Opaques	Average 6%	Average 4%
Glass	Generally less than 5%	Generally less than 10%
Others	Biotite common in cavities in about 10% of flows. Apatite invariably present, 0.5% to 1.5%	Apatite invariably present
Crystallization	Holocrystalline or nearly so; diktytaxitic textures common in rocks from the Cascade Range; nonporphyritic, or may contain up to 4% of olivine phenocrysts	Most flows have a marked platy groundmass caused by strong alignment of plagioclase microlites

fault fractures to the surface. Not all the lava came from dikes, however; much of it was erupted from low lava cones, shield volcanoes, and short fissures surmounted by cinder cones, features totally unknown among the tholeiitic basalts of the Columbia River region.

The average chemical composition of these basalts is listed in Table 5; their average mode is given in column 1 of Table 6, and the properties of their minerals are listed in Table 4; note the contrast with corresponding features for the Columbia River tholeiites, listed in Tables 1, 2 and 3. Chemically the most obvious characteristic of these Oregon basalts is richness in alumina and corresponding diminution in iron oxides. This and other features relate them to the high-alumina basalts as defined by *Kuno* [1960], but the Oregon rocks are slightly higher in alkalis and lower in alumina than Kuno's examples from Japan.

The high-alumina basalts of the Oregon plateaus and adjacent Cascade Range are chiefly of Pliocene and Quaternary age (Fig. 2), but in the central part of the area some flows are as old as Miocene [*Walker*, 1960]. Along the Ochoco–Blue Mountain axis the Miocene Picture Gorge basalt was folded and extensively eroded before the high-alumina basalt broke forth.

The first flows of high-alumina basalt to appear are generally nonporphyritic or sparsely porphyritic lavas of uniform nature, but at some volcanic centers they are associated with rocks of more siliceous composition, and also with basalts rich in phenocrysts of both plagioclase and olivine. Most of the differentiates are strongly porphyritic, of variable composition, and show other distinctive features. The undifferentiated high-alumina basalt, on the other hand, shows great uniformity in petrographic and chemical character, is present in enormous volume, and is either nonporphyritic or else contains less than four per cent of phenocrysts, all of which are olivine. Each of the 56 rocks whose modes are averaged in column 1 of Table 6 conform to this 'cut-off point' regarding phenocrysts, and as a group they are nearly as uniform in mineral composition as the Yakima tholeiites.

Tectonic relations—The high-alumina basalts clearly are associated with regions of block faulting (Fig. 1). *Fuller* [1931b; *Fuller* and *Waters*, 1929], who first pointed out the distinction between the olivine-rich (high-alumina) basalts of the Oregon plateaus and the more siliceous olivine-poor (tholeiitic) basalts of the Columbia River region, was also the first to note that the olivine basalt feeders are aligned along major fault blocks that define the Basin Ranges of southern Oregon. This close relation between normal faulting and eruption of the high-alumina magma has been confirmed by later work in many other parts of the Oregon plateaus. Many volcanic centers are on or close to major faults. Hundreds of closely spaced

TABLE 5—*Average chemical composition of high-alumina magma type, Oregon plateaus and Cascade Range*

Compound	Content
	pct
$Si O_2$	49.15
$Al_2 O_3$	17.73
$Fe_2 O_3$	2.76
$Fe O$	7.20
$Mg O$	6.91
$Ca O$	9.91
$Na_2 O$	2.88
$K_2 O$	0.72
$H_2 O^+$	0.40
$H_2 O^-$	0.25
$Ti O_2$	1.52
$P_2 O_5$	0.26
$Mn O$	0.14
$C O_2$	0.06
No. of analyses	21

TABLE 6—*Modes: high-alumina basalts, Oregon Plateaus and Cascade Range*

Mineral	Undifferentiated high-alumina basalt	Platy olivine basalt differentiate
	pct	pct
Plagioclase	51.3	58.3
Clinopyroxene	20.1	21.9
Hypersthene		1.5
Olivine	18.0	5.0
Opaques	5.7	4.4
Apatite	0.6	0.7
Glass	4.2	8.1
No. of specimens	56	18

cinder cones cluster along the low fault scarps that form a northwest-southeast en echelon belt between Three Sisters and Burns, transverse to the dominant north-south faulting of central and southern Oregon (Fig. 1). Farther north, on the Columbia Plateau, the Simcoe volcanic area of high-alumina basalt was built above zones of parallel fractures that break the underlying Yakima basalt [*Sheppard*, 1960]. The same relation between faulting and the extrusion of high-alumina basalts is also apparent in the Cascade Range. The faults that define the Hood River and Metolius fault blocks have localized many volcanic centers. Thus a close connection between Basin and Range faulting and the rise of the high-alumina magma is obvious and direct.

Differentiates of high-alumina basalt; Oregon plateaus—The bulk of the late Cenozoic lava fields of southern Oregon consist of uniform high-alumina basalt, but at many places two additional varieties of basalt, and more rarely rhyolite or rhyodacite tuffs and lavas, also appear. These have stratigraphic and petrologic relations that indicate they are differentiates of the high-alumina basalt.

Among these differentiates are strongly porphyritic basalts containing abundant large phenocrysts of plagioclase and olivine. In these rocks the total quantity of phenocrysts is generally more than 15 pct and may exceed 50 pct.

The groundmass between the phenocrysts, however, is similar in composition to typical high-alumina basalt. Plagioclase phenocrysts are much more abundant than olivine in these strongly porphyritic rocks, but where the total quantity of phenocrysts falls to five or ten per cent, olivine generally predominates. These strongly porphyritic rocks are probably crystal accumulates, enriched in plagioclase and olivine by crystal settling.

Platy olivine basalts, which range into olivine andesites, are considerably more voluminous than the crystal accumulates. They also seem to be crystallization differentiates of the original high-alumina basalt, but along the 'liquid line of descent' [*Bowen*, 1928]. They are more siliceous, richer in plagioclase, and poorer in olivine than undifferentiated high-alumina basalt. Many contain a little hypersthene in addition to clinopyroxene. Small plagioclase microlites are generally well oriented by flow, giving the rock a platy structure. Table 6, column 2 gives the average mode for 18 flows of this platy basalt. The platy olivine basalts and olivine andesites are found mainly in small lava cones rising above widespread sheets of undifferentiated high-alumina basalt. Such small volcanic centers occur in many parts of the southern Oregon volcanic plateaus but are closely spaced in the areas of maximum faulting.

Pliocene-Quaternary rhyolites and rhyolitic tuffs are present in small volume on the Oregon plateaus (older rhyolites such as those associated with the Oligocene–lower Miocene John Day formation are widespread.) Most, but not all, of the young rhyolite masses are found in areas where large and complicated volcanic edifices composed mostly of both undifferentiated and platy olivine basalt have grown above the surface of the plateaus. Four of these major volcanic centers that have received geologic study are Steens Mountain [*Fuller*, 1931b], the Mount Newberry shield volcano [*Williams*, 1935], the Medicine Lake highland [*Anderson*, 1941], and the Simcoe volcanic area [*Sheppard*, 1960]. The last three are near the western edge of the plateaus, not far from the Cascade Range. The Steens Mountain volcanic succession is exposed on a high fault block in southeastern Oregon. Although the volcanic record in each of these centers starts with typical high-alumina olivine basalt of the kind that underlies the huge hypersthene andesite volcanoes of the Cascade Range to the west, they have little else in common with the Cascade stratovolcanoes. The differentiates of the plateau centers are notably more alkalic than those of the Cascade volcanoes, and hypersthene andesites are few or absent. Rhyolite is present in all of them, though in small volume. A few of the latest olivine basalts from these centers are decidedly more alkalic than the normal high-alumina basalt of the province as a whole [see analyses in *Fuller*, 1931b, p. 121, specimen IV, and in *Sheppard*, 1960, p. 141, specimen RS-129]. Thus the alkalic tendency is not confined entirely to the more silicic differentiates.

Differentiates of high-alumina basalt; Cascade Range—The western slope of the Cascade Range is carved from early Tertiary volcanic rocks. These disappear eastward beneath a cover of closely spaced Pliocene to Recent volcanic cones and intracone lavas that occupy the crest and eastern slope of the range. The area dominated by young volcanoes is known as the High Cascades, and the part underlain by deeply eroded older volcanic rocks is called the Western Cascades (see Fig. 1).

Building of the High Cascades in central and northern Oregon began in Pliocene time with the outpouring of great volumes of undifferentiated high-alumina basalt. These flows are continuous with those that underlie the adjacent Oregon plateaus. In the High Cascades, unlike many parts of the plateau area, eruptive activity quickly became centralized into great numbers of closely spaced volcanoes, and a wider range of differentiates was formed. Large and small shield volcanoes of platy olivine basalt and olivine andesite overlap one another in the crestal part of the range, and towering among them are the huge stratovolcanoes of hypersthene andesite such as Mount Jefferson, Mount Hood, and Mount Adams. Satellitic domes and spines of dacite have also arisen, and some of the stratovolcanoes ejected large amounts of acidic pyroclastics.

Thus one of the chief differences between the Cascade Range of north-central Oregon and the plateaus to the southeast is the abundance of large volcanic cones of andesitic composition in the Cascades and their almost complete absence on the Oregon plateaus. Yet the close association of high-alumina basalt with the andesitic volcanoes of the Cascades, and the complete transitions that can be found from high-alumina basalt through platy olivine basalt, olivine andesite, and olivine-hypersthene andesite into the typical hypersthene andesites of the stratovolcanoes indicates a close genetic connection; evidently the andesites must have been derived from the high-alumina basalt by some process of fractionation or contamination. Moreover we are faced with the problem of why this production of andesite occurred on such a grand scale in the Cascade Range whereas fractionation of the same magma on the Oregon plateaus—where it poured forth on an even more voluminous scale—generally went no farther than the development of relatively small amounts of platy and porphyritic basalts, or, in the major volcanic centers, to rhyolites and alkalic basalts with little or no hypersthene andesite.

This is not the place for a full discussion of the difficult problem of the origin of orogenic andesites. *Byers* [1961] has published a recent review of this subject with some examples drawn from the Cascade area. But I think one point should be made: Whatever the process that produced hypersthene andesites and dacites in areas that had first erupted high-alumina basalts, these changes must have taken place not at the depth (probably in the mantle) where high-alumina basalt was being generated, but at higher levels probably not far beneath the volcanoes. The evidence is that undifferentiated high-alumina basalt was always available; it reached the surface from nearby vents throughout the entire period when the andesitic volcanoes were developing. Field relations, for example in the region of Mount Hood and the Columbia Gorge, show that the hypersthene andesite volcano superstructures rest on flows of undifferentiated high-alumina basalt, but that other flows of the same kind of basalt are contemporaneous with, and still others are younger than the andesite. Clearly uniform high-alumina basalt was being manufactured at depth throughout the period of andesite gestation, but in rising to the surface much of it bypassed the shallow reservoirs where differentiation to andesite had occurred. Some andesitic volcanoes, however, appear to have received new accessions of basalt magma from the mantle; they put forth flows of platy olivine andesite and olivine basalt after having erupted more acidic lavas. Moreover, the youngest flows in this part of the Cascade Range, for example the Recent flows that form the Big Lava Bed north of Columbia River in the Hood River quadrangle, are typical high-alumina olivine basalt.

The same relation holds for the major volcanic centers on the Oregon plateaus. Dikes of high-alumina basalt cut the differentiates, and the youngest flows are generally of undifferentiated basalt lava.

Miocene Intrusive Rocks of the Washington Cascades—Before leaving the andesite problem some mention should be made of the Miocene to lower Pliocene intrusive complexes of the Cascade Range in central Washington. Western Cascade rocks of Eocene to early Miocene Age lie unconformably beneath the Pliocene and younger high-alumina basalts and andesitic volcanoes just discussed in both Oregon and southern Washington. They consist of lava flows, tuffs, tuff-breccias, mudflows, and volcanic sedimentary rocks that are divisible into several formations [*Peck*, 1960; *Waters*, 1961a]. Basaltic andesite and hypersthene-augite andesite predominate, but the sequence contains great amounts of dacitic and rhyolitic pyroclastics, especially in its Oligocene and lower Miocene parts. Nearly all of the rocks of the Western Cascades have undergone a mild but widespread alteration of zeolite facies type.

Invading these altered rocks along the core of the Cascade Range in central Washington are numerous small granodiorite plutons and great swarms of closely related sills, dikes, and stocks. These were emplaced under such shallow cover that in many places they broke through to the surface in volcanic eruptions [*Fuller*, 1925; *Cater*, 1960; *Fiske, Hopson* and *Waters*, in press]. Pyroclastics from these explosive eruptions were carried eastward by streams and now form a large part of the hypersthene-hornblende andesitic and rhyodacitic debris that makes up the Ellensburg formation in the eastern foothills of the Cascades [*Waters*, 1961a]. The lower part of the Ellensburg formation, as we have already noted, rests upon the Yakima basalt, and some of the flows of iron-and-titania-rich basalt, which are the youngest representatives of the Columbia River basalts, are interbedded with it. Thus an interesting facet of Cascade geology is that while undifferentiated Yakima basalt was pouring forth in great volume on either side of the Cascade Range, plutonic rocks of granodioritic composition were working their way upward to the surface within the core of the orogen. What relation, if any, do these two events have to one another? Is the production of shallow epizonal masses of granodiorite in orogenic zones merely one aspect of the andesite problem? And what is the relation between the formation of both andesites and shallow plutons, and the simultaneous production of great quantities of undifferentiated basalt (both tholeiitic and high-alumina types) within the orogen and in adjacent areas?

Petrogenetic implications—The huge areas of volcanic rocks in the Pacific Northwest are mostly unmapped, but even though the record is incomplete some deductions about the magmatic evolution seem tenable.

(1) Three different kinds of tholeiitic basalt, and at least one kind of high-alumina basalt, have risen directly from the mantle (but perhaps from different depths within it?) and have poured forth over the surface in huge quantities. Each must have its source in a zone capable of supplying magma of uniform composition, in large quantities, and over a vast area, during a long period of time.

(2) Where the magma rose quickly and poured out in voluminous fissure eruptions, no differentiation occurred despite the large volumes of magma involved. The siliceous Yakima tholeiitic basalt is the most striking example. On the other hand, where intrusive activity became sufficiently localized beneath certain

places to build complex volcanic centers on the surface, the magma became stored and differentiated in underground reservoirs beneath the volcanoes.

(3) Differentiation in these shallow reservoirs produced a variety of derivative rocks. High-alumina basalt appears to have been trapped underground in many shallow hearths beneath both the Oregon plateaus and the High Cascades. Differentiation (and contamination?) proceeded within some of these; in general the most complete differentiation occurred in reservoirs that lay beneath the largest volcanic centers. The important point that I wish to stress, however, is that differentiation took place in these shallow reservoirs, not at the sites where the magma was generated. The evidence for this is that undifferentiated high-alumina basalt rose to the surface from nearby vents all through the periods of time in which derivative magmas were being erupted.

(4) The presence of many such differentiation centers on the Oregon plateaus, and their absence from the large basin of Yakima basalt east of the Cascade Range, probably reflects the difference in tectonic conditions under which the two magmas rose to the surface. On the Oregon plateaus active block faulting helped to form shallow reservoirs. Gradual basining of the Columbia River plateau, and the presence of huge dikes cutting brittle underlying rock, did not provide opportunity for the Yakima magma to be trapped underground. But under different tectonic conditions the Yakima basalt might have been stored and differentiated. Possibly, indeed, Yakima basalt was trapped as it rose beneath the core of the Cascade Range, and may have played a role in the development of the contemporaneous shallow granodiorite plutons, sill swarms, and related explosive eruptions of rhyodacitic debris that are confined entirely to the orogen.

Finally, in a more philosophic vein, we turn to the following problem:

The question of a single parent basalt—Pervading much of the literature on basalts is the underlying assumption that there must be *one parental basalt magma* and that all the other basalt magma types (and perhaps all varieties of igneous rocks as well) are derived from it. Many workers have wrestled with the problem of making tholeiite from alkali-olivine basalt, or

vice versa. It cannot be said that the schemes proposed have achieved outstanding success. Most of them depend upon postulated crystal sorting of pyroxenes having rather special compositions, but it is a matter of common observation that pyroxene phenocrysts are not common in basalts except in rare varieties such as ankaramites. Olivine and plagioclase are the common phenocrysts in each of the three conventional basalt magma types: tholeiites, high-alumina basalts, and alkali-olivine basalts. Moreover, each of these three magmas, on making its first appearance within a province, generally pours forth in large volume as nonporphyritic flows, or in flows with sparse to moderate amounts of phenocrystic olivine. Both the volume relations and the nonporphyritic texture belie the assumption of a long period of crystal sorting and differentiation before extrusion. Laboratory studies have not shown a solution; instead they seem to indicate that there is a thermal barrier between tholeiite and alkali-olivine basalt. Some workers, taking a different tack, postulate that tholeiite is formed by assimilation of sial in alkali-olivine basalt, but this hypothesis runs into the embarrassing absence of sial on the ocean floors; how, then, account for the great volumes of tholeiitic basalt in Hawaii and other oceanic islands? The problem gets even more perplexing as additional areas are studied and we realize that there are tholeiites and tholeiites, as well as basalts and basalts, not to mention granites and granites! The Yakima tholeiite is not the same as the Hawaiian tholeiite, and both show significant, even if minor, differences from the tholeiite of the Deccan plateaus of India, or of the Hebridean region in Scotland (see Table 2).

Very likely the search for a single parent basalt of uniform composition has its roots in the role that Bowen (and also Daly) ascribed to basalt magma in their theories of the evolution of the igneous rocks. But there is nothing in Bowen's theory of crystallization differentiation that requires only one species of parent basalt. Indeed the striking fact about the chemistry of igneous rocks is that *the convergence which they show is forward toward granite, not backward toward a single variety of basalt*. This is brought out in the memoir by *Tuttle* and *Bowen* [1958], and is strikingly documented by the graphs of differentiation index that *Thornton* and *Tuttle* [1960] have plotted for many petrographic provinces. I suggest that the key to the origin of the different basalt magma types should be sought elsewhere than in a Don Quixote assault on the windmill of a single parental basalt. Perhaps, like Sancho Panza, we shall do better to eat and sleep until means can be found to study the partial fusion of peridotite, eclogite, and other possible ingredients of the mantle under conditions that can take into account the phase modifications imposed by earth pressures at the depths where basalt magma is generated. For this we also need additional seismic evidence of the actual depths from which each of the three kinds of magmas begin their ascent to the surface.

Acknowledgments—Opportunities provided as an employee of the United States Geological Survey to do field work in parts of these areas, and financial support as a National Science Foundation Senior Postdoctoral Fellow during 12 months in Oregon, are gratefully acknowledged. Charles B. Hunt and Clifford A. Hopson have read the manuscript and made useful suggestions.

REFERENCES

ANDERSON, C. A., Volcanoes of the Medicine Lake Highland, California, *Univ. Calif. Publ., Bul. Geol. Sci., 25,* 347–422, 1941.

BOWEN, N. L., *The Evolution of the Igneous Rocks,* Princeton University Press, Princeton, 322 pp., 1928.

BYERS, F. M., Petrology of three volcanic suites, Umnak and Bogoslof Islands, Aleutian Islands, Alaska, *Geol. Soc. Amer. Bul., 72,* 93–128, 1961.

CATER, F. W., Chilled contacts and volcanic phenomena associated with the Cloudy Pass batholith, Washington, *U. S. Geol. Surv., Prof. Pap. 400-B,* B471–B473, 1960.

FISKE, R. S., C. A. HOPSON, and A. C. WATERS, Geology of Mount Rainier National Park, Washington, *U. S. Geol. Surv., Prof. Paper 444* (in press).

FULLER, R. E., *The geology of the northeastern part of the Cedar Lake quadrangle, with special reference to the deroofed Snoqualmie batholith,* unpublished M.S. thesis, University of Washington, Seattle, 96 p., 1925.

FULLER, R. E., The aqueous chilling of basaltic lava on the Columbia River Plateau, *Amer. J. Sci., 5th ser., 21,* 281–300, 1931a.

FULLER, R. E., The geomorphology and volcanic sequence of Steens Mountain in southeastern Oregon, *Univ. Washington (Seattle) Publ. Geol., 3,* 1–130, 1931b.

FULLER, R. E., and A. C. WATERS, The nature and origin of the horst and graben structure of southern Oregon, *J. Geol., 36,* 204–238, 1929.

JUNG, DIETER, Untersuchungen am Tholeyit von Tholey (Saar), *Beiträge zur Mineralogie und Petrographie, 6,* 147–181, 1958.

KENNEDY, W. Q., The parent magma of the British Tertiary province, *Mem. Geol. Surv., Summary of Progress for 1930, pt. 2,* 62–67, 1931.

KENNEDY, W. Q., Trends of differentiation in basaltic magmas, *Amer. J. Sci., 5th ser., 25,* 239–256, 1933.

KUNO, H., K. YAMASAKI, C. IIDA, and K. NAGASHIMA, Differentiation of Hawaiian magmas, *Japanese J. Geol. and Geogr., 28,* 179–218, 1957.

KUNO, H., High-alumina basalt, *J. Petrol., 1,* 121–145, 1960.

LINDGREN, W., The gold belt of the Blue Mountains of Oregon, *U. S. Geol. Survey 22nd Ann. Rept., pt. 11,* 551–776, 1901.

LOWRY, W. D., and E. M. BALDWIN, Late Cenozoic geology of the lower Columbia River valley, Oregon and Washington, *Geol. Soc. Amer. Bul., 63,* 1–24, 1952.

MERRIAM, J. C., A contribution to the geology of the John Day Basin, *Univ. Calif. Publ., Bul. Dept. Geol. Sciences, 2,* 269–314, 1901.

PECK, D. L., Cenozoic volcanism in the Oregon Cascades, *U. S. Geol. Survey, Prof. Pap. 400-B,* B308–310, 1960.

RUSSELL, I. C., A geological reconnaissance in central Washington, *U. S. Geol. Surv. Bul. 108,* 108 pp., 1893.

RUSSELL, I. C., Geology and ground water resources of Nez Perce County, Idaho, *U. S. Geol. Surv. Water-Supply Pap. 53 and 54,* 141 pp., 1901.

SHEPPARD, R. A., *Petrology of the Simcoe Mountains area,* Washington, unpublished Ph.D. dissertation, Johns Hopkins University, Baltimore, 153 pp., 1960.

SMITH, G. O., Geology and water resources of a portion of Yakima County, Washington, *U. S. Geol. Surv. Water-Supply Pap. 55,* 68 pp., 1901.

SMITH, G. O., Anticlinal mountain ridges in central Washington, *J. Geol. 11,* 166–177, 1903.

SNAVELY, P. D., R. D. BROWN, A. F. ROBERTS, and W. W. RAU, Geology and coal resources of the Centralia-Chehalis district, Washington, *U. S. Geol. Surv. Bul. 1053,* 159 pp., 1958.

SUKHESWALA, R. N. and A. POLDERVAART, Deccan basalts of the Bombay area, India, *Geol. Soc. Amer. Bul. 69,* 1475–1494, 1958.

THAYER, T. P., Petrology of the later Tertiary and Quaternary rocks of the north-central Cascade Mountains in Oregon, with notes on similar rocks in western Nevada, *Geol. Soc. Amer. Bul., 48,* 1611–1652, 1937.

THORNTON, C. P. and O. F. TUTTLE, Chemistry of igneous rocks, I, Differentiation index, *Amer. J. Sci., 258,* 664–684, 1960.

TILLEY, C. E., Some aspects of magmatic evolution, *Geol. Soc. London, Q. J., 106,* 37–61, 1950.

TUTTLE, O. F. and N. L. BOWEN, Origin of granite in the light of experimental studies, *Geol. Soc. Amer. Mem. 74,* 153 pp., 1958.

WALKER, G. W., Age and correlation of some un-named volcanic rocks in south-central Oregon, *U. S. Geol. Surv. Prof. Pap. 400-B*, B298–300, 1960.

WALKER, F., and A. POLDERVAART, Karroo dolerites of the Union of South Africa, *Geol. Soc. Amer. Bul. 60*, 591–706, 1949.

WATERS, A. C., Keechelus problem, Cascade Mountains, Washington, *Northwest Sci., 35*, 39–57, 1961a.

WATERS, A. C., Stratigraphic and lithologic varia-tions in the Columbia River basalt, *Amer. J. Sci., 259*, 583–611, 1961b.

WILKINSON, W. D., W. D. LOWRY and E. M. BALDWIN. Geology of the Saint Helens quad-rangle, Oregon, *Oregon Geology and Mineral Industries Bul. 31*, 39 pp., 1946.

WILLIAMS, H., Newberry volcano of central Ore-gon, *Geol. Soc. Amer. Bul., 46*, 253–304, 1935.

51

Reprinted from *Science* **215**:1463–1468 (1982)

The Columbia River Basalts

Peter R. Hooper

Knowledge of the dynamic processes controlling the formation and evolution of the earth's crust has greatly increased over the last two decades. Today we are aware that the 10-kilometer-thick crust beneath the oceans is composed primarily of basalt, which is being extruded in a continuing process along a worldwide (containing more SiO_2, Na_2O, and K_2O than basalt, but less MgO and FeO), while the heavier solid residue continues deeper into the mantle. Rocks of andesite composition, which are less dense than basalts, resist later subduction and accrete on the surface to form the continents.

Summary. Between 17 million and 6 million years ago, 200,000 square kilometers of the American Northwest were flooded by basaltic lava that erupted through fissures in the crust up to 150 kilometers long. Larger individual eruptions covered over a third of the Columbia Plateau in a few days. The lavas represent partial melts of the earth's mantle that were only slightly modified by near-surface, upper crustal processes. The abundant chemical and mineralogical data now available offer an opportunity to study mantle composition and the processes involved in the evolution of the earth's crust.

network of ocean ridge-rift systems such as the Mid-Atlantic Ridge and the East Pacific Rise (*1*). Basalt is generated by the melting of the less refractory part of the upwelling, ultramafic mantle that underlies the crust. Basaltic (mafic) crust so formed is displaced by still newer crust and moves laterally away from the ridges until it descends again to the mantle along subduction zones, where the crust plunges beneath continental margins (as along the western coast of South America), or under island arc systems such as those that flank the western Pacific from the Aleutian Islands to Indonesia.

In the subduction process the wet basaltic ocean crust undergoes a second partial melting as it descends to zones of higher temperature and pressure. The least refractory portion melts again and moves toward the surface as lava with the average composition of andesite

The total effect of this continuing process of plate tectonics (*1*) is a progressive differentiation of the earth's mantle by a thermal convection system that has operated since the earth's origin more than 4.5 billion years ago. It results in the less refractory and less dense components moving toward the earth's exterior and the more dense components migrating nearer to the earth's center. The less dense crust floats on the earth's mantle. The continental crust, which is both less dense and thicker than the oceanic crust, is largely exposed above sea level.

Formation of new oceanic crust along the length of an ocean ridge-rift system involves the addition of very large volumes of basaltic rock, much of which reaches the ocean floor as lava fed by the deep fissures of the rift system. This activity is normally hidden beneath the oceans. Sporadically through geologic time, similar basalt eruptions have pene-

trated the continental crust, perhaps at the beginning of a new convection cycle when the hot, upwelling part of the convection cell in the mantle moves under a continent, where its activity may eventually break the continent apart to form new oceanic crust between the separated continental fragments. Alternatively, a smaller secondary convection system may develop behind an island arc (back-arc spreading).

Although of shorter duration than ocean floor eruptions, production of continental flood basalt may be equally rapid, providing large volumes of very fluid basaltic liquid which reach the surface along fissures in the crust. The fluidity of the basaltic lava and its unusually (*2, 3*) high rate of eruption result in sheets of basalt that cover many thousands of square kilometers and build large basalt plateaus. The basalt plateaus contrast with the familiar volcanic cones formed by the more viscous and less voluminous andesitic lavas. Major continental basalt plateaus formed in this way are found in South Africa (Karroo), South America (Parana), and India (Deccan), to name a few of the larger, better known, examples. The rocks of these plateaus provide us with one of the more direct ways of studying the composition of the earth's mantle, from which they are derived by partial melting, and allow us to examine the many complex processes which lead to the formation and growth of the earth's crust.

The Columbia River Basalt

The Columbia Plateau was formed by basalt that erupted from fissures during an 11-million-year period in the Miocene Epoch (between 17 million and 6 million years ago) (Fig. 1) (*4*). More than 95 percent of the enormous volume of basalt accumulated in the first 3.5 million years (Fig. 2) and covers an area of approximately 200,000 km^2 with an average thickness of more than 1 km; this

The author is professor of geology, Washington State University, Pullman 99164.

area (Fig. 3) includes most of eastern Washington, a large part of northern Oregon, and significant parts of west central Idaho—a total area larger than the state of Washington. Other continental flood basalt provinces are even larger. The Columbia River basalts make up one of the youngest basalt plateau provinces, and although they are unusually well-preserved from weathering and erosion by a semiarid climate, they are deeply dissected by the steep-walled canyons of the great rivers of the Columbia-Snake system. The accessibility and freshness of the Columbia River basalts make them more amenable to detailed study than many of the larger, but older, basalt plateaus in more remote parts of the world, or their ocean floor equivalents.

Like other continental flood basalt provinces, the Columbia River Basalt Group was long regarded as a monotonous sequence of indistinguishable basalt flows. The development of rapid, precise techniques for the analysis of major and trace elements in rocks has changed this view. It is now technically and economically feasible to analyze many samples from a group of flows and to make the analyses available rapidly enough for field geologists to apply them directly as they map. Most Columbia River basalt flows or small groups of flows may now be identified by their chemical composition (5) (Fig. 4). Furthermore, a fluxgate magnetometer small enough to be carried on the field geologist's belt permits the magnetic polarity of each flow to be determined in the field. We know that the earth's magnetic field reverses itself periodically. Such magnetic reversals, and the striped magnetic patterns imprinted on the ocean crust parallel to and mirrored on either side of the ocean ridge-rift systems, provided one of the vital clues to our understanding of the creation of new oceanic crust by plate tectonics (6). In a vertical sequence of horizontal basalt flows accumulated over a period of 11 million years, we would therefore expect, and indeed find, many such magnetic reversals. The magnetostratigraphic rock units bracketed by these reversals (7) have been mapped (Fig. 1) and provide an unusually precise time correlation across the Columbia Plateau.

Together with the older arts of field mapping and microscope petrography, the new techniques of rapid analysis and magnetic polarity have established a detailed stratigraphic succession of the Columbia River Basalt Group (Fig. 1) (8). With a comprehensive stratigraphic succession established, it has been possible to clarify the physical and chemical evolution of the basalt magma during the 11 million years of volcanic activity, to correlate individual flows with their feeder dikes (fissures filled with solidified lava), to trace the flows back to their source, and to reconstruct the magnitude of the eruptions.

	Formation	Member	Magnetic Polarity	K / Ar Dates
		Lower Monumental	N	6 my
		Ice Harbor	N, R	8 my
		Buford	R	
	Saddle	Elephant Mountain	N, T	
	Mountains	Pomona	R	12 my
		Esquatzel	N	
	Basalt	Weissenfels Ridge	N	
		Asotin	N	
		Wilbur Creek	N	
		Umatilla	N	13.5 my
	Wanapum	Priest Rapids	R_3	
		Roza	T, R_3	
	Basalt	Frenchman Springs	N_2	
		Eckler Mountain	N_2	14.5 my
	Grande Ronde Basalt		N_2	
	Picture Gorge Basalt (R_2 / N_1)		R_2	
			N_1	
			R_1	16.5 my
	Imnaha Basalt	T	N_0	
		R_0 ?		17.0 my

(Left margin vertical labels: Columbia River Basalt Group; Yakima Basalt Subgroup; Basalt.)

Fig. 1. Stratigraphic succession of the Columbia River Basalt Group (4, 8). The Picture Gorge Basalt is limited to the John Day Basin in the south central part of the province and was fed by its own dike swarm (Fig. 3). N, normal magnetic polarity; R, reversed magnetic polarity; T, transitional magnetic polarity; and my, million years.

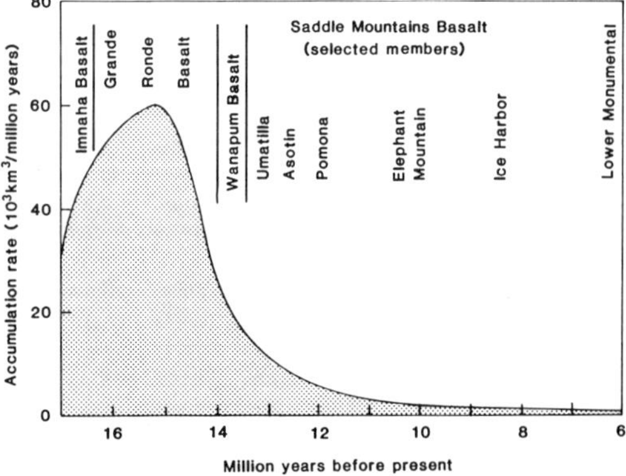

Fig. 2. Rate of basalt accumulation as a function of time.

History of the Volcanic Activity

The maximum observed thickness of the Columbia River basalts is 1500 meters, but basaltic rocks have been reported from drill cores as deep as 3000 m near the center of the plateau; the total thickness of the basalts, based on estimates of the greatest known thickness for each formation, is more than 2500 m. The Columbia River basalts fill a shallow basin; they are thickest at the center (Pasco Basin) and wedge out toward the margin of the basin (8). Rough calculations imply a total basalt volume of more than 200,000 km^3, and the number of flows ultimately identified on the plateau will probably be between 120 and 150. Individual flows range in thickness up to 120 m, with an average thickness of 15 to 30 m, and their areal extent varies from small spatter cones at source vents to major flows that cover a significant part of the Columbia Plateau, with volumes as large as 700 km³. The flows were erupted from north-northwest to south-southeast fissures concentrated in the southeast part of the plateau, where dikes cut through both older lava flows and the surrounding prebasaltic rocks (*3, 9*).

Volcanic activity began 16.9 ± 0.03 million years ago (*4*) at the southeastern extremity of the province, where the Snake River now separates Oregon and Idaho (Fig. 3). There, Imnaha Basalt lava (Fig. 1), containing large crystals of olivine, plagioclase feldspar, and occasionally clinopyroxene, poured from the fissures and filled two separate basins, which lie to the north and to the south of the Seven Devils–Wallowa Mountains divide. Some of the last Imnaha Basalt flows moved farther north into the present Clearwater drainage near Lewiston (Fig. 3), but they probably never reached the central part of the Columbia Plateau. The lava filled an older, deeply dissected surface that drained by way of a series of deep valleys to the west or southwest from the high ground of the still-rising Idaho granite batholith in the east.

There were 20 to 25 eruptions of Imnaha Basalt during a single magnetic polarity epoch lasting an estimated 500,000 years, or approximately one eruption every 20,000 years. During the transition from normal to reversed magnetic polarity about 16.5 million years ago, the nature of the erupting magma changed (Fig. 1). The later Grande Ronde Basalt is almost totally lacking in large crystals (phenocrysts) and is more siliceous than is the earlier Imnaha Basalt. With only minor compositional changes, the Grande Ronde Basalt poured out for the next 2 million years (Fig. 1), a period during which three more magnetic polarity reversals occurred (*8*). The center of activity had moved north to the present junction of the states of Oregon, Washington, and Idaho, and well over 60 flows were erupted at a greater average rate than is found either before or after this period; major eruptions occurred at the rate of one every 10,000 years at the peak of activity (Fig. 2). The Grande Ronde Basalt flooded virtually the whole Columbia Plateau, and some of its flows passed through the rising Cascade Range to reach the Pacific coast. The Grande Ronde constitutes more than 85 percent by volume of the whole Columbia River Basalt Group.

Then, about 14.5 million years ago, there was a hiatus in the volcanic activi-

ty, marking the end of the Grande Ronde Basalt period. Sediments were deposited on the lower ground to the west (Pasco Basin), and a prolonged period of weathering of the Grande Ronde Basalt occurred in the southeast. The lack of sediment deposition in the southeast reflects the gradual rise of this corner of

the Columbia Plateau, which had probably begun well before the onset of volcanism. Later flows of Grande Ronde and younger basalts did not reach the higher topographic surface in significant volume in the southeast part of the plateau; although there are feeder dikes, which are associated with small volcanic cones

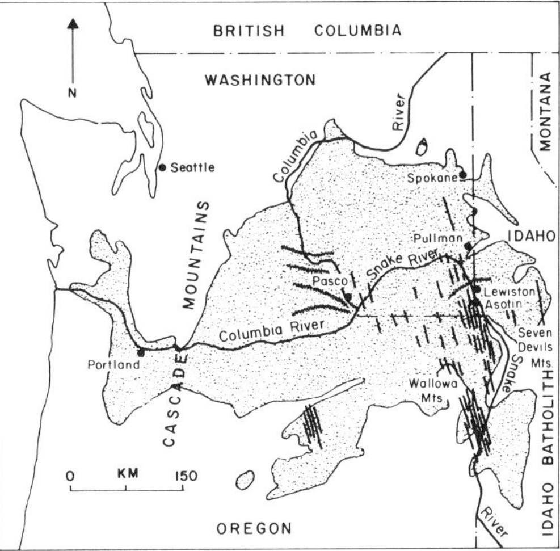

Fig. 3. Map of the area covered by the Columbia River basalts in the American Northwest. Heavy lines represent the orientation, approximate position, and concentration of the feeder dikes. Hatched lines represent the anticlinal ridges.

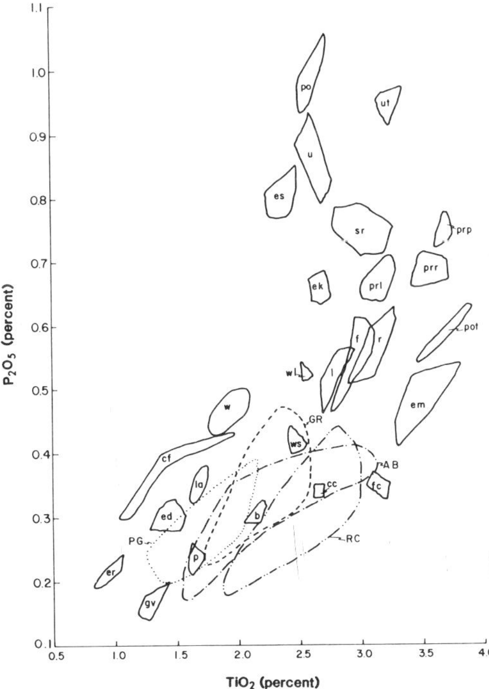

Fig. 4. Plot of percentages of P_2O_5 as a function of percentages of TiO_2 of various Columbia River Basalt flows (*5*). *RC*, Imnaha Basalt, Rock Creek type (88 analyses); *AB*, Imnaha Basalt, American Bar type (71); *PG*, Picture Gorge Basalt (40); *GR*, Grande Ronde Basalt (82). Wanapum and Saddle Mountains Basalt flows shown individually as follows: *po*, Powatka flow (21); *u*, Umatilla (20); *ut*, high TiO_2 Umatilla (5); *es*, Shumaker Creek (8); *sr*, Sopher Ridge (21); *ek*, Looking Glass (10); *prl*, Priest Rapids, Lolo type (25); *prr*, Priest Rapids, Rosalia type (10); *prp*, Priest Rapids, Pullman type (4); *pot*, Potlatch (5); *r*, Roza (10); *f*, Frenchman Springs (20); *l*, Lower Monumental (7); *wl*, Wiessenfels Ridge, Lewiston Orchards type (4); *ws*, Wiessenfels Ridge, Slippery Creek type (7); *em*, Elephant Mountain (22); *w*, Wilbur Creek (15); *cf*, Cricket Flat (8); *er*, Robinette Mountain (4); *ed*, Dodge (18); *la*, Lapwai (11); *gv*, Grangeville (12); *p*, Pomona (10); *b*, Buford (9); *cc*, Icicle Creek (5); and *fc*, Feary Creek (6).

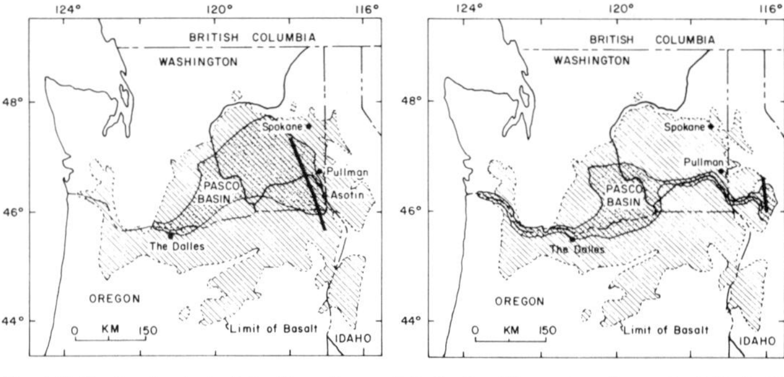

Fig. 5 (left). Areal extent of the Roza flow and its feeder dike system (heavy line) (2, 3, 8).
Fig. 6 (right) Areal extent of the Pomona flow and its feeder dike system (heavy line) (8, 12).

in a linear system, the bulk of the lava poured out farther north or west where the fissures intersected a lower land surface.

The hiatus in activity was followed by the eruption of flows of Wanapum Basalt (Fig. 1). Relatively few flows (9 to 12) erupted, but many were of exceptionally large volume and are compositionally distinct both from each other and from the flows of Grande Ronde Basalt. These great Wanapum Basalt flows are among the best known on the plateau, and one, the Roza flow, is described in detail below.

Eruption of the Wanapum Basalt represents a significant decrease in the rate that magma reached the surface (Fig. 2). A further decline in activity is seen in the final phase of the Columbia River basalt activity, the eruption of the Saddle Mountains Basalt (Fig. 1). Although its extrusion lasted for more than 7.5 million years (13.5 million to 6 million years ago), the Saddle Mountains Basalt (8) forms less than 1 percent of the total volume of the Columbia River basalt (Fig. 2), and many small, compositionally divergent flows were erupted at relatively long intervals. The progressive deformation and erosion of the Columbia Plateau, although probably no more rapid at this time than earlier, is more easily seen because of the greater length of time between eruptions (10). Most Saddle Mountains Basalt flows are clearly unconformable on older flows, typically filling erosional valleys or structural basins. Although one group (Ice Harbor Member) was erupted near the center of the plateau, most were erupted near the eastern margin. The tendency of the Saddle Mountains Basalt flows to follow, fill, and often block the river valleys and canyons resulted in the formation of lakes and subsequently the production of pillow-palagonite complexes (altered basaltic glass), which formed when a flow

plunged into and was chilled by the water. The lava was more dense than the unconsolidated sediments at the bottom of the lakes and often burrowed beneath the sediment to produce lava-sediment mixtures (11, 12). One Saddle Mountains Basalt flow, the Pomona flow, is described below.

The Roza Flow

The Roza (2, 3) is perhaps the best known flow on the Columbia Plateau. Well exposed near the surface across about 40,000 km² of the plateau (Fig. 5), it is easily recognized by its large, regularly distributed plagioclase phenocrysts. It is usually composed of two separate cooling units that probably erupted within a few hundred years of each other, with a total volume of about 1500 km³. The flow was erupted from a series of closely spaced fissures lying in a belt about 5 km wide and 200 km long and oriented in a north-northwest to south-southeast direction near the east side of the plateau (Fig. 5). In this zone some 20 vents or vent areas are indicated by the presence of pumice, some cinders, and spatter or glassy fragments which are highly oxidized and often welded together because they were still hot and partly molten when they landed. Some vents form visible cones today, like Big Butte in the southeast corner of Washington. More often, the smaller vents were covered by later flows and are now exposed only by the accidents or erosion or road building.

The Roza flow erupted on ground that had tilted gently to the west since the previous eruption, so that the lava flowed but a few miles to the east, failing to reach the sites of Asotin and Pullman (Fig. 5). Downslope to the west, however, it flowed for more than 300 km across the Pasco Basin to the Cascades, and

along the Columbia River Gorge as far as The Dalles (Fig. 5). Assuming (2) that most of the lava erupted from a 100-km length of the linear vent system and that each cooling unit had a volume of about 700 km³, and using the known temperature and viscosity of such a lava, we can calculate that the flow moved over 300 km in just a few days at an average rate of about 5 km per hour. The Roza flow is approximately 30 m thick across large parts of the central plateau. Where the lava plunged into lakes in the Pasco area, almost pure glass formed, devoid of crystals other than intratelluric plagioclase phenocrysts, suggesting that the time between eruption and arrival of the flow at the lakeshore 150 km to the west was so short that little cooling and associated crystallization occurred (2).

The enormity of this event is hard to visualize. A lava front about 30 m high, over 100 km wide, and at a temperature of 1100°C, advanced at an average rate of 5 km per hour. The lava poured from the fissures at the rate of 1 km³ per kilometer of fissure per day for about 7 days, an eruptive rate that is orders of magnitude higher than that calculated for recent activity on the Hawaiian volcanoes of Mauna Loa and Kilauea. Over the larger time span of some millions of years, however, the average rate of magma production appears similar to that of the Hawaiian volcanic center or the Iceland ridge-rift system (2, 3).

The Pomona Flow

A member of the Saddle Mountains Basalt Formation, the Pomona flow (8, 12) was erupted about 12 million years ago from a fissure on the most easterly limit of the province in north central Idaho (Fig. 6). It is easily identified both chemically and petrographically, containing numerous small phenocrysts of plagioclase, clinopyroxene, and olivine. The lava pouring from the fissure filled a local structural basin to the south, then flowed westward down the channel of the proto–Clearwater River. Large and small remnants of the flow plastered along the walls of the present Snake River canyon between Asotin and the Pasco Basin testify to its having followed and filled an earlier canyon, similar to the present one.

On reaching the Pasco Basin, more than 200 km from its feeder dike, the Pomona flow filled the river canyon and spread out over the basin floor (Fig. 6). It then continued down an ancestral Columbia River near the location of the present Columbia Gorge. A flow with the same chemical and mineralogical composition, the same magnetic polarity, and in the same apparent stratigraphic position relative to other Columbia River basalt flows is present west of Portland (*13*). Now that traces of the Pomona flow have been found in the Columbia Gorge (*14*), there can be little doubt that this same flow traveled substantially more than 550 km from north-central Idaho to the Pacific coast. It is probable that the deep-sided canyons would have created an effect not unlike a lava tunnel, in which the heated walls allow the molten lava to pour between them with negligible loss of heat. The alternative explanation, that eruption of identical basalt occurred at the same time on two sides of the active andesitic volcanoes of the Cascade Range, is much less credible (*13*).

Tectonic Framework for the Columbia River Basalt Eruption

What combination of tectonic events could have triggered this massive outpouring of basalt over the Columbia Plateau? Evolution of this part of the western American Cordilleran is still poorly understood. We know that in late Precambrian and early Paleozoic time (700 million to 400 million years ago) the western margin of the North American continent ran close to the present western border of Idaho and that crustal blocks farther west of this line, such as the Permian-Triassic rocks of the Blue Mountain Province, the rocks of the northern Cascades, and others farther west, are exotic, having moved into their present positions after their formation farther south (*15*). There is some geophysical evidence to suggest that a narrow wedge-shaped area lying just north of the Blue Mountains between these exotic blocks is underlain by oceanic, not continental, crust. It may be more than a coincidence that the main focus of the Columbia River basalt eruption lies near the apex of that wedge close to the ancient margin of the continent.

The Columbia River basalt was erupted between two active ridges. On the west the Cascade Mountains were rising by a combination of volcanic activity and upwarping. On the east the low-density granitic rocks of the Idaho batholith continued to rise along the old margin of the North American continent. In between, the ground over which the basalt flowed was tilted from east to west (*10*). Throughout the 11 million years of volcanic activity, this region was subjected to east-west tension and north-south

compression, as evidenced by the north-south dike system and the east-west anticlinal ridges (Fig. 3). It is probable that the east-west tension was the cause of a decrease in pressure at depth and of the partial melting of the upper mantle in this region.

Along the southern and southwestern margins of the Columbia Plateau and farther southwest in Oregon, horizontal movement along the vertical northwest-southeast and northeast-southwest fault planes with clockwise rotation of the intervening crustal blocks is well documented (*16*) and could have resulted from a similar stress regime. The overriding of the East Pacific ridge-rift system by the North American continental plate in the early Miocene period and the creation of a "soft margin" to the North American continent, heated by the subduction of recently formed and still hot ocean crust as suggested by Atwater (*17*), is a likely means of developing of such a stress regime.

Origin and Evolution of the Columbia River Basalts

The relatively large silica content and small magnesium/iron ratio of continental flood basalt, compared to ocean ridge basalt, has often been assumed to indicate contamination of the rising lava with silica-rich and potassium-rich continental crust. The rather small $^{87}Sr/^{86}Sr$ ratios (which will increase when older continental crust is mixed with basalt lava) of the Columbia River basalts and the failure of such major elements as silicon and potassium to vary sympathetically with an increase in the strontium isotope ratio in the basalts (with the possible exception of the very small volumes of Saddle Mountains Basalt) (*18*) makes it unlikely that crustal contamination has played a significant role in their origin. The high rate of eruption and great length of the fissure systems for each flow also suggest that crystal fractionation (removal of early crystals from residual liquids) in large chambers just below the surface is not the principal mechanism that created the compositional differences between flows. Nor is there any obvious pattern of compositional change with time.

Alternative mechanisms to account for the origin of the Columbia River basalts and the compositional differences between the flows include variation in the composition of the mantle source for each batch of lava erupted, different degrees of partial melting for one or more source compositions, and crystal fractionation at some depth well below the

surface. Theoretically, each of these possible mechanisms can be tested by the relative behavior of various chemical elements. In practice, the exercise is complex, and despite the wealth of data now available for the whole Columbia River basalt event, no wholly satisfactory solution has yet been offered.

Some elements, referred to as the incompatible elements, are almost totally excluded from the crystalline structures of the early precipitating minerals; phosphorus, barium, zirconium, lanthanum, and thorium are examples. Incompatible elements are highly concentrated in melts formed by small degrees of partial melting and are progressively diluted as the degree of partial melting increases. Conversely, they are progressively concentrated in residual liquids as the early crystalline minerals are removed. Throughout both these processes the ratios between the various incompatible elements should not vary significantly. The ratios do show substantial variation in the Columbia River basalts, and this has led to the suggestion that each major basalt flow represents a separate partial melt of a heterogeneous mantle source.

A partial melt of normal mantle rock will inevitably have a high magnesium/iron ratio and high concentrations of such elements as nickel and chromium. Once the new liquid is removed from the still crystalline mantle residue, then the magnesium/iron ratio and abundances of nickel and chromium will be reduced rapidly by the precipitation of relatively small volumes of such minerals as olivine and pyroxene. A possible explanation, then, for the unusually low magnesium/iron ratio and low concentrations of nickel and chromium in the Columbia River basalts is that they have suffered considerable crystal fractionation of olivine and pyroxene at some depth. However, simple computer models, in which various proportions of observed or possible early mineral phases are removed from one basalt flow composition to produce another flow composition, fail to provide quantitative support for this concept. Furthermore, one would expect greater variation than is in fact observed in magnesium/iron ratios and nickel and chromium values, depending on how much and what type of crystal fractionation occurred. This has led to the suggestion that crystal fractionation is minimal and that the mantle source is an iron-rich pyroxenite (*19*). Unfortunately, all the evidence provided by samples of mantle material brought to the surface in volcanic eruptions implies that such iron-rich material is extremely scarce in the mantle. To solve this basic dilemma, it has recently been suggested that the primary magma forms large chambers at the base of the crust, where crystal fractionation removes large quantities of olivine, clinopyroxene, and plagioclase, while the magma is renewed by more primary liquid from below. Such a chamber would serve both to buffer the variation in the magnesium/iron ratios and concentrations of nickel and chromium and to reduce them substantially (*20*).

That many problems remain concerning the origin of the Columbia River basalts is certain. But the wealth of chemical and mineral data now available on these rocks has provided a clear picture of the development of this immense volcanic outburst. The tectonic events associated with the eruptions are well understood and must now be reconciled with the larger concepts of plate movement and the causes of magma generation at depth. We have learned that the variation in composition between flows resulted from processes at depth rather than processes at or just below the surface, and current evidence suggests that the larger chemical differences between oceanic and continental basalts may be more closely related to the thickness of the overlying crust than to crustal contamination. Studies may now focus on the relative merits of models that emphasize the partial melting at depth of a heterogeneous, iron-rich mantle source, and more complex models in which magma reservoirs at the base of the crust fractionate early refractory minerals but are periodically tapped by a volcanic eruption and refilled by more partial melt from below.

References and Notes

1. Z, Ben-Avraham, *Am. Sci.* **69**, 291 (1981).
2. H. R. Shaw and D. A. Swanson, in *Proceedings of the Second Columbia River Basalt Symposium*, E. H. Gilmour and D. Stradling, Eds. (Eastern Washington State College, Cheney, 1970), p. 271.
3. D. A. Swanson, T. L. Wright, R. T. Helz, *Am. J. Sci.* **275**, 877 (1975).
4. Potassium-argon dates of the Columbia River basalts are from N. D. Watkins and A. K. Baksi [*Am. J. Sci.* **274**, 148 (1974)]; E. H. McKee, D. A. Swanson, and T. L. Wright [*Geol. Soc. Am. Abstr. Program* **9**, 463 (1977)]; and E. H. McKee, P. R. Hooper, and W. D. Kleck [*Isochron/West* **31**, 31 (1981)].
5. Details of the composition of Columbia River basalt flows are contained in P. R. Hooper *et al.*, *Major Element Analyses of Columbia River Basalt* [open-file report, Geology Department, Washington State University (1976), part 1]; S. P. Reidel, V. E. Camp, and M. E. Ross [*ibid.*, part 2 (1981)]; T. L. Wright, D. A. Swanson, R. T. Helz, and G. R. Byerly [*U.S. Geol. Surv. Open-File Report 79–711* (1979)]; and T. L. Wright, K. Black, D. A. Swanson, and T. O'Hearn [*ibid. 80–921* (1980)].
6. F. J. Vine and D. H. Mathews, *Nature (London)* **199**, 947 (1963).
7. S. R. Choiniere and D. A. Swanson, *Am. J. Sci.* **279**, 755 (1979); P. R. Hooper. C. R. Knowles, N. D. Watkins, *ibid.*, p. 737; D. A. Swanson *et. al.*, *U.S. Geol. Surv. Misc. Invest. Map I-1139* (1980).
8. Details of the stratigraphic succession of the Columbia River basalts may be found in D. A. Swanson, T. L. Wright, P. R. Hooper, and R. D. Bentley [*U.S. Geol. Surv. Bull. 1457-G* (1979);] and C. W. Myers and S. M. Price [*Rockwell Int. Rep. RHO-BWI-ST-4* (1979)].
9. W. H. Taubeneck, in *Proceedings of the Second Columbia River Basalt Symposium*, E. H. Gilmour and D. Stradling, Eds. (Eastern Washington State College, Cheney, 1970), p. 73.
10. P. R. Hooper and V. E. Camp, *Geology* **9**, 323 (1981).
11. G. Byerly and D. A. Swanson, *Geol. Soc. Am. Abstr. Program* **10**, 98 (1978).
12. V. E. Camp, *Geol. Soc. Am. Bull.* **92** (part 1), 669 (1981).
13. P. E. Snavely, N. S. MacLeod, H. C. Wagner, *ibid.* **84**, 387 (1973); M. H. Beeson, R. Pertta, J. Pertta, *Oreg. Geol.* **41**, 159 (1979).
14. J. L. Anderson, *Oreg. Geol.* **42**, 195 (1980).
15. G. A. Davis, J. W. H. Monger, B. C. Burchfiel, *Soc. Econ. Paleontol. Mineral. Pacific Coast Paleogeog. Symp.* **2**, 1 (1978).
16. R. W. Simpson and A. Cox, *Geology* **5**, 585 (1977); M. E. Beck, *Am. J. Sci.* **2**, 694 (1976); ___ and C. D. Barr, *Geology* **7**, 175 (1979); R. D. Bentley, J. Powell, J. L. Anderson, S. M. Farooqui, *Geol. Soc. Am. Abstr. Program* **12**, 97 (1980).
17. T. Atwater, *Geol. Soc. Am. Bull.* **81**, 3513 (1970).
18. R. T. Helz, "Chemical and experimental study of the Ice Harbor Member of the Yakema Basalt Subgroup: Evidence for intracrustal storage and contamination," paper presented at the Northwest annual meeting of the American Geophysical Union, Bend, Oregon, 1979, Abstr.; D. O. Nelson, *Geol. Soc. Am. Abstr. Program* **12**, 144 (1980).
19. T. L. Wright and R. T. Helz, *Abstracts with Program, International Discussion/Symposium on Deccan Volcanism and Related Basalt Provinces, Field Guide* (Bombay and Khandala, India, 1979), p. 61.
20. K. G. Cox, *J. Petrol.* **21**, 629 (1980); M. J. O'Hara and R. E. Matthews, *Q. J. Geol. Soc. London* **138**, 237 (1981).
21. The systematic survey of the Columbia River basalts in the last 10 years has involved the close cooperation of many geoscientists. I am particularly indebted to the work of D. A. Swanson, T. L. Wright, V. E. Camp, S. P. Reidel, and to the research team at Rockwell International, Richland, Washington. I am also grateful for the many helpful suggestions for the improvement of the manuscript by T. L. Wright, G. G. Goles, and R. D. Bentley.

52

Reprinted from pages 119-120, 137-147, 148, 149, 150, 151, 152, 153, and 154 of
Geol. Soc. America Mem. **144:**119-154 (1975)

Late Cenozoic Basaltic Volcanism and Development of the Rio Grande Depression in the Southern Rocky Mountains

PETER W. LIPMAN

AND

HARALD H. MEHNERT

*U.S. Geological Survey
Denver Federal Center
Denver, Colorado 80225*

ABSTRACT

In the Southern Rocky Mountains, upper Cenozoic basalt flows were erupted widely in areas characterized in middle Tertiary time by predominantly intermediate-composition volcanism. Initiation of basaltic volcanism coincided approximately with the beginning of extensional block faulting that resulted in development of the Rio Grande depression, a major rift structure that separates the stable platforms of the High Plains and the Colorado Plateau. Time relations are especially clear along the San Luis Valley segment of the Rio Grande depression in southern Colorado and northern New Mexico, where 16 basalt flows have been dated by K-Ar methods. Along the west margin of the San Luis Valley, silicic alkalic basalt flows as old as 26 m.y. rest unconformably on a pediment cut on middle Tertiary andesitic and related rocks (35 to 27 m.y. old), and similar basalt 20 to 0.24 m.y. old interfingers with and overlies volcaniclastic alluvial fan deposits (equivalent to Santa Fe Group) that accumulated in the subsiding depression.

Basalt erupted during late Cenozoic block faulting varies in composition with distance from the axis of the northern Rio Grande depression. Tholeiitic rocks are largely confined to the depression, and the basalt types become more alkalic to the west and east. Relatively silicic alkalic basalt, including both undersaturated and saturated types, occurs throughout the region, but very undersaturated alkalic basalt flows were erupted only on the Colorado Plateau and the High Plains. The lateral change from tholeiitic to alkalic basaltic volcanism probably reflects different conditions of magma generation in the mantle that are related to changes in crustal thickness and thermal gradients across the depression.

The compositions and compositional ranges of basalt of the Southern Rocky Mountain region are similar to those of many Pacific islands, but the proportions of basalt types are markedly different. These relations are tentatively interpreted as reflecting origin of both continental and oceanic basalt types under similar *P-T* conditions, with generation of the Southern Rocky Mountain basalt from lithospheric mantle and the oceanic basalt from relatively shallow asthenosphere.

[*Editors' Note:* Material has been omitted at this point.]

GEOGRAPHIC VARIATIONS IN BASALT COMPOSITION

A few years ago, Lipman (1969) suggested, on the basis of fragmentary data taken mainly from published literature, that upper Cenozoic basalt flows in the Southern Rocky Mountain region vary systematically in petrology with distance from the Rio Grande depression. Basalt and basaltic andesite of alkalic affinities erupted east and west of the depression, whereas tholeiitic basalt accumulated within the depression late in its evolution. The lateral change from tholeiitic to alkalic volcanism was interpreted to reflect different conditions of magma generation in the mantle. More recent studies of basaltic rocks in the region, while generally supporting these interpretations, have shown that the detailed geographic, petrologic, and age relations between rifting and basaltic volcanism are more intricate than originally recognized (Leeman and Rogers, 1970; Renault, 1970; Kudo and others, 1971; Stormer, 1972a, 1972b; Hedge and Lipman, 1972; Lipman and others, 1973; Aoki and Kudo, 1975).

About 200 major-element analyses are now available for upper Cenozoic basaltic rocks of the Southern Rocky Mountain region, including 25 new analyses (Table 3). These analyses define major compositional variations both transversely and longitudinally along the Rio Grande depression.

The transverse variations are especially evident from plots of K_2O or total alkalis versus SiO_2, which show that at any given SiO_2 content, alkali contents of the basaltic rocks are higher on each side of the depression than within it, as observed previously (Lipman, 1969; Lipman and others, 1973). The newly available analyses also indicate that, within geographically restricted areas, total alkalis tend to increase progressively away from the depression. So, basalt of the western San Juan volcanic field is distinctly higher in alkalis than otherwise petrographically similar basalt from eastern parts of the field, and basalt from central areas shows intermediate values (Fig. 8A). A similar pattern seems valid for the upper Cenozoic basalt of northeastern New Mexico, where flows in the Cimarron Mountains area on the east flank of the Sangre de Cristo Mountains are less alkalic than those farther east on the High Plains (Fig. 8B).

These overall transverse variations in alkali contents across the Southern Rocky Mountain region, approximately symmetrical with respect to the Rio Grande depression, seem to persist despite the widely differing ages (from early Miocene to Holocene) of the basalt samples plotted. Thus, in the southern San Juan field, the Pleistocene Brazos Basalt (of Doney, 1968) of the Tusas Mountains is compositionally similar to nearby Miocene basalt of the Hinsdale Formation. However,

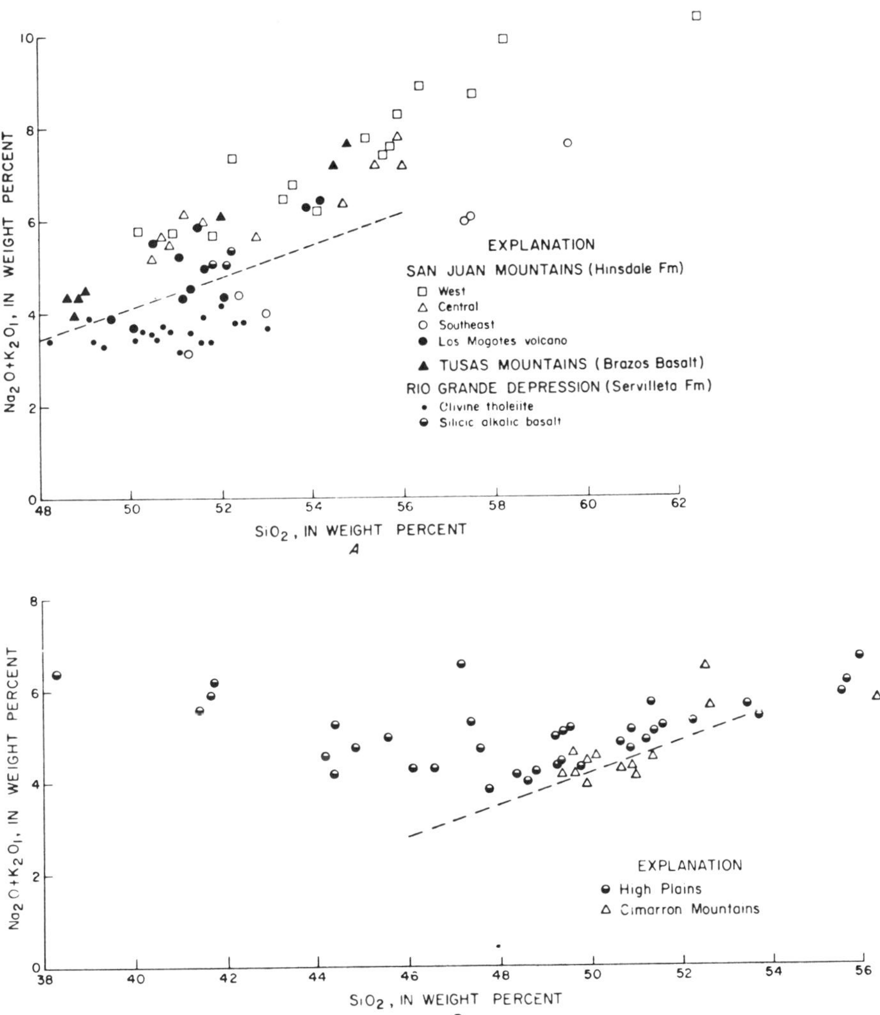

Figure 8. Plots of total alkalis and SiO₂ for selected basalt samples of the Southern Rocky Mountain region. Data are from Aoki (1967a, 1967b), Aoki and Kudo (1975), Baldwin and Muehlberger (1959), Larsen and Cross (1956), Lipman (1969), Stormer (1972a), and Table 3. All analyses plotted have been recalculated to 100 percent without H_2O or CO_2. Dashed line separates fields of Hawaiian tholeiitic and alkalic basalt (Macdonald and Katsura, 1964). A. Basalt of the San Juan Mountains and the Rio Grande depression. B. Basalt east of the Rio Grande depression (Cimarron Mountains and High Plains areas).

as emphasized by Stormer (1972b) and as documented earlier in this paper, distribution both of activity and compositional type of basalt tended to be episodic and asymmetrical to the Rio Grande depression within more restricted time intervals. For example, no analogs of the extremely low-silica alkali-olivine basalt (basanite) of the High Plains have been recognized within the San Juan field to the west of the depression, although similarly alkalic basalt does occur west of the depression in the Mount Taylor area of north-central New Mexico (Lipman and Moench, 1972) and in the Hopi Buttes region of northeastern Arizona (Williams, 1936).

Distributions of the basalt types with respect to the Rio Grande depression are also complicated by the presence within the depression of subordinate amounts of silicic alkalic basalt that petrographically resembles the dominant basalt types to the east and west (Lipman, 1969, p. 1349). Further, the andesitic to rhyolitic central volcanoes of the southern San Luis Valley have major-element compositional trends that follow continuations of trends of the silicic alkalic basalt, rather than those of the tholeiitic basalt. Thus, significant alkalic basaltic and related volcanism, in addition to the tholeiitic activity, emanated from within the depression during its evolution. Probably more significant than the ubiquitous distribution of the alkalic activity is the confinement of olivine tholeiite, low in K_2O, P_2O_5, TiO_2, U, Th, Ba, Rb, and Sr, to the northern and central parts of the depression. This tholeiite rock type is represented by the basalt of the Servilleta Formation within the southern San Luis Valley and the Espanola Basin; farther south in the central part of the depression, similar tholeiitic basalt occurs around the flanks of the Jemez Mountains and in the hills northwest of Albuquerque (Aoki and Kudo, 1973). However, basalt samples from the Cimarron Mountains east of the depression, described as tholeiitic on the basis of a normative classification (Kudo and others, 1971), have notably higher concentrations of the elements listed above than olivine tholeiites of the depression (Fig. 8B).

A similar pattern of transverse compositional variations of basalts in and adjacent to the central and northern Rio Grande depression is also strikingly evident from variations in Th and U contents and Th/K and U/K ratios, which all consistently increase away from the depression (Lipman and others, 1973). The Servilleta olivine tholeiite, which is volumetrically the predominant basalt type in the northern and central parts of the depression, has the lowest K, Th, and U contents of any basalt in the region (Fig. 9). The silicic alkalic basalt, which is the dominant type in the Southern Rocky Mountain region adjacent to the Rio Grande depression, not only has higher K, Th, and U contents than the tholeiitic basalt, but the Th/K and U/K ratios also increase with distance from the Rio Grande depression (Fig. 9). The lowest ratios of these elements occur in the basalt along the southeast margin of the San Juan field on the west margin of the Rio Grande depression and in the sparse similar silicic alkalic basalt interlayered with the tholeiitic basalt farther south within the depression (Fig. 9A). These ratios are higher in the silicic alkalic basalt of the Cimarron and Tusas Mountains, respectively, east and west of the Rio Grande depression; they are even higher in the otherwise petrologically similar basalt of the Mount Taylor area and basalt of the High Plains at even greater distances from the Rio Grande depression. Extremely high Th and U contents, as much as 30 and 8 ppm, respectively, and also high Th/K and U/K ratios, occur only at great distances from the Rio Grande depression: on the High Plains to the east and at Mount Taylor on the southeastern Colorado Plateau (Fig. 9).

South of Albuquerque, New Mexico, where the Rio Grande depression merges with the Basin and Range province, these transverse compositional relations no longer can be recognized. Upper Cenozoic basalt flows both within the depression and on each side seem to be mainly alkalic olivine basalt and basaltic andesite

(Lipman, 1969, p. 1350; Kudo and others, 1971, p. 201; Aoki and Kudo, 1975; Hoffer, 1971), similar to typical upper Cenozoic basalt in the central Basin and Range province to the west (Leeman and Rogers, 1970). Thus, a distinct longitudinal compositional variation of the basalt types within the Rio Grande depression seemingly can be related to changes in the structure of the depression: olivine tholeiite and a transverse compositional gradient of basalt are present only as far south as the depression constitutes a well-defined extensional structure separating the relatively stable regions of the Colorado Plateau and the High Plains.

Petrologically diversified basalt fields that contain tholeiitic types seem closely associated with major structural discontinuities around margins of the region of late Cenozoic extensional (basin-range) faulting in the western United States. Thus, in addition to their occurrence along the Rio Grande depression, basaltic rocks

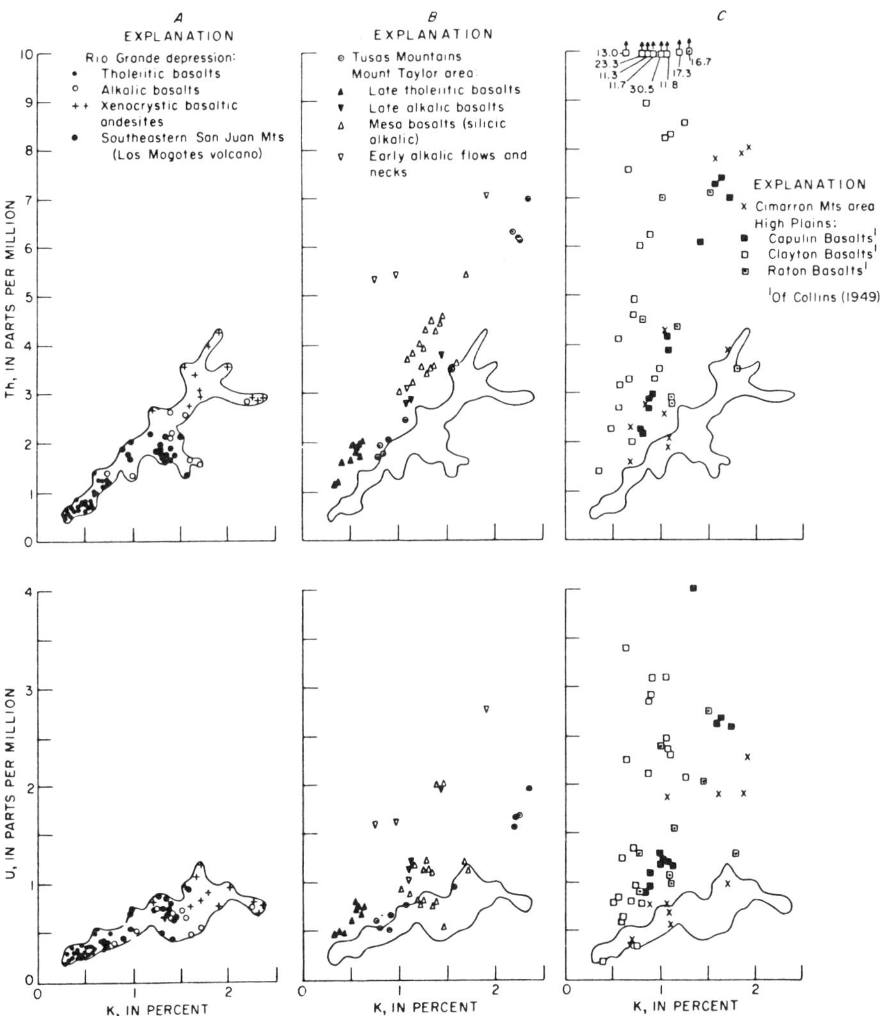

Figure 9. Plots of K, Th, and U contents of basalt of the Southern Rocky Mountain region (from Lipman and others, 1973). A. Basalt of the Rio Grande depression and the southeastern San Juan Mountains. Empirically estimated field boundary of the Rio Grande depression basalt is repeated in B and C. B. Basalt west of the Rio Grande depresssion (Tusas Mountains and Mount Taylor area). C. Basalt east of the Rio Grande depression (Cimarron Mountains and High Plains areas).

of tholeiitic affinities occur around the boundary between the Basin and Range province and the Colorado Plateau of northern New Mexico in the Mount Taylor field (Lipman and Moench, 1972) and in the San Francisco Mountain and Verde River areas of central Arizona (Robinson, 1913; McKee and Anderson, 1971, 1970, written commun.). In the block-faulted Shivwits Plateau area along the west side of the Colorado Plateau, the most abundant basalts (Grand Wash and Hurricane types of Best and Brimhall, 1970) are chemically transitional toward tholeiitic types (Table 1) and have textural features such as subophitic diktytaxitic texture, coarse grain size, and groundmass glass similar to tholeiitic basalts of the Mount Taylor and Rio Grande areas (Best and others, 1966, 1969). Near the western margin of the Basin and Range province, tholeiitic basalt occurs in the Mojave Desert and Owens Valley areas of California (Wise, 1969; Leeman and Rogers, 1970; W. P. Leeman, 1968, written commun.). Upper Cenozoic tholeiitic basalt flows are abundant at the northern edge of the Basin and Range province in the basalt plateaus of northwestern Nevada, northeastern California, and southeastern Oregon (Powers, 1932; Waters, 1962; LeMasurier, 1968; Gunn and Watkins, 1970) and in the Snake River Plain-Yellowstone National Park area (Stone, 1967; Christiansen and Blank, 1969). Tholeiitic basalt flows are seemingly sparse in the interior of the Basin and Range province; of the five analyses of tholeiites plotted in a recent paper on basalt from the Basin and Range province (Leeman and Rogers, 1970), one sample is from the Mount Taylor field, and the others are from marginal regions in northwestern Nevada and eastern California (W. P. Leeman, 1968, written commun.).

Very undersaturated alkalic basalt and basanite also occur mainly around margins of the Basin and Range province, in many of the same basaltic fields as the tholeiites (Mojave Desert—Wise, 1969; western Colorado Plateau—Best and others 1966, 1969; Mount Taylor—Lipman and Moench, 1972), and also in the tectonically stable regions beyond (Hopi Buttes in the middle of the Colorado Plateau—Williams, 1936; High Plains of northeastern New Mexico—Stobbe, 1949; Baldwin and Muehlberger, 1959; Aoki, 1967b). We know of only a single area of Cenozoic basanitic basalt in the interior Basin and Range province: the southern Pancake Range in Nevada (Vitaliano and Harvey, 1965; Trask, 1969; Scott and Trask, 1971). The basanitic basalt samples plotted in the paper by Leeman and Rogers (1970) are from the localities listed above (W. P. Leeman, 1968, written commun.).

In these areas and in most other Cenozoic volcanic areas in the Basin and Range province, however, the dominant basalt type is silicic alkalic basalt similar to the most widespread basalt type of the Southern Rocky Mountain region. Petrologically diverse basaltic suites that probably reflect magma generation over a sizable depth range in the upper mantle seem to have formed most readily in the western United States along major tectonic discontinuities (that is, boundaries of structural provinces). This relation may indicate that the discontinuities extend through the sialic crust and perhaps that they affected the entire lithospheric plate and influenced magma generation at upper mantle depths.

CONDITIONS OF MAGMA GENERATION

The compositional variations among upper Cenozoic basalt flows of the Southern Rocky Mountain region could have resulted from three major types of processes: (1) intermediate- to low-pressure fractional crystallization of the magma prior to eruption, (2) contamination interactions with sialic crust during rise of the magma to the surface, and (3) variable conditions of magma generation and high-pressure fractionation at depth within the upper mantle. By combining mineralogic, elemental, and isotopic compositional data, each of these possibilities can be evaluated under favorable conditions. These alternatives for basalt of the Southern Rocky Mountains and adjacent areas have been discussed at length elsewhere (Lipman, 1969; Doe and others, 1969; Kudo and others, 1971; Stormer, 1972a; Hedge and Lipman, 1972) and will only be summarized briefly here. For most basalt of the Southern Rocky Mountains, low-pressure fractionation and crustal contamination seem to have been relatively unimportant; the major compositional variations are seemingly due to differing degrees and (or) pressures of melting at depth within the upper mantle.

Low-Pressure Fractional Crystallization

An important constraint on any interpretation of basaltic magma genesis in the Southern Rocky Mountain region is that few of the basalt suites appear to have been affected significantly by low-pressure differentiation involving fractional crystallization and separation of phenocrysts. This sort of crystal-liquid fractionation seems to have played a significant part in producing major compositional variations in oceanic-island basaltic shield volcanoes such as Kilauea, Hawaii, where much of the observed variation can be accounted for by olivine fractionation and mixing of different batches of such fractionated lava (Powers, 1955; Murata and Richter, 1966; Wright and Fiske, 1971; Wright, 1971).

With few exceptions, olivine is the only phenocryst phase of the Southern Rocky Mountain basalt flows, and addition or subtraction of olivine cannot account for

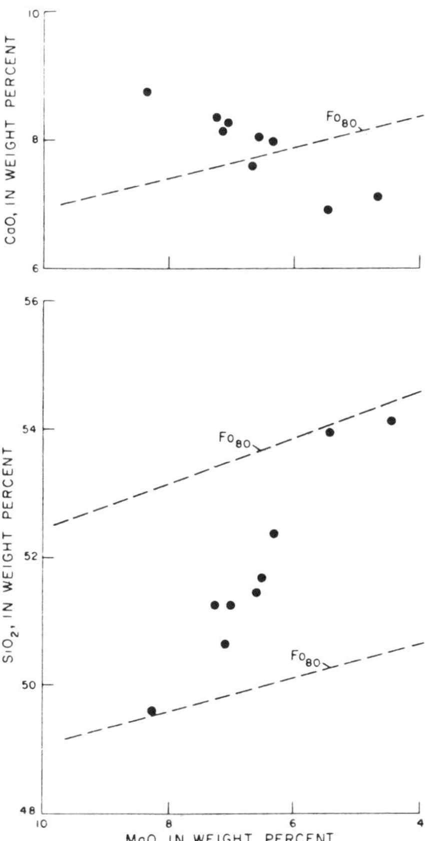

Figure 10. MgO variation diagrams for basalt of Los Mogotes volcano (data from Larsen and Cross, 1956, Table 25, nos. 3, 5; Lipman, 1969, Table 1, no. 2; Table 3). Control lines are for olivine composition of Fo_{80}, the approximate optically determined composition for phenocrysts from Los Mogotes basalt.

observed compositional variations among closely related flows. This relation is especially clear for lava flows of the small 5-m.y.-old Los Mogotes shield volcano in the southeastern San Juan field (Fig. 1). A sequence of at least 12 flows from this center, exposed where the Conejos River has cut across the south flank of the shield, displays a well-defined compositional trend from basaltic andesite, containing about 54 percent SiO_2 at the base, upward to typical silicic alkalic basalt, with about 50 percent SiO_2 at the top. Olivine is the only phenocryst phase throughout, yet the compositional trends are clearly unrelated to olivine control lines, as is readily evident on MgO variation diagrams (Fig. 10). This is in striking contrast to closely related basaltic suites from oceanic-island volcanoes. The necessary interpretation for the Los Mogotes sequence, especially when isotopic and minor-element data that preclude significant crustal contamination are considered, is that high-pressure fractionation is responsible for all major variations in even this limited sequence from a single volcano. Similar arguments are applicable to other compositionally variable basaltic sequences of localized age and geographic distribution, such as the tholeiitic basalt of the Servilleta Formation, which also contains only olivine phenocrysts. Also, an analyzed segregation vein from a Servilleta flow (Aoki, 1967a, Table 1), which clearly represents a residual liquid concentrated during crystallization of the basalt at the surface, defines a composition trend that is discordant to the trends of the overall suite of bulk samples (Aoki, 1967a). In a detailed petrologic study of the basalt of the High Plains area in northeastern New Mexico, Stormer (1972a) similarly concluded that the major compositional variations among the different basalt suites could not have resulted from low-pressure differentiation (or crustal contamination) and must have been due to high-pressure fractionation.

Effects of Crustal Contamination

Compelling evidence from concentration and isotopic data shows that the differences between the three major compositional basalt types of the region—the alkali-olivine basalt, silicic alkalic basalt, and olivine tholeiite—do not result from variable crustal contamination.

First, variations in major-oxide and minor-element concentrations among these main basalt types are such that none could have been generated from another simply by contamination by sialic crust of any plausible composition. For example, addition of sialic material to olivine tholeiite magma, which could feasibly produce concentrations of most major oxides similar to those in the silicic alkalic basalt, cannot account for variations in minor elements such as Sr. Sr content is about twice as high in the silicic alkalic basalt as in the olivine tholeiite (Table 1). Contamination by any likely material would lower Sr concentrations in basalt flows, rather than increase them. It is similarly difficult to derive silicic alkalic basalt by contamination of alkali-olivine basalt, which, though lower in SiO_2, is about the same in alkalis and is much higher in Th and U—elements that would be further increased by crustal contamination. More severe difficulties come in attempting to generate other basalt types by contamination models, for example, olivine tholeiite from alkali-olivine basalt, or vice versa.

The interpretation that crustal contamination was of limited significance in producing the variations among basalt flows of the Southern Rocky Mountains and adjacent regions is supported by Sr isotopic data, which show low and relatively uniform isotopic ratios (Kudo and others, 1971; Leeman, 1970; Hedge and Lipman, 1972). With the exception of one suspect sample, the range in 31 samples from the Southern Rocky Mountain region is only 0.7035 to 0.7050 (Fig. 11)—only about twice the standard deviation of the analytical precision (Hedge and Lipman, 1972).

Furthermore, no correlation is evident between isotopic ratio and geographic position or age or bulk-rock elemental composition. Basalt with low Sr concentrations (or high Rb/Sr ratios) should be especially sensitive to contamination effects, but the range in isotopic ratios of the Sr-poor tholeiitic rocks within the Rio Grande depression (0.7042 to 0.7048) is bracketed by that of Sr-rich alkali-olivine basaltic rocks of the High Plains (0.7035 to 0.7049). Anomalously high Sr isotopic ratios have been observed in a few basalt samples with low Sr concentrations in central New Mexico (Kudo and others, 1971), mainly from a single isotopically heterogeneous flow (Laughlin and others, 1972a). These have reasonably been interpreted as reflecting minor crustal contamination, but the other Southern Rocky Mountain basalt flows that are isotopically homogeneous have been considered essentially unmodified by crustal contamination (Hedge and Lipman, 1972).

In contrast to the Southern Rocky Mountain basalt flows, upper Cenozoic basalt farther south in New Mexico and in adjacent parts of southern Arizona commonly has distinctly lower Sr^{87}/Sr^{86} ratios (Fig. 11), indicating a differing composition of the source region of these magmas (Hedge and Lipman, in prep.).

Special interpretive problems are offered by the petrographically distinctive xenocrystic basaltic andesite types, which have been interpreted in recent years as resulting from one or more processes, including crustal contamination (Doe and others, 1969), high-pressure crystallization (Nicholls and others, 1971), or mixing of mafic and silicic magmas (U.S. Geological Survey, 1971, p. A42). The major-oxide compositions of these rocks are compatible with formation by crustal contamination of the widespread silicic alkalic basalt type. The Sr concentrations of these rocks are too high for the Sr isotopic ratios to be sensitive to the relatively modest degree of contamination required to account for the elemental concentration differences. Nevertheless, Pb isotopic compositions of three xenocrystic basaltic andesite samples from the Southern Rocky Mountains differ from associated

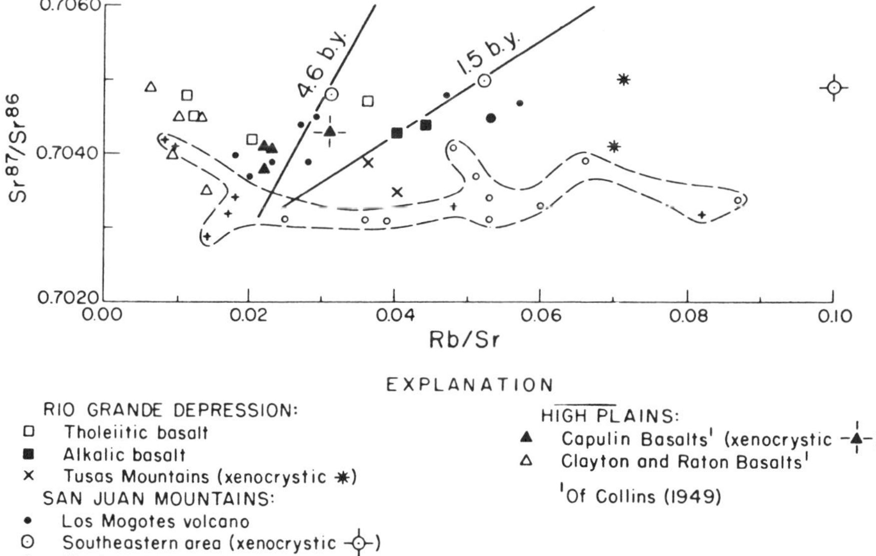

EXPLANATION

RIO GRANDE DEPRESSION:
□   Tholeiitic basalt
■   Alkalic basalt
×   Tusas Mountains (xenocrystic ✳)
SAN JUAN MOUNTAINS:
•   Los Mogotes volcano
○   Southeastern area (xenocrystic -⊕-)
●   Western area

HIGH PLAINS:
▲   Capulin Basalts[1] (xenocrystic -▲-)
△   Clayton and Raton Basalts[1]

[1]Of Collins (1949)

Figure 11. Sr^{87}/Sr^{86} plotted against Rb/Sr ratio for upper Cenozoic basalt from the Southern Rocky Mountain region (from Hedge and Lipman, in prep.). Upper Cenozoic basalt from the Basin and Range province in southern New Mexico and Arizona also plotted (enclosed by dashed line); data from Damon and others (1969, small crosses) and Leeman (1970, small circles).

nonxenocrystic basalt samples and are suggestive of crustal contamination (Doe and others, 1969), although the isotopic variability of seemingly uncontaminated volcanic rocks from the same region is a factor that weakens this argument. Also, the xenocrystic basaltic andesite types commonly have high Rb/Sr ratios that plot separately from Rb-Sr trends defined by groups of nonxenocrystic basalt types of the same area (Hedge and Lipman, 1972). These Rb/Sr relations are more readily interpreted in terms of contamination or magma-mixing models than by the high-pressure crystallization hypothesis, but the origin of these distinctive mafic rocks remains uncertain.

An important further insight into the nature of the source material of the Southern Rocky Mountain basalt is afforded by the Sr and Pb isotopic data just discussed. Figure 11 shows that many of these basaltic rocks do not contain enough Rb relative to Sr to have produced their observed Sr isotopic ratios, even given the entire 4.6 b.y. of the Earth's history. That is, they plot to the left of an isochron for the age of the Earth (Fig. 11). Because a liquid produced by partial melting will have a Rb/Sr ratio greater than that of the residual fraction, the Rb/Sr ratio of the source region for the basalt must have been even lower than the ratios observed in the samples. Similar anomalies have been noted for oceanic ridge basalt (Tatsumoto, 1966) and for a group of Cenozoic basalt types in western Nevada and eastern California (Hedge and Noble, 1971). Such basalt would seem to have been derived from a source that had been depleted in Rb by previous magma generation, as suggested by Gast (1968).

Some understanding of the nature of such a previous fractionation event may be offered by Pb isotope studies of Tertiary volcanic rocks of the San Juan field, which suggest that the source region for these rocks underwent a major geochemical fractionation about 1.7 to 1.8 b.y. ago, the approximate time of major sialic crustal formation in this area (B. R. Doe, 1970, written commun.). A similar geochemical event in the Northern Rocky Mountains, approximately contemporaneous with crustal evolution there about 2.8 b.y. ago, has been inferred from Pb isotopic data for the Absaroka Volcanic Supergroup (Peterman and others, 1970). These relations suggest that major fractionation of the mantle source region of the upper Cenozoic basalt of the Southern Rocky Mountains occurred at the time of development of sialic crust in this area 1.7 to 1.8 b.y. ago and that the composition of the underlying mantle has survived, with only slight compositional modification, to the present.

High-Pressure Fractionation

Crystallization experiments on simplified mafic systems and on actual basaltic materials at high pressures have demonstrated that basaltic magmas can be generated by partial melting of probable mantle constituents and that the composition of the basaltic melt is largely determined by the depth (pressure) of melting and the proportion of melt relative to residuum (Green and Ringwood, 1967; Green, 1968; Ito and Kennedy, 1967; Kushiro, 1968, among others). The nature of the basaltic melt is also strongly influenced by the presence of volatiles within the region of melting (Wyllie, 1971; Kushiro, 1972) and by the degree to which the melt equilibrates with mantle materials as it rises (O'Hara and Yoder, 1967). These experimental results generally suggest that tholeiitic basalt should result when the partial melting is proportionately large and occurs at shallow depth, whereas progressively more alkalic and less saturated melts are to be expected when the melting is of smaller proportion and occurs at greater depth. The role of volatile constituents is thought to be especially significant when the proportion of melt generated is small. These relations have been widely cited to explain the origin of compositionally diverse basalt suites in the western United States (Wise, 1969; Lipman, 1969; Leeman and Rogers, 1970; Baker and Ridley, 1970; Kudo and others, 1971; Stormer, 1972a; Condie and Barsky, 1972) and also to account for basalt variations in oceanic regions (Kuno, 1959, 1968; Jackson and Wright, 1970).

Admittedly, detailed extrapolation of experimentally determined pressure and depth relations for basalt generation can only approximate natural situations because the composition, proportion of melting, and volatile content of the parental mantle material are uncertain. Nevertheless, localization within the northern Rio Grande depression of olivine tholeiite that is thought to represent melting or last equilibration at relatively low pressures (less than 35 km depth, according to Green and Ringwood, 1967) suggests that generation of tholeiitic magmas may be possible within continental regions only where crustal attenuation or other major structural discontinuities permit local upwelling of mantle material and fractionation of basaltic magma at atypically shallow crustal levels. The volumetrically and areally predominant silicic alkalic basaltic rocks are thought to have been generated at greater depths, probably with a lesser degree of partial melting. Accordingly, the alkali-olivine basalt and basanite are interpreted as having been generated at even greater depth, with a still smaller fraction of partial melt and a relatively high volatile content.

Available heat-flow and deep magnetic sounding data indicate that the Southern Rocky Mountain area is one of relatively high heat flow (similar to other regions of crustal extension) between regions of lower heat flow on the Colorado Plateau and the High Plains (Roy and others, 1972; Porath, 1971; Edwards and others, 1973; Decker, 1973). Available seismic data indicate that the Southern Rocky Mountain region is characterized by sialic crust 40 to 55 km thick (Jackson and Pakiser, 1965); this seems too thick to permit generation of tholeiitic magmas in the underlying mantle, but existing data are inadequate to determine whether crustal structure varies locally across the northern and central Rio Grande depression. A longitudinal seismic profile down the axis of the depression would probably be required.

The Sr and Pb isotopic data for basalts of the Southern Rocky Mountain region indicate that any mantle upwelling along the Rio Grande depression must have been limited, because the source region for generation of all the basalt types seems to have been within that part of the lithospheric upper mantle of the American plate that has been little modified geochemically since formation of the sialic crust in this region about 1.7 b.y. ago. This lithospheric source region for the basalt magmas contrasts with inferred zones of magma generation in oceanic regions, where the lithosphere is much thinner and magma generation probably occurs at its base or within the underlying asthenosphere (Fig. 12).

Considering their contrasting environments of magma generation, the similarities in compositional range between the basalt in the Southern Rocky Mountains and basalt in ocean basins are remarkable. The most significant contrast—the volumetric prevalence of silicic alkalic basalt in the Southern Rocky Mountains, in contrast with the dominant tholeiitic basalt of ocean floors and the lower parts of oceanic-island volcanoes—probably reflects this difference in environment of magma generation. Diapiric upwelling, bringing hot deeper mantle material to a level where melting could begin by intersection of the goetherm with the peridotite solidus, presumably was slower within lower parts of continental lithosphere than within the less viscous oceanic asthenosphere and retarded both the volume of magma generated and the proportion of melting that could occur at any level within an upwelling system.

The upper Cenozoic basalt flows of southern New Mexico and Arizona that are relatively nonradiogenic in Sr (Fig. 11) may represent an intermediate site of magma generation in relation to the lithosphere-asthenosphere boundary. These flows, which occur in an early formed part of the Basin and Range province (Christiansen and Lipman, 1972), were erupted in regions previously characterized by middle Tertiary andesitic volcanic activity seemingly derived from mantle source regions but characterized by notably higher Sr^{87}/Sr^{86} ratios (Damon and others, 1969). This apparent change in Sr isotopic composition of a mantle source region could have resulted from gradual late Cenozoic upwelling of asthenospheric mantle with isotopic characteristics similar to source regions of oceanic basaltic magmas to sufficiently shallow levels in the southern Basin and Range province that it could provide a source for the younger basaltic magmas of this region. Similar asthenospheric upwelling does not seem to have been sufficient to influence magma generation farther north in more recently activated parts of the Basin and Range province or in the restricted zone of extensional faulting along the Rio Grande depression in the Southern Rocky Mountains. In both regions the upper Cenozoic volcanic rocks are relatively radiogenic in Sr, as are earlier Tertiary and Mesozoic igneous rocks of the same areas (Kistler and Peterman, 1972; Hedge and Lipman, 1972).

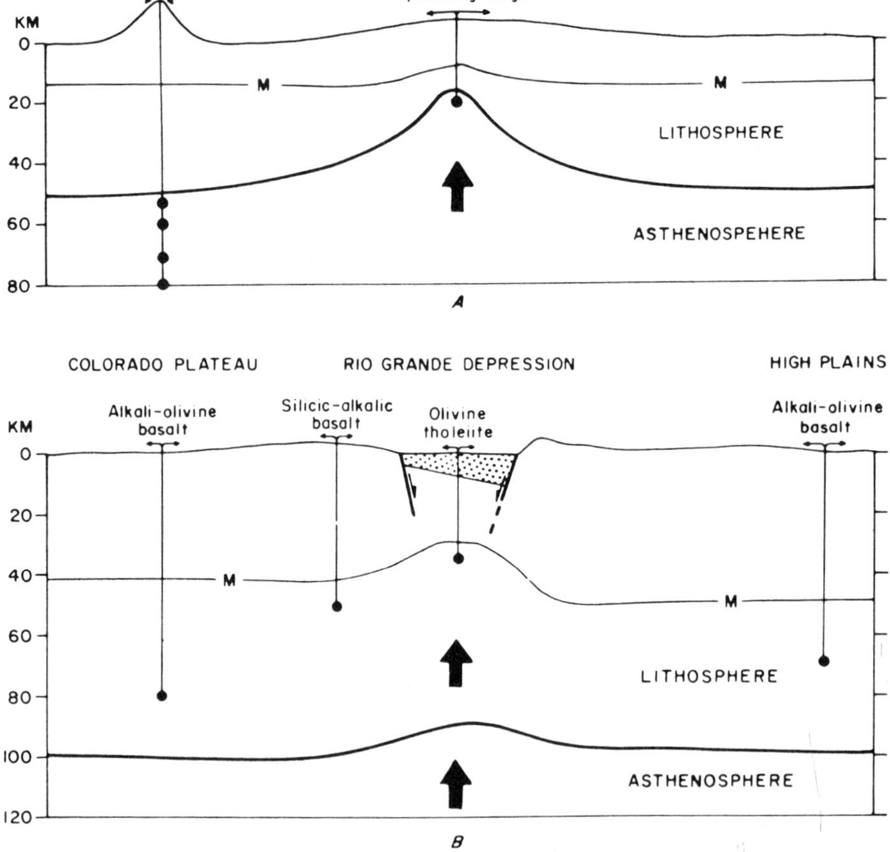

Figure 12. Diagrammatic cross sections contrasting environments of basaltic magma generation along a spreading ocean ridge (A) and in the Southern Rocky Mountain region (B). Dots indicate inferred sites of melting; large arrows indicate varying amounts of upwelling. Vertical exaggeration, ~2×.

[*Editors' Note:* Only the references cited in the preceding excerpt are reproduced here.]

REFERENCES CITED

Aoki, Ken-ichiro, 1967a, Petrography and petrochemistry of latest Pliocene olivine-tholeiites of Taos area, northern New Mexico, U.S.A.: Contr. Mineralogy and Petrology, v. 14, no. 3, p. 191-203.

——1967b, Alkaline and calc-alkaline basalts from Capulin Mountain, northeastern New Mexico, U.S.A.: Japanese Assoc. Mineralogists, Petrologists and Econ. Geologists Jour., v. 58, no. 4, p. 143-151.

Aoki, K., and Kudo, A. M., 1975, Major element variations of late Cenozoic basalts of New Mexico: New Mexico Univ. Pubs. Geology, no. 8 (in press).

Baker, I., and Ridley, W. I., 1970, Field evidence and K, Rb, Sr data bearing on the origin of the Mt. Taylor volcanic field, New Mexico, U.S.A.: Earth and Planetary Sci. Letters, v. 10, no. 1, p. 106-114.

Baldwin, Brewster, and Muehlberger, W. R., 1959, Geologic studies of Union County, New Mexico: New Mexico Bur. Mines and Mineral Resources Bull. 63, 171 p.

Best, M. G., and Brimhall, W. H., 1970, Late Cenozoic basalt types in the western Grand Canyon region, *in* Hamblin, W. K., and Best, M. G., eds., The western Grand Canyon district: Utah Geol. Soc., Guidebook to Geology of Utah, no. 23, p. 57-74.

Best, M. G., Hamblin, W. K., and Brimhall, W. H., 1966, Preliminary petrology and chemistry of late Cenozoic basalts in the western Grand Canyon region: Brigham Young Univ. Research Studies Geology Ser., v. 13, p. 109-123.

Best, M. G., Brimhall, W. H., and Hamblin, W. K., 1969, Late Cenozoic basalts on the western margin of the Colorado Plateaus, Utah and Arizona: Brigham Young Univ. Research Studies Geology Ser., Rept. 69-1, 39 p.

Christiansen, R. L., and Blank, H. R., Jr., 1969, Volcanic evolution of the Yellowstone Rhyolite Plateau and the eastern Snake River Plain, *in* Symposium on volcanoes and their roots: Oxford, England, Internat. Assoc. Volcanic Chemistry Earth's Interior, p. 220-221.

Christiansen, R. L., and Lipman, P. W., 1972, Cenozoic volcanism and plate tectonic evolution of the western United States—[Pt.] II, late Cenozoic: Royal Soc. London Philos. Trans., ser. A. v. 271, no. 1213, p. 249-284.

Damon, P. E., and collaborators, 1969, Correlation and chronology of ore deposits and volcanic rocks: U.S. Atomic Energy Comm. Progress Rept. COO-689-120, 137 p.

Decker, E. R., 1973, Geothermal studies in the Southern Rocky Mountain region, 1971-73: Geol. Soc. America, Abs. with Programs (Rocky Mountain Sec.), v. 5, no. 6, p. 475-476.

Doe, B. R., Lipman, P. W., Hedge, C. E., and Kurasawa, Hajime, 1969, Primitive and contaminated basalts from the Southern Rocky Mountains, USA: Contr. Mineralogy and Petrology, v. 21, no. 2, p. 142-156.

Doney, H. H., 1968, Geology of the Cebolla quadrangle, Rio Arriba County, New Mexico: New Mexico Bur. Mines and Mineral Resources Bull. 92, 114 p.

Edwards, C. L., Reiter, M. A., and Weidman, C., 1973, Geothermal studies in New Mexico and southern Colorado [abs.]: EOS (Am. Geophys. Union Trans.), v. 54, p. 463.

Gast, P. W., 1968, Trace element fractionation and the origin of theoleiitic and alkaline magma types: Geochim. et Cosmochim. Acta, v. 32, no. 10, p. 1057-1086.

Green, D. H., 1968, Origin of basaltic magmas, *in* Hess, H. H., and Poldervaart, A., eds., Basalts—The Poldervaart treatise on rocks of basaltic composition, Vol. 2: New York and London, Interscience Pubs., Inc., p. 835-862.

Green, D. H., and Ringwood, A. E., 1967, The genesis of basaltic magmas: Contr. Mineralogy and Petrology, v. 15, no. 2, p. 103-190.

Gunn, B. M., and Watkins, N. D., 1970, Geochemistry of the Steens Mountain Basalts, Oregon: Geol. Soc. America Bull., v. 81, no. 5, p. 1497-1516.

Hedge, C. E., and Lipman, P. W., 1972, Upper Cenozoic basalts of the Southern Rocky Mountain region: P. 2, Strontium isotopes and the geochemistry of Sr, Rb, and Ba: Geol. Soc. America, Abs. with Programs (Cordilleran Sec.), v. 4, no. 3, p. 169.

Hedge, C. E., and Noble, D. C., 1971, Upper Cenozoic basalts with high Sr^{87}/Sr^{86} and Sr/Rb ratios, southern Great Basin, western United States: Geol. Soc. America Bull., v. 82, no. 12, p. 3503-3510.

Hoffer, J. M., 1971, Mineralogy and petrology of the Santo Tomas-Black Mountain basalt field, Potrillo volcanics, south-central New Mexico: Geol. Soc. America Bull., v. 82, no. 3, p. 603-611.

Ito, Keisuke, and Kennedy, G. C., 1967, Melting and phase relations in a natural peridotite to 40 kilobars: Am. Jour. Sci., v. 265, no. 6, p. 519-538.

Jackson, E. D., and Wright, T. L., 1970, Xenoliths in the Honolulu volcanic series, Hawaii: Jour. Petrology, v. 11, p. 405-430.

Jackson, W. H., and Pakiser, L. C., 1965, Seismic study of crustal structure in the Southern Rocky Mountains, in Geological Survey research 1965: U.S. Geol. Survey Prof. Paper 525-D, p. D85-D92.

Kistler, R. W., and Peterman, Z. E., 1972, Variations in Sr, Rb, K, Na, and initial Sr^{87}/Sr^{86} in Mesozoic granitic rocks in California: Geol. Soc. America, Abs. with Programs (Cordilleran Sec.), v. 4, no. 7, p. 562-563.

Kudo, A. M., Aoki, K. I., and Brookins, D. G., 1971, The origin of Pliocene-Holocene basalts of New Mexico in the light of strontium-isotopic and major-element abundances: Earth and Planetary Sci. Letters, v. 13, no. 1, p. 200-204.

Kuno, H., 1959, Origin of Cenozoic petrographic provinces of Japan and surrounding area: Bull. Volcanol., v. 20, p. 37-76.

——1968, Differentiation of basalt magmas, in Hess, H. H., and Poldervaart, A., eds., Basalt—The Poldervaart treatise on rocks of basaltic composition, Vol. 2: New York and London, Interscience Pubs., Inc., p. 623-688.

Kushiro, Ikuo, 1968, Compositions of magmas formed by partial zone melting of the Earth's upper mantle: Jour. Geophys. Research, v. 73, no. 2, p. 619-634.

——1972, Effect of water on the composition of magmas formed at high pressures: Jour. Petrology, v. 13, no. 2, p. 311-334.

Larsen, E. S., Jr., and Cross, Whitman, 1956, Geology and petrology of the San Juan region, southwestern Colorado: U.S. Geol. Survey Prof. Paper 258, 303 p.

Laughlin, A. W., Brookins, D. G., and Carden, J. R., 1972a, Variations in the initial strontium ratios of a single basalt flow: Earth and Planetary Sci. Letters, v. 14, no. 1, p. 79-82.

Leeman, W. P., 1970, The isotopic composition of strontium in late-Cenozoic basalts from the Basin-Range province, western United States: Geochim, et Cosmochim. Acta, v. 34, no. 8, p. 857-872.

Leeman, W. P., and Rogers, J.J.W., 1970, Late Cenozoic alkali-olivine basalts of the Basin-Range province, USA: Contr. Mineralogy and Petrology, v. 25, p. 1-24.

LeMasurier, W. E., 1968, Crystallization behavior of basalt magma, Santa Rosa Range, Nevada: Geol. Soc. America Bull., v. 79, no. 8, p. 949-972.

Lipman, P. W., 1969, Alkalic and tholeiitic basaltic volcanism related to the Rio Grande depression, southern Colorado and northern New Mexico: Geol. Soc. America Bull., v. 80, no. 7, p. 1343-1353.

Lipman, P. W., and Moench, R. H., 1972, Basalts of the Mount Taylor volcanic field, New Mexico: Geol. Soc. America Bull., v. 83, no. 5, p. 1335-1344.

Lipman, P. W., Bunker, C. M., and Bush, C. A., 1973, Potassium, thorium, and uranium contents of upper Cenozoic basalts of the Southern Rocky Mountain region, and their relation to the Rio Grande depression: U.S. Geol. Survey Jour. Research, v. 1, no. 4, p. 387-401.

Macdonald, G. A., and Katsura, T., 1964, Chemical composition of Hawaiian lavas: Jour. Petrology, v. 5, pt. 1, p. 82-133.

McKee, E. H., and Anderson, C. A., 1971, Age and chemistry of Tertiary volcanic rocks in north-central Arizona and relation of the rocks to the Colorado Plateaus: Geol. Soc. America Bull., v. 82, no. 10, p. 2767-2782.

Murata, K. J., and Richter, D. H., 1966, The settling of olivine in Kilauean magma as shown by lavas of the 1959 eruption: Am. Jour. Sci., v. 264, no. 3, p. 194-203.

Nicholls, J., Carmichael, I.S.E., and Stormer, J. C., Jr., 1971, Silica activity and P total in igneous rocks: Contr. Mineralogy and Petrology, v. 33, no. 1, p. 1-20.

O'Hara, M. J., and Yoder, H. S., Jr., 1967, Formation and fractionation of basic magmas at high pressures: Scottish Jour. Geology, v. 3, pt. 1, p. 67-117.

Peterman, Z. E., Doe, B. R., and Prostka, H. J., 1970, Lead and strontium isotopes in rocks of the Absaroka volcanic field, Wyoming: Contr. Mineralogy and Petrology, v. 27, no. 2, p. 121-130.

Porath, H., 1971, Magnetic variation anomalies and seismic low-velocity zone in the western United States: Jour. Geophys. Research, v. 76, no. 11, p. 2643-2648.

Powers, H. A., 1932, The lavas of the Modoc Lava Bed quadrangle, California: Am. Mineralogist, v. 17, no. 7, p. 253-294.

——1955, Composition and origin of basaltic magma of the Hawaiian Islands: Geochim. et Cosmochim. Acta, v. 7, nos. 1-2, p. 77-107.

Renault, Jacques, 1970, Major-element variations in the Potrillo, Carrizozo, and McCartys basalt fields, New Mexico: New Mexico Bur. Mines and Mineral Resources Circ. 113, 22 p.

Robinson, H. H., 1913, The San Franciscan volcanic field, Arizona: U.S. Geol. Survey Prof. Paper 76, 213 p.

Roy, R. F., Blackwell, D. D., and Decker, E. R., 1972, Continental heat flow, in Robertson, E. C., ed., The nature of the solid earth: McGraw-Hill Book Co., p. 506-543.

Scott, D. H., and Trask, N. J., 1971, Geology of the Lunar Crater volcanic field, Nye County, Nevada: U.S. Geol. Survey Prof. Paper 599-I, 22 p.

Stobbe, H. R., 1949, Petrology of volcanic rocks of northeastern New Mexico: Geol. Soc. America Bull., v. 60, no. 6, p. 1041-1095.

Stone, G. T., 1967, Petrology of upper Cenozoic basalts of the western Snake River Plain, Idaho [Ph.D. thesis]: Boulder, Colo., Univ. Colorado, 392 p.

Stormer, J. C., Jr., 1972a, Mineralogy and petrology of the Raton-Clayton volcanic field, northeastern New Mexico: Geol. Soc. America Bull., v. 83, no. 11, p. 3299-3322.

——1972b, Ages and nature of volcanic activity on the southern High Plains, New Mexico and Colorado: Geol. Soc. America Bull., v. 83, no. 8, p. 2443-2448.

Tatsumoto, M., 1966, Genetic relations of oceanic basalts as indicated by lead isotopes: Science, v. 153, no. 3740, p. 1094-1101.

Trask, N. J., 1969, Ultramafic xenoliths in basalt, Nye County, Nevada, in Geological Survey research 1969: U.S. Geol. Survey Prof. Paper 650-D, p. D43-D48 [1970].

U.S. Geological Survey, 1971, Geological Survey research 1971: U.S. Geol. Survey Prof. Paper 750-A, 418 p. [1972].

Vitaliano, C. J., and Harvey, R. D., 1965, Alkali basalt from Nye County, Nevada: Am. Mineralogist, v. 50, nos. 1-2, p. 73-84.

Waters, A. C., 1962, Basalt magma types and their tectonic associations—Pacific Northwest of the United States, in The crust of the Pacific Basin: Am. Geophys. Union Geophys. Mon. 6, p. 158-170.

Williams, Howel, 1936, Pliocene volcanoes of the Navajo-Hopi country: Geol. Soc. America Bull., v. 47, no. 1, p. 111-172.

Wise, W. S., 1969, Origin of basaltic magmas in the Mojave Desert area, California: Contr. Mineralogy and Petrology, v. 23, no. 1, p. 53-64.

Wright, T. L., 1971, Chemistry of Kilauea and Mauna Loa lava in space and time: U.S.

Wright, T. L., and Fiske, R. S., 1971, Origin of the differentiated and hybrid lavas of Kilauea Volcano, Hawaii: Jour. Petrology, v. 12, no. 1, p. 1-65.

Wyllie, P. J., 1971, Role of water in magma generation and initiation of diapiric uprise in the mantle: Jour. Geophys. Research, v. 76, no. 5, p. 1328-1338.

Reprinted from pages 45–47 and 57–66 of *Geol. Soc. Australia Jour.*
23 (Pt. 1):45–67 (1976)

GEOCHEMISTRY AND PETROGENESIS OF THE NEWER BASALTS OF VICTORIA AND SOUTH AUSTRALIA

By A. J. IRVING and D. H. GREEN

The Pliocene-Recent Newer Basalts province contains a large variety of magma types: quartz tholeiite, olivine tholeiite, olivine basalt, alkali olivine basalt, hawaiite, K-rich hawaiite, nepheline basanite, transitional olivine analcimite and olivine nephelinite (normatively low-K basanite to low-K nepheline mugearite), nepheline hawaiite, K-rich nepheline hawaiite, and nepheline mugearite. Despite some spatial restriction of the different magmas, there is no well-defined compositional progression with time. Many of the alkaline lavas contain Cr-diopside lherzolite-series xenoliths and must be of direct upper-mantle derivation.

Based on systematics of Mg-values (100 Mg/Mg + Fe^{2+} ratios) coupled with data for the olivine-liquid equilibrium and the results of other experimental studies, it is deduced that most of the basanitic magmas are primary melts of garnet-peridotite mantle, essentially unmodified by crystal fractionation. The nepheline hawaiites and nepheline mugearites are considered to be fractionated liquids derived from basanites by progressive removal of mainly olivine and kaersutite *at mantle pressures*. The hawaiites may be similarly related to alkali olivine-basalt parental liquids. The olivine tholeites and olivine basalts cannot be primitive mantle melts and must have undergone fractionation of olivine and perhaps pyroxene. The quartz-tholeiite and K-rich hawaiitic magmas are probably products of low-pressure fractionation processes.

A synthesis of chemical data for Tertiary-Quaternary basaltic lavas from throughout eastern Australia reveals that nepheline basanite is the most widespread mantle-derived magma type, and that the high-pressure lineage nepheline basanite → nepheline hawaiite → nepheline mugearite → nepheline benmoreite is much more important than previously recognized.

INTRODUCTION

The Tertiary volcanics of Victoria and southeast South Australia were extruded in two distinct episodes. The first, in the Palaeocene to Early Miocene, is represented by the Older Volcanics, which occur as eroded plugs and flow remnants chiefly in Gippsland (eastern Victoria), but also sporadically to the north and west of Melbourne (see Singleton & Joyce, 1969; Edwards, 1938*a*). Post-Miocene volcanic activity is represented by the Newer Volcanics, which were erupted chiefly from Late Pliocene to Recent over large areas of Victoria (the Western District Plains, Werribee Plains, and Central Highlands) and in a small region in the far southeast of South Australia (see Fig. 1).

The Newer Volcanics comprise over 99 per cent basaltic rocks, with minor trachytic rocks, and accordingly the term Newer Basalts has been commonly used, and will be used here, to denote the post-Miocene basaltic rocks of the province. The Newer Basalts represent a continental mixed tholeiitic-alkaline basaltic suite, with a wide variety of basalt types in space and time, and an abundance of xenoliths and megacrysts of high-pressure origin (see Irving, 1974*a, b*). In attempting to derive a petrogenetic scheme for the genesis and evolution of basaltic magmas in a single province, we have relied not so much on field and petrographic aspects, as on major-element geochemistry and the application of data from experimental studies. Trace-element data for a selection of the samples studied here have also been obtained, and their implications for petrogenesis will be discussed in a separate paper.

Previous studies of the Newer Volcanics have collectively encompassed most of the province in varying detail. Petrological studies include Howchin (1901), Jutson (1905), Stanley (1909, 1910), Grayson & Mahony (1910), Skeats & Summers (1912), Fenner (1921), Orr (1927), Mahony (1931), Skeats & James (1937), Edwards (1938*b*), Hills (1938), Edwards & Crawford (1940), Coulson (1938, 1941, 1954), Gill (1947), Condon (1951), Solomon (1952), Yates (1954), Hanks (1955), and Spencer-Jones *in* Boutakoff

A. J. IRVING & D. H. GREEN

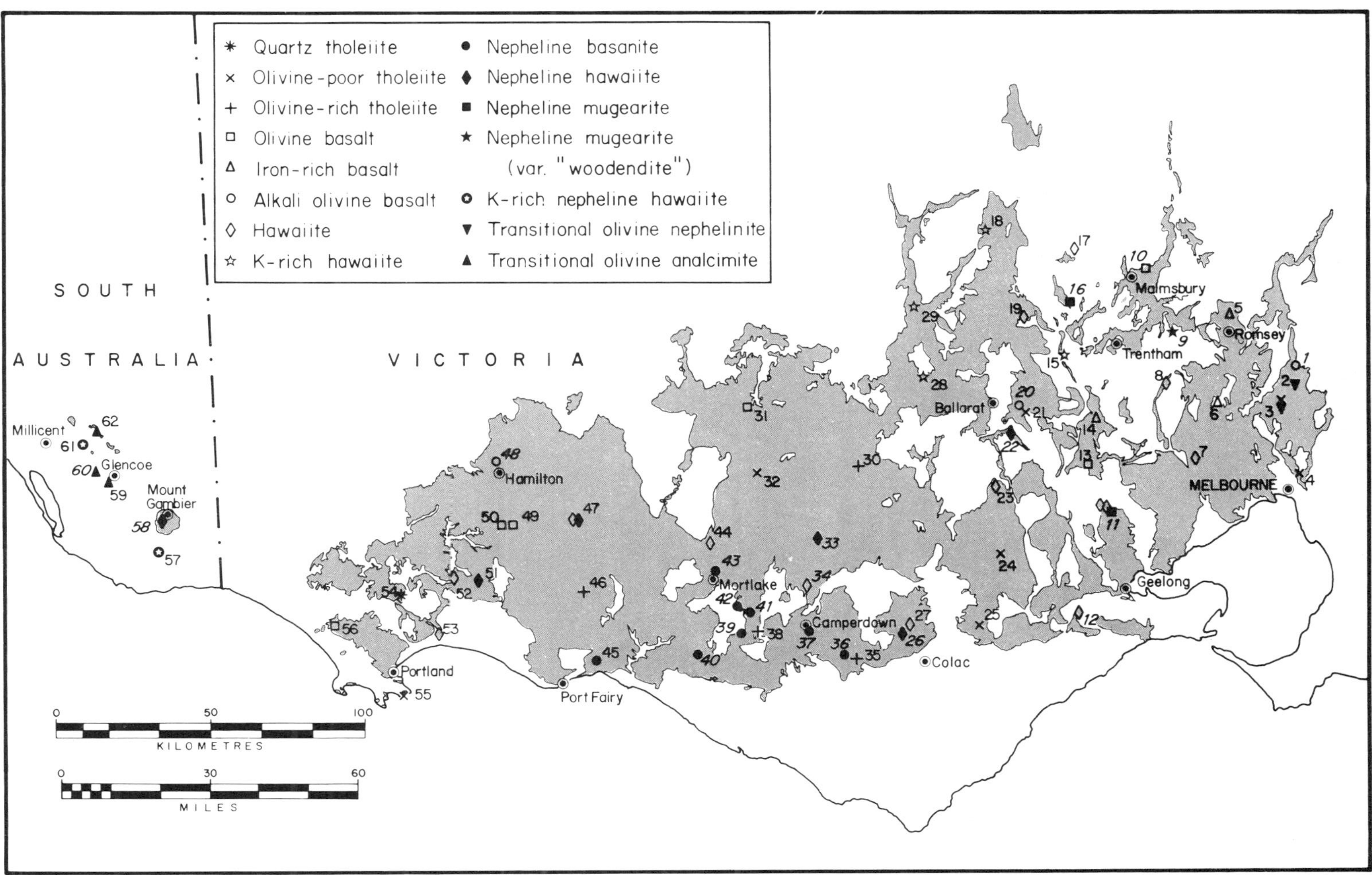

(1963). The most extensive contribution of chemical data was that of Edwards (1938b).

Geomorphological studies, such as those of Hills (1938), Gill (1950), Ollier & Joyce (1964), Ollier (1967), Singleton & Joyce (1971), have provided much of the knowledge of relative ages of lavas within the province. The stratigraphic relationships between certain lavas and Pleistocene aeolian calcarenite dunes have also been used to evaluate relative ages (e.g. Sprigg, 1952; Coulson, 1938; Boutakoff, 1963). The rather limited K-Ar dating of the Newer Basalts (McDougall *et al.*, 1966; Rahman & McDougall, 1972; Wellman, 1974) has been confined mainly to lava flows rather than eruptive centres, but indicates ages ranging from 4.5 to 0.57 m.y. Evidence of more recent activity in South Australia is contained in aboriginal legends (Hossfeld, 1950), and $^{14}C-$ dating of material from tuffs suggests ages of 5000 to 6000 years at Tower Hill (Gill, 1967), and 1400 to 4800 years at Mount Gambier (Gill, 1955; Dury & Langford-Smith, 1968).

In his classic study of the Newer Volcanics of central Victoria, Edwards (1938b) recognized a group of trachytic rocks and a group of basaltic rocks. The trachytic rocks are confined to the Macedon-Trentham district, where they occur as isolated, steep-sided plugs and domes (e.g. Blue Mountain, Babbingtons Hill, Hanging Rock, The Camel's Hump). They include sölvsbergites, phonolites, and anorthoclase trachytes, and appear on geomorphological grounds (e.g. Ollier, 1967) to be older than nearby basaltic lavas; this is con-firmed by recent K-Ar and Rb-Sr dating which indicate an age of 6.3 m.y. (Dasch *et al.*, 1972). Thus aside from the unusual 'woodendite' from Racecourse Hill, Woodend, these acidic to intermediate members of the Newer Volcanics appear to be temporally distinct from the basaltic rocks. They can be also distinguished by their limited geographic extent and the lack of a continuous chemical spectrum between the two groups (as shown by the analyses of Edwards, 1938b). It is thus doubtful whether the trachytic volcanics represent classical products of shallow-level crystal fractionation from Newer Basalt parent magmas (as postulated by Edwards), and it appears probable that they are quite distinct[1]. Both these and the unusual hypersthene trachyandesites associated with more typical basaltic rocks in the Mount Gisborne district (Edwards & Crawford, 1940) have been excluded from this study.

Edwards divided the basaltic rocks of central Victoria into several types (e.g. Ballan type, etc.) based largely on criteria of petrography and field relationships, with lesser regard to chemical features. Unfortunately, these petrographic types do not necessarily have particularly well defined chemical compositions. Many of the flows sampled are quite oxidized, and Edwards' use of 'iddingsite' as a key mineral in several types undoubtedly served to obscure primary chemical features.

Fig. 1. Distribution of magma types within the Newer Basalts province (stippled). Cr-diopside lherzolite series xenoliths are present at localities numbered in italics. *Key to localities:*

1. Mt Frazer	22. Mt Bunninyong	*43.* Mt Shadwell
2. Bald Hill	23. Mt Mercer	44. Mt Flat Top (Mondilibi)
3. Mt Ridley	24. Gow's Hill	45. Tower Hill
4. Merri Creek (flow)	25. Mt Gellibrand	46. Hawkesdale (flow)
5. Melbourne Hill	*26.* Red Hill	47. Mt Rouse
6. Mt Holden	27. Warrion Hill	*48.* Mt Bainbridge
7. Mt Cottrill	28. Mt Callender	49. Mt Napier
8. Mt Bullengarook	29. Mt Mitchell	50. Flow west of Mt Napier
9. Racecourse Hill (Woodend)	*30.* Mt Widderin (flow)	51. Mt Eccles (Lake Surprise Crater)
10. Green Hill	31. The Bald Hill	52. Flow near Lake Condah
11. The Anakies	32. Mt Hamilton	53. Tyrendarra flow
12. Mt Moriac	*33.* Mt Elephant	54. Mt Eckersley
13. Mt Wallace	*34.* Mt Kurweeton	55. Cape Grant (flow)
14. Mt Gorong	35. Stoneyford (flow)	56. Mt Kincaid
15. Leonard's Hill	*36.* Mt Porndon	57. Mt Schank
16. Mt Franklin	*37.* Mt Leura	*58.* Mt Gambier (Brown Lake)
17. Mt Consultation	38. Marida Yallock (flow)	59. The Bluff
18. Mt Mooloort	*39.* Terang (flow)/Mt. Terang	*60.* Mt Watch
19. Smeaton Hill	40. Mt Warrnambool	61. Mt Burr
20. Mt Warrenheip	*41.* Noorat (flow)/Mt Noorat	62. Mt McIntyre
21. Dunnstown (flow)	42. Kolora (flow)	

---

[1] The $^{87}Sr/^{86}Sr$ initial ratio of 0.7103 $\pm$ 0.0024 measured for nine sölvsbergites and soda trachytes (Dasch *et al.*, 1972) is inconsistent with their derivation from any of the basaltic lavas of the province with $^{87}Sr/^{86}Sr$ ratios of 0.7035-0.7045: Dasch & Green, 1975; Stueber, 1969; Stuckless & Irving, 1976).

[Editors' Note: Material has been omitted at this point.]

SPATIAL AND TEMPORAL DISTRIBUTION OF NEWER BASALTS MAGMAS

Members of most of the chemically distinctive lava types are geographically grouped in sub-provinces (Fig. 1). Aside from the apparently extensive tholeiite and olivine basalt plains, good examples are the volcanic hills composed of basanite, of K-rich hawaiite, and of olivine tholeiite. The olivine nephelinites occur at the eastern, and olivine analcimites at the western, extremities of the province. The restriction of the very young basanitic volcanoes and maars to the southern margin of the province coincides approximately with a zone of anomalously high electrical conductivity centred beneath Colac at depths less than 50 km, which may represent a region of partial melting in the upper mantle or a receding thermal high related to the volcanism (Bennett & Lilley, 1973). The nepheline mugearite and K-rich nepheline hawaiite cones are both confined to restricted parts of the province, and there is some tendency for hawaiite volcanoes to be concentrated in the southeastern section.

The geochronology of the Newer Basalts is only imperfectly known. As outlined above, there are relatively few isotopic ages (especially for the volcanic hills) and much of the available evidence derives from geomorphological considerations (e.g. Ollier & Joyce, 1964; Ollier, 1967); however, some generalized relationships can be deduced. In broad terms, there is the suggestion of a time sequence from a tholeiitic flood plain or shield through superposed hawaiitic volcanoes to basanitic volcanoes, although very few of the investigated centres contain more than one basalt type. In detail, however, this sequence is invalid. Among the very old volcanoes of the Portland-Hamilton area, there is the quartz tholeiite of

Mount Eckersley, but also the alkali olivine basalt of Mount Bainbridge. Also, the olivine tholeiites of the Stony Rises and in particular Mount Hamilton are almost certainly as young as most of the alkaline lavas (see Ollier, 1967). Again, the hawaiites and nepheline hawaiites of Mounts Gambier, Rouse, and Eccles and the Tyrendarra flow are probably younger than most of the basanites, and, as mentioned above, the almost identical lavas of Mounts Burr and Schank appear to be widely separated in time.

Thus, in summary, it appears that in the interval from about 4.5 m.y. to sub-historic times the province has developed by recurring eruptions of a variety of subalkaline and alkaline magmas, which although to a certain degree spatially restricted, did not form a well-defined compositional progression with time.

CRITERIA FOR RECOGNITION OF PRIMARY AND DERIVATIVE MAGMAS

It is generally agreed by petrologists that basaltic magmas are generated by processes of partial melting in the Earth's mantle. An objective in the study of basaltic rocks is to extrapolate from features of erupted lavas to processes of magma generation and evolution. Such an extrapolation is at first sight intractable, because an erupted lava is likely to have had a very complex history, beginning with the partial melting process itself, and involving subsequent crystal fractionation, contamination, or both at all levels up to the surface. There are, however, a number of characteristics in some lavas which provide constraints on the extent of these processes.

According to the weight of current isotopic and other trace-element geochemical evidence, Cr-diopside-spinel lherzolite-series xenoliths in basalts represent accidental, modified samples of the Earth's upper mantle (see Frey & Green, 1974). It can thus be argued that basalts containing these dense fragments must have moved directly and rapidly from the mantle to the surface without significant modification by crystal fractionation (*i.e.* settling) processes at shallow crustal levels (Green, 1969, 1970*a*; Maaløe, 1973). Xenoliths of other ultramafic rocks, rich in aluminous pyroxenes, as well as aluminous pyroxene megacrysts, may also imply mantle derivation for the lavas containing them (see Irving, 1974*a*, 1974*b*). Thus, an erupted basalt liquid containing any of these high-pressure inclusions must be either (a) a primary partial melt of the mantle, or (b) a derivative of such a primary melt pro-

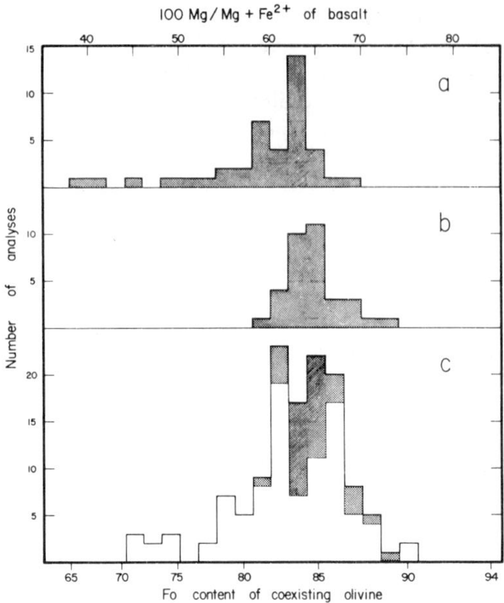

Fig. 4. Histograms of Mg-values and compositions of coexisting olivines (for $K_D^{OL-LIQ} = 0.33$):

(a) Newer Basalts magmas lacking Cr-diopside lherzolite-series xenoliths. Fe_2O_3/FeO ratios adjusted as discussed in text.

(b) Newer Basalts magmas containing Cr-diopside lherzolite-series xenoliths. Multiple analyses were averaged for Racecourse Hill, The Anakies (east) (2 groups) Mts Franklin, Warrenheip, Elephant, Porndon (2 groups), Leura (2 groups), Shadwell (2 groups), Gambier, and Watch. Fe_2O_3/FeO ratios adjusted as discussed in text. Data for all analysed samples before and after adjustment are illustrated by Green *et al.* (1974, fig. 1).

(c) Data of (b) combined with data for 95 lherzolite-bearing basaltic magmas from other eastern Australian Tertiary-Quaternary provinces (Fe_2O_3/FeO adjusted to 0.15). Most of the latter data are unpublished (*see* Green *et al.*, 1974), the remainder being from Binns *et al.* (1970), Wass (1971), Lovering & White (1969), Cundari (1973), and Table I.

duced by crystal fractionation processes *at high pressures.*

Possibly the most important equilibrium pertinent to processes of partial melting in the upper mantle is that between olivine, the most abundant upper mantle phase, and basaltic liquid. Experimental studies at zero pressure (Grove *et al.*, 1973) and 1 atmosphere (Roeder & Emslie, 1970) indicate a value of 0.33 for the distribution coefficient,

$$K_D, \; OL-LIQ = (\frac{Fe}{Mg}) \, OL / (\frac{Fe}{Mg}) LIQ.$$ An alternative way of expressing this relation is
$$X_{OL} = 100 X_{LIQ} \div [X_{LIQ} + K_D (100 - X_{LIQ})],$$
where $X = 100 \, Mg/Mg+Fe^{2+}$. A summary of data for Mg-value of liquids (X_{LIQ}) and Fo content of coexisting olivines (X_{OL}) was given by Cawthorn *et al.* (1973). Data from experimental studies on terrestrial and Ti-poor lunar compositions at pressures up to 30 kb also yield K_D^{OL-LIQ} near 0.33, thus demonstrating that pressure has a negligible effect on the equilibrium.

Recent estimates of the chemical composition of the undepleted Earth's upper mantle by Carter (1970), Ringwood (1966), and Nicholls (1967) have Mg-values of 87.4 to 89.3[1], and the olivines of these peridotitic assemblages would have similar values. If compositions in the range Fo_{86} to Fo_{90} are considered, this would imply equilibrium with basaltic liquids with Mg-values between 66 and 75 (for $K_D, \, ^{OL-LIQ}=0.33$). Melts which are primary on the basis of this criterion commonly contain high-pressure inclusions; but since the presence of inclusions at the surface depends ultimately only on the vigour of the final eruption, it is possible to have primary melts that lack inclusions. Mg-values below 66 would imply some history of crystal fractionation in the basalt. Because of the necessarily rather subjective estimation of pre-eruptive Fe_2O_3 contents, the 100 $Mg/Mg+Fe^{2+}$ ratio is far from an ideal parameter, but in the absence of data on such partition relationships as Mg/Ni and Mg/Co, it is the most definitive index of ferromagnesian crystal fractionation processes.

In Figure 4 the Mg-values of basaltic magmas containing and lacking Cr-diopside lherzolite xenoliths are compared. Also shown are the compositions of olivine coexisting with liquid for $K_D, \, ^{OL-LIQ}= 0.33$. The lherzolite-bearing magmas can be consistently interpreted as primary liquids or slightly fractionated liquids derived from peridotite containing olivine of composition Fo_{87-90}. The lherzolite-

---

[1] Hutchison *et al.* 1970) have listed average compositions of French spinel lherzolites in basalts (Mg-value = 91.0-92.6) and South African peridotite xenoliths in kimberlite (Mg-value = 94.7). These and the estimate of Harris *et al.* (1967) (with an Mg-value of 91.0) are considered to be depleted compositions (see Frey & Green, 1974).

free magmas show a greater spread of Mg-values, consistent with equilibrium with olivine of composition Fo_{65-85}. On this basis, it is deduced that the majority of the lherzolite-free basalts have undergone some crystal fractionation, and, in particular, the quartz tholeiite and olivine tholeiites cannot be interpreted as primary liquids but have probably lost olivine.

The widespread presence in erupted lavas of xenoliths of crustal material suggests that the possibility of contamination by such material should always be entertained, although it can be argued that any lava containing inclusions of high-pressure (>10 kb) origin is unlikely to have had the opportunity for crustal contamination, even if crustal xenoliths were incorporated en route to the surface. Isotopic parameters such as $^{87}Sr/^{86}Sr$ ratio can provide more definitive evidence of the extent of contamination processes: the $^{87}Sr/^{86}Sr$ ratios for 14 examples (of all major chemical types) from the Newer Basalts are within the normal range for uncontaminated modern basalts (Dasch & Green, 1975; Stueber, 1969; Stuckless & Irving, 1976).

GENESIS OF THE NEWER BASALTS MAGMAS

The detail of interpretations regarding the development of the Newer Basalts province as a whole is necessarily limited by its complexity in both space and time, and discussion is here confined to likely processes and physical conditions involved in the evolution of the major basalt types. Observations made on the lavas themselves and their inclusions have been integrated with evidence from relevant experimental studies.

Basanitic Magmas

On the basis of criteria detailed above, the only examples of the Newer Basalts which could have developed as direct upper mantle melts are the majority of the basanites, including some of the transitional olivine analcimites (low-K 'basanites'). The Mount Frazer alkali olivine basalt (Mg-value = 68.6) could also be primary, but in view of the very high normative olivine content of this sample the possibility of contamination with xenocrystal olivine must be considered. Some of the basanites, notably those from Mount Leura, cannot be primary melts on the grounds of low Mg-values. The more magnesian basanites commonly contain megacrysts which attest to crystallization and possible crystal fractionation of these or related magmas at high pressures. The extent of fractionation may not

have been very great, and many of the basanites are probably only slightly removed in composition from their parental mantle melts.

Recent experimental studies (Green, 1973a) on a Mount Leura basanite (2650) with 10% added olivine (Fo_{90}) indicate that the Victorian basanite magmas may be derived by small degrees (about 5%) of partial melting of hydrous mantle peridotite (containing about 0.1–0.3% water) at pressures of 25–30 kb (*i.e.* depths of 80-100 km). At these pressures residual mantle peridotite contains garnet as well as olivine, orthopyroxene, and clinopyroxene Green, 1973b). Data for the rare-earth elements in basanites from Victoria (Frey & Green, 1974; F. A. Frey, unpubl.) and elsewhere (Kay & Gast, 1974) affirm that such magmas must form by very small degrees of partial melting of peridotite containing garnet as a residual phase.

The relatively high contents of Ti, K, and P characteristic of basanites are also consistent with such small degrees of melting and these elements (especially K and P) are particularly enriched in the Victorian basanites. In the source peridotite these 'incompatible' elements are contained in early-melting accessory phases such as amphibole, mica, apatite, ilmenite, and titanoclinohumite, and it is notable that pargasite, phlogopite, and apatite are present in some Victorian Cr-diopside lherzolite xenoliths (Frey & Green, 1974). The basanites from Mount Leura, which have exceptionally high K, Ti, and P, have a higher Fe-Mg ratio than those of, say, Mount Porndon and Mount Noorat, and some part of their enrichment in incompatible elements may have been produced by small amounts of fractionation (major olivine with some clinopyroxene and/or orthopyroxene) from a more typical basanite parent.

The very low K contents of the transitional olivine analcimites virtually exclude source regions containing mica. Coupled with the very high Ti and P contents of these magmas, this suggests that apatite and ilmenite or perhaps titanoclinohumite were the important accessory phases at subsolidus conditions in their source regions. Such apparent regional variation in the accessory mineralogy, minor element chemistry and isotope chemistry of upper mantle peridotite also emerges from studies of Cr-diopside lherzolites (*e.g.* Frey & Green, 1974; Frey & Prinz, 1971 and unpubl. data, Dasch & Green, 1975). Both the olivine analcimites and the more fractionated olivine nephelinites transitional to basanites probably originated by slightly lower degrees of partial melting, possibly at deeper levels than the more typical basanites (*cf.* Bultitude & Green, 1971).

Hawaiitic and Mugearitic Magmas

All the analysed samples of hawaiite, nepheline hawaiite, and nepheline mugearite have Mg-values less than 66 and are thus considered to be derivative magmas rather than primary mantle melts. Over two-thirds of the investiated examples of these magma types are associated with Cr-diopside lherzolite xenoliths, implying that, at least in these cases, the crystal fractionation processes responsible were operative at mantle pressures. Such a high-pressure origin cannot be excluded for the remainder of these lavas.

Nepheline hawaiites and nepheline mugearites

In terms of Newer Basalt lavas now observed at the surface, the only candidates as primary melts parental to the hawaiitic and mugearitic magmas appear to be basanite, transitional olivine analcimite, and possibly a more primitive alkali olivine basalt type. The spatial restriction of the transitional olivine analcimites and their very low K contents would seem to exclude them as parents of widespread significance; however, several lines of evidence suggest a link at high pressures between the more typical basanites, the nepheline hawaiites, and the nepheline mugearites.

A lherzolite-bearing flow of nepheline hawaiite (2154) occurs at Noorat, immediately southeast of the complex basanite volcano Mount Noorat, from which it is possibly derived. More definitive evidence of a link between the basanite, nepheline hawaiite, and nepheline mugearite magmas is provided by the chemistry of the clinopyroxene megacrysts found within examples of each. General similarities are evident amongst all these megacrysts; however the occurrence at one nepheline mugearite centre (Mt Franklin) of clinopyroxene megacrysts with compositions representing an extension of the chemical trend defined by clinopyroxenes in the basanites at Mount Noorat suggests a continuous compositional variation in liquids precipitating the clinopyroxenes and a genetic relationship between these two magma types in particular (*see* Irving, 1974*b*).

Nepheline mugearite 2102 from The Anakies (east) has been the subject of an extensive experimental investigation (Irving & Green, 1972, and unpubl. data). From this study, it was concluded that the nepheline mugearite and nepheline hawaiite magmas could be derived from hydrous basanitic parental liquids

by high pressure (12-20 kb) crystal fractional processes involving the removal particularly of kaersutitic amphibole, with olivine, apatite, aluminous clinopyroxene, and biotite in varying proportions. Further evidence for the important role of amphibole in producing this magmatic lineage is provided by trace element data, and in particular by the observed decrease in K/Rb ratio from basanites through nepheline hawaiites to nepheline mugearites (A. J. Irving, unpubl.). Particular differences in contents of incompatible minor and trace elements (*e.g.* K, Ti, P, Ba, Sr, Rb, etc.) between otherwise similar magmas (compare P_2O_5 contents (0.6% and 0.9%) of the nepheline mugearites from The Anakies (east) and Mt Franklin) are interpreted as largely inherited chemical traits of parent magmas. Even though the contents of these elements can be modified by crystal fractionation processes, large differences in concentration are more probably determined by variations in accessory mineralogy and partial melting processes in the primitive upper mantle source regions.

Hawaiites

In calculating fractionation trends from basanite to nepheline mugearite, we have found that the effect of kaersutite subtraction was largely to decrease the Ca:Na ratio in the derivative liquids without significantly altering the normative nepheline content. It is thus predictable that just as the nepheline hawaiites may be related to the basanites by a fractionation model involving chiefly amphibole control, so the hawaiites of the Newer Basalts (which differ primarily in lower normative *ne*) may be related to an alkali olivine basalt parent. A role for aluminous orthopyroxene in such fractionation may be indicated by the presence of megacrysts of this mineral in the hawaiite at The Anakies (west) and in the Kyogle hawaiite (Wilkinson & Binns, 1969). Although, with the possible exception of the Mount Frazer lava, alkali olivine basalt magmas of demonstrable direct upper mantle derivation are not present among the Newer Basalts, all the observed alkali olivine basalt magmas can be related to a more primitive type by schemes involving minor olivine and pyroxene fractionation at upper mantle levels.

K-rich hawaiitic magmas

The K-rich hawaiites of the Ballarat region and the K-rich nepheline hawaiites of South Australia are possibly low-pressure fractionation products of other (high-pressure) hawaiite

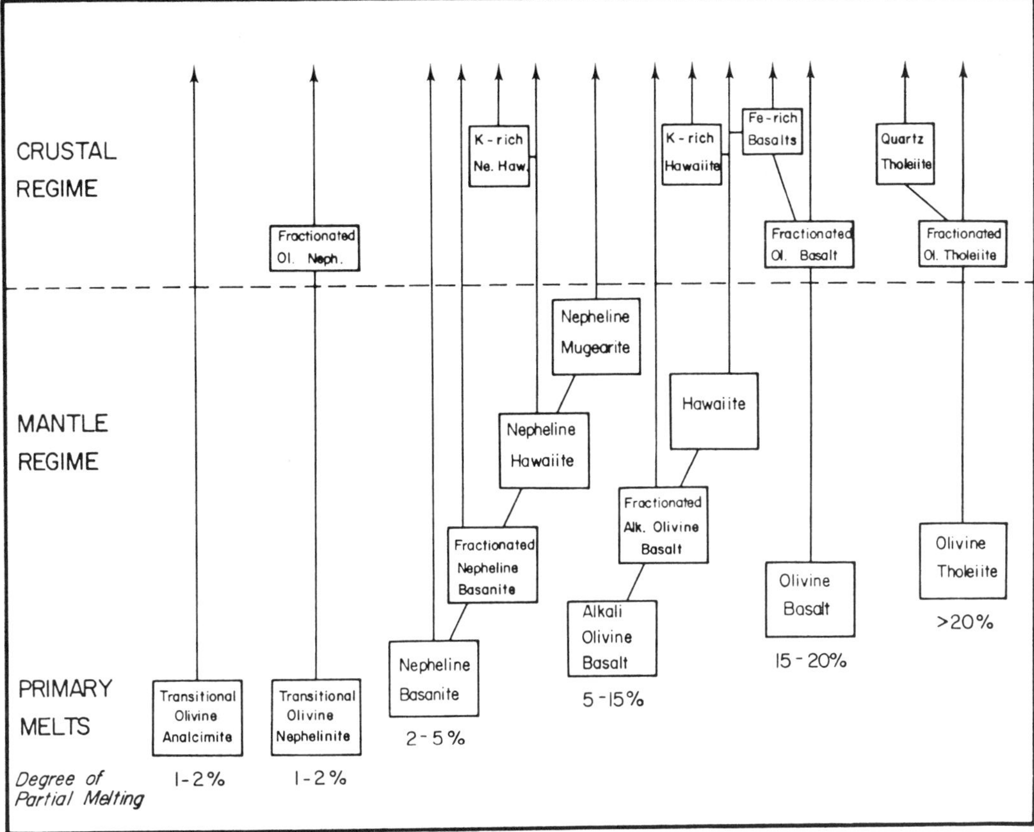

Fig. 5. Schematic summary of primary melts and magmatic lineages for the **Newer Basalts** magmas. Approximate degrees of partial melting are after Green (1970*b*, 1973*a*) and Kay & Gast (1974). Vertical scale has only partial depth significance.

magmas respectively (which in each case are present in the near vicinity). All these lavas are relatively iron-rich (with Mg-values less than 60), and are also very noticeably enriched in K, Ti, and P. There is no conclusive evidence as to the pressures at which these magmas might have fractionated, although the systematic absence of high-pressure inclusions is consistent with a relatively low-pressure origin. Experimental data (Irving & Green, unpubl.) suggest that the fractionation at pressures less than 5 kb (even under wet conditions) of mugearitic and hawaiitic magmas will be controlled by olivine and eventually clinopyroxene and plagioclase over a temperature interval of at least 120°C, which would result primarily in a marked iron-enrichment trend accompanied by enhanced incompatible element contents in derivative liquids. The three iron-rich basalts show some similarities to the K-rich hawaiites, yet differ in terms of lower K and Mg-value. It is pro-

bable that these lavas represent examples of more extreme (and perhaps multiple) low-pressure fractionation of magmas which may have ranged in bulk composition from olivine basalt to hawaiite.

Tholeiite and Olivine Basalt Magmas

The tholeiitic and olivine basalt magmas of the Newer Basalts are too high in Fe/Mg to be primary magmas produced by relatively large degrees (greater than 20%) of partial melting of mantle peridotite (*cf.* Green & Ringwood, 1967). The tholeiites of Gows Hill, Mount Gellibrand, Mount Hamilton, Dunnstown, Cape Grant, Merri Creek, Mount Ridley and Mount Eckersley, which have Mg-values of 50.6–62.9 and are poor or lacking in normative olivine (0.7–8%), are most likely fractionates of more olivine-rich tholeiite magmas at very low pressures, probably less than 5 kb (*cf.* Green *et al.*, 1967; Green, 1970*b*). The olivine basalts from Bald Hill, Mount Kincaid, and

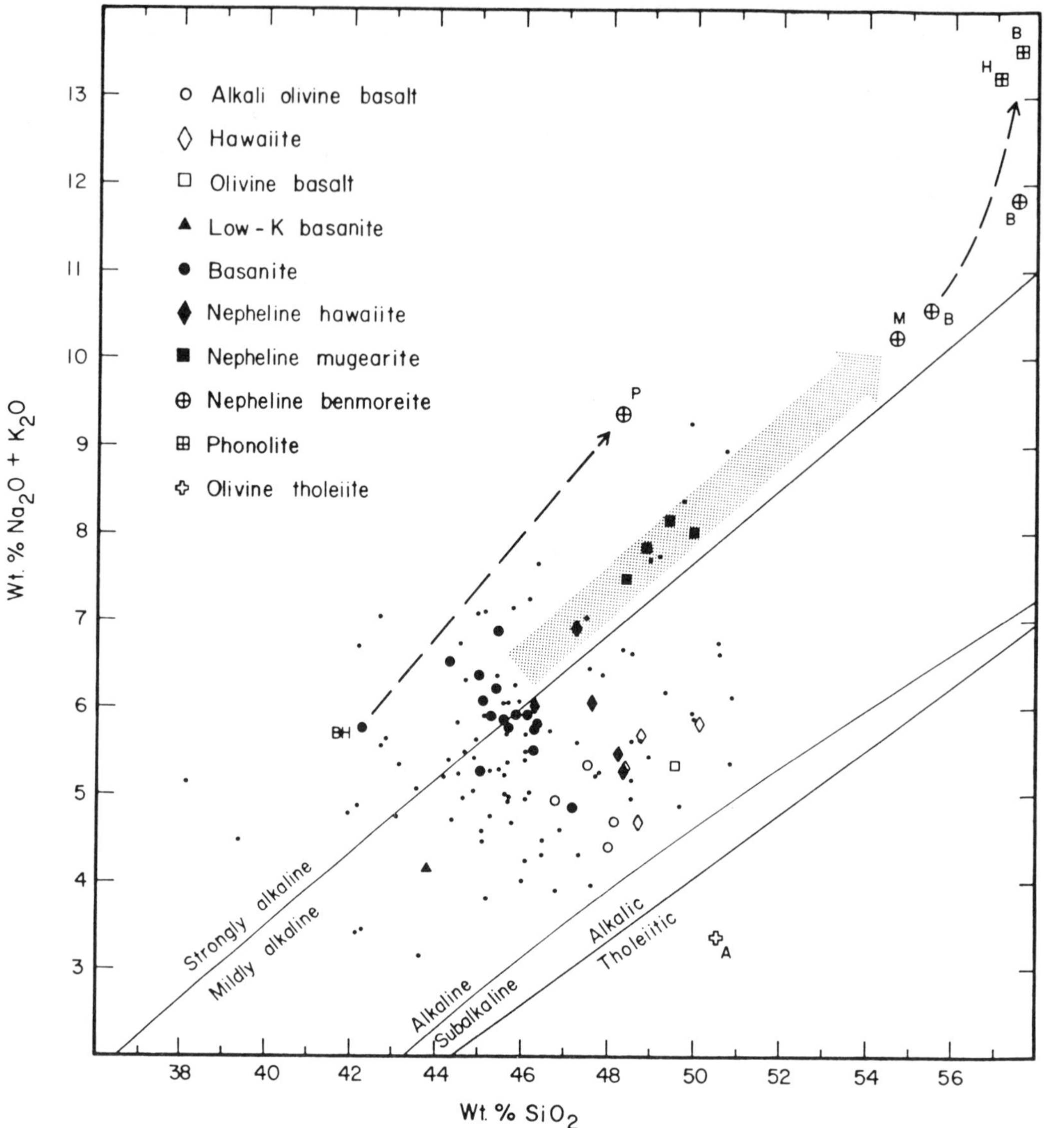

Fig. 6. Alkalies *versus* silica plot for Cr-diopside lherzolite-bearing magmas from eastern Australia and elsewhere. Data for the Newer Basalts magmas (large symbols) averaged as in Fig. 4b (low-K basanite = transitional olivine analcimite). Small dots refer to 95 other eastern Australian magmas (volatile-free), as in Fig. 4c. Other abbreviations and data sources are A = Andover, Tasmania (Sutherland, 1974), BH = Black Head, New Zealand (Price & Taylor, 1973), P = Pigroot, New Zealand (Price & Green, 1972), M = Mt Mitchell, Queensland (Green *et al.*, 1974), B = Bokkos, Nigeria (Price & Irving, unpubl.), H = Heldburg, E. Germany (Price & Irving, unpubl.). Dividing lines as for Fig. 2, with alkalic/tholeiitic dividing line after MacDonald & Katsura (1964).

Mount Wallace and the olivine tholeiites transitional to them from Marida Yallock, Hawkesdale, Mount Widderin, and Stoneyford have, in general, slightly higher Mg-values (58.6–63.6) and much greater normative olivine contents (13.9–19.8). Such differences are explicable in terms of different conditions of melting of source peridotite and thus different normative olivine/hypersthene ratios for parental magmas, with the further complication of variable olivine or (olivine + pyroxene) fractionation at low pressures.

Summary

A schematic summary of the suggested parental melts and magmatic lineages for the Newer Basalts is presented in Fig. 5.

COMPARISON WITH OTHER CAINOZOIC VOLCANICS OF EASTERN AUSTRALIA

The Newer Basalts province contains a wide range of lava compositions from quartz tholeiite to basanite. A particular feature of the province is the development of hawaiitic and mugearitic lavas, yet very young examples of even more evolved alkaline magmas such as benmoreite, trachyte, phonolite, and alkali rhyolite are apparently lacking, in contrast to their occurrence among Tertiary-Quaternary volcanics elsewhere in eastern Australia. In addition, extremely undersaturated lavas, like the olivine nephelinites and olivine melilitites of Tasmania and the Victorian Older Volcanics, appear to be absent from the Newer Basalts (see Sutherland, 1969*b*).

In Figure 6 we compare the (Na_2O+K_2O) *vs.* SiO_2 variation for lherzolite-bearing Newer Basalts magmas with that for over 90 lherzolite-bearing magmas from elsewhere in eastern Australia. With one remarkable exception, an olivine tholeiite from Tasmania (Sutherland, 1974), the lherzolite-bearing magmas plot within the alkaline field (as defined by Irvine & Baragar, 1971, or Macdonald & Katsura, 1964) or within the strongly alkaline field (as defined by Saggerson & Williams, 1964). Basanite is the most widespread mantle-derived magma type. The trend from basanite through nepheline hawaiite to nepheline mugearite as a high-pressure magma series, which is particularly discernible among the Newer Basalts, may be extrapolated to the distinctive nepheline benmoreite from Mount Mitchell, Southeast Queensland (Green *et al.*, 1974). The

unusual mafic nepheline benmoreite (or phonolite) from Pigroot in the East Otago province of New Zealand (Price & Green, 1972) may be similarly related to basanites of the same province. We have suggested on the basis of experimental and other evidence that fractionation of kaersutitic amphibole is largely responsible for this high-pressure lineage, and the observed association of kaersutite megacrysts with many magmas of this series throughout eastern Australia (see Wass & Irving, 1976) is supporting evidence. As demonstrated in Nigeria and Eastern Germany (Irving & Price, 1974), continued fractionation can produce high-pressure phonolites, which may also be represented in eastern Australia. As studies following that of Binns *et al.* (1970) are beginning to reveal, such evolved members of high-pressure alkaline magma series are much more important throughout eastern Australia than previously recognized.

ACKNOWLEDGEMENTS

We wish to thank Professor A. J. R. White, Dr I. McDougall, Mr R. Overton, and Dr T. H. Green for collecting several of the samples, and Dr R. A. Binns for permission to use an unpublished analysis. XRF analyses of most samples were made at the Australian National University with the expert assistance of P. H. Beasley; 15 XRF analyses were obtained at Macquarie University through the generosity of Dr R. H. Flood. Special thanks are due to E. Kiss for his highly skilled efforts in carrying out the majority of the wet chemical determinations, and to A. Powell for preparation of numerous thin sections. The critical comments of Professor D. S. Coombs are greatly appreciated.

REFERENCES

BENNETT, D. J., & LILLEY, F. E. M., 1973: Electrical conductivity structure in the southeast Australian region. *Geophys. J. Roy. astr. Soc.,* 37, pp. 191-206.

BINNS, R. A., DUGGAN, M. B., & WILKINSON, J. F. G., 1970: High pressure megacrysts in alkaline lavas from northeastern New South Wales. *Am. J. Sci., 269,* pp. 132-168.

BOUTAKOFF, N., 1963: The geology and geomorphology of the Portland area. *Mem. geol. Surv. Vict., 22.*

BULTITUDE, R. J., & GREEN, D. H., 1971: Experimental study of crystal-liquid relationships at high pressure in olivine nephelinite and basanite compositions. *J. Petrol., 12,* pp. 121-147.

CARTER, J. L., 1970: Mineralogy and chemistry of the earth's upper mantle based on the partial fusion-partial crystallisation model. *Bull. geol. Soc. Am., 81,* pp. 2021-2034.

CAWTHORN, R. G., FORD, C. E., BIGGAR, G. M., BRAVO, M. S., & CLARKE, D. B., 1973: Determination of the liquid composition in experimental samples: discrepancies between microprobe analysis and other methods. *Earth planet. Sci. Lett., 21,* pp. 1-5.

CONDON, M. A., 1951: The geology of the lower Werribee River, Victoria. *Proc. Roy. Soc. Vict., 63,* pp. 1-24.

COOMBS, D. S., & WILKINSON, J. F. G., 1969: Lineages and fractionation trends in undersaturated volcanic rocks from the East Otago volcanic province (New Zealand) and related rocks. *J. Petrol., 10*, pp. 440-501.

COULSON, A., 1938: The basalts of the Geelong district. *Proc. Roy. Soc. Vict., 50*, pp. 251-257.

COULSON, A., 1941: The volcanoes of the Portland district. *Proc. Roy. Soc. Vict., 53*, pp. 394-402.

COULSON, A., 1954: The volcanic rocks of the Daylesford district. *Proc. Roy. Soc. Vict., 65*, pp. 113-124.

CUNDARI, A., 1973: Petrology of the leucite-bearing lavas in New South Wales. *J. geol. Soc. Aust., 20*, pp. 465-492.

DASCH, E. J., & GREEN, D. H., 1975: Strontium isotope geochemistry of lherzolite inclusions and host basaltic rocks, Victoria, Australia. *Am. J. Sci., 257*, pp. 461-469.

DASCH, E. J., WELLMAN, P., & MILLAR, D. J., 1972: Age and strontium isotope geochemistry of highly differentiated volcanic rocks from the 'Newer Volcanics', southeastern Australia (abstr.). *Abstracts with Programs, Geol. Soc. Am. Mtgs., 4*, pp. 483-484.

DURY, G. J., & LANGFORD-SMITH, T., 1968: Australian geochronology: Checklist 3. *Aust. J. Sci., 30*, pp. 304-306.

EDWARDS, A. B., 1938a: Petrology of the Tertiary Older Volcanic rocks of Victoria. *Proc. Roy. Soc. Vict., 51*, pp. 73-98.

EDWARDS, A. B., 1938b: The Tertiary volcanic rocks of central Victoria. *Q.J. geol. Soc. Lond., 94*, pp. 243-320.

EDWARDS, A. B., & CRAWFORD, W., 1940: The Cainozoic volcanic rocks of the Gisborne district, Victoria. *Proc. Roy. Soc. Vict., 52*, pp. 281-311.

FENNER, C., 1921: The craters and lakes of Mount Gambier, South Australia. *Trans. Roy. Soc. S. Aust., 45*, pp. 169-205.

FREY, F. A., & GREEN, D. H., 1974: The mineralogy, geochemistry and origin of lherzolite inclusions in Victorian basanites. *Geochim. cosmochim. Acta, 38*, pp. 1023-1059.

FREY, F. A., & PRINZ, M., 1971: Ultramafic nodules from San Carlos, Arizona: mineralogy and chemical composition (abstr.). *Abstracts with Programs, Geol. Soc. Am. Mtgs., 3*, pp. 573-574.

GILL, E. D., 1947: Ecklin Hill—a volcano in the Western District of Victoria. *Vict. Naturalist, 64*, pp. 130-134.

GILL, E. D., 1950: An hypothesis relative to the age of some Western District volcanoes, Victoria. *Proc. Roy. Soc. Vict., 60*, pp. 189-194.

GILL, E. D., 1955: Radiocarbon dates for Australian archaeological and geological samples. *Aust. J. Sci., 18*, pp. 49-52.

GILL, E. D., 1967: Evolution of the Warrnambool-Port Fairy coast and the Tower Hill eruption, Western Victoria, in JENNINGS, J. N., & MABBUTT, J. A. (eds.), *Landform Studies from Australia and New Guinea.* ANU Press, pp. 340-364.

GRAYSON, H. J., & MAHONY, D. J., 1910: The geology of the Camperdown and Mount Elephant districts. *Mem. geol. Surv. Vict., 9.*

GREEN, D. H., 1969: The origin of basaltic and nephelinitic magmas in the earth's mantle. *Tectonophysics, 7*, pp. 409-422.

GREEN, D. H., 1970a: A review of experimental evidence on the origin of basaltic and nephelinitic magmas. *Phys. Earth planet. Interiors, 3*, pp. 221-235.

GREEN, D. H., 1970b: The origin of basaltic and nephelinitic magmas. *Trans. Leicester lit. phil. Soc., 44*, pp. 26-54.

GREEN, D. H., 1973a: Conditions of melting of basanite magma from garnet peridotite. *Earth planet. Sci. Lett., 17*, pp. 456-465.

GREEN, D. H., 1973b: Experimental melting studies on a model upper mantle composition at high pressure under water-saturated and water-undersaturated conditions. *Earth planet. Sci. Lett., 19*, pp. 37-53.

GREEN, D. H., & RINGWOOD, A. E., 1967: The genesis of basaltic magmas. *Contr. Miner. Petrol., 15*, pp. 103-190.

GREEN, D. H., EDGAR, A. D., BEASLEY, P., KISS, E., WARE, N. G., 1974: Upper mantle source for some hawaiites, mugearites and benmoreites. *Contr. Miner. Petrol., 48*, pp. 33-43.

GREEN, T. H., GREEN, D. H., & RINGWOOD, A. E., 1967: The origin of high-alumina basalts and their relationships to quartz tholeiites and alkali basalts. *Earth planet. Sci. Lett., 2*, pp. 41-51.

GROVE, T. L., WALKER, D., LONGHI, J., STOLPER, E., & HAYS, J. F., 1973: Petrology of rock 12002 and origin of picritic basalts at Oceanus Procellarum. *Proc. Fourth Lunar Sci. Conf., Geochim. cosmochim. Acta, Suppl. 4, Vol. 1*, pp. 995-1011.

HANKS, W., 1955: Newer Volcanic vents and lava fields between Wallan and Yuroke, Victoria. *Proc. Roy. Soc. Vict., 57*, pp. 1-16.

HARRIS, P. G., REAY, A., & WHITE, I. G., 1967: Chemical composition of the upper mantle. *J. geophys. Res., 72*, pp. 6359-6369.

HARVEY, M., & JOPLIN, G. A., 1941: A note on some leucite bearing rocks from New South Wales with special reference to an ultrabasic occurrence at Murrumburrah. *J. Proc. Roy. Soc. N.S.W., 74*, pp. 419-422.

HILLS, E. S., 1938: The age and physiographic relationships of the Cainozoic volcanic rocks of Victoria. *Proc. Roy. Soc. Vict., 51*, pp. 112-139.

HOSSFELD, P. S., 1950: The late Cainozoic history of the southeast of South Australia. *Trans. Roy. Soc. S. Aust., 73*, pp. 232-279.

HOWCHIN, W., 1901: Notes on the extinct volcanoes of Mount Gambier and Mount Schank, South Australia. *Trans. Roy. Soc. S. Aust., 25*, pp. 54-62.

HUTCHISON, R., PAUL, D. K., & HARRIS, P. G., 1970: Chemical composition of the upper mantle. *Miner. Mag., 37*, pp. 726-729.

IRVINE, T. N., & BARAGAR, W. R. A., 1971: A guide to the chemical classification of the common volcanic rocks. *Canad. J. Earth Sci., 8*, pp. 523-548.

IRVING, A. J., 1974a: Pyroxene-rich ultramafic xenoliths in the Newer Basalts of Victoria, Australia. *N. Jb. Miner. Abh., 120*, pp. 147-167.

IRVING, A. J., 1974b: Megacrysts from the Newer Basalts and other basaltic rocks of southeastern Australia. *Bull. geol. Soc. Am., 85*, pp. 1503-1514.

IRVING, A. J., & GREEN, D. H., 1972: Experimental study of phase relationships in a high-pressure mugearitic basalt as a function of water content (abstr.). *Abstracts with Programs, Geol. Soc. Am. Mtgs., 4*, pp. 550-551.

IRVING, A. J., & PRICE, R. C., 1974: Mantle-derived phonolitic rocks from Nigeria, Eastern Germany and New Zealand (abstr.). *EOS, Trans. Am. geophys. Union, 55*, p. 487.

JOYCE, E. B., 1971: A study of maars in the post-Miocene volcanics of southeastern South Australia (abstr.). *Abstr. ANZAAS 43rd Congress, Brisbane*, p. 72.

JUTSON, J. T., 1905: Notes on the volcanic history of Mount Shadwell. *Vict. Naturalist, 22*, p. 8.

KAY, R. W., & GAST, P. W., 1974: The rare earth content and origin of alkali-rich basalts. *J. Geol., 81*, pp. 653-682.

KESSON, S. E., 1972: Basic alkaline rocks. Unpub. Ph.D. thesis, Australian National University, Canberra.

LOVERING, J. F., & WHITE, A. J. R., 1969: Granulitic and eclogitic inclusions from basic pipes at Delegate, Australia. *Contr. Miner. Petrol., 21*, pp. 9-52.

MAALØE, S., 1973: Temperature and pressure relations of ascending primary magmas. *J. geophys. Res., 78*, pp. 6877-6886.

MACDONALD, G. A., & KATSURA, T., 1964: Chemical composition of Hawaiian lavas. *J. Petrol., 5*, pp. 82-133.

MAHONY, D. J., 1931: Alkaline Tertiary rocks near Trentham and at Drouin, Victoria. *Proc. Roy. Soc. Vict., 43*, pp. 123-129.

MCDOUGALL, I., ALLSOPP, H. D., & CHAMALAUN, F. H., 1966: Isotopic dating of the Newer Volcanics of Victoria, Australia, and geomagnetic polarity epochs. *J. geophys. Res., 71*, pp. 6107-6118.

MCINERNY, K., 1929: The building stones of Victoria, Part II. The igneous rocks. *Proc. Roy. Soc. Vict., 41*, pp. 121-159.

MORGAN, W. R., 1968: The geology and petrology of Cainozoic basaltic rocks in the Cooktown area, north Queensland. *J. geol. Soc. Aust., 15*, pp. 65-78.

NICHOLLS, G. D., 1967: Geochemical studies in the ocean as evidence for the composition of the mantle, in RUNCORN, S. K. (ed.), *Mantles of the Earth and Terrestrial Planets*. Interscience, pp. 285-304.

OLLIER, C. D., 1967: Landforms of the Newer Volcanic Province of Victoria, in JENNINGS, J. N., & MABBUTT, J. A. (eds), *Landform Studies from Australia and New Guinea*. A.N.U. Press, pp. 315-339.

OLLIER, C. D., & JOYCE, E. B., 1964: Volcanic physiography of the Western Plains of Victoria. *Proc. Roy. Soc. Vict., 77*, pp. 357-376.

ORR, D., 1927: An olivine-anorthoclase-basalt from Daylesford. *Proc. Roy. Soc. Vict., 40*, pp. 34-44.

PRICE, R. C., & GREEN, D. H., 1972: Lherzolite nodules in a 'mafic phonolite' from north-east Otago, New Zealand. *Nature phys. Sci., 235*, pp. 133-134.

PRICE, R. C., & TAYLOR, S. R., 1973: The geochemistry of the Dunedin Volcano, East Otago, New Zealand: rare earth elements. *Contr. Miner. Petrol., 40*, pp. 195-205.

RAHMAN, A., & MCDOUGALL, I., 1972: Potassium-argon ages on the Newer Volcanics of Victoria. *Proc. Roy. Soc. Vict., 85*, pp. 61-69.

RINGWOOD, A. E., 1966: The chemical composition and origin of the Earth, in HURLEY, P. M. (ed.), *Advances in Earth Science*. M.I.T. Press, pp. 287-356.

ROEDER, P. L., & EMSLIE, R. F., 1970: Olivine-liquid equilibrium. *Contr. Miner. Petrol., 29*, pp. 275-289.

SAGGERSON, E. P., & WILLIAMS, L. A. J., 1964: Ngurumanite from southern Kenya and its bearing on the origin of rocks in the northern Tanganyika alkaline district. *J. Petrol., 5*, pp. 40-81.

SINGLETON, O. P., & JOYCE, E. B., 1969: Cainozoic volcanicity in Victoria. *Geol. Soc. Aust. spec. Publ. 2*, pp. 145-154.

SKEATS, E. W., & JAMES, A. V. G., 1937: Basaltic barriers and other surface features of the Newer Basalts of Western Victoria. *Proc. Roy. Soc. Vict., 49*, pp. 245-278.

SKEATS, E. W., & SUMMERS, H. S., 1912: The geology and petrology of the Macedon district. *Geol. Surv. Vict. Bull. 24*.

SOLOMON, M., 1952: The volcanic deposits of southeast South Australia as sources of quarry stone. *Min. Rev. (Adelaide), 93*, pp. 133-137.

Sprigg, R. C., 1952: The geology of the South-East Province, South Australia, with special reference to Quaternary coastline migrations and modern beach developments. *Geol. Surv. S. Aust. Bull., 29.*

Stanley, E. R., 1909: Complete analysis of the Mount Gambier Basalt, with petrographical descriptions. *Trans. Roy. Soc. S. Aust., 33,* pp. 82-100.

Stanley, E. R., 1910: Lherzolite and olivine from Mount Gambier. *Trans. Roy. Soc. S. Aust., 34,* pp. 63-68.

Stuckless, J. S., & Irving, A. J., 1976: Strontium isotope geochemistry of megacrysts and host basalts from southeastern Australia. *Geochim. cosmochim. Acta, 40,* pp. 209-213.

Stueber, A. M., 1969: Abundances of K, Rb, Sr and Sr isotopes in ultramafic rocks and minerals from western North Carolina. *Geochim. cosmochim. Acta, 33,* pp. 543-553.

Sutherland, F. L., 1969a: A review of the Tasmanian Cainozoic volcanic province. *Geol. Soc. Aust. spec. Publ., 2,* pp. 133-144.

Sutherland, F. L., 1969b: A comparison of the Cainozoic volcanic provinces of Victoria and Tasmania. *Proc. Roy. Soc. Vict., 82,* pp. 179-186.

Sutherland, F. L., 1974: High-pressure inclusions in tholeiitic basalt and the range of lherzolite-bearing magmas in the Tasmanian volcanic province. *Earth planet. Sci. Lett., 24,* pp. 317-324.

Wass, S. Y., 1971: Studies on some basaltic rocks in New South Wales. Unpub. Ph.D. thesis, University of Sydney, Sydney.

Wass, S. Y., & Irving, A. J., compilers, 1976: XENMEG: a catalogue of occurrences of xenoliths and megacrysts in volcanic rocks of eastern Australia. *Spec. Publ. Aust. Mus.,* in press.

Wellman, P., 1974: Potassium-argon ages on the Cainozoic volcanic rocks of eastern Victoria, Australia. *J. geol. Soc. Aust., 21,* pp. 359-376.

Wilkinson, J. F. G., 1962: Mineralogical, geochemical and petrogenetic aspects of an analcite-basalt from the New England District of New South Wales. *J. Petrol., 3,* pp. 192-214.

Wilkinson, J. F. G., 1968: Analcimes from some potassic igneous rocks and aspects of analcime-rich assemblages. *Contr. Miner. Petrol., 18,* pp. 252-269.

Wilkinson, J. F. G., & Binns, R. A., 1969: Hawaiite of high pressure origin from northeastern New South Wales. *Nature, Lond. 222,* pp. 553-555.

Yates, H., 1954: The basalts and granitic rocks of the Ballarat district. *Proc. Roy. Soc. Vict., 66,* pp. 63-101.

*Dr A. J. Irving,**
Dr R. H. Green,
Research School of Earth Sciences,
Australian National University,
P.O. Box 4, Canberra, A.C.T. 2600.

**Present address:*
The Lunar Science Institute,
3303 NASA Road 1,
Houston, Texas, U.S.A. 77058.

Part VII

CORRELATION OF BASALT TYPE AND TECTONIC ENVIRONMENT

Editors' Comments
on Papers 54 Through 58

54   SCHMIDT
Excerpt from *Petrology of the Volcanic Rocks*

55   JAKEŠ and WHITE
*Major and Trace Element Abundances in Volcanic Rocks of
Orogenic Areas*

56   ROGERS
Excerpts from *Criteria for Recognizing Environments of Formation of Volcanic Suites; Application of These Criteria to Volcanic Suites in the Carolina Slate Belt*

57   PEARCE and CANN
Excerpts from *Tectonic Setting of Basic Volcanic Rocks Determined Using Trace Element Analyses*

58   PEARCE and GALE
*Identification of Ore-Deposition Environment from Trace-
Element Geochemistry of Associated Igneous Host Rocks*

One principal impetus for the classification of basalts has always
been a correlation of rock type with conditions of magma formation.
The portion of the paper by Kennedy (1933) reproduced as Paper 1
clearly indicates efforts in this direction. The obvious intention has
been to find some property of a basalt that is unique to a particular
tectonic setting, thus allowing ancient rocks to indicate the tectonic
conditions of the place and time at which they formed. This type of
correlation can, at least in theory, be done without any effort to
understand the reasons for the development of particular rocks. In
this part, we discuss only the relationships between rock type and
environment. Although many of the references cited herein discuss
the petrologic evolution of basalts, our treatment of this topic has
been restricted to Part II.

There are two approaches to the topic. An apparently simple-
minded, but actually very complex, method is to list as many
compositional, mineralogical, textural, structural, stratigraphic, and

so forth properties as one can find for different basalt assemblages of known modern environments and attempt to find some property, or set of properties, that distinguishes rocks of one environment from those of all others. These distinguishing properties are then applied to the ancient rocks. This qualitative multivariate approach has the advantage of not omitting any information that might be useful and has the disadvantage of requiring the handling of tremendous amounts of data. The approach also assumes uniformitarianism (i.e., lack of variation) in the production of basalt compositions through time, which may not be a valid concept.

Far too many papers have used this qualitative approach to attempt even a listing of them. The papers cited here, and those referenced, contain useful summaries or particularly pertinent information that has not been included in other summaries.

One of the earliest, and most obvious, efforts was to distinguish the basalts of oceanic regions from those of continents. At one time, alkali basalts of oceanic islands were regarded as separated from SiO_2-saturated tholeiites by the andesite line (see diagram from Schmidt, Paper 54). In the 1950s and 1960s, however, the abundance of tholeiitic basalts in the ocean basins began to be recognized, with some important early papers being those of Powers (1955) on Hawaii and Engel and Engel (1964) on MORB. Further information on the distinctive character of ocean-margin and continental volcanism also accumulated during this period, with some of the more significant papers being the study of the Pacific Northwest of the United States by Waters (Paper 50) and work in the Aleutians summarized by Coats (1962). Green and Poldervaart (1955) attempted an early synthesis of variation in basalt types.

Later studies have, of course, added much more information. The papers we find most useful for determining the descriptive nature of basalts from different tectonic environments are as follows:

Mid-ocean ridge basalts: Kay, Hubbard, and Gast (1970), Glassley (1974), Pearce (1975), and Hart (1976). Hawkins (1977) and Wood et al. (1981) discussed the great similarity between basalts of open-ocean ridges and those of back-arc basins;

Oceanic islands: Aumento (1968), Schilling (1975), and Zielinski (1975);

Intra-oceanic island arcs: Jakeš and Gill (1970), Donnelly et al. (1971), Garcia (1978), Morrison (1980), and Perfit et al. (1980);

Continental margin subduction zones: Jakeš and White (Paper 55), and Miyashiro (1974);

Continental plateau basalts: Wright, Grolier, and Swanson (1973);

Continental rift zones: Mutschler, Finn, and Ludington (1978).

We should also call attention to Condie (1976), who summarized characteristics of volcanic rocks of different modern environments and applied and compared the distinctions to possible Archean equivalents, and Pearce (1976), who summarized major element compositions of many types of basalts. Jakeš and White (Paper 55) summarized the characteristics of subduction-related volcanic rocks from various settings. An attempt to synthesize all the possible discriminating properties of volcanic rock suites has recently been made by Rogers (Paper 56).

The other approach to identification of basalt suites is to use a limited amount of information—commonly, trace elements—in a quantitative fashion. The major thrust of this line of investigation has been to determine relationships between elements that are regarded as immobile during posteruptive processes including metamorphism, hydrothermal activity, and weathering. Whether all of the elements are as immobile as claimed is still an open question; for example, Smith and Smith (1976) have clearly shown that strontium, an element at one time regarded as immobile, is greatly affected by posteruptive processes.

One of the earliest major efforts to use immobile elements for basalt discrimination was by Chayes (1965). Chayes particularly noted the tendency of intra-oceanic basalts to have higher TiO_2 contents than basalts from continental margins. More recent work has been conducted largely by British geochemists (e.g., Pearce and Cann, 1973; Pearce, 1975; Floyd and Winchester, 1978; Pearce and Norry, 1979; Wood, Joron, and Treuil, 1979; and Thompson et al., 1980). Elements used include Ti, Cr, Sr, Y, Zr, Nb, and the recently proposed Th-Hf-Ta triad. The classic paper in this approach is by Pearce and Cann (1973), and we have reproduced some significant diagrams from it as Paper 57. A paper that synthesizes much of this work is by Pearce and Gale (Paper 58).

A word of caution should be inserted concerning the use of immobile-element diagrams in the absence of other information. Many of the distinctions in these diagrams are for rock types rather than for identifiable tectonic situations. For example, several diagrams identify fields of calc-alkali basalts, and there is a tendency to assume that this term is equivalent to subduction-zone basalts; in fact, the correspondence may not be that close. Furthermore, many diagrams refer to within-plate basalts, which could include oceanic islands, continental volcanic centers, and plumes. Thus, a position on a diagram may not uniquely specify a tectonic environment of formation. In addition to this caution, we should reiterate two other problems mentioned earlier—namely, the problem of posteruptive alteration of volcanic rocks and the hopeful assumption that, say, a tholeiite from a

Paleozoic island arc has the same composition as one from a modern arc.

Finally, in our opinion, field petrologists interested in the environment of formation of volcanic suites need to pay more attention to experimental results such as those summarized in Part II. These studies allow us to place constraints on conditions of basalt magma genesis and later modification, and these restrictions are invaluable in relating basalt compositions to tectonic environments. The summary paper by Green (1969) and the book by Yoder (1976) are outstanding efforts in this regard.

REFERENCES

Aumento, F., 1968, The Mid-Atlantic Ridge Near 45°N: II. Basalts from the Area of Confederation Peak, Canadian *Jour. Earth Sci.* **5:**1–21.

Chayes, F., 1965, Titania and Alumina Content of Oceanic and Circumoceanic Basalt, *Mineralog. Mag.* **34:**126–131.

Coats, R. R., 1962, Magma Types and Crustal Structure in the Aleutian Arc, in *The Crust of the Pacific Basin,* G. A. MacDonald and H. Kuno, eds., American Geophysical Union Monograph 6, Washington, D.C., pp. 92–109.

Condie, K. C., 1976, Trace-Element Geochemistry of Archean Greenstone Belts, *Earth Sci. Rev.* **12:**393–417.

Donnelly, T. W., J. J. W. Rogers, P. Pushkar, and R. L. Armstrong, 1971, Chemical Evolution of the Igneous Rocks of the Eastern West Indies—An Investigation of Thorium, Uranium, and Potassium Distributions, and Lead and Strontium Isotopic Ratios, *Geol. Soc. America Mem. 130,* pp. 181–224.

Engel, A. E. J., and C. G. Engel, 1964, Composition of Basalts from the Mid-Atlantic Ridge, *Science* **144:**1330–1333.

Floyd, P. A., and J. A. Winchester, 1978, Identification and Discrimination of Altered and Metamorphosed Volcanic Rocks Using Immobile Elements, *Chem. Geology* **21:**291–306.

Garcia, M. D., 1978, Criteria for the Identification of Ancient Volcanic Arcs, *Earth Sci. Rev.* **14:**147–165.

Glassley, W., 1974, Geochemistry and Tectonics of the Crescent Volcanic Rocks, Olympic Peninsula, Washington, *Geol. Soc. America Bull.* **85:**785–794.

Green, D. H., 1969, The Origin of Basaltic and Nephelinitic Magmas in the Earth's Mantle, *Tectonophysics* **7:**409–422.

Green, J., and A. Poldervaart, 1955, Some Basaltic Provinces, *Geochim. et Cosmochim. Acta* **7:**177–188.

Hart, S. R., 1976, Chemical Variance in Deep Ocean Basalts, *Initial Reports of the Deep Sea Drilling Project* **34:**301–335.

Hawkins, J. W., Jr., 1977, Petrologic and Geochemical Characteristics of Marginal Basin Basalts, in *Island Arcs, Deep Sea Trenches, and Back-Arc Basins,* M. Talwani and W. C. Pitman, III, eds., Maurice Ewing Series, vol. 1, American Geophysical Union, Washington, D.C., pp. 355–365.

Jakeš, P., and J. B. Gill, 1970, Rare Earth Elements and the Island Arc Tholeiitic Series, *Earth and Planetary Sci. Letters* **9:**17–28.

Kay, R., N. J. Hubbard, and P. W. Gast, 1970, Chemical Characteristics and Origin of Oceanic Ridge Volcanic Rocks, *Jour. Geophys. Research* **75:**1585-1613.

Kennedy, W. Q., 1933, Trends of Differentiation in Basaltic Magmas, *Am. Jour. Sci.,* 5th series, **25:**239-256.

Miyashiro, A., 1974, Volcanic Rock Series in Island Arcs and Active Continental Margins, *Am. Jour. Sci.* **274:**321-355.

Morrison, G. W., 1980, Characteristics and Tectonic Setting of the Shoshonite Rock Association, *Lithos* **13:**97-108.

Mutschler, F. E., D. D. Finn, and S. Ludington, 1978, Magmatism and Related Ore Deposits of Extensile Continental Environments—Computer Exercises in Petrochemical Pattern Recognition and Prediction, *Program and Abstracts of 1978 International Symposium on the Rio Grande Rift, Santa Fe, New Mexico, Los Alamos Scientific Laboratory,* pp. 64-65.

Pearce, J. A., 1975, Basalt Geochemistry Used to Investigate Past Tectonic Environments on Cyprus, *Tectonophysics* **25:**41-67.

Pearce, J. A., 1976, Statistical Analysis of Major-Element Patterns in Basalts, *Jour. Petrology* **17:**15-43.

Pearce, J. A., and J. R. Cann, 1973, Ophiolite Origin Investigated by Discriminant Analysis Using Ti, Zr and Y, *Earth and Planetary Sci. Letters* **12:**339-349.

Pearce, J. A., and M. J. Norry, 1979, Petrogenetic Implications of Ti, Zr, Y, and Nb Variations in Volcanic Rocks, *Contr. Mineralogy and Petrology* **69:**33-47.

Perfit, M. R., D. A. Gust, A. E. Bence, R. J. Arculus, and S. R. Taylor, 1980, Chemical Characteristics of Island-Arc Basalts—Implications for Mantle Sources, *Chem. Geology* **30:**227-256.

Powers, H. A., 1955, Composition and Origin of Basaltic Magma of the Hawaiian Islands, *Geochim. et Cosmochim. Acta* **7:**77-107.

Schilling, J. G., 1975, Azores Mantle Blob—Rare Earth Evidence, *Earth and Planetary Sci. Letters* **25:**103-115.

Smith, R. E., and S. E. Smith, 1976, Comments on the Use of Ti, Zr, Y, Sr, K, P, and Nb in Classification of Basaltic Magmas, *Earth and Planetary Sci. Letters* **32:**114-120.

Thompson, R. N., M. A. Morrison, D. P. Mattey, A. P. Dickin, and S. Moorbath, 1980, An Assessment of the Th-Hf-Ta Diagram as a Discriminant for Tectonomagmatic Classifications and in the Detection of Crustal Contamination of Magmas, *Earth and Planetary Sci. Letters* **50:**1-10.

Wood, D. A., V. -L. Joron, and M. Treuil, 1979, A Re-appraisal of the Use of Trace Elements to Classify and Discriminate between Magma Series Erupted in Different Tectonic Settings, *Earth and Planetary Sci. Letters* **45:**326-336.

Wood, D. A., D. P. Mattey, V.- L. Joron, N. G. Marsh, J. Tarney, and M. Treuil, 1981, A Geochemical Study of 17 Selected Samples from Basement Cores Recovered at Sites 447, 448, 449, 450, and 451, Deep Sea Drilling Project Leg 59, *Initial Reports of the Deep Sea Drilling Project* **59:**743-752.

Wright, T. L., M. J. Grolier, and D. A. Swanson, 1973, Chemical Variation Related to the Stratigraphy of the Columbia River Basalt, *Geol. Soc. America Bull.* **84:**371-386.

Yoder, H. S., Jr., 1976, *Generation of Basaltic Magma,* National Academy of Sciences, Washington, D.C., 265p.

Zielinski, R. A., 1975, Trace Element Evolution of a Suite of Rocks from Reunion Island, Indian Ocean, *Geochim. Cosmochim. Acta* **39:**713-734.

Reprinted from page 128 of *U.S. Geol. Survey Prof. Paper 280-B,* pp. 127–175 (1957)

PETROLOGY OF THE VOLCANIC ROCKS

Robert George Schmidt

[*Editors' Note:* Only Figure 11 is reproduced here.]

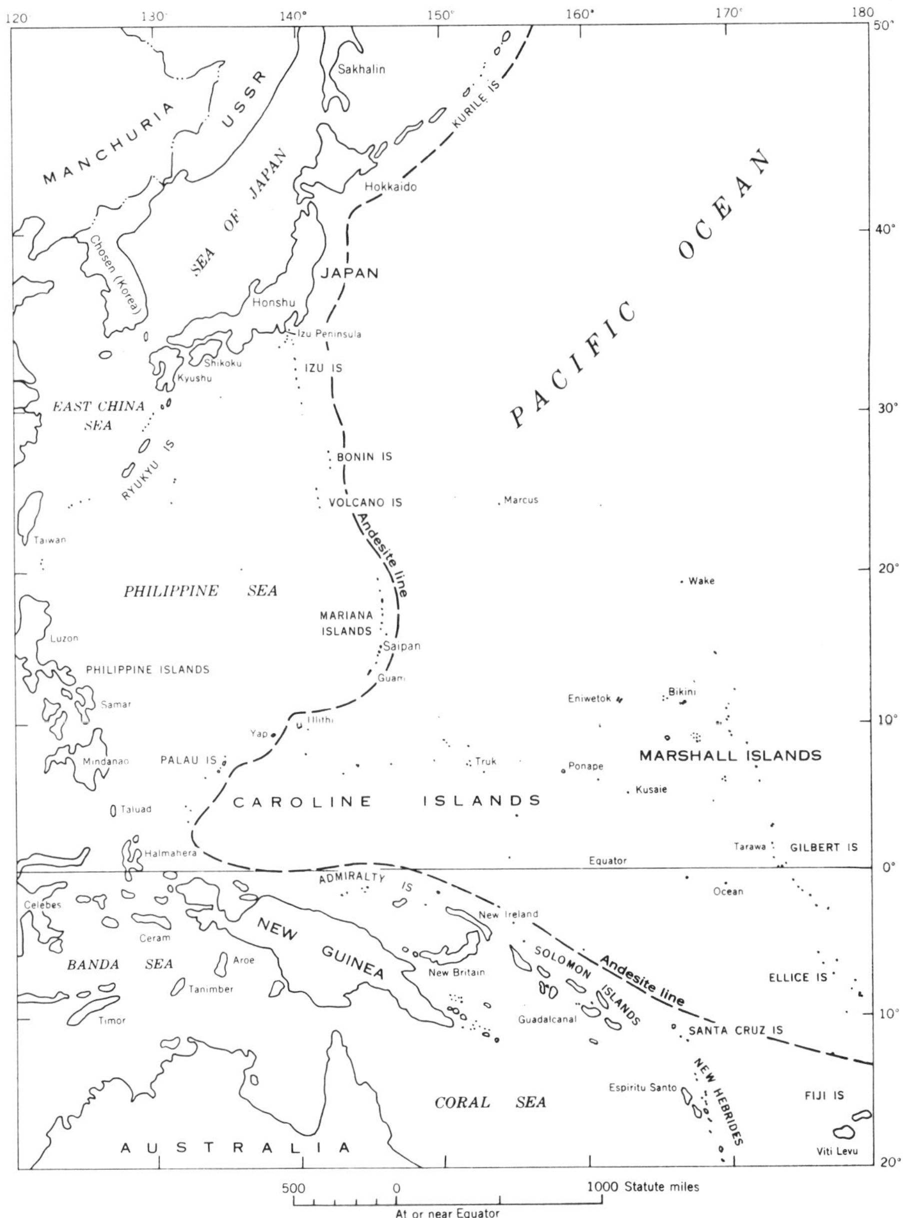

FIGURE 11.—Index map of the western north Pacific Ocean.

55

P. JAKEŠ *The Lunar Science Institute, 3303 Nasa Road 1, Houston, Texas 77058*

A. J. R. WHITE *Department of Geology, The Australian National University, Canberra, A. C. T., Australia*

Major and Trace Element Abundances in Volcanic Rocks of Orogenic Areas

ABSTRACT

Major and trace element abundances in volcanic rocks of island arcs vary regularly in both the lateral and in the stratigraphic sequence. The earliest manifestations of island arc evolution are tholeiitic rocks occurring on the oceanic side of recent island arcs. These tholeiites are characterized by chemical features such as low K_2O content and relatively high iron enrichment, that is, high FeO/MgO in medium SiO_2 (59 percent) rocks. The tholeiites are followed stratigraphically or laterally by calc-alkaline rocks and finally by shoshonites or alkaline rocks, showing progressive increase in K_2O content, increase in K_2O/Na_2O ratio, and decreasing iron enrichment. Trace elements of the potassium type are the most sensitive indicators of compositional variation showing differences between tholeiites and shoshonites by two orders of magnitude. Trace elements and major elements suggest a continuous gradational sequence from tholeiites through calc-alkaline rocks to shoshonites. Rare-earth elements show two diverse abundance patterns: (1) a primitive chondritic-like pattern is characteristic for island arc tholeiitic association; and (2) a fractionated light rare-earth elements enriched pattern is typical for calc-alkaline and shoshonitic and alkaline associations.

INTRODUCTION

The mechanism of continental growth (Hamilton, 1970) by the addition of primary but relatively highly fractionated sialic material has been attributed to andesitic volcanism in island arcs (Wilson, 1952; Taylor and White, 1965, 1966; Taylor, 1967; Dickinson, 1968; Green and Ringwood, 1968; Ringwood, 1969; and others). Island arcs have, in general, been considered to be piles of andesites with subordinate basalts or dacites, and of sediments derived from them. Chemical diversity of island-arc lavas has recently been pointed out (Kuno, 1950, 1966; Sugimura, 1968; Jakeš and White, 1969, 1970; Donnelly and others, 1972; Jakeš and Gill, 1970), and chemical features such as K_2O content have been correlated with the seismic evidence on island arc structure (Dickinson, 1968; Dickinson and Hatherton, 1967).

Seismic data (Sykes, 1966; Isacks and others, 1968) together with morphology, gravity, and heat-flow measurements support the model in which large volumes of oceanic crust are consumed under island arcs (Isacks and others, 1968). Island-arc volcanism is a result of the tectonic interaction of oceanic lithosphere and continental lithosphere. In this paper, we discuss the composition of recent island-arc volcanic rocks in relation to their geographical as well as their stratigraphic relations and point out differences in composition between volcanic rocks in arcs and those in continental margins.

ISLAND-ARC ROCK ASSOCIATIONS

We here classify volcanic rocks in island arcs on the basis of chemical composition following the suggestions of Taylor and White (1966). Such classification is used to avoid mineralogical and textural features which reflect a mode of emplacement, type of eruption, and subsequent cooling history. Comparison of recent island-arc rocks and those of older orogenic areas (even those slightly metamorphosed) thus can be made, and the classification also avoids genetic terms (for example, spilitic pillow lava).

Principal divisions of the island-arc volcanic rocks are based on major-element chemistry, particularly the K_2O versus SiO_2 relation, total alkali contents, K_2O/Na_2O ratio, or "iron enrichment" (in AMF diagrams). These features define a rock association (Jakeš and Gill, 1970). In each association, there is a wide range of SiO_2 content which is used to subdivide the association (Table 1). Basalts have a lower SiO₂ content than 52 percent; andesites range from 52 to 62 percent with a subgroup of low-Si (basaltic) andesites with 52 to 56 percent, and dacites with SiO₂ higher than 63 percent. In this classification, the name of the association is used along with the name of the rock, for example, tholeiitic dacite, or shoshonitic basalt. Boundaries between each rock association are as artificial as boundaries between each rock type within a particular association. In the island arcs, there is a continuum from one arc association to another, but it is the regularity of change in time and space that is discussed here.

VARIATIONS OF ROCK COMPOSITION ACROSS ISLAND ARCS

Sugimura (1961, 1968) reviewed the work of Yamasaki, Kuno, Moore, Rittman, and others on the variation of rocks across island arcs. In his 1961 paper, Sugimura related K₂O content and depth of earthquake foci, but Kuno's (1966) scheme of parental basalt variation from tholeiitic through high-alumina to alkali-olivine, and its relation to depth of earthquake foci have been widely accepted. Dickinson and Hatherton (1967) and Dickinson (1968) have developed the idea further. In those island arcs where the seismic zone dips beneath the arc with depth to the Benioff zone increasing from the trench toward the continent, recent volcanics show an increase in K₂O at a given SiO₂ content. This correlation between rock

chemistry and depth to the Benioff zone must be emphasized; the distance of the respective centers from the trench is not a critical feature, for this is simply related to the dip of the Benioff zone, which ranges from 25° for the Indonesian part of New Guinea, through 45° in the Tonga-Kermadec region, to vertical at Bougainville (Denham, 1969).

The slope of the K₂O versus SiO₂ plot also increases across an island arc toward the continent (Fig. 1). This is supported by observations of Dickinson (1968) and Gill (1970), but there are exceptions. For example, in the high-K calc-alkaline rocks described by Jakeš and Smith (1970), the gradient of K₂O versus SiO₂ is relatively low, and shoshonites may have very flat or even negative gradients (Joplin, 1968; Ruxton, 1966).

Na₂O does not change as much as K₂O in sections across island arcs with a given SiO₂ content, and consequently the K₂O/Na₂O ratio increases with increasing K₂O from approximately 0.1 in the low-K rocks on the oceanic side of the arc, to 1.0 in high-K rocks (shoshonites) on the continental side but may reach 3 or 4 (western Kamchatka; Erlikh, 1968).

Rocks of each of the three associations (that is, tholeiitic, calc-alkaline, shoshonitic) show a wide range of SiO₂, and this permits evaluation of Fe-Mg variations. Island-arc tholeiites are similar to those of tholeiites of continental and oceanic areas but the "iron enrichment" is less than that of abyssal tholeiites (Jakeš and Gill, 1970). The degree of "iron enrichment"

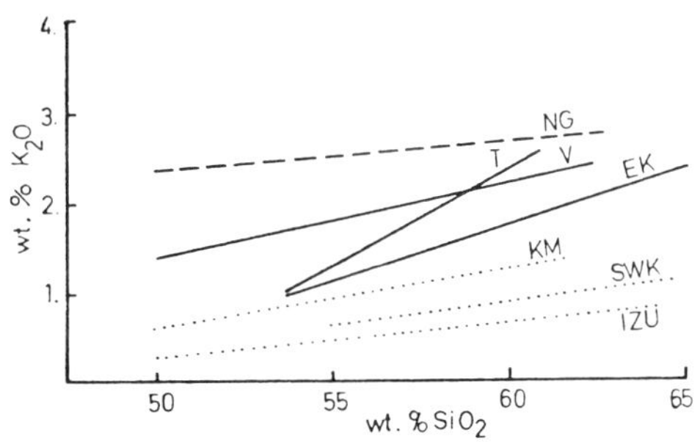

Figure 1. Correlation of SiO₂ versus K₂O wt percent in volcanic rocks of island arcs. Dotted lines = tholeiitic associations: KM = Karkar and Manam Islands, New Guinea, SWK = South West Kurile Islands, IZU = Izu islands and peninsula; full lines = calc alkaline associations: EK = Eastern Kamchatka, T and V = Mount Trafalgar and Mount Victory, Cape Nelson Eastern Papua; dashed line = shoshonitic association New Guinea Highlands.

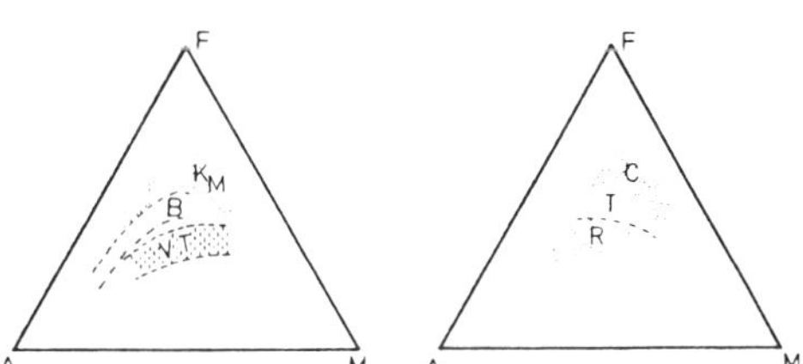

Figure 2. A (Na₂O + K₂O), M (MgO), F (FeO + .9 Fe₂O₃) generalized plot for volcanic rocks of island arcs. Left triangle = examples from Melanesia: tholeiitic association: KM = Karkar and Manam Island calc-alkaline associations: B = Bougainville, and V T = Mount Victory and Trafalgar Eastern Papua. Shoshonitic association of New Guinea plots within V-T field. In right triangle volcanic rocks from Guadalcanal (Solomon Islands), of different age: C = Cretaceous, T = Miocene, R = Recent (B. Hackman, 1969, personal commun.).

in an island-arc suite decreases as the depth of the underlying Benioff zone increases. Rocks farthest from the oceanic side of the arc (for example, high-K calc-alkaline or shoshonitic rocks) have virtually no "iron enrichment" (Fig. 2).

The rocks of island arcs are generally high in Al_2O_3, irrespective of SiO_2 content. The typical calc-alkaline, low- and medium-K rocks are fairly constant in Al_2O_3, usually not less than 16.0 percent and not more than 17.5 percent over most of the SiO_2 range. In island-arc tholeiites, Al_2O_3 is in the 14 to 17.5 percent range. In high-K rocks, there is a fluctuation between low (14 percent) and extremely high values (19 percent).

In Kuno's (1966) scheme of variation of composition in Cenozoic island-arc lavas, there is a substantial difference in TiO_2 between tholeiitic basalts (low values) and alkali basalts (high values). If rocks of the same SiO_2 content are compared, there is a wide spread of values in calc-alkaline rocks and the shoshonitic association appears to have higher titanium contents than both tholeiitic and calc-alkaline.

Although the ratio of ferrous to ferric iron is affected by near-surface oxidation, it appears that the proportion of more oxidized rocks increases away from the trench.

In each association (tholeiitic, calc-alkaline, shoshonitic), rocks with a wide range of SiO_2 values occur; however, proportions of rocks having a particular SiO_2 content are different for each association. Basaltic rocks with less than 54 percent SiO_2 are the most abundant rocks of the tholeiitic association, basaltic andesites with about 56 percent SiO_2 are very frequent in low-K calc-alkaline associations,

TABLE 1. NOMENCLATURE AND CHARACTERISTICS OF PRINCIPAL VOLCANIC ROCKS FROM ISLAND ARCS IN THE RANGE 50 TO 65 PERCENT SiO_2

	— Increasing K_2O at given SiO_2 ⟶	
	— Increasing K_2O/Na_2O ⟶	
	— Decreasing "iron enrichment" ⟶	

| SiO_2 wt. percent | Rock association | | |
	Tholeiitic	Calc-alkaline (low-K, high-K)	Shoshonitic
50	Tholeiitic	Calc-alkaline	Shoshonitic
51	basalt	basalt	basalt
52	(tholeiite)	(high-alumina basalt)	
53			(absarokite)
54			
55	Tholeiitic	Low-Si andesite	
56	andesite		(shoshonite)
57	(icelandite)	(basaltic andesite)	Shoshonitic
58			andesite
59		Andesite	(banakite)
60			
61			
62			
63	Tholeiitic	Dacite	Shoshonitic
64	dacite		dacite
65			(latite, quartz latite)

Tholeiitic association: Basaltic rocks with high Al_2O_3 (14 to 18 percent) and less than 4 percent of $Na_2O + K_2O$ occur most frequently. K_2O is low and usually does not exceed 1.2 percent for SiO_2-rich rocks (58 percent). The slope of K_2O versus SiO_2 is only 0 to 0.3 percent K_2O per 5 percent SiO_2. The K_2O/Na_2O ratios are low, usually less than 0.35; the FeO/MgO ratio varies widely and "iron enrichment" (in AMF diagram) is characteristic in medium SiO_2 members (icelandites or tholeiitic andesites). In most rocks, Fe_2O_3/FeO does not exceed 0.5. Tholeiitic association generally corresponds to Kuno's (1950) "pigeonitic" series, whereas "hypersthenic" series corresponds to calc-alkaline series.

Calc-alkaline association: Al_2O_3 content is high (15 to 19 percent), although lower values (15 to 16.5 percent) occur in high SiO_2 (>63 percent) rocks and wider variation in Al_2O_3 content is characteristic of high-K rocks. Total alkali content is moderate to high and increases gently with increasing SiO_2. FeO/MgO ratio is almost constant, and there is little or no "iron enrichment." The K_2O/Na_2O ratio varies between 0.35 to 0.75. K_2O varies from relatively low (1 percent) to high (3 percent) at 58 percent SiO_2, and accordingly calc-alkaline rocks can be divided into low-K and high-K types. Slopes of K_2O versus SiO_2 variation are steeper in K-rich rocks and within the whole association vary from 0.3 to 0.8 percent K_2O per 5 percent SiO_2.

Shoshonitic association: The shoshonitic association is still poorly documented partly because of its confusion with alkaline basalt series. There is a high, but very variable Al_2O_3 content (14.5 to 20.0 percent) in all shoshonites. A high content of total alkali elements (>5 percent) irrespective of SiO_2 is primarily due to high K_2O. The K_2O/Na_2O ratio is high, fluctuating around unity. As in calc-alkaline associations, the FeO/MgO ratio is almost constant and rocks do not show any clearly defined "iron enrichment." The slope of K_2O versus SiO_2 correlation is very steep in the lower SiO_2 region (48 to 52 percent SiO_2), but in the range 52 to 60 percent SiO_2, the slope approaches zero, and in some rocks may be negative (Jakeš and Smith, 1970; Ruxton, 1966).

and the frequency of types richer in SiO_2 increases with increasing K_2O. However, in shoshonitic rocks, the more basic members with 54 percent SiO_2 are again most prominent.

VARIATIONS IN STRATIGRAPHIC SEQUENCE OF VOLCANIC ROCKS IN ISLAND ARCS

The complete variation in lava composition across island arcs, from tholeiitic to calc-alkaline and finally to shoshonitic associations, is a feature of island arcs of an advanced stage of evolution (old island arcs) with a clearly definable seismic plane dipping toward the continent (for example, Kamchatka, Japan, Sunda Island, New Guinea, New Zealand). Jakeš and White (1969) and Gill (1970) suggested that in regions where the seismic plane is not clearly developed or where it is vertical or dips toward the oceanic side (for example, Solomon Islands, New Hebrides, Fiji), a similar compositional variation is shown in the stratigraphy of the volcanic pile. In general, tholeiitic rocks are the earliest manifestation of island-arc evolution (Baker, 1968) and are followed

in time by calc-alkaline rocks and finally by shoshonites (Gill, 1970).

Such progressive change of character of volcanism is also supported by detailed study of the evolution of one volcano over a period of time. Aramaki (1963) has observed an increase in total alkalis (at given SiO_2 contents) of the Asama lavas with increasing age. Gorshkov (1962) pointed out similar evolutionary trends of Kurile-Kamchatka lavas. Erlikh's (1968) data show an increase of K_2O content and decrease of "iron enrichment" with decreasing age in the lavas of Kamchatka. Jakeš and Smith (1970) demonstrate that the increase of K_2O in calc-alkaline rocks of Cape Nelson correlates with decreasing age and trends of K_2O versus SiO_2 in the most recent lavas resemble those of shoshonites.

Gill's (1970) account of the geochemical evolution of Fiji provides an example of the parallelism between the lateral and the stratigraphic progression of lava compositions. In Fiji, the increase of slope for K_2O versus SiO_2 trendline correlates with the decreasing age of the volcanic rocks. Similar conclusions have

TABLE 2A. MAJOR ELEMENT ABUNDANCES IN VOLCANIC ROCKS OF ISLAND ARCS*

| | Island arc tholeiites | | | Calc-alkaline rocks | | | | | | Shoshonitic | | |
	Basalt	Tholeiitic andesite	Tholeiitic dacite	High-Al basalt	Low-Si andesite	Low-K andesite	Andesite	High-K andesite	Dacite	Shoshonite	Latite	Abyssal tholeiite
	1	2	3	4	5	6	7	8	9	10	11	12
SiO_2	51.57	57.40	79.20	50.59	54.54	59.05	59.64	58.52	66.80	53.74	59.27	49.34
TiO_2	.80	1.25	.23	1.05	1.13	.69	.76	.76	.23	1.05	.56	1.49
Al_2O_3	15.91	15.60	11.10	16.29	16.26	17.07	17.38	16.20	18.24	15.84	15.90	17.04
Fe_2O_3	2.74	3.48	.52	3.66	2.31	3.90	2.54	2.93	1.25	3.25	2.22	1.99
FeO	7.04	5.01	.90	5.08	5.40	2.57	2.72	3.28	1.02	4.85	3.19	6.82
MnO	.17	..	..	.17	.12	.15	.09	.09	.06	.11	.10	.17
MgO	6.73	3.38	.36	8.96	6.97	3.25	3.95	4.14	1.50	6.36	5.45	7.19
CaO	11.74	6.14	2.06	9.50	7.50	7.09	5.92	5.59	3.17	7.90	5.90	11.72
Na_2O	2.41	4.20	3.40	2.89	3.64	3.80	4.40	3.64	4.97	2.38	2.67	2.73
K_2O	.44	.43	1.58	1.07	1.49	1.27	2.04	2.67	1.92	2.57	2.68	.16
P_2O_5	.11	.44	..	.21	.23	.20	.28	.25	.09	.54	.41	.16
H_2O	.45	..	..	.81	1.31	.64	1.08	1.47	.26	1.09	1.44	..

* Examples are from Melanesia.
1. Tholeiitic basalt, Lake Dakataua, Talasea, New Britain (Lowder and Carmichael, 1970).
2. Tholeiitic andesite, Viti Levu, Fiji (Gill, 1970).
3. Tholeiitic dacite, Saipan (Taylor, 1969).
4. High-Al basalt, north slope of Mt. Trafalgar, Cape Nelson, East Papua (Jakeš and Smith, 1970).
5. Low-Si andesite, south slope of Mt. Trafalgar, Cape Nelson, East Papua (Jakeš and Smith, 1970).
6. Low-K andesite, Mau Quarry, Viti Levu, Fiji (Jakeš, 1970).
7. Andesite, lava dome of Mt. Lamington, East Papua (Jakeš and White, 1969).
8. High-K andesite, northwestern slope of Mt. Trafalgar, Cape Nelson, East Papua (Jakeš and Smith, 1970).
9. Dacite, Savo volcano, Guadalcanal, Solomon Islands (Jakeš and White, 1969).
10. Shoshonite, Gumnach River, Mt. Hagen, New Guinea Highlands (Jakeš and White, 1969).
11. Latite, Tambul, Mt. Giluwe, New Guinea Highlands (Jakeš and White, 1969).
12. Abyssal tholeiite (Engel, 1965).

been reached by Donnelly and others (1970), who showed that the tholeiitic volcanism of the Lesser Antilles gives way with time to calc-alkaline volcanism. Similarly Hackman (1970, personal commun.) has found a decrease in the "iron enrichment" in volcanic rocks of Guadalcanal (Solomon Islands) from Miocene to Holocene lavas. Jolly recently (1971) presented another example of high-K rocks (shoshonitic) with little or no iron enrichment relative to magnesium following the calc-alkaline association in Lesser Antilles.

The concept of progressive evolution of island arcs from tholeiitic to tholeiitic plus calc-alkaline is supported by the fact that island arcs in which only tholeiitic or low-K calc-alkaline rocks occur (for example, Tonga-Kermadec, Izu-Marianas, central Kurile Islands, South Sandwich Islands) are associated with younger rocks (generally post-Mesozoic). Island arcs, where tholeiitic, calc-alkaline, and high-K calc-alkaline rocks occur together, contain pre-Miocene or pre-Cretaceous geologic elements (for example, Solomon Islands, New Hebrides, Aleutian Islands). When shoshonitic rocks occur (Kamchatka, Japan, Sunda Islands, or New Guinea) the age of basement rocks and associated sediments is usually pre-Mesozoic. In these highly developed arcs, there may be suites of acidic calc-alkaline rocks (here called Andean).

This variation suggests that the basement of island arcs is formed from arc tholeiitic rocks, very often of submarine emplacement, along with complementary ultramafic rocks as suggested from geophysical evidence (Kuno, 1968). In ideal cases, the tholeiitic rocks are associated with deep-water sediments, and are overlain by calc-alkaline rocks associated with shallow-water sediments derived essentially from the tholeiitic and later calc-alkaline rocks (orogenic graywackes). It should be emphasized that lower-K-volcanic activity in any stage usually does not cease. The further stages of arc evolution are characterized by the presence of higher-K (calc-alkaline or shoshonitic) rocks with concurrent eruptions of lower-K rocks on the oceanic side. In the latest stage, mainly acidic (andesite-dacite) calc-alkaline rocks appear.

TRACE ELEMENT VARIATIONS IN ISLAND-ARC ROCKS

Each rock association in an island arc also has characteristic abundances of trace elements. Trace elements show a positive or negative correlation with SiO_2 content and comparison between rock associations has, therefore, been made (where possible) between rocks of the same level of SiO_2 content.

In general, the potassium-type elements (for example, Rb, Ba, Sr, and Pb in island-arc rocks) show the largest variations in a geographic (from the oceanic to the continental side of an arc) (*compare* Hart and others, 1970) and stratigraphic (from older to younger units) sense (Table 2). Tholeiitic rocks of island arcs have very low Rb values, ranging from 2 to 30 ppm over a range of SiO_2 from 49 to 75 percent. Rocks with low SiO_2 have an average of 5

TABLE 2B.　TRACE ELEMENT ABUNDANCES IN VOLCANIC ROCKS FROM ISLAND ARCS

	Island arc tholeiites			Calc-alkaline association			Shoshonitic association		
	Basalt	Andesite	Dacite	Basalt	Andesite	Dacite	Basalt	Andesite	Dacite
SiO_2	52%	58%	63%	52%	58%	63%			
Rb	5.0	6.0	15	10	30	45	75	100	120
Ba	75	100	175	115	270	520	1,000	850	900
Sr	200	220	90	330	385	460	700	850	850
K/Rb	1,000	890	870	340	430	380	200	200	200
La	1.1	2.4	5.5	9.6	11.9	14	14	18	..
Ce	2.6	..	15	19	24	19	28	35	..
Y	..	..	23	20	21	20	..	..	..
Yb	1.4	2.4	2.7	2.7	1.9	1.4	2.1	1.2	..
La/Yb	1.0	1.0	1.9	3.5	6.2	10	6.6	15	..
Th	.5	.31	1.6	1.1	2.2	1.7	2	2.8	
U	.15	.34	.85	.2	.7	.6	1.0	1.3	
Th/U	1.6	.9	1.88	5.9	3.2	2.7	2.0	2.1	
Ni	30	20	1	25	18	5	20	..	..
V	270	175	19	255	175	68	200	..	..
Cr	50	15	4	40	25	13	30	..	..
Zr	70	70	125	100	110	100	50	150	200
Hf	1.0	1.0	2.6	2.6	2.3	3.8	1.0	3.2	3.8

Data in p.p.m.

Data in Table 2B were compiled from different sources and do not correspond to rocks in Table 2A. Trace element values are preferred values which emerge from compilation of data of other authors, namely: Prinz, 1968; Taylor, 1968; Taylor et al., 1969; Jakeš and Gill, 1970; Gill, 1970; and unpublished data by these authors.

ppm of Rb which is still twice the value of abyssal tholeiites (Kay and others, 1970). Low-K calc-alkaline rocks (55 percent SiO_2) have a higher Rb content (10 ppm) than rocks of the arc tholeiitic series with the same SiO_2 content. The abundance of Rb increases through medium- to high-K calc-alkaline rocks, and Taylor (1969) gives 31 ppm as an average value for andesite (59 percent SiO_2) with 1.6 percent K_2O. In the shoshonitic association, the content of Rb is high, with 80 ppm in rocks with 54 percent SiO_2 (for example, Nicholls and Carmichael, 1969; Jakeš and White, 1970). In more siliceous rocks, the Rb content increases to values of more than 100 ppm; Jolly (1971) gives values to 150 ppm.

Potassium-rubidium ratios have been discussed in detail by Jakeš and White (1970). They suggested that the highest K/Rb ratio in a given association decreases from 1,000 in island-arc tholeiites through calc-alkaline rocks with values of about 500, to 250 in shoshonites.

Barium behaves in a way similar to that of rubidium. Values between 50 and 100 ppm are characteristic of low-K (0.2 to 0.4 percent K_2O) basaltic rocks of the island-arc tholeiite association (Jakeš and Gill, 1970; Gill, 1970; A. Ewart, 1970, personal commun.; see Table 2). In typical calc-alkaline rocks (59 percent SiO_2), values of about 800 ppm are common (Jakeš and Smith, 1970). In calc-alkaline rocks with SiO_2 near 63 percent, Ba values are 1,000 ppm. Shoshonitic rocks with about 54 percent SiO_2 have Ba contents near 600 ppm, but more siliceous types (59 percent) have Ba contents in excess of 1,000 (Nicholls and Carmichael, 1969). In some high-K rocks (for example, in Eastern Papua), the crystallization of a mica phase distorts the pattern of Ba as well as Rb variation, and even a decrease of Ba with increasing SiO_2 can be observed (Jakeš and Smith, 1970).

The lead content increases from the tholeiites to high-K calc-alkaline types (Table 2). Values near 4 ppm for low-K calc-alkaline andesites and 10 to 15 ppm for high-K calc-alkaline andesites are typical. Data on the Pb contents of shoshonitic rocks are limited, but it appears that they are similar to high-K andesites (a shoshonitic rock from the New Guinea Highlands has 16 ppm at 59 percent SiO_2). The "least fractionated" or more basic shoshonitic rocks have very low Pb values (3 to 5 ppm).

The complexity of Sr variation and its abundances result from the strong fractionation of Sr by feldspar. Kolbe and Taylor (1966),

Hall (1967), Rhodes (1969), and others found a decrease of Sr with increasing SiO_2 in calc-alkaline plutonic rocks. Similar trends are observed in island-arc rocks within the tholeiitic association, but in typical island-arc calc-alkaline rocks, such trends are not clear (see Baker, 1968) or in high-K calc-alkaline rocks (for example, Eastern Papua, Jakeš and Smith, 1970), the converse is found; Sr increases with increasing SiO_2. There is a wide range in the Sr abundances of island-arc tholeiites, but the data of Lowder and Carmichael (1970), Gill (1970), and Prinz (1968) suggest values of 200 to 250 ppm at 53 percent SiO_2. This is twice the characteristic value of the oceanic tholeiites (Kay and others, 1970), and hence Sr may be a useful parameter in delineating island-arc and oceanic tholeiites. In the low-K rocks of the Izu Peninsula (Taylor and White, 1966), and New Hebrides (A. Ewart, 1970, personal commun.) Sr abundances at 56 to 58 percent SiO_2 are generally in the 350 to 450 ppm range. In high-K calc-alkaline rocks and shoshonites, Sr contents are invariably high, and values of 800 ppm for the shoshonitic association are characteristic.

The total content of *rare earth elements* (REE) in island-arc volcanic rocks is relatively low and usually does not exceed 100 ppm (Taylor, 1969). The most important fact emerging from new data (Jakeš and Gill, 1970) is that there are two types of distribution patterns in recently erupted island-arc associations.

Chondritic or "primitive" REE patterns characterize tholeiitic rocks. New data on the basement rocks from the Solomon Islands, Eastern Papua, Macquarie Ridge, on recent tholeiitic volcanoes from New Britain, and recent data of Masuda, 1966; Taylor, 1968; Gill, 1970; and A. Ewart and A. L. Graham (1970, personal commun.) demonstrate that tholeiitic rocks with primitive REE patterns are not confined to mid-oceanic ridges (Schilling, 1969). Thus, genetic constraints, based on REE, applied to rocks of the mid-oceanic ridges should also be applied to rocks of the outer part of the island arcs.

Patterns strongly enriched in light REE (fractionated) occur in calc-alkaline and shoshonitic rocks of island arcs. Within island-arc rocks with this type of REE pattern, there is only a weak correlation between K_2O and the La/Yb ratio (Jakeš and Gill, 1970), suggesting a relatively sharp division between the two associations.

Data on abundances of large highly charged

cations (Zr, Th, U) also show a regular pattern of variation. There is a positive correlation between Zr and SiO_2, but at a given SiO_2 content, the Zr content of island-arc tholeiites is less than that of calc-alkaline rocks; however, Gill (1970) found a decrease in Zr abundances in passing from calc-alkaline rocks to shoshonitic rocks, whereas Nicholls and Carmichael (1969) found relatively high abundances of Zr in shoshonites, indicating the opposite trends.

Data given by Gill (1970), Taylor (1969), and Donnelly and others (1971) suggest that the absolute abundances of both Th and U (although extremely variable) and Th/U ratios increase from tholeiites to shoshonites.

Cr and Ni are easily affected by removal or addition of ferromagnesian phases such as olivine. The abundances of Cr, Ni, and V show an inverse correlation with SiO_2. There is a wide range of absolute abundances even for one rock association within a particular region (Baker, 1968), but, in general, ferromagnesian trace-element concentrations in island-arc rocks (except in a few rocks of cumulative origin) are low. The abundances in calc-alkaline rocks are normally very low, but in rocks with less than 54 percent SiO_2, Cr and Ni are sometimes high. In the tholeiitic association, such enrichment is considered to result from olivine and clinopyroxene accumulation. The variability in Cr and Ni content is illustrated by rocks from Izu (Taylor and White, 1966) or high-K rocks from Papua (Jakeš and Smith, 1970). In both areas, there are higher than the average concentrations of these elements.

In contrast to Cr and Ni abundances, V contents are much the same in all three associations and 120 to 175 ppm at SiO_2 58 percent are typical; higher values seem to be typical of low-K rocks.

ISLAND-ARC CALC-ALKALINE ROCKS VERSUS ANDEAN CALC-ALKALINE ROCKS

Calc-alkaline volcanism is also associated with active continental margins (for example, South American Andes), and significant volumes of calc-alkaline rock were also erupted in the "intra-continental" orogenic chains (for example, Carpathians in central Europe).

McBirney (1969) and Forbes and others (1969) have shown differences in composition of continental and island-arc andesites, pointing out the differences in Na_2O, K_2O, and FeO contents, but they found an overlap of both. Yoder (1969), using examples from the West Indies, Solomon Islands, and Paricutin, mentioned differences in $MgO/FeO + Fe_2O_3$ of island-arc calc-alkaline rocks. Hamilton (1969, 1970) pointed out the similarities of Andean volcanic rocks and Mesozoic batholithic masses in the western United States, and accepted the view that continental-margin volcanism and island-arc volcanism differ, although Hamilton considers that both of these types are related to the Benioff zone.

In major element chemistry, the continental margin (Andean) calc-alkaline rocks have equivalents in high-K calc-alkaline rocks of island arcs, and thus differences advocated by Forbes and others (1969) and Yoder (1969) are not necessarily real. Generalized differences are given in Table 3, for instance in continental areas, the rocks with higher SiO_2 contents (more than 63 percent) have K_2O/Na_2O ratios close to 1; whereas in the island-arc rocks, the K_2O/Na_2O ratio would be constant within the whole range of SiO_2 values (50 to 60 percent).

The distinction between Andean and island-arc volcanism is also seen from trace-element data (Zeil and Pichler, 1967) and El-Hinnawi and others (1969).

If rocks with the same SiO_2 and K_2O are compared, the abundances of trace elements of the potassium type (Ba, Sr, Rb) and Zr, Th, and U are higher in the Andean type (Siegers and others, 1969); whereas K/Rb ratios are lower (Table 3). V, Ni, and Cr contents are variable but generally low in both calc-alkaline associations; however, they seem to be slightly lower in the Andean associations (Jakeš, unpub. data).

TABLE 3. GENERALIZED DIFFERENCES BETWEEN CALC-ALKALINE VOLCANIC ROCKS OF ISLAND ARCS AND THOSE OF CONTINENTAL MARGINS

	Continental margin (Andean)	Island arc
Range of SiO_2	56 to 75 percent	50 to 66 percent
FeO + Fe_2O_3/MgO	higher than 2.0	lower than 2.0
K_2O/Na_2O	0.60 to 1.1	less than 0.8
Trace elements at same K_2O and SiO_2	higher Rb, Ba, Sr, Th, U, Zr; lower K/Rb (230), Th/U	lower Rb, Ba, Sr, Th; Zr higher K/Rb (400), Th/U
Phenocrysts	biotite, hornblende, clinopyroxene, orthopyroxene, rare quartz, garnet, cordierite	clinopyroxene, orthopyroxene, hornblende (rare biotite); no quartz, garnet, or cordierite
Sequence of phenocryst crystallization (hbe-andesites)	hornblende→clinopyroxene→orthopyroxene	clinopyroxene→hornblende→clinopyroxene

The island-arc calc-alkaline rocks grade into low-K rocks (island-arc tholeiites) on the oceanic side, and into high-K rocks (shoshonitic on the continental side of the arcs. On the other hand, Andean calc-alkaline volcanic rocks invariably have high-K contents, and distinct spatial variations in lava composition do not occur. However, in some Andean type andesites (Carpathians), there is slight increase of K_2O at a given SiO_2 content in a stratigraphical sequence (Kuthan, 1968).

In island arcs, calc-alkaline rocks with SiO_2 > 63 percent are rare (<10 percent) and rocks with less than 56 percent SiO_2 are moderately abundant ($\cong 30$ percent), whereas in continental areas, the calc-alkaline association is characterized by the presence of andesites, dacites, and rhyolites. Rocks with less than 56 percent SiO_2 are rare in continental areas, or are absent (for example, Northern Andes, Katsui, 1969; Alaskan volcanoes, Forbes and others, 1969).

DISCUSSION OF ISLAND-ARC ASSOCIATION

Tholeiitic, calc-alkaline, and shoshonitic associations occurring in island-arc areas are closely related in space and time. These associations are defined on the basis of major-element chemistry but also, in regions with a well-developed seismic plane, by the depth to the Benioff zone beneath the volcano. Although each association has a typical major- and trace-element abundance (Table 2), there are compositional gradations between them, and therefore all boundaries, if drawn at all, are artificial. In this sequence, the most important variations are the increase of K_2O content at given SiO_2 content, the increase of K_2O/Na_2O ratio, the decrease of "iron enrichment," and the increase in the concentration of the potassium type (Rb, Ba) and of other large cations (Zr, Th, U). The REE pattern changes abruptly from "primitive" in tholeiitic rocks to one enriched in light REE (fractionated) in calc-alkaline and shoshonitic rocks.

If it is assumed that rocks of island arcs have their source at depths corresponding to that of the underlying Benioff zone (Dickinson and Hatherton, 1967), then their source material is the upper mantle or the upper part of a descending slab of oceanic lithosphere at depths ranging from 80 to 250 km.

Some restrictions on hypotheses of evolution of island-arc volcanic sequences are imposed by major element chemistry. Assuming an homogeneous source material, potassium abundances suggest that the amount of fractional or partial melting in the mantle or in the descending slab must decrease with depth from 15 to 25 percent to produce island-arc tholeiites to as low as 2 to 5 percent to produce calc-alkaline and shoshonitic rocks.

The fact that rocks of the same SiO_2 content in the tholeiitic association differ in trace-element abundances (Rb, Ba, Sr) from the rocks with the same SiO_2 content in the calc-alkaline or in the shoshonitic association suggests that simple fractionational crystallization of one parent magma is not likely to produce these and other differences. The transitional character of all associations in true island-arc areas, must be pointed out once again, and thus any petrogenetic scheme is insufficient if it can be applied only to a single association or group of rocks.

Although we do not attempt to treat trace-element abundances quantitatively (Gast, 1968; Shaw, 1970) because of large fluctuations and essentially unknown values of distribution coefficients, trace-element abundances provide a number of constraints on the character of the source material.

Rare-earth element abundances of island-arc lavas suggest that:

1. Tholeiites of island arcs come from relatively primitive, unfractionated, or little-fractionated material. Fractional crystallization of olivine and minor pyroxene could have occurred, since this will not change the over-all primitive pattern. But fractional (selective) melting of amphibole from parageneses containing phases which do not substantially discriminate between light and heavy REE (olivine, orthopyroxene, spinel) could also create similar primitive patterns. The extensive fractionation of plagioclase in tholeiites probably did not occur, although a slight increase in Eu in some samples could indicate fractional melting involving plagioclase.

2. Because the REE are strongly fractionated in calc-alkaline and shoshonitic rocks compared to tholeiite, these rocks may be derived from more primitive material by a very small degree of partial melting. Also the residuum or cumulate of calc-alkaline and shoshonitic rocks must have contained minerals which strongly discriminate light REE in favor of the

melt. Garnet and clinopyroxene appear to be most effective.

The main distribution coefficients for large cations and abundances of these trace elements in island-arc rocks indicate that tholeiitic rocks could have been in equilibrium with olivine and pyroxene, and calc-alkaline rocks with olivine, clinopyroxene, and garnet. Equilibrium of tholeiitic rocks with amphibole, because of the K/Rb ratio of tholeiites, calc-alkaline rocks and shoshonites with biotite (because of high Rb, Ba, K), or plagioclase (because of Sr) is unlikely. Instead, these minerals are believed to determine the geochemical characteristics of island-arc rocks. Amphibole and mica are stable in some high-pressure–high-temperature subsolidus parageneses, but if melting occurs in the mantle parageneses or in the thermally retarded descending slab (Minear and Töksoz, 1970), these phases enter the melt. It is suggested that if the temperature exceeds the limit of stability of any of these mineral phases, then their breakdown provides the necessary water, and the magma inherits some chemical characteristics of the mineral concerned. Griffin and Murthy (1969) advocated a fractional melting process involving various hydrous phases to explain the geochemical characteristics of basaltic rocks in oceanic areas. To explain the decreasing K/Rb ratios across island arcs, Jakeš and White (1970) suggested that more mica relative to amphibole is involved in partial melting with increasing depth.

We present two extreme possibilities for evolution of island-arc rocks, but believe that both models (A and B) are complementary. Model A does not consider any extensive contribution from upper mantle material to the magmas appearing in the island arcs, and the magmas are derived from the slab of hydrated oceanic crust underthrusting the island arc. In model B, this underthrust oceanic crust provides "water" or more probably a water-rich siliceous melt, and variations in lava compositions reflect different degrees of partial melting, as well as fractionation of the magmas during their ascent.

We assume that if some material of the underthrust slab (oceanic crust) undergoes partial melting and is depleted in a low melting fraction during its descent, then the residual material cannot contribute to the production of further lavas and, therefore, the melts produced at greater depths are believed to come from "undepleted" oceanic crust. The amount of material in the productive upper part of the descending slab is large enough to allow for this assumption. It is also suggested that material with lower melting point than that of hydrous oceanic tholeiite is more likely to contribute to melts produced along the upper part of the seismic plane. These materials could be alkaline rocks of decomposed seamounts or sediments dragged along the seismic zone.

REFERENCES CITED

Aramaki, S., 1963, Geology of Asama volcano: Tokyo Univ. Fac. Sci., sec. II, v. 14, p. 229–443.

Baker, P. E., 1968, Comparative volcanology and petrology of the Atlantic island-arcs: Bull. Volcanol., v. 32, p. 189–206.

Denham, D., 1969, Distribution of earthquakes in the New Guinea-Solomon Island region: Jour. Geophys. Research, v. 74, p. 4290–4299.

Dickinson, W. R., 1968, Circum-Pacific andesite types: Jour. Geophys. Research, v. 73, p. 2261–2269.

Dickinson, W. R., and Hatherton, T., 1967, Andesitic volcanism and seismicity around the Pacific: Science, v. 157, p. 801–803.

Donnelly, T. W., Rogers, P., Pushkar, P., and Armstrong, R. L., 1972, Chemical evolution of the igneous rocks of the eastern West Indies: An investigation of thorium, uranium and potassium distribution, and lead and strontium isotopic ratio: Geol. Soc. America (in press).

Erlikh, E. N., 1968, Petrohimiya Kainozoiskoi Kurilo-Kamchatskoi vulkanicheskoi provintsii: Moskva, Nauka, p. 277.

El-Hinnawi, E. E., Pichler, H., and Zeil, W., 1969, Trace element distribution in Chilean ignimbrites: Contr. Mineralogy and Petrology, v. 24, p. 50–62.

Engel, C. G., Engel, A. E., and Havens, R. G., 1965, Chemical characteristics of oceanic basalts and the upper mantle: Geol. Soc. America Bull., v. 76, p. 719–725.

Forbes, R. B., Ray, D. K., Katsura, T., Matsumoto, H., Haramura, H., and Furst, M. J., 1969, The comparative chemical composition of continental vs. island arc andesites in Alaska: Oregon Dept. Geology and Mineral Industries Bull., v. 65. p. 111–120.

Gast, P. W., 1968, Trace element fractionation and the origin of tholeiitic and alkaline magma types: Geochim. et Cosmochim. Acta, v. 32, p. 1057–1086.

Gill, J. B., 1970, Geochemistry of Viti Levu, Fiji, and its evolution as an island arc: Contr. Mineralogy and Petrology, v. 27, p. 179–203.

Gorshkov, G. S., 1962, Petrochemical features of volcanism in relation to the types of the earth's crust: Am. Geophys. Union Mon., v. 6, p. 110–115.

Green, T. H., and Ringwood, A. E., 1968, Genesis of the calc-alkaline igneous rock suite: Contr. Mineralogy and Petrology, v. 18, p. 105-162.

Griffin, W. L., and Murthy, V. R., 1969, Distribution of K, Rb, Sr, and Ba in some minerals relevant to basalt genesis: Geochim. et Cosmochim. Acta, v. 33, p. 1389–1414.

Hall, A., 1967, The variation of some trace elements in the Rosses granite complex, Donegal: Geol. Mag., v. 104, p. 99-109.

Hamilton, W., 1969, The volcanic central Andes, a modern model for Cretaceous batholiths and tectonics of western North America: Oregon Dept. Geology and Mineral Industries Bull., v. 65, p. 175–184.

—— 1970, Mesozoic California and the underflow of Pacific mantle: Geol. Soc. America Bull., v. 80, p. 2409–2430.

Hart, S. R., Brooks, C., Krogh, T. E., Davis, G. L., and Nava, D., 1970, Ancient and modern volcanic rocks: a trace element model: Earth and Planetary Sci. Letters, v. 10, p. 17–28.

Isacks, B., Oliver, J., and Sykes, L. B., 1968, Seismology and the new global tectonics: Jour. Geophys. Research, v. 73, p. 5855–5899.

Jakeš, P., 1970, Analytical and experimental geochemistry of volcanic rocks from island arcs [Ph.D. thesis]: Canberra, Australian National Univ., 134 p.

Jakeš, P., and Gill, J. B., 1970, Rare earth elements and the island arc tholeiitic series: Earth and Planetary Sci. Letters, v. 9, p. 17–28.

Jakeš, P., and Smith, I. E., 1970, High potassium calc-alkaline rocks from Cape Nelson, eastern Papua: Contr. Mineralogy and Petrology, v. 28, p. 259–271.

Jakeš, P., and White, A.J.R., 1969, Structure of the Melanesian arcs and correlation with distribution of magma types: Tectonophysics: v. 8, p. 223–236.

—— 1970, K/Rb ratios of rocks from island arcs: Geochim. et Cosmochim. Acta, v. 34, p. 849–856.

—— 1971, Hornblendes from calc-alkaline volcanic rocks of island arcs and continental margins: Am. Mineralogist (in press).

Jolly, W. T., 1971, Potassium-rich igneous rocks from Puerto Rico: Geol. Soc. America Bull., v. 82, p. 399–408.

Joplin, G. A., 1968, The shoshonite association: A review: Geol. Soc. Australia Jour., v. 15, p. 275–294.

Katsui, Y., 1969, Andesites from the Andes and Antarctica: Oregon Dept. Geology and Mineral Industries Bull., v. 65, p. 193.

Kay, R., Hubbard, N. J., and Gast, P. W., 1970, Chemical characteristics and origin of oceanic ridge volcanic rocks: Jour. Geophys. Research, v. 75, p. 1585–1610.

Kolbe, P., and Taylor, S. R., 1966, Major and trace element relationships in granodiorites and granites from Australia and South Africa: Contr. Mineralogy and Petrology, v. 12, p. 202–222.

Kuno, H., 1950, Petrology of Hakone volcano and the adjacent areas, Japan: Geol. Soc. America Bull., v. 61, p. 957–1020.

—— 1966, Lateral variation of basalt magma type across continental margins and island arcs: Bull. Volcanol., v. 29, p. 195–222.

—— 1968, Origin of andesite and its bearing on the island arc structure: Bull. Volcanol., v. 32–1, p. 141–176.

Kuthan, M., 1968, Young volcanic rocks of the Carpathians in Slovakia, in Regional geology of Czechoslovakia: Ustredni Ustav Geologicky, Praha, p. 628–667.

Lowder, G. G., and Carmichael, I.S.E., 1970, The volcanoes and caldera of Talasea, New Britain geology and petrology: Geol. Soc. America Bull., v. 81, p. 17–38.

Masuda, A., 1966, Lanthanides in basalts of Japan with three distinct types: Geochem. Jour., v. 1, p. 11–26.

McBirney, A., 1969, Andesitic and rhyolitic volcanism of orogenic belts: Am. Geophys. Union Mon., v. 13, p. 501–506.

Minear, J. W., and Töksoz, M. N., 1970, Thermal regime of a downgoing slab and new global tectonics: Jour. Geophys. Research, v. 75, p. 1397–1419.

Nicholls, J., and Carmichael, I.S.E., 1969, A commentary on the absarokite-shoshonite-banakite series of Wyoming, U.S.A.: Schweizer. Mineralog. u. Petrog. Mitt., v. 49, p. 47–64.

Prinz, M., 1968, Geochemistry of basaltic rocks: Trace elements, in Hess, H., and Poldervaart, A., eds., Basalts: New York, Interscience, p. 271–323.

Rhodes, J. M., 1969, The geochemistry of a granite-gabbro association at Hartley, New South Wales [Ph.D. thesis]: Canberra, Australian National Univ., 403 p.

Ringwood, A. E., 1969, Composition and evolution of the upper mantle: Am. Geophys. Union Mon., v. 13, p. 1–17.

Ruxton, B. P., 1966, A late Pleistocene to Recent rhyodacite-trachybasalt-basaltic latite volcanic association in northeast Papua: Bull. Volcanol., v. 29, p. 347–369.

Schilling, J. G., 1969, Red Sea floor origin, rare earth evidence: Science, v. 165, p. 1357–1359.

Shaw, D. M., 1970, Trace element fractionation during anatexis: Geochim. et Cosmochim. Acta, v. 37, p. 237–242.

Siegers, A., Pichler, H., and Zeil, W., 1969, Trace element abundances in the "Andesite" formation of northern Chile: Geochim. et Cosmochim. Acta, v. 33, p. 882–887.

Sugimura, A., 1961, Regional variation of the

K_2O/Na_2O ratios of volcanic rocks in Japan and environs: Geol. Soc. Japan Jour., v. 67, p. 292–300.

Sugimura, A., 1968, Spatial relations of basaltic magmas in island arcs, *in* Hess, H. H., and Poldervaart, A., eds., Basalts: New York, Interscience, p. 573–575.

Sykes, L. R., 1966, The seismicity and deep structure of island arcs: Jour. Geophys. Research, v. 71, p. 2981–3006.

Taylor, S. R., 1967, The origin and growth of continents, Tectonophysics, v. 4, p. 17–34.

—— 1968, Geochemistry of andesites, *in* Ahrens, L. H., ed., Origin and distribution of the elements: New York-London-Paris, Pergamon Press, p. 559–583.

—— 1969, Trace element chemistry of andesites and associated calc-alkaline rocks: Oregon Dept. Geology and Mineral Industries Bull., v. 65, p. 43–63.

Taylor, S. R., and White, A.J.R., 1965, Geochemistry of andesites and the growth of continents: Nature, v. 208, p. 271–273.

—— 1966, Trace element abundances in andesites: Bull. Volcanol., v. 29, p. 174–194.

Taylor, S. R., Capp, A. C., and Graham, A. L., 1969, Trace element abundances in andesites. II. Saipan, Bougainville and Fiji: Contr. Mineralogy and Petrology, v. 23, p. 1–26.

Wilson, J. T., 1952, Orogenesis as the fundamental geological process: Am. Geophys. Union Trans., v. 33, p. 444–449.

Yoder, H. S., 1969, Calc-alkalic andesites: Experimental data bearing on origin of their assumed characteristics: Oregon Dept. Geology and Mineral Industries Bull., v. 65, p. 111–120.

Zeil, W., and Pichler, H., 1967, Die Kanozoische Rhyolith-Formation im mittleren Abschnitt der Anden: Geol. Rundschau, v. 57, p. 48–81.

MANUSCRIPT RECEIVED BY THE SOCIETY APRIL 2, 1971

REVISED MANUSCRIPT RECEIVED JULY 26, 1971

THE LUNAR SCIENCE INSTITUTE CONTRIBUTION No. 65 (NASA CONTRACT TO UNIVERSITIES SPACE RESEARCH ASSOCIATION No. NSR 09-051-001)

AUTHOR'S PRESENT ADDRESS: GEOLOGICAL INSTITUTE, CZECHOSLOVAK ACADEMY OF SCIENCES, SUCHDOL 2, PRAHA 6, CZECHOSLOVAKIA (JAKEŠ)

56

CRITERIA FOR RECOGNIZING ENVIRONMENTS OF FORMATION OF VOLCANIC SUITES; APPLICATION OF THESE CRITERIA TO VOLCANIC SUITES IN THE CAROLINA SLATE BELT

John J. W. Rogers

[*Editors' Note:* Only Tables 1 through 6 are reproduced here.]

TABLE 1. MID-OCEAN RIDGE BASALT (MORB) AND BACK-ARC BASIN

SiO_2 range	— restricted (46 to 52%)	
eruptive style	— subaqueous; some pillows, no pyroclastics	
mineralogy	— olivine and plagioclase phenocrysts	
K_2O	0.1-0.2%	
Rb	2 ppm	lower than in all
Ba	15 ppm	other basalt types
Sr	100 ppm	
TiO_2	1.5%	
Zr	100 ppm	higher than in
Cr	300 ppm	subduction-zone basalts
Ni	100 ppm	
REE pattern	— light REE depletion	
$Zr-TiO_2$	— linear relationship	
FeO*/MgO	— < 1.5 (lower than in arc or continental-margin basalts)	

TABLE 2. OCEANIC ISLAND

Two suites (possibly transitional)

Subalkaline suite

basalt similar to MORB but with higher:

TiO_2	1.5%
K_2O	0.4%
Ba	50-100 ppm

REE pattern shows light REE enrichment

Alkaline suite

SiO_2 range	— mostly basalt; some rocks up to 65%	
mineralogy	— alkaline; normative, and some modal, nepheline	

Alkali olivine basalts

higher than other basalts in:

TiO_2	2.7%	
K_2O	1.2%	higher in most trace elements
Ba	300 ppm	than other basalts
Sr	400 ppm	

405

TABLE 3.   EARLY INTRA-OCEANIC ISLAND ARC

SiO_2 range	- 50% to 75%; andesite rare; may be bimodal (basalt-dacite or spilite-keratophyre)
eruptive style	- subaqueous; some pillows; some non-explosive pyroclastics
mineralogy	- clinopyroxene, plagioclase, and some olivine phenocrysts in mafic rocks

Mafic rocks

K_2O	0.5%	
Rb	5 ppm	higher than in MORB; similar
Ba	100 ppm	to calcalkaline arc basalts
Sr	200 ppm	
TiO_2	0.8%	
Zr	50 ppm	lower than in MORB; similar
Cr	50 ppm	to calcalkaline arc basalts
Ni	50 ppm	
REE pattern	- flat	
Zr-TiO	- linear relationship	
FeO*/ MgO	- >1.5	

Silicic rocks

impoverished in lithophilic elements

TABLE 4.   MATURE ISLAND ARC (CALCALKALINE)

SiO_2 range	- unimodal; andesite common; >68% rare
eruptive style	- subaerial; flows and pyroclastics
mineralogy	- clinopyroxene in andesites and basalts

Mafic rocks

similar to early arc basalts except:

light REE enriched (contrasting to flat pattern in early basalts)

no linear relationship between Zr and TiO_2

Andesites

K_2O	1.0%	
Rb	15 ppm	lower than in continental-
Ba	300 ppm	margin andesites
Sr	400 ppm	
TiO_2	0.7%	
Zr	100 ppm	lower than in continental-
Cr	30 ppm	margin andesites
Ni	5 ppm	

Note: Calcalkaline suite shows steady increase in lithophilic element content toward the more silicic rocks.

TABLE 5. CONTINENTAL MARGIN (CALCALKALINE)

SiO_2 range	-	mostly 60% to 75%; minor basalts
eruptive style	-	subaerial; commonly pyroclastic, some flows
mineralogy	-	amphibole in intermediate rocks

Mafic rocks

    similar to intra-oceanic-arc calcalkaline basalts

Andesites

K_2O	2.0%	
Rb	50 ppm	higher than in intra-
Ba	500 ppm	oceanic arc andesites
Sr	600 ppm	

TiO_2	0.9%	
Zr	200 ppm	higher than in intra-
Cr	60 ppm	oceanic arc andesites
Ni	30 ppm	

Note: Rate of increase in abundances of lithophilic elements with increasing SiO_2 is higher than in intra-oceanic arc suites.

TABLE 6. CONTINENTAL

great diversity of rock types:
    mafic to silicic
    flows and pyroclastics
    subalkaline and alkaline
    peraluminous and peralkaline

Plateau basalts

SiO_2 range	- 48% to 52%; rare silicic horizons
eruptive style	- subaerial; flows, rare pyroclastics
mineralogy	- quartz-normative to nepheline-normative basalts

compositionally similar to calcalkaline arc basalts except higher in:

TiO_2	2.0%
Cr	200 ppm
Ni	30 ppm

Rift zones

    great variety of rock types, including:
        plateau basalts
        alkaline suite (basalts, phonolites, etc.)
        peralkaline suite
        ignimbrites (peralkaline and peraluminous)

57

TECTONIC SETTING OF BASIC VOLCANIC ROCKS DETERMINED USING TRACE ELEMENT ANALYSES

J. A. Pearce and J. R. Cann

[*Editors' Note:* Only Figures 2 through 4 are reproduced here.]

J.A. Pearce, J.R. Cann, Tectonic setting of basic volcanic rocks

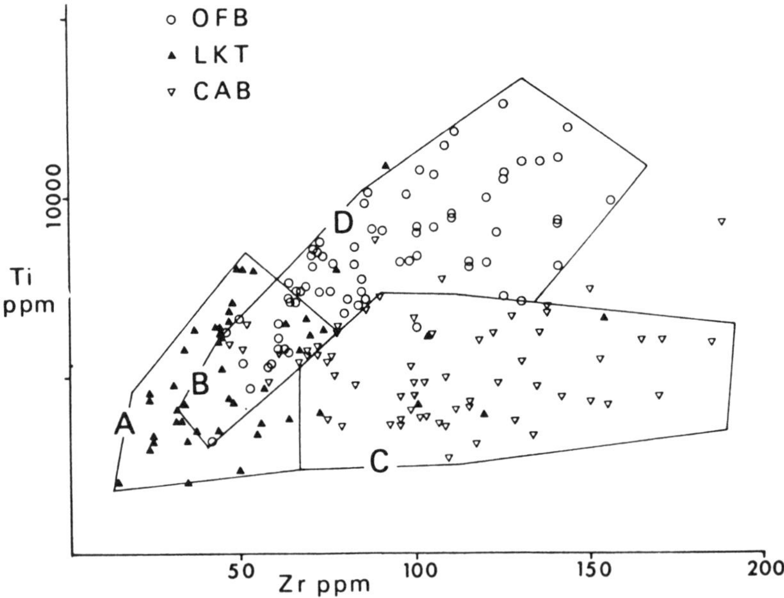

Figure 2. Discrimination diagram using Ti and Zr. Ocean-floor basalts (OFB) plot in fields D and B, low-potassium tholeiites (LKT) in fields A and B, and calc-alkali basalts (CAB) in fields C and B.

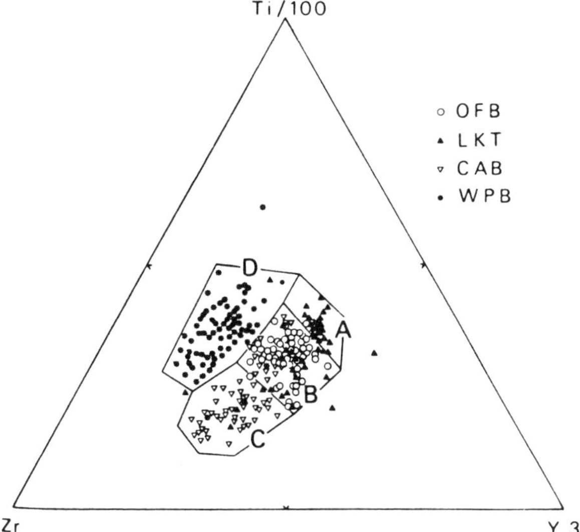

Figure 3. Discrimination diagram using Ti, Zr and Y. "Within-plate" basalts (WPB) i.e. ocean island or continental basalts—plot in field D, ocean-floor basalts (OFB) in field B, low-potassium tholeiites (LKT) in fields A and B, calc-alkali basalts (CAB) in fields C and B.

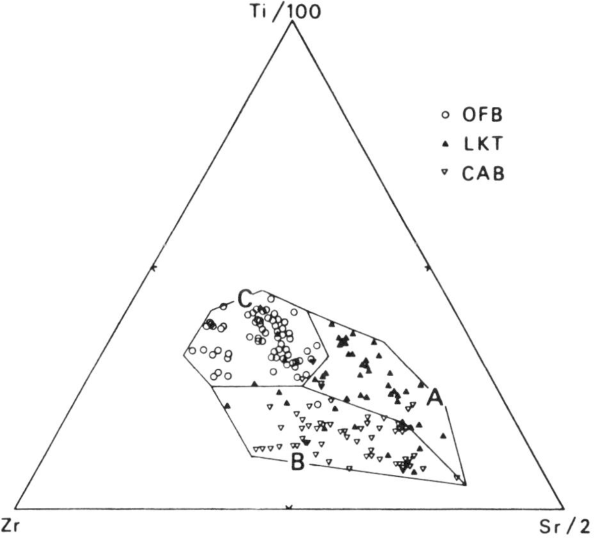

Figure 4. Discrimination diagram using Ti, Zr and Sr. Ocean-floor basalts (OFB) plot in field C, low-potassium tholeiites (LKT) in field A, and calc-alkali basalts (CAB) in field B.

Identification of ore-deposition environment from trace-element geochemistry of associated igneous host rocks

J. A. Pearce
Department of Earth Sciences, The Open University, Milton Keynes, Buckinghamshire, England

G. H. Gale
Department of Mines, Resources and Environmental Management, Winnipeg, Manitoba, Canada

550.42:546:553.067

Synopsis

The tectonic environment of formation of volcanogenic massive sulphide deposits and porphyry tin and copper deposits can be identified from the geochemical characteristics of the associated igneous rocks. The 'stable' trace-element geochemistry (involving Ti, Zr, Y, Nb, Cr and rare-earth elements) and geology of metabasalts related to 12 massive sulphide deposits indicate that the deposits studied fall into four distinct classes. (1) Cyprus-type, including Cyprus, Oman and Betts Cove, possibly formed during the early stages of back-arc basin development; (2) Løkken-type, including Løkken and York Harbour, possibly formed at back-arc basin spreading centres; (3) Joma-type, including Joma, Røros and Bidjovagge, possibly formed in a small ocean of Red Sea type; and (4) Gjersvik-type, including Gjersvik, Buchans, Noranda and Lynn Lake, possibly formed during an early stage of island arc evolution (Gjersvik), later during island arc evolution (the Buchans Kuroko-type deposit) or in a Precambrian setting (Noranda and Lynn Lake). Deposits related to major ocean ridge crests appear to be small and relatively uncommon, perhaps because relatively few favourable sites for ore deposition exist in such environments.

The environment of intrusion of acid—intermediate igneous rocks can be deduced by use of diagrams based on the element Nb (e.g. SiO_2 versus Nb). On this basis, tin-bearing granites can be classified either as within-plate magmas (Nigeria) or as magmas from evolved volcanic arc settings (e.g. Bolivia, Cornwall and Indonesia); the latter group can be further subdivided geologically into post-orogenic and back-arc extensional settings. Although the continental crust may play a significant role in the formation of tin deposits, the geochemical data presented here suggest that partial melting of tin-enriched mantle above a subduction zone may be the single most important genetic factor.

The purpose of this paper is to show how igneous geochemistry can be used to identify the tectonic environments in which volcanogenic ore deposits are formed. The paper is divided into two parts: the first deals with deposits formed in a submarine environment (i.e. volcanogenic massive sulphide deposits), and the second with deposits formed in a continental environment — in particular, tin deposits.

Many works of synthesis have already been published on the relationship between massive sulphide deposits and plate tectonics.[1–5] These use mainly geological criteria to point to a variety of possible environments of massive sulphide

deposition, at major ocean ridges, in marginal basins, in island arcs and in oceanic islands. In many cases, however, particularly in the deformed and metamorphic rocks of mountain belts, the geological evidence is ambiguous. Geochemical techniques then have a useful application. Because the host rocks are commonly metamorphosed, stable trace-element geochemistry[6] can be used to deduce their environment of eruption. Elements such as Ti, Zr, Y, Nb,[7,8,9] P,[9] Cr[10,11] and the rare earths[12] all exhibit systematic differences between basalts from different tectonic settings while remaining relatively immobile during weathering and metamorphic processes. In this paper we study the concentrations of these elements in the volcanic rocks related to mineralization in the Norwegian Caledonides, Newfoundland, Cyprus and Oman and in Precambrian terrain in Scandinavia and Canada.

Ore deposits related to continental magmatism can also be studied by such a geochemical approach. It is fairly well established that porphyry copper deposits are formed above destructive plate margins.[13] The environment of formation of tin deposits is much more enigmatic, however. The possibilities suggested[14–18] are post-orogenic, within-plate ('hot-spot') and destructive plate margin settings. Only rarely can the precise setting of tin-bearing granites be identified from geological criteria alone. Geochemical methods are developed in this paper to distinguish between granitic rocks originating in 'within-plate' and 'volcanic arc' settings. These methods are applied to some tin-bearing granitic rocks in Cornwall, Bolivia, Indonesia and Nigeria.

Massive sulphide deposits

Magmatism in the oceanic environment

If (as is widely accepted) hydrothermal circulation of sea water through oceanic crust is the dominant process in the formation of massive sulphide deposits,[19] then, given a sufficiently high geothermal gradient and permeable crust, massive sulphides could form at almost any site of submarine igneous activity. So, before studying the lavas related to these deposits, it is first necessary to geochemically categorize lavas from major ocean spreading centres, marginal basin spreading centres, continental margins, island arc seamounts, ocean islands and seismic/aseismic ridges. To do this, three main magma types are first defined: *ocean-floor*

Table 1 Mean concentrations of trace elements (ppm) in three main oceanic magma types (from literature survey made in 1973[33]); ocean island basalt column includes both tholeiitic and alkalic types

	Ocean-floor tholeiites	Island arc tholeiites	Ocean island basalt
Ti	8400	5000	13 500
Zr	83	45	211
Y	28	18	27
Nb	2.5	1.5	27
Cr	280	107	200
La	3.0	3.2	27
Ce	10	6.5	56
Yb	2.5	2.0	2.3
K	1300	2700	5450
Rb	2.5	4.7	13
Ba	8.5	60	176
Sr	120	175	415
U	0.09	0.15	0.64
Th	0.16	0.37	0.77
Cu	73	62	81
Co	51	30	60
V	230	270	240
Ga	17	15	18

basalts (which include tholeiitic and alkalic types), *within-plate basalts* (which also include tholeiitic and alkalic types) and *island arc basalts* (which include tholeiitic and calc-alkaline types). Within-plate tholeiites are distinguished from ocean-floor tholeiites by their higher concentrations of most lithophile trace elements, the few notable exceptions including Y, the heavy rare earths, Sc and Cr. Island arc tholeiites are distinguished from ocean-floor tholeiites by lower concentrations of small ion lithophile elements, including Cr; their concentrations of large ion lithophile elements are typically similar to or greater than those of ocean-floor tholeiites. Table 1 illustrates the nature of these geochemical differences.

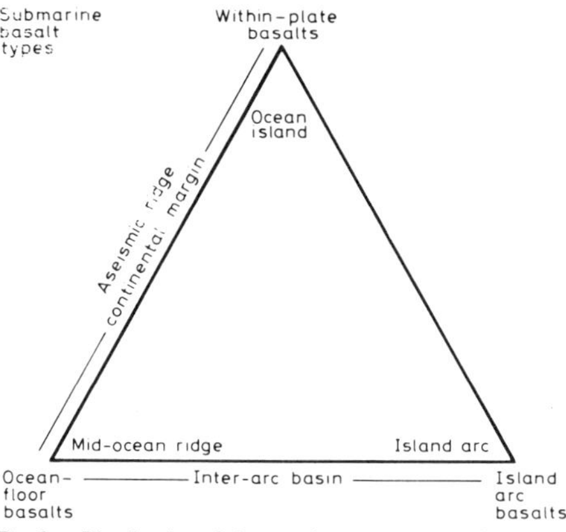

Fig. 1   Distribution of three main magma types within ocean basins — illustrated by use of triangular diagram with main magma types at apices and transitional magma types along sides

In practice, lavas do not always fall neatly into one or other of these magma types. The main reason for this appears to be the existence of transitional tectonic environments. For example, if a spreading centre in a marginal

basin directly overlies a subduction zone, its magma type might be expected to be intermediate between ocean-floor and island arc type; similarly, magmas erupted during the initial stages of sea-floor spreading could be intermediate between ocean-floor and within-plate magma types. Fig. 1 illustrates the probable relationship between magma type and tectonic setting. It emphasizes that geochemical studies will often need to be supplemented by geological evidence for an unambiguous interpretation to be made.

Geochemical characteristics of the host lavas

Stable trace-element geochemistry, coupled with geological information, is used here to characterize metavolcanic rocks related to a variety of massive sulphide deposits. We deal first with the geochemical information, which is summarized in Table 2.

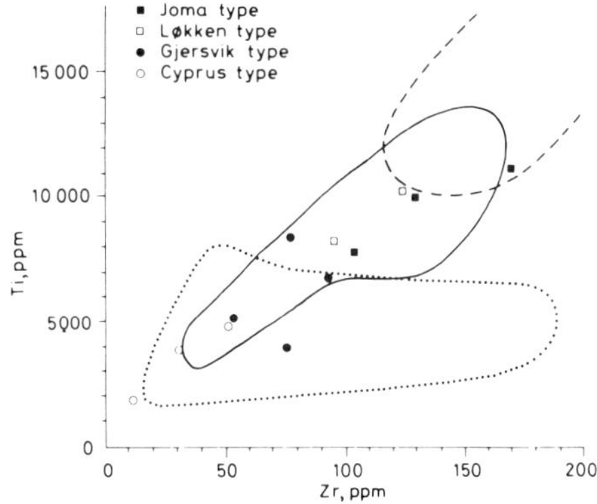

Fig. 2   Mean analyses of metabasalts from various massive sulphide deposit types plotted on Ti–Zr discrimination diagram.[8] Dotted line denotes field of island arc basalts and andesites, continuous line field of ocean-floor basalts and dashed line field of within-plate basalts

Various discriminant diagrams can be used to identify magma type. Fig. 2 is a plot of Ti against Zr[8] in which there

Table 2   Mean concentrations of 'stable' trace elements (ppm) in greenstones used in this study. Lavas from deposits 1, 2, 5, 7, 9 and 10 include samples from actual mines; remainder are samples from same geologic unit containing deposit, but not from deposit itself

Deposit		Location of lavas	Ti	Zr	Y	Nb	Cr	P₂O₅	No. of analyses
1	Cyprus deposits [20, 21]	Troodos Massif [11, 22] (axis sequence)	5900	61	28	1	100	0.08	85
		Troodos Massif [11, 22] (upper pillow lavas)	3270	31	15	1	405	0.06	45
2	Lasail, Oman [23]	Lasail [23]	3750	34	16	—	109	0.05	15
3	Betts Cove,[24] Newfoundland	Baie Verte [25]	1950	15	11	2	—	—	3
4	York Harbour, Newfoundland [26]	Bay of Islands complex [27]	10700	125	32	—	160	0.17	7
5	Løkken, Norway [28, 29]	Løkken [30]	8550	93	26	4	221	0.09	38
6	Røros, Norway [31]	Gjelsvik sequence [32]	9850	129	32	—	114	0.14	18
7	Joma, Norway [28, 29]	Joma [30]	11400	171	26	35	367	0.19	5
8	Bidjovagge, Norway	Finnmark [33]	8050	103	24	—	171	0.09	18
9	Gjersvik, Norway [28, 29]	Gjersvik [30]	8400	73	27	5	24	0.07	16
10	Buchans, Newfoundland	Buchans [34]	3900	75	—	—	74	0.13	90
11	Lynn Lake, Manitoba [36]	Rusty Lake greenstones [33, 35]	5100	54	16	2	139	0.24	24
12	Noranda, Ontario [36]	Noranda [37]	7150	92	—	—	157	0.13	78

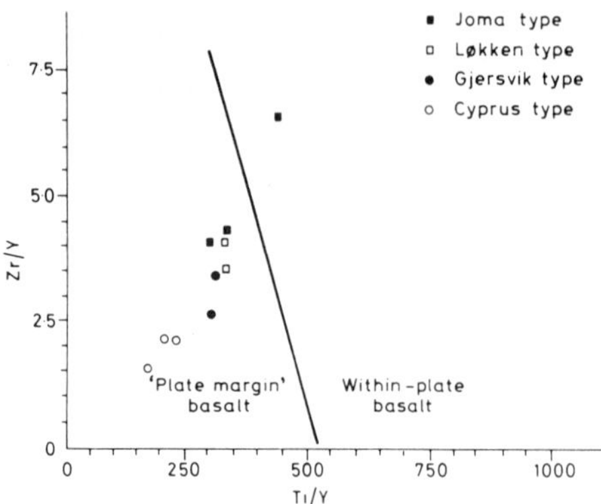

Fig. 3   Mean analyses of metabasalts from various massive sulphide deposit types plotted on Zr/Y—Ti/Y discrimination diagram (analogous to Ti—Zr—Y diagram of Pearce and Cann [8]). Diagram separates within-plate from other magma types

is a general increase in Ti and Zr from island arc through ocean-floor to within-plate magma types. Fig. 3 is a plot of Ti/Y against Zr/Y. This distinguishes within-plate from other magma types, making use of the distinctive enrichment in Ti and Zr (but not Y) of within-plate basalts; Fig. 4 is a plot of Ti and Cr[11] which separates the 'plate margin basalts' of Fig. 3 into ocean-floor and island arc basalts, since for

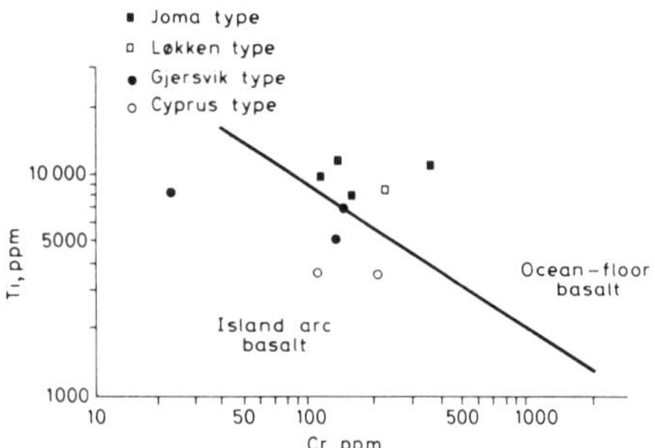

Fig. 4   Mean analyses of metabasalts from various massive sulphide deposit types plotted on log Ti—log Cr[11] discrimination diagram

any given Ti concentration island arc basalts will nearly always contain lower Cr concentrations.

Fig. 5 shows chondrite-normalized rare-earth patterns, which provide useful additional information on magma type. Ocean-floor tholeiites and island arc tholeiites show slight rare-earth element depletions; ocean-floor alkali basalts and island arc calc-alkaline basalts show slight rare-earth element enrichment; whereas within-plate basalts show moderate to strong enrichment in the light rare-earth elements. The massive sulphide deposits investigated in this paper are associated with various basalt types listed above and can be divided into four categories according to the characteristics of the host lavas. Each of these categories has been named after one of its member deposits, and its inferred tectonic environment is given in parenthesis.

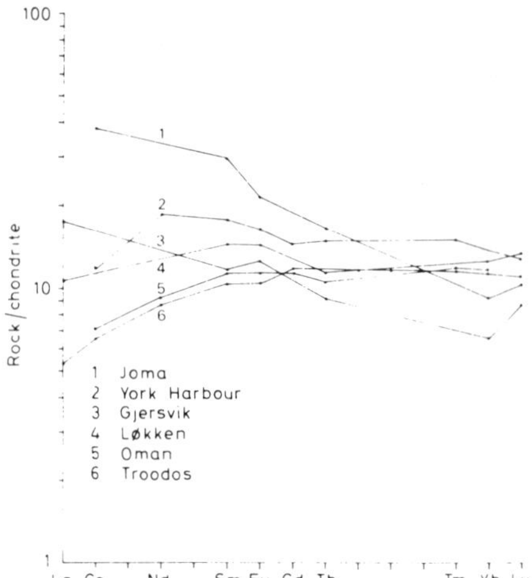

Fig. 5   Chondrite-normalized rare-earth patterns for typical samples of metabasalt from various massive sulphide deposit types. Cyprus analysis from Kay and Senechal [38]

Cyprus-type (?early back-arc) deposits   Deposits of this type were identified from the Troodos Massif, Cyprus,[20, 21] Oman [23] and Betts Cove, Newfoundland,[24] and all contained Cu > Zn. Geologically, these deposits are situated in the lava unit of fully developed ophiolite complexes, from which it is implicit that they were formed at a constructive plate margin. Basaltic pillow lavas form the main host rocks. Cyprus and Oman contain two lava units [21,22] — the lower basaltic, the upper containing rocks of komatiitic composition as well as basalts. The lower unit is thought to have been erupted at the ridge axis; the upper off-axis outside the zone of deep circulation of sea water. Ore deposits in these areas are mostly located at the upper—lower pillow lava boundary, though some (such as Lasail, Oman [23]) occur within the lower sequence and are cut by numerous feeder dykes to the overlying lavas.

Geochemically, the lavas are characterized by low concentrations of the small ion lithophile elements, including Cr, and therefore classify as island arc tholeiites in Figs. 2, 3 and 4; the light rare-earth-depleted patterns in Fig. 5 are consistent within this interpretation. Clearly, the geological and geochemical data are at variance. A consistent model can,

Table 3   Some important characteristics of Cyprus lavas

1	Two petrologically distinct lava units
2	Concentrations of Ti, Zr and Y decrease up lava sequence [22]
3	Overall, Ti, Zr and Y concentrations are low [11, 22] and there are strong light rare-earth element depletions [38]
4	Low Cr and MgO concentrations (at given Ti concentration) [11]
5	Basaltic komatiites present
6	Orthopyroxene phenocrysts in some lavas [11, 22]
7	$^{87}Sr/^{86}Sr \sim 0.7036$ in rocks apparently unaffected by isotopic exchange with sea water [39]
8	Chromite deposits
9	Underlying 'upper mantle' ultramafics very depleted [40] in their basaltic component

however, be built up by considering a transitional back-arc environment where new oceanic crust is being formed at spreading axes above subduction zones. There is some petrogenetic evidence for this hypothesis. Table 3 lists some of the ways in which lavas of the Cyprus type differ from rocks dredged and drilled from major ocean ridges. Features 3, 5 and 9 in Table 3 can be explained in terms of high degrees of partial melting under hydrous conditions; features 4, 6 and 8 can be explained by crystal fractionation under hydrous conditions;[41] feature 7 can be explained by input of water rich in radiogenic Sr derived from subducted oceanic crust; features 3—7 are common characteristics of island arc volcanic rocks. This particular type of back-arc environment is therefore an attractive possibility, although this is still at present only one of several hypotheses proposed.

Løkken-type (evolved back-arc) deposits   Løkken, Norway,[28, 29] and York Harbour, Newfoundland,[26, 27] fall into this category. These deposits contain Cu > Zn and Cu < Zn, respectively. The York Harbour deposit is found in the lavas of the Bay of Islands ophiolite complex; the Løkken deposit occurs within an inverted lava sequence with associated gabbros but no exposed ultramafics. Both Løkken and York Harbour apparently contain two basaltic lava units, one above and one below the ore deposit.

Geochemically, the lavas contain higher concentrations of all the small ion lithophile elements in comparison with Cyprus-type lavas, although the upper lava units again appear more depleted in these elements than the lower units. These lavas classify as ocean-floor basalts in Figs. 2, 3 and 4 and this is supported by light rare-earth element depletion in Fig. 5. The question still remains whether the setting was in a major or marginal ocean basin; lavas of similar character have been found in both environments. The geological setting of the Løkken lavas [42] and their high concentration of large ion lithophile elements (which cannot be accounted for solely by alteration) both favour a back-arc setting similar to the Marianas basin at the present day. The situation for York Harbour is ambiguous in view of the conflicting interpretations of Newfoundland geology,[43] and both a major or marginal basin environment are possible.

Joma-type (continental margin) deposits   This category includes Palaeozoic deposits from Joma, Norway,[26, 29] Røros, Norway,[31] and the Proterozoic deposit at Bidjovagge in Finnmark, Norway.[30] These deposits are all related to greenschist facies metabasalts and more silicic tuffs; metagabbros and serpentinites of uncertain association are also found in these areas. Metamorphosed sedimentary rocks overlie or are interbedded with the volcanics, and these include argillaceous and arenaceous rocks in the Røros area, and quartzites and limestones at Joma and in Finnmark. Geologically, the environment appears to resemble that of 'Sullivan-type' [5] deposits.

Geochemically, the Joma samples contain high Ti/Y and Zr/Y ratios (Figs. 2 and 3) and light rare-earth-element-enriched rare-earth patterns (Fig. 5), both indicative of a within plate origin. The Røros and Finnmark lavas had lower Ti and Zr concentrations, and this, coupled with high Cr concentrations, points to an ocean-floor basalt character. Geochemically, they closely resemble the Løkken type of greenstone. If the geology and the geochemistry are combined, a model can be built up to suggest that these deposits were formed at a continental margin, in a small ocean of Red Sea type. At Joma both the geology (shelf sedimentation) and geochemistry (the within-plate character) support this. At Røros and Finnmark the lavas have an ocean-floor basalt character; but, as Fig. 2 shows, they are chemically somewhat transitional towards within-plate basalts; the presence of silicic tuffs and interbedded sediments also argues against a normal oceanic ridge setting. Their geochemistry and the geological settings are therefore both consistent with their formation in a continental margin environment.

Gjersvik-type (island arc) deposits   This type of deposit, defined by the peculiar geochemical characteristics of its related basic lavas, encompasses several sub-classes of very different geological character.

With regard to the *Gjersvik type*, the Gjersvik Cu—Zn deposit itself [28, 29] lies within a pile of predominantly basic metavolcanic rocks, although some andesitic and silicic lavas, agglomerates and dykes are present. Geochemically, the lavas contain moderate concentrations of Ti, Zr and Y (Figs. 2 and 3), but very low Cr (Fig. 4), which classifies them as island arc basalts; flat rare-earth patterns are consistent with this interpretation. The nearby Zn—Cu deposit at Skorovas, described elsewhere in this volume,[44] has a similar setting, but with more andesites and fewer basalts in the volcanic pile; the geochemistry of the basic lavas from Gjersvik and Skorovas is very similar.[44] The deposits seem on this evidence to have formed during submarine volcanism in an island arc setting.

Kuroko-type deposits have been described in detail from Japan and elsewhere,[45, 46] where the geological evidence favours an island arc origin. There are, however, several geological differences between these deposits and the Gjersvik type of deposit. These differences include the nature of the lavas (Kuroko deposits are mainly related to acid pyroclastic rocks), and the ore mineralogy and texture (Kuroko deposits contain Cu, Zn and Pb minerals with barite and gypsum). Kuroko-type deposits appear to be genetically related to small acid intrusions. Nevertheless, published geochemical data on andesites from the Buchans deposit, Newfoundland,[34] closely resemble data on the andesites from the Gjersvik area.

The *Noranda type* of Zn—Cu deposit occurs within Precambrian greenstone terrain in association with acid pyroclastic rocks.[47] Geochemical data on the more basic rocks from the Noranda [37] and Lynn Lake [33, 35] regions revealed low to moderate Ti, Zr and Y concentrations (Figs. 2 and 3) and low Cr (Fig. 4) — characteristics very similar to those of basic lavas from the Gjersvik region. The Lynn Lake lavas appeared to be tholeiitic in character, the Noranda lavas more calc-alkaline. Thus, although we do not yet know sufficient about tectonic processes in the Archaean to interpret these data in terms of tectonic environment, we can say that the lava geochemistry most closely resembles that from present-day island arcs.

Tectonic setting of massive sulphide deposits

The main conclusion that can be drawn from this study is that massive sulphides can form in a variety of environments. Most of the examples used here appeared to have formed in a back-arc setting or during incipient formation of an ocean basin; Fig. 6 illustrates the possible environments of formation of the various types of massive sulphide deposit in a typical western Pacific arc—trench system.

The results also raise the important question: do all settings in which submarine volcanism takes place have an equal massive sulphide forming potential? The evidence from this paper suggests that perhaps they do not. With the possible exception of the small deposit at York Harbour, no good examples have been found of massive sulphide deposits related to the major ocean ridge types of volcanic

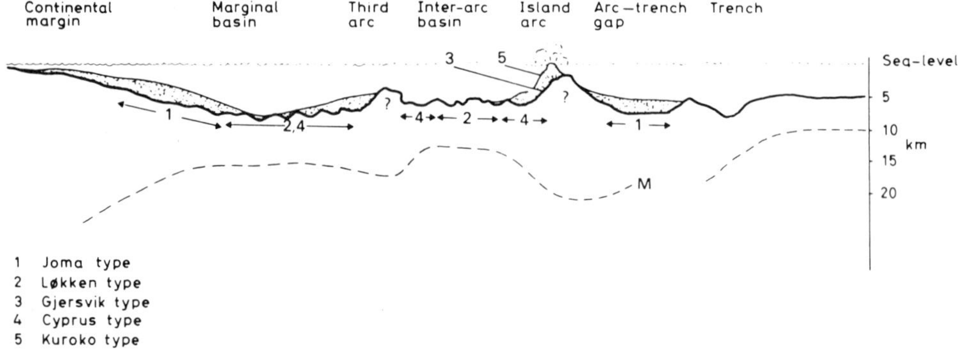

Fig. 6   Possible distribution of massive sulphide deposit types in marginal basin setting. Section taken from Karig [80]

rock. This might be explained in terms of the absence of true mid-ocean ridge ophiolites in the geologic record. A few probable examples do exist, however, in Macquarie Island,[4] the western Alps [49,50] and Støren and Stavenes [30] in the Norwegian Caledonides. These do contain tiny deposits, confirming that ore-forming processes do take place in this environment; but there are no known deposits of any major economic value.

If this apparent absence of deposits in major oceans is real, it can be explained in several ways. Magmatic fluids could be more important than has hitherto been considered; perhaps magmatic fluids in a back-arc setting contain a source of reduced sulphur derived from the breakdown of pyrite to form pyrrhotite in subducted oceanic crust. Perhaps the passage of the fluids through evaporites at some continental margins yields an important source of sulphur. Both these possibilities can be tested by thermodynamic calculations. The most probable explanation may, however, lie in the necessity to have a *site* for ore deposition. The highly faulted ocean crust that exists during early spreading and in back-arc settings could provide fault-bounded troughs where low Eh and restricted sea-water circulation would provide a favourable environment for ore deposition; these conditions are less likely to be met in an evolved ocean.

Tin deposits

Magmatism in the continental environment

Tin deposits are related to acid-intrusive rocks emplaced into continental crust — at 'hot spots' within plates, at destructive plate margins and as a result of continent—continent collision. These rocks belong to two main magma types: *volcanic arc magmas* (which include rocks of the tholeiitic, calc-alkaline and shoshonitic series) and *within-plate magmas* (which include rocks of the tholeiitic and alkalic series). There are substantial geochemical variations both within and between these magma types, whose relationship with tectonic setting is illustrated in Fig. 7.

Acid and intermediate within-plate magmas tend to have high concentrations of both large and small ion lithophile elements, with the exception of a few elements, e.g. Ti and Sr, which have not behaved incompatibly during magma genesis. The concentrations of the incompatible elements increase from tholeiitic to alkali magma types. Volcanic arc magmas also contain high conentrations of the incompatible large ion lithophile elements — concentrations which increase from tholeiitic through calc-alkaline to shoshonitic types. By contrast, however, concentrations of small ion lithophile elements are low, particularly in the island arc tholeiitic series.

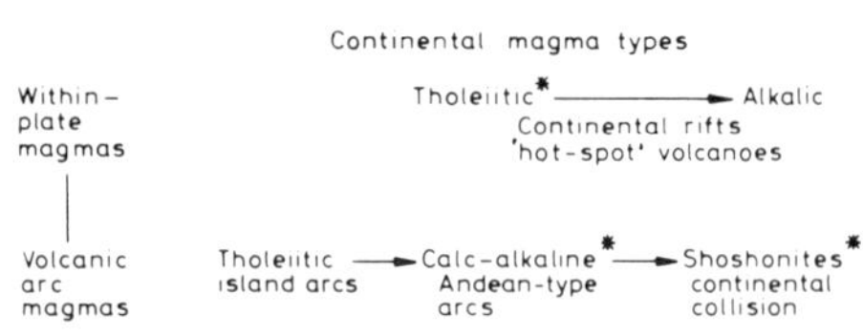

Fig. 7   Distribution of magma types within continental environment. Magma types transitional between within-plate and volcanic arc types do occur, but only in rare tectonic settings

The most effective elements for identifying whether an acid igneous rock has a within-plate or volcanic arc character are therefore those with small highly charged ions. Nb is particularly useful, and diagrams such as SiO_2—Nb, K_2O—Nb and Zr—Nb are all valuable in the identification of magma type. The SiO_2—Nb diagram presented in Fig. 8 separates low Nb volcanic arc magmas from high Nb within-plate magmas. An overlap occurs on this diagram between tholeiitic within-plate magmas and high K calc-alkaline or shoshonitic magmas from volcanic arc and post-orogenic settings. K_2O and Rb can be used to partly eliminate this overlap, provided that K-silicate alteration has not taken place. *Acid* within-plate magmas range from about 100 ppm (at Nb = 15 ppm) to 300 ppm Rb (at Nb = 500 ppm), whereas *acid* volcanic arc magmas range from about less than 50 ppm Rb (at Nb = 5 ppm) to greater than 200 ppm (at Nb > 20 ppm). This rough guide suggests that acid rocks which contain very high Rb and K_2O concentrations and which plot within the overlap field have a volcanic arc origin.

Because so little is known about geochemical and mineralogical variations in the lower crust and upper mantle it is impossible to give a precise explanation for this diagram. A few general points can, however, be made.

It is apparent from the magnitude of the Nb variation (5—500 ppm in acid rocks) that variations in the degree of

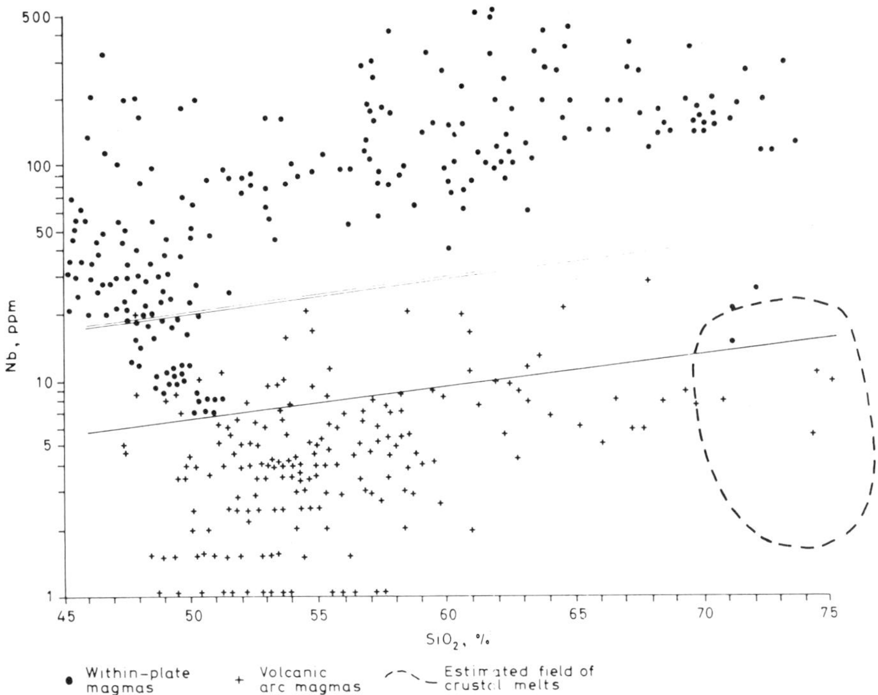

Fig. 8   Nb–SiO$_2$ diagram showing data from volcanic rocks of known tectonic settings. Single diagonal line gives lower limit for within-plate magma type; double diagonal line gives upper limit for volcanic arc magma type. In general, Nb content of within-plate magmas increases from tholeiitic to alkalic types, and Nb content of volcanic arc magmas increases from island arc to shoshonitic types. Estimate of Nb–SiO$_2$ range of crustal melts derived from sediments metamorphosed up to and including amphibolite facies superimposed on diagram

partial melting and/or crystal fractionation cannot alone explain why this element does discriminate between the various tectonic environments. The nature and composition of the source rock may be the single, most important factor.

Partition coefficients for Nb[33] between crystal and melt are probably low (<0.2) for olivine, orthopyroxene, garnet and plagioclase, moderately low (<0.6) for clino-pyroxene, moderately high (>1.0) for amphibole and high for micas (>2.5), magnetite, apatite and zircon. As a result, the presence of minor phases, particularly minerals such as phlogopite,[52,53] in the source rock and their behaviour during melting will play an important part in determining the concentration of Nb in the melt. Because of their high ionic potential, Nb^{5+} ions will not be strongly partitioned into aqueous fluids. Ions of low ionic potential, such as K$^+$, will, however, be readily transported in these fluids.

This last point may partially explain the geochemical characteristics of volcanic arc magmas. The mantle wedge above the subduction zone will be modified by aqueous fluids derived from dehydration of oceanic lithosphere.[54,55] These fluids should contain high K$^+$ concentrations and relatively low Nb^{5+} concentrations. Thus, any acid magma derived from partial melting of this mantle followed by crystal fractionation will not be enriched in Nb. By contrast, heterogeneities in within-plate settings involve pro-

cesses that affect K$^+$ and Nb^{5+} to an approximately equal extent. These could be caused by migration of small amounts of incompatible rich melt[56] or represent variations in the melting–recycling history of the mantle. In either case partial melting of mantle enriched in this way followed by crystal fractionation should produce acid melts rich in Nb. There are also other factors that could also partially explain the Nb variations — in particular, the behaviour of some of these Nb-bearing minor phases in the mantle. For example, residual mica phases will reduce the Nb content of the melt in equilibrium with them.

The arguments outlined above apply to magmas derived from the mantle; but, of course, some acid magmas may have had a crustal source. A field for crustal melts has therefore been superimposed on this diagram. This was delineated by explaining the geochemistry of concordant granitic veins in Precambrian terrains (see, for example, Bowes[57] and Drury[58]) and by carrying out petrogenetic calculations on the melting of metasedimentary rocks. Nb-bearing phases, hornblende, biotite, muscovite or phlogopite, are usually residual phases during partial melting of pelitic and siliceous sediments[59,60,61] and this constrains the melt to low Nb concentrations in most models. There are, however, some exceptions to this: partial melting of eclogite and some granulites will leave pyroxene, garnet and plagioclase as residual phases, and high Nb melts can therefore be produced.

Most granites produced by melting of continental crust at depths where amphiboles and micas are residual phases should, however, plot in or close to the area shown and will geochemically overlap (and perhaps include) tholeiitic within-plate magmas and magmas from volcanic arcs, particularly as far as their Nb and SiO_2 concentrations are concerned.

Geochemical characteristics of tin-bearing magmas

Table 4 summarizes some of the important geochemical characteristics of tin-bearing granites from Nigeria,[62,63,64] Cornwall,[17,18] Bolivia[65] and Indonesia.[66] These data have been plotted on to the SiO_2–Nb diagram (Fig. 9).

The common characteristic of these four granite provinces is their very high concentration of K_2O and Rb. There are, however, significant differences in the concentrations of other elements. The Nigerian granites contain very high concentrations of Nb, Zr, Y and the rare earths and plot in the within-plate magma field in Fig. 9. This interpretation is consistent with geological evidence for a 'hot-spot' origin for these granites.[15,61,62] The other granites contain much lower concentrations of these elements and, in Fig. 9, plot in the overlap zone between volcanic arc and within-plate magmas; the very high Rb concentrations, however, point to a volcanic arc magma type. Geological criteria are then needed to distinguish between an active volcanic arc and a post-orogenic setting. The Bolivian granites were clearly related to subduction processes; however, their setting in the Eastern Cordillera of the Andes argues in favour of an origin associated with tensional stresses and back-arc rifting rather than with the processes that produced magmas in the Western Cordillera (the volcanic arc *sensu stricto*). The Cornish granites are geochemically almost identical to the Bolivian granites, although geological opinion is divided between a post-orogenic and a volcanic arc setting.[17,18] The Indonesia granites contain concentrations of Rb, Zr and Nb similar to those of the Bolivian and Cornish granites, but significantly higher concentrations of Y and the heavy rare earths. This does not indicate any obvious difference in tectonic environment (a back-arc rifting setting is geologically likely[16]), but it does suggest significant differences in the role of garnet during the genesis of the magma.

To illustrate the composition of intrusive rocks related to porphyry copper deposits some analyses[60] from the region of the Coed-y-Brenin porphyry copper deposit[67] in Wales have been plotted in Fig. 9. They plot, as expected[68,69] in the field of volcanic arc magmas, their very low Nb concentrations favouring an island arc rather than continental margin setting.

Table 4   Mean concentrations (ppm) of some trace elements in tin-bearing granites (note that Nigeria data include all Younger Granites; tin-bearing biotite granites have slightly lower concentrations of incompatible trace elements)

	Nigeria	Bolivia	Indonesia	Cornwall
SiO_2, %	74.5	67.1	70.8	72.4
Zr	650	173	130	78
Y	220	19	64	13
Nb	170	21	15	15
La	185	41	68	33
Sm	57	5.2	11.2	7.0
Yb	34	1.6	4.7	0.9
Rb	330	310	405	580

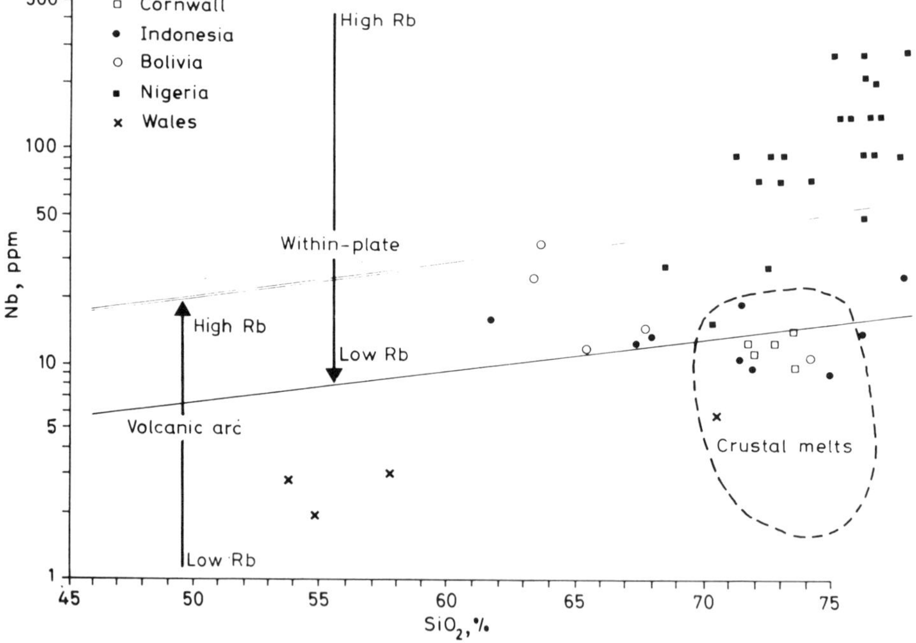

Fig. 9 SiO_2 and Nb data from some tin-bearing granites plotted on to Nb–SiO_2 diagram of Fig. 8. Also plotted for comparison are analyses of intrusive rocks from Coed-y-Brenin porphyry copper district, Wales

Source of tin-rich magmas

A number of hypotheses have been proposed to explain why certain magmas should have high tin concentrations: these include the melting of tin-rich detrital sediments [70] or granitic rocks [71] in continental crust; melting of upper mantle;[72] and evolution of fluorine-rich fluids derived from breakdown of apatite in subducted oceanic crust followed by mantle or crustal melting.[16] Here we briefly set out some of the petrological constraints on such processes, emphasizing the deep-seated processes rather than the processes that concentrate and deposit the tin at shallow crustal levels. The following factors appear to be significant.

(*a*)   As was stressed in the previous section, the only obviously distinctive features of tin-bearing granites are that they are emplaced into continental rather than oceanic crust and that they are regional rather than local phenomena. They can be formed in most tectonic settings.

(*b*)   Examination of patterns of trace-element partitioning [73] and the few available Sn analyses of mineral phases suggests that partition coefficients for Sn between crustal and melt will be low for quartz, feldspar, pyroxene, garnet or olivine, moderate for amphibole, but high ($>$ 1) for micas and other minor phases such as apatite, zircon and sphene. This means that production of tin-rich magmas by partial melting is not possible if micas and amphiboles make up a significant part of the residue. Thus, partial melting of pelitic rocks or of any rock metamorphosed in amphibolite or greenschist facies is not likely to yield tin-rich magmas; partial melting of eclogite and granulite facies metabasic rocks or partial melting of the mantle could yield tin-rich magmas, provided that the tin-bearing phases, such as phlogopite, are completely melted and the source rocks themselves contain sufficiently high tin concentrations.

(*c*)   Although there are few thermodynamic data on tin complexes, it is evident that tin has a particular affinity for fluorine (see, for example, Mitchell and Garson [16] and Barsukov [74]) and that the partitioning of tin into a melt or aqueous fluid will increase with increasing fluorine concentration. The main fluorine-bearing minerals in the lower crust and mantle are phlogopite and apatite (minerals which also accommodate tin), so the stability of these minerals is likely to be particularly important.

(*d*)   It is well established [75] that fluorine-rich melts and fluids can leach tin from mica lattices. Reaction between fluorine-rich magma and wallrock may therefore be a potent mechanism for tin enrichment.

(*e*)   The small amount of available data shows that tin concentrations in mantle-derived basalts range from 0.2 ppm (see, for example, Gill [76]) in island arc tholeiites to $>$ 5 ppm in within-plate alkali basalts. In addition, a recent compilation by Bailey and Macdonald [77] has shown that peralkaline obsidians from continental settings contain significantly higher F/Cl ratios than those from oceanic settings.

The factors listed above have implications for the origin of the four tin-bearing granite provinces discussed in this paper. The Nigerian granites may give the simplest story, since no subduction zone was involved in their genesis. These granites appear to contain concentrations of Zr, Nb and the rare earths that are much higher than those expected for crustal melts, even when the proposed enrichment or redistribution of these elements during wallrock alteration has been taken into account.[78] Furthermore, their geochemistry resembles that of acid rocks from ocean islands where no continental crust existed. This evidence supports a mantle origin for the granites — by partial melting and subsequent crystal fractionation. Initial $^{87}Sr/^{86}Sr$ ratios and lead isotope ratios [63] indicate that some crustal contamination may have taken place, although without stable

isotope data this cannot be regarded as unequivocal evidence. The tin itself could have both a mantle component (see above) and a crustal component (tin-bearing pegmatites have been recorded in the Pan-African basement [63,64]).

The other tin-bearing granites contain lower concentrations of the small ion lithophile elements and some analyses do overlap the field of crustal melts in Fig. 9; neither a crustal origin nor an origin from mantle enriched by fluids emanating from subducted oceanic crust can therefore be ruled out. We suggest the following multistage model to explain the origin of the Bolivian tin-bearing granites. This model most closely resembles that of Mitchell and Garson,[16] although there are a number of important differences. *Stage one* involves subduction of oceanic crust beneath the South American continent. During the long time-interval (Late Triassic—Tertiary)[79] between incipient subduction and emplacement of tin-bearing magmas, the composition of the mantle wedge beneath Bolivia is altered by input of small amounts of melt/aqueous fluids derived from eclogite in the subducted oceanic lithosphere. A mantle phase such as phlogopite could accommodate both tin and fluorine (and, of course, other elements, such as potassium) introduced in this way.

Stage two involves partial melting of this enriched mantle in a back-arc rifting environment. The tectonic process is probably analogous to that which causes marginal basins to form behind island arcs, but with one important difference — the time-gap between incipient subduction and rifting is much greater because of the great thickness of continental crust. This time-gap may be necessary to permit sufficient accumulation of tin in the mantle prior to partial melting. *Stage three* involves ascent of magma to the surface. Further concentration of tin can take place by crystal fractionation and by wallrock alteration. Alternatively, this basic magma may initiate partial melting of the lower crust; high tin concentrations are likely if the source rocks are in eclogite or granulite facies and are enriched in tin by volatile transfer from the underlying basic magma.

The details of this model may need to be modified for the Cornish and Indonesian provinces. For example, the melting of enriched mantle in stage two might result from continent—continent collision rather than back-arc rifting. A similar model could also explain the Nigerian province; the Nigerian Younger Granites could have been derived from mantle that had been enriched in tin during the Pan-African tectonothermal event but that had not been involved in magma genesis until the Jurassic. It must be emphasized, however, that this model is only one of several that could explain the data available. Nevertheless, it does fit several petrogenetic constraints and more and better petrogenetic data will enable it to be rejected, modified or improved.

Conclusions

(1)   Stable trace-element characteristics of metabasalts related to massive sulphide deposits indicate that such deposits have been formed in a variety of different environments.

(2)   Geology and geochemistry together indicate that the deposits studied formed at Red Sea and marginal basin types of spreading centres and in island arcs: to our knowledge no large deposit was formed at a spreading axis within a large ocean.

(3)   We speculate that within oceanic crust really favourable sites for ore deposition may only exist in highly faulted small ocean basins of Red Sea and marginal basin type.

(4)   The geochemical characteristics and geological setting of tin-bearing granites indicate that tin deposits of this type can form within continental crust at 'hot spots', in exten-

417

sional regimes above subduction zones and in continental collision environments.

(5)   We speculate that tin-rich magmas can form by a three-stage process that involves (a) a long period of enrichment of the mantle above a subduction zone in tin and other elements as a result of incipient melting of, or reactions within, subducted oceanic crust metamorphosed to eclogite facies, (b) partial melting of this phlogopite-bearing tin-enriched mantle as a result of back-arc spreading processes, continental collision or 'hot-spot' activity and (c) crystal fractionation, wallrock reactions and perhaps crustal melting as this initially basic magma ascends to the surface.

Acknowledgment

We gratefully acknowledge the assistance of the following: Dr. A. O. Brunfelt, P-R. Graff, G. Marriner, Dr. P. J. Potts, M. Sarre, B. Sundvoll and C. K. Winter for help and advice on analytical techniques; Dr. D. Alderton, Dr. J. N. Grant and Dr. R. H. Verschure for samples from tin provinces; Martin Taylor (Prospection, Ltd.) for samples of drill core from the Lasail prospect, Oman; Dr. W. R. A. Baragar, Tor Grenne, Dr. W. J. Rea and M. Steeves for unpublished data; Dr. J. R. Cann, J. Cramer, A. J. Fleet, Dr. C. Halls, Dr. D. Roberts, Professor F. M. Vokes, J. B. Wright and geologists from the various massive sulphide mines and prospects for valuable discussions; and Professor I. G. Gass, Dr. I. L. Gibson and J. B. Wright for constructive criticism of the manuscript. Most of the rare-earth analyses were carried out (by J.A.P.) at the Mineralogisk—Geologisk Museum in Oslo during tenure of a Royal Society European fellowship; X-ray fluorescence analyses were carried out at the University of East Anglia, Norwich; Bedford College, London; and NGU, Trondheim, Norway.

References

1. Sillitoe R. H. Formation of certain massive sulphide deposits at sites of sea-floor spreading. *Trans. Instn Min. Metall. (Sect B: Appl. earth sci.)*, **81**, 1972, B141—8.

2. Sillitoe R. H. Environments of formation of volcanogenic massive sulfide deposits. *Econ. Geol.*, **68**, 1973, 1321—5.

3. Hutchinson R. W. Volcanogenic sulfide deposits and their metallogenic significance. *Econ. Geol.*, **68**, 1973, 1223—46.

4. Sawkins F. J. Sulfide ore deposits in relation to plate tectonics. *J. Geol.*, **80**, 1972, 377—97.

5. Sawkins F. J. Massive sulphide deposits in relation to geotectonics. In press.

6. Cann J. R. Rb, Sr, Y, Zr and Nb in some ocean floor basaltic rocks. *Earth planet. Sci. Lett.*, **10**, 1970, 7—11.

7. Pearce J. A. and Cann J. R. Ophiolite origin investigated by discriminant analysis using Ti, Zr and Y. *Earth plant. Sci. Lett.*, **12**, 1971, 339—49.

8. Pearce J. A. and Cann J. R. Tectonic setting of basic volcanic rocks determined using trace element analyses. *Earth planet. Sci. Lett.*, **19**, 1973, 290—300.

9. Floyd P. A. and Winchester J. A. Magma type and tectonic setting discrimination using immobile elements. *Earth planet. Sci. Lett.*, **27**, 1975, 211—8.

10. Bloxam T. W. and Lewis A. D. Ti, Zr, and Cr in some British pillow lavas and their petrogenetic affinities. *Nature, phys. Sci.*, **237**, 1972, 134—6.

11. Pearce J. A. Basalt geochemistry used to investigate past tectonic environments on Cyprus. *Tectonophysics, **25**, 1975, 41—67.

12. Herrmann A. G. Potts M. J. and Knake D. Geochemistry of the rare earth elements in spilite from the oceanic and continental crust. *Contr. Mineral. Petrol.*, **44**, 1974, 1—16.

13. Sillitoe R. H. A plate tectonic model for the origin of porphyry copper deposits. *Econ. Geol.*, **67**, 1972, 184—97.

14. Halls C. Tin and plate tectonics. *Ind. Eastern Eng.*, **116**, 1974, 493—4.

15. Sillitoe R. H. Tin mineralisation above mantle hot spots. *Nature, Lond.*, **248**, 1974, 497—9.

16. Mitchell A. H. G. and Garson M. S. Relationship of porphyry copper and circum-Pacific tin deposits to palaeo-Benioff zones. *Trans. Instn Min. Metall. (Sect B: Appl. earth sci.)*, **81**, 1972, B10—25.

17. Mitchell A. H. G. Southwest England granites: magmatism and tin mineralization in a post-collision tectonic setting. *Trans.Instn Min. Metall. (Sect. B: Appl. earth sci.)*, **83**, 1974, B95—7.

18. Bromley A. V. Tin mineralization of Western Europe: is it related to crustal subduction? *Trans. Instn Min. Metall. (Sect. B: Appl. earth sci.)*, **84**, 1975, B28—30.

19. Spooner E. T. C. and Fyfe W. S. Sub-sea-floor metamorphism, heat and mass transfer. *Contr. Mineral. Petrol.*, **42**, 1973, 287—304.

20. Hutchinson R. W. and Searle D. L. Stratabound pyrite deposits in Cyprus and relations to other sulphide ores. In *Proc. IMA—IAGOD meetings '70, IAGOD volume* (Tokyo: Society of Mining Geologists of Japan, 1971), 198—205. *(Spec. Issue no. 3)*

21. Constantinou G. and Govett G. J. S. Genesis of sulphide deposits, ochre and umber of Cyprus. *Trans. Instn Min. Metall. (Sect. B: Appl. earth sci.)*, **81**, 1972, B34—46.

22. Smewing J. D. Simonian K. O. and Gass I. G. Metabasalts from the Troodos Massif, Cyprus: genetic implication deduced from petrography and trace element geochemistry. *Contr. Mineral. Petrol.*, **51**, 1975, 49—64.

23. Prospection Ltd. Personal communications, 1976.

24. Upadhyay H. D. and Strong D. F. Geological setting of the Betts Cove copper deposits, Newfoundland: an example of ophiolite mineralization. *Econ. Geol.*, **68**, 1973, 161—7.

25. Kidd W. S. F. The evolution of the Baie Verte lineament, Burlington Peninsula, Newfoundland. Ph.D. thesis, University of Cambridge, 1974.

26. Duke N. A. and Hutchinson R. W. Geological relationships between massive sulfide bodies and ophiolitic volcanic rocks near York Harbour, Newfoundland. *Can. J. Earth Sci.*, **11**, 1974, 53—69.

27. Williams H. and Malpas J. Sheeted dikes and brecciated dike rocks within transported igneous complexes, Bay of Islands, Western Newfoundland. *Can. J. Earth Sci.*, **9**, 1972, 1216—29.

28. Vokes F. M. Caledonian massive sulphide deposits in Scandinavia: a comparative review. In press.

29. Vokes F. M. and Gale G. H. Metallogeny relatable to continental drift in sothern Scandinavia. In press.

30. Gale G. H. Reconnaissance geochemical survey of Caledonian volcanics and intrusives in central and south

Norway: Part II — Analytical data. *Rep. Norges. geol. unders.* 1228A, 1974.

31. Rui I. J. Structural control and wall rock alteration at Killingdal mine, Central Norwegian Caledonides. *Econ. Geol.,* **68**, 1973, 859—83.

32. Grenne T. and Pearce J. A. Unpublished data.

33. Pearce J. A. Unpublished data.

34. Thurlow J. G. Swanson E. A. and Strong D. F. Geology and lithogeochemistry of the Buchans polymetallic sulfide deposits, Newfoundland. *Econ. Geol.,* **70**, 1975, 130—44.

35. Steeves M. A. Geology of the Rusty Lake area. *Publ. Manitoba Mines Brch.* In press.

36. Hutchinson R. W. Ridler R. H. and Suffel G. G. Metallogenic relationships in the Abitibi Belt, Canada: a model for Archaean metallogeny. *CIM Trans.,* **74**, 1971, 106—15.

37. Baragar W. R. A. Unpublished data.

38. Kay R. W. and Senechal R. G. The rare earth geochemistry of the Troodos ophiolite complex. *J. geophys. Res.,* **81**, 1976, 964—70.

39. Chapman H. J. Spooner E. T. C. and Smewing J. D. ^{87}Sr enrichment of ophiolitic rocks from Troodos, Cyprus, indicates sea water interaction. *Trans. Am. geophys. Union E⊕S,* **56**, 1975, 1074. (Abstract)

40. Menzies M. and Allen C. Plagioclase lherzolite—residual mantle relationships within two eastern Mediterranean ophiolites. *Contrib. Mineral. Petrol.,* **45**, 1974, 197—213.

41. Nicholls I. A. and Ringwood A. E. Production of silica-saturated tholeiitic magmas in island arcs. *Earth planet. Sci. Lett.,* **17**, 1972, 243—6.

42. Gale G. H. and Roberts D. Trace element geochemistry of Norwegian lower Palaeozoic basic volcanics and its tectonic implications. *Earth planet. Sci. Lett.,* **22**, 1974, 380—90.

43. Strong D. F. Plate tectonic setting of Appalachian—Caledonian mineral deposits as indicated by Newfoundland examples. *Trans. Am. Inst. Min. Engrs,* **256**, 1974, 121—8.

44. Halls C. *et al.* Geological setting of the Skorovas orebody within the allochthonous metavolcanic stratigraphy of the Gjersvik Nappe, central Norway. In *Volcanic processes in ore genesis* (London: Institution of Mining and Metallurgy and Geological Society, 1977), 128—51.

45. Horikoshi E. Volcanic activity related to the formation of the Kuroko-type deposits in the Kosaka District, Japan. *Mineral. Deposita,* **4**, 1969, 321—45.

46. Clark L. A. Volcanogenic ores: comparison of cupriferous pyrite deposits of Cyprus and Japanese Kuroko deposits. In *Proc. IMA—IAGOD meetings '70, IAGOD volume* (Tokyo: Society of Mining Geologists of Japan, 1971), 206—15. (*Spec. Issue* no. 3)

47. Sangster D. F. Precambrian volcanogenic massive sulphide deposits in Canada: a review. *Pap. geol. Surv. Can.,* 72—22, 1972, 44 p.

48. Varne R. and Rubenach M. J. Geology of Macquarie Island and its relationship to oceanic crust. In *Antarctic oceanology II: the Australian—New Zealand Sector* Hayes D. E. ed. (Washington, D.C.: American Geophysical Union, 1972), 251—66. (*Antarct. Res. Ser.* no. 19)

49. Bickle M. J. and Pearce J. A. Oceanic mafic rocks in the Eastern Alps. *Contrib. Mineral. Petrol.,* **49**, 1975, 177—89.

50. Dietrich V. J. Evolution of the Eastern Alps: a plate tectonics working hypothesis. *Geology,* **4**, 1976, 147—52.

51. Delong S. E. Hodges F. N. and Arculus R. J. Ultramafic and mafic inclusions, Kanaga Island, Alaska and the occurrence of alkaline rocks in island arcs. In press.

52. Flower M. F. J. Schmincke H.-U. and Thompson R. N. Phlogopite stability and the ^{87}Sr/^{86}Sr step in basalts along the Reykjanes Ridge. *Nature, Lond.,* **254**, 1975, 404—6.

53. Aoki K. Phlogopites and potassic richterites from mica nodules in South African kimberlites. *Contr. Mineral. Petrol.,* **48**, 1974, 1—7.

54. Ringwood A. E. The petrological evolution of island arc systems. *Jl geol. Soc. Lond.,* **130**, 1974, 183—204.

55. Gill J. B. Role of underthrust oceanic crust in the genesis of a Fijian calc-alkaline suite. *Contr. Mineral. Petrol.,* **43**, 1974, 29—45.

56. Green H. W. A CO_2 charged asthenosphere. *Nature, phys. Sci.,* **238**, 1972, 2—5.

57. Bowes D. R. Geochemistry of Precambrian crystalline basement rocks, North-West Highlands of Scotland. In *24th Int. Geol. Congr.* (Montreal: The Congress, 1972), Sect. 1, 97—103.

58. Drury S. A. The chemistry of some granitic veins from the Lewisian of Coll and Tiree, Argyllshire, Scotland. *Chem. Geol.,* **9**, 1972, 175—93.

59. Brown G. C. and Fyfe W. S. The transition from metamorphism to melting: status of the granulite and eclogite facies. In *24th Int. geol. Congr.* (Montreal: The Congress, 1972), Sect. 2, 27—34.

60. Kilinc I. A. Experimental study of partial melting of crustal rocks and formation of migmatites. In *24th Int. geol. Congr.* (Montreal: The Congress, 1972), Sect. 2, 109—13.

61. Clifford T. N. Stumpfl E. F. and McIver J. R. A sapphirine—cordierite—bronzite—phlogopite paragenesis from Namaqualand, South Africa. *Mineralog. Mag.,* **40**, 1975, 347—56.

62. Bowden P. The geochemistry of some Nigerian igneous rocks. Ph.D. thesis, Imperial College, London, 1961.

63. Bowden P. and van Breemen O. Isotopic and chemical studies on younger granites from Northern Nigeria. *Proc. conference on African geology, 1970* (London: Commonwealth Gological Liaison Office, 1971), 105—20.

64. Wright J. B. Controls of mineralization in the older and younger tin fields of Nigeria. *Econ. Geol.,* **65**, 1970, 945—51.

65. Priem H. N. A. *et al.* Isotope geochronology in the Indonesian tin belt. *Geol. Mijnbouw,* **54**, 1975, 61—70.

66. Sillitoe R. H. Halls C. and Grant J. N. Porphyry tin deposits in Bolivia. *Econ. Geol.,* **70**, 1975, 913—27.

67. Rice R. and Sharp G. J. Copper mineralization in the forest of Coed-y-Brenin, North Wales. *Trans. Instn Min. Metall. (Sect. B: Appl. earth sci.),* **85**, 1976, B1—13.

68. Rea W. J. *et al.* Igneous intrusion and metallogenesis on and around the Harlech Dome. In *Mineral exploitation and economic geology 1974* (Cardiff: University College, 1974), pap. 8, 5 p.

69. Allen P. M. *et al.* Geochemistry of some igneous rocks from the Harlech Dome and their relationship to mineralization. *Trans. Instn Min. Metall. (Sect. B: Appl. earth sci.),* **85**, 1976, B100—8.

70. Schuiling R. D. Tin belts on the continents around the Atlantic Ocean. *Econ. Geol.,* **62,** 1967, 540—50.

71. Bowden P. Origin of the younger granites of northern Nigeria. *Contr. Mineral. Petrol.,* **25,** 1970, 153—62.

72. Noble J. A. Metal provinces of the western United States. *Bull. geol. Soc. Am.,* **81,** 1970, 1607—24.

73. Jensen B. B. Patterns of trace element partitioning. *Geochim. cosmochim. Acta,* **37,** 1973, 2227—42.

74. Barsukov V. L. The geochemistry of tin. *Geochemistry,* 1957, 41—52.

75. Cramer J. Personal communication, 1975; Alderton D. and Jackson N. Personal communications, 1976.

76. Gill J. B. Geochemistry of Viti Levu, Fiji, and its evolution as an island arc. *Contr. Mineral. Petrol.,* **27,** 1970, 179—203.

77. Bailey D. K. and Macdonald R. Fluorine and chlorine in peralkaline liquids and the need for magma generation in an open system. *Mineralog. Mag.,* **40,** 1975, 405—14.

78. Alexiev E. I. Rare earth elements in younger granites of Northern Nigeria and the Cameroons and their genetic significance. *Geochemiya,* 1970, 192—8; *Geochemistry intn.,* **7,** 1970, 127—32.

79. James D. E. A plate tectonic model for the evolution of the Central Andes. *Bull. geol. Soc. Am.,* **82,** 1971, 3325—46.

80. Karig D. E. Ridges and basins of the Tonga—Kermadec island-arc system. *J. geophys. Res.,* **75,** 1970, 239—54.

420

AUTHOR CITATION INDEX

Abbott, E. W., 11
Adams, F. D., 3
Ahrens, L. H., 404
Akimoto, S., 250, 280, 334
Albee, A. L., 236
Alexiev, E. I., 420
Allègre, C. J., 205, 206, 262, 286
Allen, C., 206, 419
Allen, J. B., 151
Allen, P. M., 419
Allen, R., 37, 45, 151
Allen, R. O., 299
Allsopp, H. D., 385
Andersen, O., 151
Anderson, A. T., 221
Anderson, C. A., 113, 351, 372
Anderson, D. L., 191, 333
Anderson, J. L., 358
Andrews, P., 57
Anepohl, J. K., 11
Anhaeusser, C. R., 268
Aoki, K., 371, 372, 419
Aramaki, S., 402
Archer, P., 272
Arculus, R. J., 392, 419
Armstrong, R. L., 37, 261, 268, 391, 402
Arndt, N. T., 261
Arth, J. G., 261, 268
Atwater, T., 291, 358
Aumento, F., 279, 286, 299, 391
Auvray, B., 261

Backlund, H., 15
Baeckström, O., 15
Bailey, D. K., 420
Bailey, E. B., 3, 14, 15, 23, 70, 113, 338
Baker, I., 371
Baker, P. E., 37, 40, 50, 402
Baksi, A. K., 358
Bakumenko, I. T., 184
Baldridge, W. S., 338
Baldwin, B., 371
Baldwin, E. M., 351, 352
Ballard, R. D., 285
Bankston, D. C., 301
Baragar, W. R. A., 40, 385
Barker, D. S., 3
Barr, C. D., 358
Barsukov, V. L., 420
Barth, T. F. W., 3
Bass, M. N., 11
Batiza, R., 291
Bazarova, T. Y., 184
Beasley, P., 384
Beck, M. E., 358
Beeson, M. H., 311, 358
Bell, J. D., 3
Bell, P. M., 151, 236
Ben-Avraham, Z., 358
Bence, A. E., 206, 221, 236, 272, 291, 392
Bender, J. F., 206, 221, 236, 272, 291

Benioff, H., 316
Bennett, D. J., 383
Bentley, R. D., 358
Bergeron, M., 286
Bertrand, J. M., 261
Best, M. G., 3, 333, 371
Bickle, M. J., 419
Biggar, G. M., 236, 383
Billington, S., 333
Binns, R. A., 383, 386
Birch, F., 131
Bjornsson, S., 301
Black, K., 358
Blackwell, D. D., 373
Blais, S., 261
Blake, D. H., 37, 301
Blanchard, D. P., 222, 292
Blank, H. R., Jr., 371
Bloxam, T. W., 418
Boettcher, A. L., 71, 183, 184, 191, 192, 236, 334
Bonatti, E., 207
Bond, W. L., 236
Bottinga, Y., 205
Boutakoff, N., 383
Bowden, P., 419, 420
Bowen, N. L., 3, 13, 70, 75, 86, 120, 151, 172, 351
Bowes, D. R., 419
Boyd, F. R., 120, 131, 132, 151, 192, 236
Bravo, M. S., 236, 333, 383
Brenner, N. L., 222, 236
Brett, R., 183
Bridgewater, D., 268
Brimhall, W. H., 371
Bromley, A. V., 418
Brookins, D. G., 372
Brooks, C., 61, 241, 261, 268, 286, 323, 403
Brown, G. C., 419
Brown, I. A., 50
Brown, R. D., 351
Bruce, E., 261
Brunfelt, A. O., 299
Bryan, W. B., 205, 207, 221, 261, 285, 286, 287, 291, 299
Buch, C. A., 372
Buddington, A. F., 113
Bultitude, R. J., 183, 279, 383
Bunch, T. E., 205
Bunker, C. M., 372
Burchfiel, B. C., 11, 358
Burkholder, F. R., 268
Burnham, C. W., 70
Bush, C. A., 372
Byerly, G. R., 236, 358
Byers, F. M., Jr., 37, 351

Cameron, W. E., 236
Camp, V. E., 358
Cann, J. R., 221, 299, 392, 418
Capdevila, R., 261
Capp, A. C., 37, 301, 404
Carden, J. R., 372

Carlson, R. W., 291
Carmichael, I. S. E., 3, 37, 44, 192, 221, 323, 334, 372, 403
Carter, J. L., 205, 383
Cater, F. W., 351
Cawthorn, R. G., 70, 236, 316, 334, 383
Chamalaun, F. H., 385
Chapman, H. J., 419
Charette, M. P., 311
Chayes, F., 44, 45, 285, 391
Chipman, D. W., 236
Chodos, A. A., 236
Choiniere, S. R., 358
Christiansen, R. L., 371
Christie, P. M., 262
Church, S. E., 261, 323
Clague, D. A., 205, 311
Clark, L. A., 419
Clark, S. P., Jr., 132, 183, 184
Clarke, D. B., 236, 24, 241, 334, 383
Clifford, T. N., 419
Clough, C. T., 3, 70, 338
Coats, R. R., 37, 391
Cohen, L. H., 279
Coleman, R. G., 205
Compston, W., 249, 268
Condie, K. C., 261, 262, 316, 323, 391
Condon, M. A., 383
Constantinou, G., 418
Cooke, D. L., 268
Coombs, D. S., 384
Cooper, J. A., 268
Cordier, P. L. A., 3
Coulson, A., 384
Cox, A., 358
Cox, D. C., 23
Cox, K. G., 3, 57, 334, 338, 358
Craig, H., 57
Crawford, W., 384
Cross, W., 372
Cullers, R. L., 286
Cundari, A., 384
Cuyubamba, A., 323

Daly, R. A., 4, 13, 15
Damon, P. E., 371
Darwin, C. R., 4
Dasch, E. J., 384
Davidson, C. G., 61
Davis, B. T. C., 131, 151, 172
Davis, D. A., 23
Davis, G. A., 358
Davis, G. L., 241, 261, 268, 403
Davis, K. E., 236, 261
Dawson, J. B., 151
Deans, T., 151
Decker, E. R., 371, 373
Deer, W. A., 4, 72, 113
Delaney, J. R., 221
DeLong, S. E., 72, 222, 237, 292, 334, 419
Demitriev, L. V., 61

421

Denham, D., 402
DePaepe, P., 268
Depaolo, D. J., 261
Desmarest, N., 4
Deutsch, S., 268
Devine, J. D., 11, 273
Dewey, J. F., 221
Dick, H. J. B., 205, 241
Dickey, J. S., 205, 285, 287
Dickin, A. P., 392
Dickinson, W. R., 37, 192, 249, 268, 316, 323, 334, 402
Dietrich, V. J., 236, 419
Dixon, J. R., 222, 236, 237
Dixon, S. A., 222, 236, 237
Dodge, F. C. W., 330
Doe, B. R., 268, 339, 371, 373
Doering, W. P., 268
Doherty, P. C., 286
Doney, H. H., 371
Donnelly, T. W., 11, 37, 316, 391, 402
Dott, R. H., Jr., 323
Drever, H. I., 57, 61
Drury, S. A., 419
Duda, A., 338
Dugan, M. A., 292
Duggan, M. B., 383
Duke, N. A., 418
Dungan, M. A., 71, 221, 222
Dury, G. J., 384

Easton, J., 132
Eckstrand, O. R., 61, 262
Edgar, A. D., 384
Edwards, A. B., 384
Edwards, C. L., 371
Eggler, D. H., 70, 192
Egorov, L. S., 192
Eichelberger, J. C., 221, 323
El-Hinnawi, E. E., 402
Elsasser, W. M., 132
Elthon, D., 236
Emslie, R. F., 205, 207, 385
Engdahl, E. R., 334
Engel, A. E. J., 57, 249, 261, 280, 291, 299, 316, 334, 391, 402
Engel, C. G., 57, 249, 261, 280, 291, 299, 334, 391, 402
England, J. L., 120, 131, 151, 236
Erlank, A. J., 261
Erlikh, E. N., 402
Evans, R., 11, 273
Ewart, A., 192, 299
Ewing, A. E., 11
Ewing, M., 11, 222, 262, 280

Falvey, D. A., 323
Farooqui, S. M., 358
Faure, G., 249, 250
Fenner, C. N., 86, 384
Findlay, D. C., 61
Finger, L. W., 285
Finn, D. D., 392
Fisher, D. E., 299

Fisher, L., 291
Fisher, R. L., 299
Fiske, R. S., 286, 351, 373
Flower, M. F. J., 419
Floyd, P. A., 391, 418
Forbes, R. B., 402
Ford, C. E., 236, 334, 383
Fouque, F. A., 4
Fox, P. J., 221, 222
Frey, F. A., 205, 261, 268, 273, 285, 286, 287, 291, 292, 311, 384
Fuchs, K., 338
Fujii, T., 205, 236
Fuller, R. E., 351
Furst, M. J., 402
Fyfe, W. S., 334, 418, 419

Gale, G. H., 61, 418, 419
Gancarz, A. J., 236
Ganguly, J., 221
Gansser, A., 236
Garcia, M. D., 391
Garson, M. S., 418
Gaskell, T. F., 334
Gass, I. G., 50, 205, 280, 418
Gast, P. W., 183, 205, 206, 249, 261, 262, 268, 272, 280, 286, 291, 334, 371, 385, 392, 402, 403
Geiss, J., 132
Gill, E. D., 384
Gill, J. B., 11, 31, 40, 192, 262, 299, 316, 334, 391, 402, 403, 419, 420
Gilmour, E. H., 352, 358
Glassley, W. E., 261, 391
Glikson, A. Y., 261
Goldberg, E., 132
Goodwin, A. M., 261
Goranson, R. W., 120
Gorshkov, G. S., 402
Govett, G. J. S., 418
Graham, A. L., 37, 301, 404
Grant, J. N., 419
Grant, N. K., 268
Grayson, H. J., 384
Green, D. H., 9, 11, 57, 70, 132, 151, 158, 172, 183, 205, 236, 250, 261, 268, 279, 280, 291, 311, 334, 371, 383, 384, 385, 391
Green, H. W., II, 183, 419
Green, J., 391
Green, T. H., 132, 316, 334, 384, 403
Greenbaum, D., 221
Griffin, W. L., 250, 311, 403
Grillot, L. R., 206
Grolier, M. J., 392
Grossman, L., 183
Grove, T. L., 237, 384
Gunn, B. M., 323, 371
Gust, D. A., 392

Haas, J. L., 184
Hague, G., 45
Hajash, A., 272
Häkli, T. A., 280
Hall, A., 403
Hall, J., 4
Halls, C., 418, 419
Hamblin, W. K., 371
Hameurt, J., 261
Hamilton, W., 403
Hanks, W., 384
Hanson, G. N., 206, 236, 262, 268, 272, 291, 311
Haramura, H., 402
Harker, A., 99
Harris, P. G., 40, 120, 132, 151, 250, 268, 280, 291, 384, 385
Harrison, C. G., 71
Harrison, N. M., 261
Hart, P. J., 268
Hart, S. R., 61, 205, 236, 241, 261, 268, 286, 291, 292, 299, 391, 403
Hart, R. A., 261
Harvey, M., 384
Harvey, R. D., 373
Harwood, H. F., 44, 151
Haskin, L. A., 280, 286, 299
Haskin, M. A., 280, 286
Hatherton, T., 37, 192, 249, 316, 323, 334, 402
Havens, R. G., 57, 249, 261, 280, 291, 299, 334, 402
Hawkesworth, C. J., 261
Hawkins, J. W., Jr., 261, 391
Hayes, D. E., 71, 419
Hays, J. D., 323
Hays, J. F., 222, 236, 237, 384
Hedge, C. E., 192, 249, 250, 268, 323, 339, 371
Heier, K. S., 249
Hein, S. M., 261, 268
Hekinian, R., 205, 286
Helmke, P. A., 299
Helz, R. T., 358
Henderson, W. T., 192
Hepworth, J. V., 268
Hermance, J. F., 206
Herrmann, A. G., 418
Hess, H. H., 4, 11, 57, 249, 250, 268, 316, 334, 371, 372, 404
Hibberson, W. O., 205, 236
Higazy, R. A., 151
Hilde, T. W. C., 299
Hildreth, R. A., 192, 268
Hills, E. S., 384
Hodges, F. N., 221, 236, 419
Hoffer, J. M., 372
Holloway, J. R., 286, 334
Holmes, A., 44, 132, 151, 334
Hooper, P. R., 358
Hopson, C. A., 351
Horikoshi, E., 419
Hossfeld, P. S., 385
Hotz, P. E., 23

Houtermans, F. G., 132
Howchin, W., 385
Howie, R. A., 151
Huang, W. L., 72, 184
Hubbard, N. J., 206, 268, 272, 392, 403
Hurley, P. M., 132, 249, 250, 311, 385
Hutchinson, R. W., 418, 419
Hutchison, R., 385
Hyndman, D. W., 4

Iddings, J. P., 11, 45, 91, 98
Iida, C., 23, 30, 351
Illies, J. H., 338
Irvine, T. N., 40, 70, 206, 221, 236, 385
Irving, A. J., 385, 386
Isacks, B. L., 333, 403
Ito, K., 70, 172, 250, 279, 280, 372

Jackson, E. D., 372
Jackson, W. H., 372
Jahn, B.-M., 261, 262
Jakeš, P., 11, 37, 40, 262, 323, 334, 391, 403
Jakobsson, S. P., 292, 299
James, A. V. G., 385
James, D. E., 261, 323, 420
James, H. L., 316
Jamieson, B. G., 241, 334
Jaques, A. L., 236, 316
Jennings, J. N., 384, 385
Jensen, B. B., 420
Johannes, W., 236
Johannsen, A., 57
Johnson, R. L., 57
Johnston, R., 57, 61
Johnston, T., 11, 273
Jolly, W. T., 403
Joplin, G. A., 11, 45, 384, 403
Joron, V.-L., 392
Joyce, E. B., 385
Jung, D., 351
Jutson, J. T., 385

Kable, E. J. D., 261
Kanaev, V. F., 61
Karig, D. E., 261, 323, 420
Katsui, Y., 43
Katsura, T., 11, 23, 57, 372, 385, 402
Kausel, E. G., 268
Kay, R. W., 206, 262, 268, 272, 291, 385, 392, 403, 419
Keisuke, 372
Kennedy, G. C., 70, 120, 172, 250, 279, 280, 372
Kennedy, W. Q., 11, 14, 15, 23, 70, 338, 351, 392
Kesson, S. E., 236, 268, 385
Keyes, M. G., 24
Kidd, W. S. F., 221, 418
Kilinc, I. A., 419

Kingsley, R., 11, 273, 286
Kiss, E., 384
Kistler, R. W., 372
Kleck, W. D., 358
Klerkx, J., 268
Knake, D., 418
Knopoff, L., 268
Knowles, C. R., 358
Koehnken, P. J., 11
Kolbe, P., 403
Kopp, A. C., 192
Korzhinskii, D. S., 206
Koster van Groos, A. F., 184
Krakatoa Committee of the Royal Society, 98
Krogh, T. E., 241, 261, 268, 403
Kudo, A. M., 371, 372
Kuno, H., 11, 23, 30, 70, 132, 172, 192, 221, 222, 316, 334, 351, 372, 391, 403
Kurasawa, H., 371
Kushiro, I., 11, 71, 132, 172, 184, 192, 205, 206, 236, 250, 261, 268, 280, 286, 334, 372
Kuthan, M., 403

L.S.P.E.T., 57
Lacroix, A., 40, 90
Laird, J., 236
Lambert, I. B., 250, 334
Lancelot, J. R., 286
Langford-Smith, T., 384
Langmuir, C. H., 206, 236, 272, 291
Langseth, M. G., Jr., 206, 280
Larsen, E. S., Jr., 372
Lasaga, A. C., 237
Laughlin, A. W., 372
Lavin, O. P., 222
Leeds, A. R., 268
Leeman, W. P., 241, 291, 338, 372
Lefevre, C., 323
Leishman, J., 61
Le Maitre, R. W., 50
LeMasurier, W. E., 372
Leonard, B. F., 316
Le Pichon, X., 206, 280
Lewis, A. D., 418
Lewis, J. V., 94
Lida, C., 70
Lilley, F. E. M., 383
Lindgren, W., 351
Lippard, S. J., 339
Lofgren, G. E., 222
Logatchev, N. A., 339
Long, P. E., 222, 292
Longhi, J., 236, 237, 384
Lovering, J. F., 385
Lowder, G. G., 37, 403
Lowell, R. P., 273
Lowry, W. D., 351, 352
Ludington, S., 392
Lugmuir, G. W., 291

Ma, M.-S., 291
Maaløe, S., 385
Mabbutt, J. A., 384, 385
McBirney, A. R., 11, 192, 268, 280, 334, 403
McCall, G. J. H., 61
McCallum, I. S., 311
McCamy, K., 221
Macdonald, G. A., 11, 23, 57, 372, 385, 391
Macdonald, R., 420
Macdougall, J. D., 291
McDougall, I., 249, 268, 385
McElhinney, M. W., 236
MacGregor, I. D., 132, 151, 311
McGregor, V. R., 268
McIlver, J. R., 419
McInerny, K., 385
McKee, E. H., 358, 372
McKenzie, D. P., 280
MacLeod, N. S., 358
Mahony, D. J., 384, 385
Major, A., 132
Malpas, J., 418
Manson, V., 57, 250
Mao, H. K., 236
Marsh, B. D., 316, 323, 334
Marsh, N. G., 392
Mason, R., 268
Mason, S. L., 4
Masuda, A., 37, 268, 403
Mather, K. F., 4
Mathews, D. H., 358
Mathews, R. E., 71
Matsumoto, H., 402
Mattey, D. P., 392
Medaris, L. G., 286
Mehnert, H. H., 339
Meijer, A., 262
Melson, W. G., 236, 280, 286, 292, 299, 301
Menard, H. W., 291
Menzies, M. A., 206, 419
Mercy, E. L. P., 151
Merriam, J. C., 351
Metais, D., 45
Michael, P. G., 221
Michel-Levy, A., 4
Millar, D. J., 384
Minear, J. W., 403
Mitchell, A. H. G., 418
Mitronovas, W., 334
Miyashiro, A., 11, 222, 262, 323, 392
Modreski, P. J., 183, 192
Moench, R. H., 372
Mohr, P. A., 206, 339
Monger, J. W. H., 358
Monkman, L. G., 57
Moorbath, S., 250, 261, 262, 392
Moore, J. G., 205, 221, 285, 286, 287, 339
Moores, E. M., 61
Moorhouse, W. W., 268
Morey, G. B., 262

Morgan, W. J., 268
Morgan, W. R., 37, 385
Morrison, G. W., 392
Morrison, M. A., 392
Morse, S. A., 4
Mottl, M. J., 272
Moxham, R. L., 57
Muehlberger, W. R., 371
Muenow, D. W., 221
Muir, I. D., 23, 45, 50
Murali, A. V., 291
Murata, K. J., 372
Murthy, V. R., 250, 262, 311, 403
Mutschler, F. E., 222, 392
Myers, C. W., 358
Myers, J. S., 268
Mysen, B. O., 71, 183, 184, 206, 236, 286

Nagashima, K., 23, 30, 70, 351
Nakamura, N., 268
Naldrett, A. J., 61, 261, 262
Natland, J. W., 311
Nava, D., 241, 261, 268, 403
Nelen, J. A., 236
Nelson, S. A., 221
Nelson, W. H., 45
Neprochnov, Y. P., 61
Nesbitt, R. W., 12, 61, 262, 273
Neumann, E.-R., 339
Newton, E. T., 91
Newton, R. S., 236
Nicholls, G. D., 280, 385
Nicholls, I. A., 71, 334, 419
Nicholls, J., 221, 372, 403
Nielson, D. R., 316, 323, 334
Ninkovich, D., 323
Nishikawa, M., 172, 280, 334
Noble, D. C., 323, 371
Noble, J. A., 61, 420
Nockolds, S. R., 37, 45, 113, 151
Norry, M. J., 392
Novitsky-Evans, J. M., 11, 316
Nyquist, L. E., 262

O'Donnell, T. H., 206, 222, 236, 237
O'Hara, M. J., 70, 71, 132, 151, 158, 184, 206, 222, 236, 250, 280, 291, 292, 316, 334, 372
O'Hearn, T., 236, 358
O'Nions, R. K., 261, 268
Oliver, J., 403
Ollier, C. D., 385
Orr, D., 385
Osborn, E. F., 71
Oversby, V. M., 268
Oxburgh, E. R., 206, 280

Packham, G. H., 323
Pakiser, L. C., 372
Pankhurst, R. J., 3, 250, 268
Papike, J. J., 222
Parmentier, E. M., 206
Paster, T. P., 299

Paul, D. K., 385
Peacock, M. A., 11
Pearce, J. A., 392, 418, 419
Peck, D. L., 351
Perfit, M. R., 392
Perret, F. A., 40, 98
Pertta, J., 358
Pertta, R., 358
Peterman, Z. E., 192, 268, 372, 373
Petrova, G. N., 61
Philpotts, J. A., 268, 280
Phinney, R. A., 261
Pichler, H., 323, 402, 403, 404
Pierce, W. G., 45
Pilger, A., 339
Playfair, J., 4
Poetz, J., 286
Poldervaart, A., 4, 11, 24, 57, 249, 250, 268, 316, 334, 352, 371, 372, 391, 404
Popoff, M., 339
Porath, H., 373
Potts, M. J., 323, 418
Powell, J., 358
Powell, J. L., 205, 250, 268, 291
Powell, M. A., 222
Powers, H. A., 373, 392
Presnall, D. C., 71, 206, 222, 236, 237
Price, R. C., 268, 385
Price, S. M., 358
Priem, H. N. A., 419
Prinz, M., 384
Prostka, H. J., 373
Pushkar, P., 37, 391, 402
Pyke, D. R., 61, 261, 262

Ragland, P. C., 316
Rahman, A., 385
Ramberg, I. B., 339
Ramlal, K., 57
Raspe, R. E., 4
Rau, W. W., 351
Ray, D. K., 402
Rea, W. J., 419
Reay, A., 132, 384
Reidel, S. P., 358
Reiter, M. A., 371
Renault, J., 373
Rhodes, J. M., 71, 221, 222, 292, 403
Rice, R., 419
Richard, P., 206, 262
Richards, J. R., 268
Richey, J. E., 3, 70, 338
Richter, D. H., 372
Richter, F. M., 268
Ridler, R. H., 268, 419
Ridley, W. I., 236, 371
Riecker, R. E., 338, 339
Rikunov, L. M., 61
Ringwood, A. E., 4, 9, 11, 70, 71, 132, 151, 172, 192, 205, 206, 236, 250, 261, 268, 280, 291, 311, 316, 323, 324, 334, 371, 384, 385, 403, 419
Ritsema, A. R., 261
Roberts, A. F., 351
Roberts, D., 419
Robertson, E. C., 183, 373
Robinson, H. H., 373
Roedder, E., 71, 184
Roeder, P. L., 207, 237, 385
Rogers, J. J. W., 11, 37, 268, 316, 338, 372, 391
Rogers, P., 402
Rona, P. A., 273
Roobol, M. J., 323
Rosendahl, B. R., 291
Rosler, A., 339
Ross, M. E., 358
Rougon, D. J., 222
Rowell, J. A., 120, 151
Rowlett, H., 221
Roy, R. F., 373
Roy, S. D., 291, 311
Royce, D., 221
Rubenach, M. J., 419
Rubey, W. W., 262
Ruckmick, J. C., 61
Rui, I. J., 419
Runcorn, S. K., 132, 280, 385
Russell, I. C., 351
Ruxton, B. P., 403

Saggerson, E. P., 385
Sangster, D. F., 419
Sawkins, F. J., 418
Schairer, J. F., 71, 151, 172
Schilling, J.-G., 11, 205, 207, 222, 241, 268, 273, 286, 291, 292, 301, 392, 403
Schmincke, H.-U., 338, 419
Schmitt, R. A., 291
Schnetzler, C. C., 268, 280
Schreiber, E., 221
Schreinemakers, 79
Schrock, R. L., 222
Schuiling, R. D., 420
Schwarzer, R. R., 11, 268
Scott, D. H., 373
Searle, D. L., 418
Seeber, L., 334
Senechal, R. G., 419
Seyfried, W. E., 272
Shand, S. J., 13
Shannon, E. V., 89
Sharaskin, A. Ya., 12, 273
Sharp, G. J., 419
Shaver, R. H., 323
Shaw, D. M., 262, 268, 292, 311, 403
Shaw, H. R., 358
Shibata, T., 222, 237, 273, 292
Shido, F., 11, 222, 262
Shih, C. Y., 262
Shimizu, N., 206, 262
Siefert, F., 236
Siegers, A., 403

Sillitoe, R. H., 418, 419
Simonian, K. O., 418
Simpson, E. S. W., 23
Simpson, R. W., 358
Sims, P. K., 262
Singleton, O. P., 385
Skeats, E. W., 385
Sleep, N. H., 280
Smewing, J. D., 418, 419
Smith, A. L., 192, 323
Smith, G. O., 351
Smith, I. E., 334, 403
Smith, R. E., 392
Smith, S. E., 392
Snavely, P. E., 351, 358
Sobolev, V. S., 184
Solomon, M., 385
Sood, M. K., 4
Sorensen, H., 268, 291
Spallanzani, L., 4
Spera, F. J., 221
Spooner, E. T. C., 418, 419
Sprigg, R. C., 386
Stanley, E. R., 386
Stebbins, J., 292
Steeves, M. A., 419
Steinnes, E., 299
Steven, T. A., 339
Stewart, D. B., 172
Stice, G. D., 268, 299
Stillman, C. J., 57
Stobbe, H. R., 373
Stoiber, R. E., 316, 323, 334
Stolper, E., 222, 237, 384
Stone, G. T., 373
Stormer, J. C., Jr., 372, 373
Stradling, D., 358
Strong, D. F., 418, 419
Stuckless, J. S., 386
Stueber, A. M., 386
Stumpfl, E. F., 419
Suffel, G. G., 419
Sugimura, A., 316, 334, 403, 404
Sukheswala, R. N., 351
Summers, H. S., 385
Sun, S.-S., 12, 262, 273, 286,
  292, 311
Sutherland, D. S., 50
Sutherland, F. L., 386
Swainbank, J. G., 268
Swanson, D. A., 358, 392
Swanson, E. A., 419
Sykes, L. R., 334, 403, 404
Syono, Y., 250, 280, 334

Takahashi, E., 237
Talukdar, S. C., 11
Talwani, M., 71
Tanaka, T., 268
Tarney, J., 262, 392
Tatsumoto, M., 37, 262, 268,
  286, 292, 373
Taubeneck, W. H., 358
Taylor, S. R., 37, 192, 206, 236,
  250, 268, 291, 301, 385, 392,
  403, 404
Teall, J. J. H., 91
Thayer, T. P., 351
Thomas, H. H., 14, 15, 113, 268
Thompson, F., 261
Thompson, G., 205, 221, 273,
  280, 285, 287, 291, 292, 299,
  301
Thompson, R. N., 206, 236, 392,
  419
Thorarinsson, S., 301
Thornton, C. P., 351
Thurlow, J. G., 419
Tiercelin, J.-J., 339
Tilley, C. E., 8, 12, 23, 24, 30,
  45, 50, 72, 113, 132, 151,
  152, 172, 184, 207, 280, 292,
  316, 351
Tilton, G. R., 249, 261, 268
Töksoz, M. N., 403
Trask, N. J., 373
Treuil, M., 392
Truckle, P. H., 339
Tsuboi, S., 91, 92
Turcotte, D. L., 206, 280
Turekian, K. K., 183, 184
Turner, F. J., 3, 4, 45, 113,
  151, 334
Tuttle, O. F., 152, 351
Tyrrell, G. W., 99

U.S. Geological Survey, 373
Udintsev, G. B., 61
Ukhanov, V. V., 192
Upadhyay, H. D., 418
Upton, B. G. J., 301
Urey, H. C., 57
Uyeda, S., 334

Vail, J. R., 57
Vallier, T., 236
van Andel, T. H., 280, 285
van Breemen, O., 419
Vaniman, D. T., 222
Varne, R., 419
Vening Meinesz, F. A., 132
Verhoogen, J., 3, 4, 45, 113, 132,
  151, 334
Viljoen, M. J., 61, 262, 268
Viljoen, R. P., 61, 262, 268
Vine, F. J., 61, 358
Vinogradov, A. P., 61, 192
Vitaliano, C. J., 373
Vogt, J. H. L., 83
Vokes, F. M., 418
von Bunsen, R. W. E., 3
Von Eckermann, H., 151

Wadati, K., 316
Wadsworth, W. J., 301
Wager, L. R., 4, 72, 113
Wagner, H. C., 358
Wahl, W., 14

Wahlstrom, E. E., 4
Walker, D., 72, 222, 236, 237,
  292, 384
Walker, F., 24, 352
Walker, G. W., 352
Walther, J. V., 268
Ware, N. G., 384
Washington, H. S., 24, 91, 92,
  120
Wass, S. Y., 386
Wasserburg, G. J., 261
Waters, A. C., 24, 113, 351, 352,
  373
Watkins, N. D., 268, 358, 371
Watkinson, D. H., 184
Watson, E. B., 237
Wedepohl, K. H., 311
Weidman, C., 371
Welke, H., 250
Wellman, P., 384, 386
Werner, A. G., 4
White, A. J. R., 37, 262, 323,
  334, 385, 403, 404
White, I. G., 384
White, W. M., 11, 207, 273, 286,
  287, 292
Wiik, H. B., 184
Wilcox, R. E., 37
Wiliams, A. F., 152
Wilkinson, J. F. G., 72, 383, 384,
  386
Wilkinson, W. D., 352
Williams, A. F., 57
Williams, H., 11, 113, 268, 352,
  373, 418
Williams, L. A. J., 339, 385
Wilson, G. V., 3, 70, 338
Wilson, H. D. B., 57
Wilson, J. F., 261
Wilson, J. T., 268, 404
Wimmenauer, W., 50
Winchester, J. A., 391, 418
Windley, B. F., 261, 262, 268
Wise, W. S., 373
Wood, B. J., 221
Wood, D. A., 392
Wood, D. N., 57
Wright, T. L., 222, 236, 280, 286,
  358, 372, 373, 392
Wright, W. B., 3, 70, 338, 419
Wyllie, P. J., 4, 57, 61, 72, 152,
  183, 184, 250, 311, 334, 373

Yamasaki, K., 23, 30, 70, 351
Yates, H., 386
Yeats, R. S., 286
Yoder, H. S., Jr., 4, 8, 12, 24, 71,
  72, 113, 132, 151, 152, 172,
  184, 206, 207, 250, 268, 280,
  292, 316, 323, 372, 392, 404

Zajac, M., 11, 273
Zeil, W., 323, 402, 403, 404
Zielinski, R. A., 392
Zindler, A., 292

SUBJECT INDEX

Absarokite, 43, 396
Achondrite, 52-56
Afar, east Africa, 336
Aleutians, 31, 35-36, 321, 331, 333, 389, 398
Alkali basalts, 8, 10, 41, 65-68, 114, 118-121,
123, 127, 129-130, 147, 154, 157-158,
181-182, 191, 201, 241, 263-266, 270, 272,
276-277, 291, 305, 307, 309-310, 359-370,
389. *See also* Series, alkaline
Alkali-lime index, 33
Alnöites, 146, 148
Amphibole, 185, 309-310, 316, 324, 326, 380,
401-402. *See also* Fractionation,
amphibole
Amphibolite, melting of, 114
Analcimite, 374-375, 377, 379
Andes Mountains, South America, 49, 321-323,
400
Andesite, 16-17, 29, 40-41, 68, 110, 112, 172-173,
185, 189-190, 298, 315, 337, 348-349,
394-398, 406
Andesite line, 389, 393
Ankaramite, 20, 23, 41, 52-56, 60
Apatite, 265-266, 379-380
Archean greenstone belts, 251-260, 263, 267
Arcs, continental margin and intra-oceanic,
320-323. *See also* Island arcs
Australia, southeastern, 338, 374-383
Azores, 271, 281, 283, 304

Banakite, 43, 396
Barberton Mountains, South Africa, 11, 52-56,
258
Basalt tetrahedron, 7-9, 67, 188
Basanite, 9, 20, 23, 41, 67, 158, 181, 362,
374-375, 377, 379-380, 383
Basin and Range Province, U.S.A., 362-364, 367,
370
southern Oregon, 340-341, 345-348
Benioff zones, 248, 314, 318, 320-323, 325,
328-333, 395-396, 401
Benmoreite, 41-42, 47, 374
Bimodal suites, 337
Biotite, 47, 149, 380. *See also* Phlogopite
Bowen trend, 66, 69
Buffer, sanidine-eclogite, 326

Calcalkaline basalt, 345-350, 408-409. *See also*
High-alumina basalt; Series, calcalkaline
Carbon dioxide, effect of, 43, 65-66, 68, 174-183,
185-191, 203, 214
Carbonatite, 50, 68, 146, 148, 157, 182, 190
Caribbean, 10, 34, 39-40, 49, 264-265, 400. *See
also* Lesser Antilles
Cascade Range, northwest U.S.A., 103, 110-112,
322, 340-350
Chondrite, 52-56
Circum-Pacific, 28-31, 35-36, 38, 264-265, 322,
330-331, 333, 395, 397-399, 400
Clinopyroxene, 70, 121, 128, 130-131, 149, 180
214, 306, 310, 380
Colorado Plateau, U.S.A., 363-364, 369

Columbia River Plateau, northwest U.S.A., 338,
340-350, 353-358
Comendite, 46
Contamination
crustal, 241, 245-248, 327, 364-368
oceanic sediment, 251
Continental basalts, 241, 270, 336, 389-390, 407.
See also Tholeiite, continental
Continental margins, 316, 394, 400, 407, 411
Cotectic
high pressure, 227
low pressure, 214
Crust
hydrated oceanic, 402
intermediate, 315
undepleted oceanic, 402
Cusp, mantle solidus, 193, 200-203, 236

Dacite, 29, 41, 172-173, 394-398
Deccan trap, 16, 22, 87, 89, 91
Diabase, 14, 16, 34, 89-90, 93-94, 248
Diagram
alkali-silica, 20-21, 25, 29, 54, 361, 382
AMF, 25, 395
potash-silica, 31, 36, 328, 330, 395
Differentiation. *See* Fractionation
Dolerite. *See* Diabase

East African rift, 50, 258, 297, 337
Eclogite, 53, 114-119, 134-136, 138-140, 147-150,
322. *See also* Fractionation, eclogite
Eclogite nodules, 133, 138, 141-142, 145-146,
148-149
Eifel area, Germany, 337
Ethiopian plateau, 337
Ethiopian rift valley, 336-337. *See also* East
African rift

FAMOUS area, mid-Atlantic ridge, 194-196, 198,
209, 271, 281-287
Fenner trend, 66, 69, 85-100
Flood basalt, 40, 336. *See also* Plateau basalt
Fluid inclusions, peridotite nodules, 68
Fractionation
amphibole, 324, 374, 380, 383
closed-system, 272
constant oxygen pressure, 101-113
constant total composition, 101-113
eclogite, 118-119, 142, 144-149, 154, 243-244,
289
garnet, 68, 114, 118
high-pressure, 119-121, 137-142, 146, 149-150,
288-289, 364, 366, 368-370, 374, 377-378,
380, 383
low-pressure, 119-121, 142-146, 149-150, 155,
219, 223, 233, 243-244, 274-275, 284, 288,
294, 305, 311, 318, 357-358, 364-366, 369,
374, 380-381, 401
magnetite, 66, 69, 85-88, 305, 311
olivine, 68, 153-154, 197, 233-234, 274-275,
324, 374, 380
omphacite, 114, 118

open-system, 272, 288
orthopyroxene, 67, 154
plagioclase, 111, 113, 274-275
polybaric, 67

Garnet. *See* Fractionation, garnet
Geothermal gradient, 190, 193, 196-197, 202
Grande Ronde Basalt, Columbia River basalts,
    354-355

Haplobasalt, 73-84
Haplodiorite, 73-84
Harzburgite, 153, 170-171, 201, 223-224, 236, 279
Hawaii, U.S.A., 8, 17-23, 48, 52-53, 56, 183, 243,
    264-265, 272, 305-311, 322, 350
Hawaiite, 20, 23, 41, 47, 305, 311, 338, 374-375,
    377, 380-381, 383
High-alumina basalt, 8-9, 27-30, 41, 110-112,
    121, 123, 127, 129, 154, 158, 314, 318,
    345-350, 396-397. *See also* Calcalkaline
    basalt
High Plains, U.S.A., 362-363, 366, 369
Hydrogen, effect of, 68, 185-191
Hydrothermal activity, ridges, 272
Hypersthene, 20-21. *See also* Orthopyroxene

Icelandite, 41, 396
Ilmenite, 379
Immobile elements, 390, 408-417
Imnaha Basalt, Columbia River basalts, 354-355
Incompatible elements, 122, 147, 327, 384
    absarokite-shoshonite-banakite series, 43
    Archean greenstone belts, 253-255
    Columbia River basalts, 357
    continental margin volcanics, 407
    FAMOUS area, 204, 281-285, 287
    Hawaii, 305, 307-308, 310
    island arc basalt, 36, 39, 251, 323, 406, 414.
        See also Incompatible elements, orogenic
        volcanics
    mid-Atlantic ridge, 288, 290-291
    minerals, 146
    MORB, 193, 251, 274, 279, 281, 293, 295, 304,
        405, 410
    Newer Basalts, Australia, 379-380
    Ninetyeast Ridge, 294-303
    oceanic island basalts, 405, 410
    orogenic volcanics, 34, 394, 398-402. *See also*
        Incompatible elements, island arc basalts
    Rio Grande rift, 362-363, 367
    tholeiitic basalts, 242-249
Indian Ocean, 294-303, 322
Indonesia, 36, 49, 321, 331, 395, 410, 416
Iron enrichment, 10, 46-47, 85-100, 395, 397
Island arcs, 31, 34-36, 40, 314-315, 318, 320-324,
    389, 394-402, 406, 411
Isofracts, 75-76

Japan, 25, 27-31, 36, 38, 314, 321, 325, 328,
    332, 346, 397-398

Kamchatka, U.S.S.R., 36, 395, 397-398
Kilauea, Hawaii, 21, 243, 365

Kilbourne's Hole, New Mexico, U.S.A., 199-200
Kimberlite, 52-56, 68, 133, 138, 140-142,
    145-150, 153, 157, 190
Komatiite, 10, 52-56, 58-60
    Archean, 58-62, 257-258
    basaltic, 58-62, 252, 257-258
    peridotitic, 58-62, 252, 257-258
    pyroxenitic, 62
Korea, 27-30

Late Yakima-Ellensburg magma, Columbia River
    basalts, 342-345
Lead isotopes, 263-264, 279, 367-369
Lesser Antilles, Caribbean, 31, 35-36, 39-40,
    322, 398
Lherzolite, 69, 137, 153, 193, 201. *See also*
    Peridotite
Lherzolite nodules. *See* Peridotite nodules
LIL elements. *See* Incompatible elements
Limburgite, 52-56
Liquid immiscibility, 66, 69
Low velocity zone, 190, 278
Lunar rocks, 52-56

Madagascar, 52-55
Magma. *See also* Partial melting
    parental, 67, 350-351
    primary. *See* Primary magmas
    separation from source, 66, 121, 128-131
Magma mixing, 69, 208-209, 215-216, 220-221,
    272, 283, 288
Magnesium-iron ratio, 104, 185, 189, 191,
    194-195, 198, 200, 231-232, 282, 287-288,
    304, 374, 378, 405-406
Magnetite. *See* Fractionation, magnetite
Major elements, 15, 53-56, 104-105, 246, 390
    absarokite-shoshonite-banakite series, 42
    continental margin volcanics, 407
    continental volcanics, 407
    Hawaiian basalts, 306-307, 310
    island arc basalts, 406
    komatiite, 52-56, 59-60, 62
    MORB, 32, 274-276, 294-297, 300, 302-303, 405
    Newer Basalts, Australia, 382-383
    Ninetyeast Ridge, 294-300, 302-303
    oceanic island basalts, 405
    orogenic volcanics, 27-29, 394-398, 400-401
    Rio Grande rift, 360-361, 365
Manchuria, 27-30
Mantle, 240
    depleted in incompatible elements, 251, 279,
        293
    hydrated, 34, 274, 316, 379, 413
Mantle convection, 125-126
Mantle diapir. *See* Mantle plume
Mantle heterogeneity, 240-241, 248-249, 266,
    272, 279, 288, 291-292, 357
Mantle plume, 201, 267, 271-272, 278, 290, 337,
    369, 390
Martinique, 39-40
Massive sulphide deposits, 410-414
Mauna Loa, Hawaii, 243
Melilite basalt, 186, 190

"Metamorphic facies," upper mantle, 67
Metamorphism, low-grade, 253–254
Metatholeiite, 52–56
Mid-Atlantic ridge, 32, 271, 281–282, 287–291,
  304, 322
Moho, 114
Mont Pelée, Martinique, 39
MORB (mid-ocean ridge basalt), 8, 32, 34, 68–70,
  183, 186, 193–204, 208–221, 252, 255–257,
  266, 271–272, 274–279, 282, 293, 389, 405,
  411. *See also* Ocean-floor basalt
  N-type, 293
  P-type, 293
Mugearite, 17, 23, 28–29, 41, 47, 305, 338,
  374–375, 377, 380, 383
Mull, Scotland, 14, 22, 66, 95, 110–112, 336–337
  non-porphyritic central type basalt 14, 66,
    336–337
  plateau type basalt, 14, 66, 336–337

Neodymium isotopes, 258–259
Nephelinite, 9, 41, 67, 158, 181, 186, 189–190,
  263, 374–375, 377, 379
New South Wales, Australia, 49–50, 264
New Zealand, 382–383, 397
Newer Basalts, Victoria and South Australia,
  374–383
Ninetyeast Ridge, Indian Ocean, 272, 294–303

Ocean-floor basalts, 264, 408–410, 413. *See also*
  MO-RB
Oceanic basalts, 240–241, 270–271, 389. *See also*
  MORB; Ocean-floor basalts
Oceanic island basalts, 270–271, 389–390, 405,
  410–411
Oceanite, 22, 60
Oceanographer fracture zone, 208–221
Olivine, 18, 27, 154, 167–171, 194. *See*
  Fractionation, olivine
Olivine basalt, 7–8, 14–15, 18–19, 121, 129–130,
  375, 377, 381–382
Omphacite. *See* Fractionation, omphacite
Ophiolite, 39, 194, 196
Orthopyroxene, 47, 67–68, 121, 128, 130–131,
  140–141, 150, 154, 180, 230, 274–25, 306.
  See also Fractionation, orthopyroxene
Oxygen, effect of, 66, 68, 101–113, 154

Pacific Northwest, U.S.A., 340–350, 389
Palisades sill, 22, 89, 94
Papua, western Pacific, 36, 329, 397, 399
Partial melting, 65, 124–125, 172, 379
  constant total composition, 153
  crust, 324, 402
  cusp on solidus, 193, 200–203, 236
  degree of, 68, 198–201, 243, 251, 264, 274–277,
    279, 281, 285, 289, 291, 306, 326, 332,
    357, 365, 369, 401–402
  depth of, 66, 68, 365, 401–402
  dynamic, 271, 287, 289
  eclogite, 116–120, 322, 324. *See also* Eclogite
  equilibrium, 193, 203–204, 289–290
  fractional, 193, 203–204, 289–290

garnet peridotite, 67, 114, 117, 119–120, 124,
  133, 139–141, 147, 173
pyrolite, 128. *See also* Pyrolite
Partitioning, crystal-liquid, 265, 289, 326,
  378–379, 415
Peridotite. *See* Lherzolite; Harzburgite; Partial
  melting
  melting point, 277
  spinifex texture, 257
Peridotite nodules, 68, 122, 130, 133, 138, 149,
  153–154, 199–200, 374, 377
Phlogopite, 186, 190, 265–266, 305, 308–310, 324,
  326. *See also* Biotite
Phonolite, 337
Phosphorous, 266, 308, 355, 379–381
Picrite, 41, 52–56, 60, 67, 70, 121, 127,
  129–131, 133, 138, 150, 157–158, 223–224,
  234–236, 389. *See also* Primary magmas
Picture Gorge magma, Columbia River basalts,
  342–345, 354
Pillow lavas, 32, 294, 406
Plagioclase, 79–80, 274. *See also* Fractionation,
  plagioclase
Plateau basalt, 14–15, 18, 336–337, 389, 407.
  See also Flood basalt
Pomona flow, Columbia River basalts, 356
Potash, 266, 315–316, 325–326, 330–332, 379–381,
  397–398
Potash-depth relations, subduction zones, 190,
  315–316, 320–323, 325–334, 394–395
Potassic mafic lavas, 44, 146–150, 157
Primary magmas, 7–8, 13–14, 67, 69–70, 119–120,
  133, 153, 195–197, 223–224, 231–233, 236,
  328, 381
Primary phase field
  clinopyroxene, 73–75, 214
  forsterite, 167–171
  orthopyroxene, 169, 181, 197–198
  plagioclase, 75–79
Pyrolite, 68, 121–124, 126–127, 129, 131, 203,
  305, 307–308
Pyroxenite, 50, 116, 154

Queensland, Australia, 49, 382

Rare-earth elements, 34, 253, 256–258, 264, 266,
  274–279, 281–284, 287–290, 293–294,
  297–298, 300–304, 394, 398–399, 401,
  405–406, 412
Rhine Valley, Germany, 50, 263
Rhodesia, 53, 245, 249, 254
Rhyolite, 17, 29, 41, 337
Rift basalt, 336–338, 359–370, 389, 407
Rio Grande rift, western U.S.A., 337–338, 359–370
Ross Island, Antarctica, 263–265
Roza flow, Columbia River basalts, 356

Saddle Mountains Basalt, Columbia River
  basalts, 354–356
San Juan Mountains, Colorado, U.S.A., 360–363,
  366, 368
Scotland, 14, 22, 53, 66, 95, 110–112, 247–248,
  336–337

Seawater interactions, 32, 253, 272, 294, 296
Sea-floor spreading, 248, 314
Series
   absarokite-shoshonite-banakite, 42–45
   alkaline, 7, 9, 13, 17–23, 25, 27–30, 41,
      44, 46–48, 306, 314, 318, 336–337,
      359–360, 362, 374–375, 377–383, 405,
      407
   Andean, 39, 400–401. *See also* Series, high-K
      calcalkaline; Series, shoshonitic calcal-
      kaline, 10, 13, 31, 34–35, 41, 44, 46–50,
      103, 154, 315, 324, 390, 394–402, 406–408
   high-K calcalkaline, 324, 394–402. *See also*
      Series, Andean; Series, shoshonitic hy-
      persthenic, 10
   island arc tholeiite, 8, 31, 35–36, 40, 314–315,
      318, 324, 397, 399, 406, 410, 412–413. *See*
      also Series, primitive island arc; Series,
      low-K tholeiite
   low-K tholeiiten, 39, 408–409. *See also* Series,
      primitive island arc; Series, island arc
      tholeiite
   pigeonitic, 10, 21, 29, 31
   primitive island arc, 10, 34, 315. *See also*
      Series, island arc tholeiite; Series, low-K
      tholeiite
   shoshonitic, 10, 36, 39, 43–44, 46–50, 324,
      394–402, 414. *See also* Series, high-K
      calcalkaline; Series, Andean subalkaline,
      7, 25, 38, 41, 82–84, 270, 314, 405, 407
   tholeiite, 8–10, 14–18, 20–22, 27–30, 41,
      44, 46–48, 65–68, 103, 110–114, 118–120,
      147, 154, 270–272, 318, 336, 340–345, 349,
      359–360, 374–376, 381–382, 389, 394–402
Silica saturation, 18–20, 175–181, 187–189, 193
Skaergaard intrusion, Greenland, 38, 66, 103,
   110–111
Snake River Plain, western U.S.A., 322–323, 364
Soret effect, 66, 69
Spilite, 36
Spinel, 130, 137
Strontium isotopes, 246–248, 263–265, 279–283,
   285, 293, 322–323, 357
Subduction zones, 314–316, 320–323, 332–333,
   389–390, 394–402
Systems, experimental
   dry, 73–84, 101–108, 133, 142, 159–172, 188,
      193, 196, 203, 214–215
   plus carbon dioxide, 176–177, 182
   plus water, 162–169, 173–174
   plus water and carbon dioxide, 174–176, 188

Tasmania, Australia, 46, 48, 382
Textures, 62, 217–218, 295
Thermal divides, 114, 136, 154

Tholeiite. *See also* Series, tholeiite
   continental, 246–249, 340–350, 359–370. *See*
      also Continental basalt
   island arc. *See* Series, island arc tholeiite
   oceanic. *See* MORB
   olivine, 8, 18, 41, 67–68, 121, 123, 127–130,
      133, 158, 173, 191, 201, 223, 318, 374–375,
      377
   quartz, 41, 121, 123, 128, 157–158, 168, 172,
      223, 374–375
Tin deposits, 410, 414–417
Titanium, 218–219, 231–232, 306, 355, 379–381,
   390
Trachyandesite, 337
Trachybasalt, 41, 47
Trachyte, 17, 29, 41, 44, 47, 50, 338, 374, 376
Transition, solid state, 67, 69, 115–116, 315
Transition metals, trace, 36, 39, 43, 145, 147, 253,
   255–256, 274, 287, 289, 293–303, 310, 357,
   398, 400, 405–407
"Transitional" basalt, 8, 272, 305–307, 309–310,
   411
Tristan da Cunha, 47–48
Tristanite, 41, 47
Tschermak's molecule, 176, 189

Underplating, 324

Valley of Ten Thousand Smokes, Alaska, U.S.A.,
   99
Viti Levu, Fiji, 31, 35–36, 333, 397
Victoria, Australia, 46, 48–49, 374–383
Volatile transfer, 285

Wallrock reaction, 122, 153, 244–249, 287–289,
   327, 331
Wanapun Basalt, Columbia River basalts,
   354–356
Watchung Basalt, New Jersey, U.S.A., 86, 94
Water, effect of, 43, 65–66, 68, 159–172,
   174–175, 178, 181–183, 185–191, 203,
   276–277, 324
Weathering, submarine. *See* Seawater inter-
   actions
West Greenland-Baffin Island, 245–246
Within-plate basalts, 390, 409, 411, 413–414
Wyoming, U.S.A., 43–44, 46, 49, 364

Xenoliths, peridotite. *See* Peridotite nodules

Yakima magma, Columbia River basalts, 22,
   342–345, 350, 354

Zonation, magma chamber, 285
Zone refining. *See* Wallrock reaction

About the Editors

PAUL C. RAGLAND was born in 1936 in Lubbock, Texas. His undergraduate degree was from Texas Tech and both graduate degrees were from Rice University. The PhD was conferred in 1962, at which time he joined the faculty at the University of North Carolina, Chapel Hill. He remained there until 1978, when he became chairman at The Florida State University in Tallahassee, where he is currently professor of geology. The majority of his current research deals with the petrochemistry of intrusive igneous rocks, particularly diabase dikes and granitic plutons.

JOHN J. W. ROGERS received the B.S. and Ph.D. in geology from the California Institute of Technology in 1952 and 1955, respectively, and the M.S. from the University of Minnesota in 1952. He was a member of the faculty at Rice University, Houston, Texas from 1954 to 1974, where he served at different times as chairman of the geology department and Master of Brown College. He has been W. R. Kenan, Jr., Professor of Geology at the University of North Carolina at Chapel Hill since 1975. His current major research activities are studies of continental crustal evolution in southern India and in North Africa and the Middle East.